Universitext

T0171845

T.S. Blyth

Lattices and Ordered Algebraic Structures

 Springer

T.S. Blyth
Emeritus Professor
School of Mathematics and Statistics
Mathematical Institute
University of St Andrews
North Haugh
St Andrews
KY16 9SS
UK

Mathematics Subject Classification (2000): 06-01, 20-01, 06A05, 06A06, 06A12, 06A15, 06A23, 06A99, 06B05, 06B10, 06B15, 06C05, 06C10, 06C15, 06D05, 06D10, 06D15, 06D20, 06E05, 06E10, 06E15, 06E20, 06E25, 06E30, 06F05, 06F15, 06F20, 06F25, 12J15, 20M10, 20M15, 20M17, 20M18, 20M19, 20M99

British Library Cataloguing in Publication Data
Blyth, T. S. (Thomas Scott)
  Lattices and ordered algebraic structures
  1. Lattice theory   2. Ordered algebraic structures
  I. Title
  511.3′3

Library of Congress Cataloging-in-Publication Data
Blyth, T. S. (Thomas Scott)
  Lattices and ordered algebraic structures / T.S. Blyth.
    p. cm. — (Universitext)
  Includes bibliographical references and index.

  1. Ordered algebraic structures.   2. Lattice theory.   I. Title.
  QA172.B58   2005
  511.3′3—dc22   2004056612

ISBN 978-1-84996-955-0   e-ISBN 978-1-84628-127-3
Springer Science+Business Media
springeronline.com

© Springer-Verlag London Limited 2010

Printed in the United States of America
12/3830-543210 Printed on acid-free paper

Preface

The notion of an order plays an important rôle not only throughout mathematics but also in adjacent disciplines such as logic and computer science. The purpose of the present text is to provide a basic introduction to the theory of ordered structures. Taken as a whole, the material is mainly designed for a postgraduate course. However, since prerequisites are minimal, selected parts of it may easily be considered suitable to broaden the horizon of the advanced undergraduate. Indeed, this has been the author's practice over many years.

A basic tool in analysis is the notion of a continuous function, namely a mapping which has the property that the inverse image of an open set is an open set. In the theory of ordered sets there is the corresponding concept of a residuated mapping, this being a mapping which has the property that the inverse image of a principal down-set is a principal down-set. It comes therefore as no surprise that residuated mappings are important as far as ordered structures are concerned. Indeed, albeit beyond the scope of the present exposition, the naturality of residuated mappings can perhaps best be exhibited using categorical concepts. If we regard an ordered set as a small category then an order-preserving mapping $f : A \to B$ becomes a functor. Then f is residuated if and only if there exists a functor $f^+ : B \to A$ such that (f, f^+) is an adjoint pair.

Residuated mappings play a central rôle throughout this exposition, with fundamental concepts being introduced whenever possible in terms of natural properties of them. For example, an order isomorphism is precisely a bijection that is residuated; an ordered set E is a meet semilattice if and only if, for every principal down-set $x^{\downarrow}$, the canonical embedding of $x^{\downarrow}$ into E is residuated; and a Heyting algebra can be characterised as a lattice-based algebra in which every translation $\lambda_x : y \mapsto x \wedge y$ is residuated. The important notion of a closure operator, which arises in many situations that concern ordered sets, is intimately related to that of a residuated mapping. Likewise, Galois connections can be described in terms of residuated mappings, and vice versa. Residuated mappings have the added advantage that they can be composed to form new residuated mappings. In particular, the set $\operatorname{Res} E$ of residuated mappings on an ordered set E forms a semigroup, and here we include descriptions of the types of semigroup that arise.

A glance at the list of contents will reveal how the material is marshalled. Roughly speaking, the text may be divided into two parts though it should be stressed that these are not mutually independent. In Chapters 1 to 8 we deal with the essentials of ordered sets and lattices, including boolean algebras, p-algebras, Heyting algebras, and their subdirectly irreducible algebras. In Chapters 9 to 14 we provide an introduction to ordered algebraic structures, including ordered groups, rings, fields, and semigroups. In particular, we include a characterisation of the real numbers as, to within isomorphism, the only Dedekind complete totally ordered field, something that is rarely seen by mathematics graduates nowadays. As far as ordered groups are concerned, we develop the theory as far as proving that every archimedean lattice-ordered group is commutative. In dealing with ordered semigroups we concentrate mainly on naturally ordered regular and inverse semigroups and provide a unified account which highlights those that admit an ordered group as an image under a residuated epimorphism, culminating in structure theorems for various types of Dubreil-Jacotin semigroups.

Throughout the text we give many examples of the structures arising, and interspersed with the theorems there are bundles of exercises to whet the reader's appetite. These are of varying degrees of difficulty, some being designed to help the student gain intuition and some serving to provide further examples to supplement the text material. Since this is primarily designed as a non-encyclopaedic introduction to the vast area of ordered structures we also include relevant references.

We are deeply indebted to Professora Doutora Maria Helena Santos and Professor Doutor Herberto Silva for their assistance in the proof-reading. The traditional free copy is small recompense for their labour. Finally, our thanks go to the editorial team at Springer for unparalleled courtesy and efficiency.

St Andrews, Scotland   T.S. Blyth

Contents

1

Ordered sets; residuated mappings

1.1 The concept of an order

The reader will recall that a **binary relation** on a non-empty set E is a subset R of the cartesian product set $E \times E = \{(x,y) \mid x,y \in E\}$. We shall say that $x, y \in E$ are R-**related** whenever $(x,y) \in R$, this often being written in the equivalent form $x\,R\,y$. In general there are many properties that binary relations may satisfy on a given set E. In particular, for example, the reader will be familiar with the notion of an **equivalence relation** on E, namely a binary relation R that is

(1) **reflexive** $[(\forall x \in E)\ (x,x) \in R]$;

(2) **symmetric** $[(\forall x, y \in E)$ if $(x,y) \in R$ then $(y,x) \in R]$;

(3) **transitive** $[(\forall x, y, z \in E)$ if $(x,y) \in R$ and $(y,z) \in R$ then $(x,z) \in R]$.

If we define the **dual** of R to be the relation R^d given by

$$(x,y) \in R^d \iff (y,x) \in R,$$

then we may state (2) in the equivalent form $R = R^d$.

Here we shall be particularly interested in the situation where property (2) is replaced by the property

(2′) **anti-symmetric** $[(\forall x, y \in E)$ if $(x,y) \in R$ and $(y,x) \in R$ then $x = y]$,

which may be expressed as $R \cap R^d = \mathrm{id}_E$ where id_E denotes the relation of equality on E.

Definition If E is a non-empty set then by an **order** on E we mean a binary relation on E that is reflexive, anti-symmetric, and transitive.

We usually denote an order by the symbol $\leqslant$. Variants include $\preceq$ and $\sqsubseteq$. It is traditional to write the expression $(x,y) \in\ \leqslant$ in the equivalent form $x \leqslant y$ which we read as 'x is less than or equal to y'.

Thus $\leqslant$ is an order on E if and only if

(1) $(\forall x \in E)\ x \leqslant x$;

(2′) $(\forall x, y \in E)$ if $x \leqslant y$ and $y \leqslant x$ then $x = y$;

(3) $(\forall x, y, z \in E)$ if $x \leqslant y$ and $y \leqslant z$ then $x \leqslant z$.

Definition By an **ordered set** $(E; \leqslant)$ we shall mean a set E on which there is defined an order $\leqslant$.

Other common terminology for an order is a **partial order**, and for an ordered set is a **partially ordered set** or a **poset**.

According to Birkhoff [13] the defining properties of an order occur in a fragmentary way in the work of Leibniz (circa 1690). The present formulation emerged from the work of Peirce [91], Schröder [101], and Hausdorff [62].

Example 1.1 On every set the relation of equality is an order.

Example 1.2 On the set $\mathbb{P}(E)$ of all subsets of a non-empty set E the relation $\subseteq$ of set inclusion is an order.

Example 1.3 On the set $\mathbb{N}$ of natural numbers the relation $|$ of divisibility, defined by $m|n$ if and only if m divides n, is an order.

Example 1.4 If $(P; \leqslant)$ is an ordered set and Q is a subset of P then the relation $\leqslant_Q$ defined on Q by

$$x \leqslant_Q y \iff x \leqslant y$$

is an order on Q. We often write $\leqslant_Q$ simply as $\leqslant$ and say that Q *inherits the order $\leqslant$ from* P.

Thus, for example, the set $\operatorname{Equ} E$ of equivalence relations on E inherits the order $\subseteq$ from $\mathbb{P}(E \times E)$.

Example 1.5 The set of even positive integers may be ordered in the usual way, or by divisibility.

Example 1.6 If $(E_1; \leqslant_1), \ldots, (E_n; \leqslant_n)$ are ordered sets then the cartesian product set $\underset{i=1}{\overset{n}{\times}} E_i$ can be given the **cartesian order** $\leqslant$ defined by

$$(x_1, \ldots, x_n) \leqslant (y_1, \ldots, y_n) \iff (i = 1, \ldots, n)\ x_i \leqslant_i y_i.$$

More generally, if $\big((E_\alpha; \leqslant)\big)_{\alpha \in A}$ is a family of ordered sets then we can order the cartesian product set $\underset{\alpha \in A}{\times} E_\alpha$ by defining

$$(x_\alpha)_{\alpha \in A} \leqslant (y_\alpha)_{\alpha \in A} \iff (\forall \alpha \in A)\ x_\alpha \leqslant y_\alpha.$$

Note that here we have used the same symbol $\leqslant$ for each of the orders involved.

Example 1.7 Let E and F be ordered sets. Then the set $\operatorname{Map}(E, F)$ of all mappings $f : E \to F$ can be ordered by defining

$$f \leqslant g \iff (\forall x \in E)\ f(x) \leqslant g(x).$$

In particular, if we let $\mathbf{n} = \{1, 2, \ldots, n\}$ and consider a real $n \times n$ matrix $A = [a_{ij}]$ to be the mapping $f : \mathbf{n} \times \mathbf{n} \to \mathbb{R}$ given by $f(i, j) = a_{ij}$ then we can order the set of such matrices by

$$A \leqslant B \iff (\forall i, j)\ a_{ij} \leqslant b_{ij}.$$

We say that elements x, y of an ordered set $(E; \leqslant)$ are **comparable** if either $x \leqslant y$ or $y \leqslant x$. We denote this symbolically by writing $x \nparallel y$. If all pairs of elements of E are comparable then we say that E forms a **chain**, or that $\leqslant$ is a **total order**. In contrast, we say that $x, y \in E$ are **incomparable**, and write $x \parallel y$, when $x \nleqslant y$ and $y \nleqslant x$. If all pairs of distinct elements of E are incomparable then clearly $\leqslant$ is equality, in which case we say that E forms an **antichain**.

Example 1.8 The sets $\mathbb{N}$, $\mathbb{Z}$, $\mathbb{Q}$, $\mathbb{R}$ of natural numbers, integers, rationals, and real numbers form chains under their usual orders.

Example 1.9 In Example 1.2, the singleton subsets of $\mathbb{P}(E)$ form an antichain under the inherited inclusion order.

EXERCISES

1.1. Let $(P_1; \leqslant_1)$ and $(P_2; \leqslant_2)$ be ordered sets. Prove that the relation $\leqslant$ defined on $P_1 \times P_2$ by

$$(x_1, y_1) \leqslant (x_2, y_2) \iff \begin{cases} x_1 <_1 x_2, \\ or \ x_1 = x_2 \ and \ y_1 \leqslant_2 y_2 \end{cases}$$

is an order (the **lexicographic order** on $P_1 \times P_2$). Show also that $\leqslant$ is a total order if and only if $\leqslant_1$ and $\leqslant_2$ are total orders.

1.2. Let P_1 and P_2 be disjoint sets. If $\leqslant_1$ is an order on P_1 and $\leqslant_2$ is an order on P_2 prove that the following defines an order on $P_1 \cup P_2$:

$$x \leqslant y \iff \begin{cases} x, y \in P_1 \ and \ x \leqslant_1 y, \\ or \ x, y \in P_2 \ and \ x \leqslant_2 y. \end{cases}$$

The resulting ordered set is called the **ordered disjoint union** of P_1 and P_2 and is denoted by $P_1 \dot\cup P_2$.

1.3. Let P_1 and P_2 be disjoint sets. If $\leqslant_1$ is an order on P_1 and $\leqslant_2$ is an order on P_2 prove that the following defines an order on $P_1 \cup P_2$:

$$x \leqslant y \iff \begin{cases} x, y \in P_1 \ and \ x \leqslant_1 y, \\ or \ x, y \in P_2 \ and \ x \leqslant_2 y, \\ or \ x \in P_1 \ and \ y \in P_2. \end{cases}$$

The resulting ordered set is called the **vertical sum**, or the **linear sum**, of P_1 and P_2 and is denoted by $P_1 \oplus P_2$.

1.4. Let E be an ordered set in which every chain and every antichain is finite. Prove that E is finite.

Theorem 1.1 *If R is an order on E then so is its dual R^d.*

Proof Clearly, if R satisfies the properties of being reflexive, anti-symmetric and transitive then so also does R^d. $\square$

In what follows we shall denote the dual of an order $\leqslant$ on E by the symbol $\geqslant$ which we read as 'greater than or equal to'. Then the ordered set $(E; \geqslant)$ is called the **dual** of $(E; \leqslant)$ and is often written as E^d.

As a consequence of Theorem 1.1 we can assert that to every statement that concerns an order on a set E there is a dual statement that concerns the corresponding dual order on E. This is the basis of the useful

Principle of Duality *To every theorem that concerns an ordered set E there is a corresponding theorem that concerns the dual ordered set E^d. This is obtained by replacing each statement that involves $\leqslant$, explicitly or implicitly, by its dual.*

In what follows we shall make several applications of the Principle of Duality. By way of illustration, if $(E; \leqslant)$ is an ordered set then by a **top element** or **maximum element** of E we mean an element $x \in E$ such that $y \leqslant x$ for every $y \in E$. A top element, when it exists, is unique. In fact, if x, y are both top elements of E then on the one hand $y \leqslant x$ and on the other $x \leqslant y$ whence, by the anti-symmetric property of $\leqslant$, we have $x = y$. The dual notion is that of a **bottom element** or **minimum element**, namely an element $z \in E$ such that $z \leqslant y$ for every $y \in E$. By the above and the Principle of Duality, we can assert immediately that a bottom element, when it exists, is unique. An ordered set that has both a top element and a bottom element is said to be **bounded**.

In what follows we shall use the notation $x < y$ to mean $x \leqslant y$ and $x \neq y$. Note that the relation $<$ thus defined is transitive but is not an order since it fails to be reflexive; moreover, $x < y$ and $y < x$ are incompatible. We denote the dual of the relation $<$ by the symbol $>$.

Definition In an ordered set $(E; \leqslant)$ we say that x is **covered by** y (or that y **covers** x) if $x < y$ and there is no $a \in E$ such that $x < a < y$. We denote this by using the notation $x \prec y$.

Many ordered sets can be represented by means of a **Hasse diagram**. In such a diagram we represent elements by points and interpret $x \prec y$ by

i.e. we join the points representing x and y by an increasing line segment.

Example 1.10 Let $E = \{1, 2, 3, 4, 6, 12\}$ be the set of positive divisors of 12. If we order E in the usual way, we obtain a chain. If we order E by divisibility, we obtain the Hasse diagram

Example 1.11 Ordered by set inclusion, $\mathbb{P}(\{a, b, c\})$ has Hasse diagram

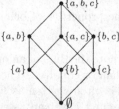

Clearly, the Hasse diagram for the dual of an ordered set E is obtained by turning that of E upside-down.

EXERCISES

1.5. Draw the Hasse diagrams for all possible orders on sets of $3, 4, 5$ elements.

1.6. Draw the Hasse diagram for the set of positive divisors of 210 when ordered by divisibility.

1.7. If p, q are distinct primes and m, n are positive integers, draw the Hasse diagram for the set of positive divisors of $p^m q^n$, ordered by divisibility.

1.8. Let P_1 and P_2 be the ordered sets with Hasse diagrams

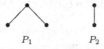

P_1 P_2

Draw the Hasse diagrams of $P_1 \times P_2$ and $P_2 \times P_1$ under the cartesian order (Example 1.6). Comment on the result.

1.9. With P_1 and P_2 as in Exercise 1.8, draw the Hasse diagrams of $P_1 \times P_2$ and $P_2 \times P_1$ under the lexicographic order (Exercise 1.1), and those of the vertical sums $P_1 \oplus P_2$ and $P_2 \oplus P_1$ (Exercise 1.3).

1.2 Order-preserving mappings

Definition If $(A; \leqslant_1)$ and $(B; \leqslant_2)$ are ordered sets then we say that a mapping $f : A \to B$ is **isotone** (or **order-preserving**) if

$$(\forall x, y \in A) \quad x \leqslant_1 y \Rightarrow f(x) \leqslant_2 f(y);$$

and is **antitone** (or **order-inverting**) if

$$(\forall x, y \in A) \quad x \leqslant_1 y \Rightarrow f(x) \geqslant_2 f(y).$$

Example 1.12 If E is a non-empty set and $A \subseteq E$ then $f_A : \mathbb{P}(E) \to \mathbb{P}(E)$ given by $f_A(X) = A \cap X$ is isotone. If X' is the complement of X in E then the assignment $X \mapsto X'$ defines an antitone mapping on $\mathbb{P}(E)$.

Example 1.13 Given $f : E \to F$ consider the induced **direct image map** $f^\to : \mathbb{P}(E) \to \mathbb{P}(F)$ defined for every $X \subseteq E$ by $f^\to(X) = \{f(x) \mid x \in X\}$ and the induced **inverse image map** $f^\leftarrow : \mathbb{P}(F) \to \mathbb{P}(E)$ defined for every $Y \subseteq F$ by $f^\leftarrow(Y) = \{x \in E \mid f(x) \in Y\}$. Each of these mappings is isotone.

We shall now give a natural interpretation of isotone mappings. For this purpose we require the following notions.

Definition By a **down-set** (or **hereditary subset**) of an ordered set $(E; \leqslant)$ we shall mean a subset D of E with the property that if $x \in D$ and $y \in E$ is such that $y \leqslant x$ then $y \in D$. We include the empty subset of E as a down-set. By a **principal down-set** we shall mean a down-set of the form $x^\downarrow = \{y \in E \mid y \leqslant x\}$. Dually, we define an **up-set** to be a subset U such that if $x \in U$ and $y \in E$ is such that $y \geqslant x$ then $y \in U$; and a **principal up-set** to be an up-set of the form $x^\uparrow = \{y \in E \mid y \geqslant x\}$.

Example 1.14 In the chain $\mathbb{Q}^+$ of positive rationals the set $\{q \in \mathbb{Q}^+ \mid q^2 \leqslant 2\}$ is a down-set that is not principal.

Example 1.15 If A and B are down-sets of an ordered set E then clearly so also are $A \cap B$ and $A \cup B$. This is not true in general for principal down-sets. For example, in

$$\begin{array}{cc} c\bullet & \bullet d \\ & \\ a\bullet & \bullet b \end{array}$$

we have $c^{\downarrow} \cap d^{\downarrow} = \{a, b\} = a^{\downarrow} \cup b^{\downarrow}$.

Isotone mappings are characterised by the following properties.

Theorem 1.2 *If E, F are ordered sets and if $f : E \to F$ is any mapping then the following statements are equivalent:*

(1) *f is isotone;*

(2) *the inverse image of every principal down-set of F is a down-set of E;*

(3) *the inverse image of every principal up-set of F is an up-set of E.*

Proof (1) $\Rightarrow$ (2): Suppose that f is isotone. Let $y \in F$ and let $A = f^{\leftarrow}(y^{\downarrow})$. If $A \neq \emptyset$ let $x \in A$. Then for every $z \in E$ with $z \leqslant x$ we have $f(z) \leqslant f(x) \leqslant y$ whence $z \in A$. Thus A is a down-set of E.

(2) $\Rightarrow$ (1): For every $x \in E$ we have $x \in f^{\leftarrow}[f(x)^{\downarrow}]$. By (2) this is a down-set of E, so if $y \in E$ is such that $y \leqslant x$ we have $y \in f^{\leftarrow}[f(x)^{\downarrow}]$. It follows that $f(y) \leqslant f(x)$ and therefore f is isotone.

(1) $\Leftrightarrow$ (3): This follows from the above by the Principle of Duality. $\square$

1.3 Residuated mappings

In view of the above natural result, we now investigate under what conditions the inverse image of a principal down-set is also a principal down-set. The outcome will be a type of mapping that will play an important role in the sequel.

Theorem 1.3 *If E, F are ordered sets then the following conditions concerning $f : E \to F$ are equivalent:*

(1) *the inverse image under f of every principal down-set of F is a principal down-set of E;*

(2) *f is isotone and there is an isotone mapping $g : F \to E$ such that $g \circ f \geqslant \mathrm{id}_E$ and $f \circ g \leqslant \mathrm{id}_F$.*

Proof (1) $\Rightarrow$ (2): If (1) holds then it follows from Theorem 1.2 that f is isotone. In symbolic form, (1) becomes

$$(\forall y \in F)(\exists x \in E) \; f^{\leftarrow}(y^{\downarrow}) = x^{\downarrow}.$$

Now for every given $y \in F$ this element x is clearly unique, so we can define a mapping $g : F \to E$ by setting $g(y) = x$. Since $f^{\leftarrow}$ is isotone it follows that so is g. For this mapping g we have $g(y) \in g(y)^{\downarrow} = x^{\downarrow} = f^{\leftarrow}(y^{\downarrow})$, so $f[g(y)] \leqslant y$ for all $y \in F$ and therefore $f \circ g \leqslant \mathrm{id}_F$; and $x \in f^{\leftarrow}[f(x)^{\downarrow}] = g[f(x)]^{\downarrow}$ so that $x \leqslant g[f(x)]$ for all $x \in E$ and therefore $g \circ f \geqslant \mathrm{id}_E$.

$(2) \Rightarrow (1)$: If (2) holds then on the one hand we have

$$f(x) \leqslant y \Rightarrow x \leqslant g[f(x)] \leqslant g(y),$$

and on the other we have

$$x \leqslant g(y) \Rightarrow f(x) \leqslant f[g(y)] \leqslant y.$$

It follows from these observations that $f(x) \leqslant y$ if and only if $x \leqslant g(y)$ and therefore $f^{\leftarrow}(y^{\downarrow}) = g(y)^{\downarrow}$ from which (1) follows. $\qquad \square$

Definition A mapping $f : E \to F$ that satisfies either of the equivalent conditions of Theorem 1.3 is said to be **residuated**.

We note in particular that if $f : E \to F$ is a residuated mapping then an isotone mapping $g : F \to E$ which is such that $g \circ f \geqslant \mathrm{id}_E$ and $f \circ g \leqslant \mathrm{id}_F$ is in fact *unique*. To see this, suppose that g and $g^{\star}$ are each isotone and satisfy these properties. Then $g = \mathrm{id}_E \circ g \leqslant (g^{\star} \circ f) \circ g = g^{\star} \circ (f \circ g) \leqslant g^{\star} \circ \mathrm{id}_F = g^{\star}$. Similarly, $g^{\star} \leqslant g$ and therefore $g = g^{\star}$.

We shall denote this unique g by f^{+} and call it the **residual** of f.

It is clear from the above that $f : E \to F$ is residuated if and only if, for every $y \in F$, there exists

$$f^{+}(y) = \max f^{\leftarrow}(y^{\downarrow}) = \max\{x \in E \mid f(x) \leqslant y\}.$$

Moreover, $f^{+} \circ f \geqslant \mathrm{id}_E$ and $f \circ f^{+} \leqslant \mathrm{id}_F$.

Example 1.16 Simple calculations reveal that if $f : E \to F$ then the direct image map $f^{\to} : \mathbb{P}(E) \to \mathbb{P}(F)$ is residuated with residual $f^{\leftarrow} : \mathbb{P}(F) \to \mathbb{P}(E)$.

Example 1.17 If E is any set and $A \subseteq E$ then $\lambda_A : \mathbb{P}(E) \to \mathbb{P}(E)$ defined by $\lambda_A(X) = A \cap X$ is residuated with residual λ_A^{+} given by $\lambda_A^{+}(Y) = Y \cup A'$.

Example 1.18 For $m \in \mathbb{N}\setminus 0$ define $f_m : \mathbb{N} \to \mathbb{N}$ by $f_m(n) = mn$. Then f_m is residuated with $f_m^{+}(p) = \lfloor \frac{p}{m} \rfloor$ where $\lfloor q \rfloor$ denotes the integer part of $q \in \mathbb{Q}$.

Example 1.19 Every bounded operator f on a Hilbert space H induces a residuated mapping on the set of closed subspaces of H, namely that given by $M \mapsto [f^{\to}(M)]^{\perp\perp}$.

Example 1.20 If S is a semigroup, define a multiplication on $\mathbb{P}(S)$ by

$$XY = \begin{cases} \{xy \mid x \in X, y \in Y\} & \text{if } X, Y \neq \emptyset; \\ \emptyset & \text{otherwise.} \end{cases}$$

Then multiplication by a fixed subset of S is a residuated mapping on $\mathbb{P}(S)$.

Example 1.21 If R is a commutative ring with a 1 then multiplication by a fixed ideal of R is a residuated mapping on the ordered set $(I(R); \subseteq)$ of ideals.

The notion of a residuated mapping has its roots in investigations by Certaine [37], Ward and Dilworth [112], and Dilworth [43] into multiplicative ideal theory which relates directly to Example 1.21.

EXERCISES

1.10. For each integer $n \geqslant 1$ let $\mathbf{n}$ denote the chain $1 < 2 < 3 < \cdots < n$. Prove that a mapping $f : \mathbf{n} \to \mathbf{m}$ is residuated if and only if it is isotone and $f(1) = 1$.

1.11. Let E be a bounded ordered set with bottom element 0 and top element 1. Given $e \in E$, let $\alpha_e, \beta_e : E \to E$ be defined by

$$\alpha_e(x) = \begin{cases} 0 & \text{if } x = 0; \\ e & \text{otherwise,} \end{cases} \qquad \beta_e(x) = \begin{cases} 0 & \text{if } x \leqslant e; \\ 1 & \text{otherwise.} \end{cases}$$

Prove that α_e and β_e are residuated and determine their residuals.

1.12. Let E be the ordered set given by the Hasse diagram

Show that the mapping $f : E \to E$ given by

$$f(x_{ij}) = \begin{cases} x_{0,j+1} & \text{if } i = 0, j \neq n; \\ x_{01} & \text{if } i = 0, j = n; \\ x_{i-1,j+1} & \text{if } i \neq 0, j \neq n; \\ x_{i-1,1} & \text{if } i \neq 0, j = n. \end{cases}$$

is residuated and determine f^+.

For every non-empty set E the residuated mappings on $\mathbb{P}(E)$ are completely described in the following result.

Theorem 1.4 *Let E be a non-empty set and let R be a binary relation on E. Then the mapping $\xi_R : \mathbb{P}(E) \to \mathbb{P}(E)$ given by the prescription*

$$\xi_R(A) = \{y \in E \mid (\exists x \in A)\ (x, y) \in R\}$$

is residuated. Moreover, every residuated mapping $f : \mathbb{P}(E) \to \mathbb{P}(E)$ is of this form for some binary relation R on E.

Proof Let $i : \mathbb{P}(E) \to \mathbb{P}(E)$ be the antitone mapping that sends each subset of E to its complement. Consider the isotone mapping $\xi_R^+ = i \circ \xi_{R^d} \circ i$. It is readily verified (draw pictures!) that $\xi_R \circ \xi_R^+ \leqslant \text{id}$ and $\xi_R^+ \circ \xi_R \geqslant \text{id}$, whence ξ_R^+ is the residual of ξ_R.

To see that every residuated mapping $f : \mathbb{P}(E) \to \mathbb{P}(E)$ is of this form for some binary relation R on E, consider the relation R_f defined on E by

$$(x, y) \in R_f \iff y \in f(\{x\}).$$

Observe that $\xi_{R_f}(\{x\}) = \{y \in E \mid (x, y) \in R_f\} = f(\{x\})$, so that f and ξ_{R_f} agree on singletons.

Now if $k : \mathbb{P}(E) \to \mathbb{P}(E)$ is any residuated mapping then, since it is isotone, for every non-empty subset A of E we have $k(A) = k\left(\bigcup_{x \in A} \{x\} \right) = \bigcup_{x \in A} k(\{x\})$. In fact, if $B = \bigcup_{x \in A} k(\{x\})$ then clearly $k(A) \supseteq B$. On the other hand, $k(\{x\}) \subseteq B$ for every $x \in A$ and so $\{x\} \subseteq k^+(B)$ whence $A = \bigcup_{x \in A} \{x\} \subseteq k^+(B)$ and therefore $k(A) \subseteq B$. The resulting equality, applied to both f and ξ_{R_f}, together with the fact that f and ξ_{R_f} agree on singletons now gives $f(A) = \bigcup_{x \in A} f(\{x\}) = \bigcup_{x \in A} \xi_{R_f}(\{x\}) = \xi_{R_f}(A)$ whence we obtain $f = \xi_{R_f}$. $\quad\square$

Particular properties of residuated mappings are the following.

Theorem 1.5 *If $f : E \to F$ is residuated then*
$$f \circ f^+ \circ f = f \quad \text{and} \quad f^+ \circ f \circ f^+ = f^+.$$

Proof Since f is isotone, it follows from Theorem 1.3 that $f \circ f^+ \circ f \geqslant f \circ \mathrm{id}_E = f$, and that $f \circ f^+ \circ f \leqslant \mathrm{id}_F \circ f = f$, from which the first equality follows. The second is established similarly. $\quad\square$

Theorem 1.6 *If $f : E \to F$ and $g : F \to G$ are residuated mappings then so is $g \circ f : E \to G$, and $(g \circ f)^+ = f^+ \circ g^+$.*

Proof Clearly, $g \circ f$ and $f \circ g$ are isotone. Moreover,
$$(f^+ \circ g^+) \circ (g \circ f) \geqslant f^+ \circ \mathrm{id}_F \circ f = f^+ \circ f \geqslant \mathrm{id}_E;$$
$$(g \circ f) \circ (f^+ \circ g^+) \leqslant g \circ \mathrm{id}_F \circ g^+ = g \circ g^+ \leqslant \mathrm{id}_G.$$
Thus, by the uniqueness of residuals, $(g \circ f)^+$ exists and is $f^+ \circ g^+$. $\quad\square$

Corollary *For every ordered set E the set $\mathrm{Res}\, E$ of residuated mappings $f : E \to E$ forms a semigroup, as does the set $\mathrm{Res}^+ E$ of residual mappings $f^+ : E \to E$.* $\quad\square$

EXERCISES

1.13. If $f, g : E \to E$ are residuated prove that $f \leqslant g \iff g^+ \leqslant f^+$. Deduce that the semigroups $\mathrm{Res}\, E$ and $\mathrm{Res}^+ E$ are anti-isomorphic.

1.14. If $f : E \to E$ is residuated prove that $f = f^+ \iff f^2 = \mathrm{id}_E$.

1.15. If $f : E \to F$ is residuated prove that the following are equivalent:
$$(1)\ f^+ \circ f = \mathrm{id}_E; \quad (2)\ f \text{ is injective}; \quad (3)\ f^+ \text{ is surjective}.$$

1.16. If E has a top element 1 prove that the mapping $\Theta : \mathrm{Res}\, E \to E$ given by $\Theta(f) = f(1)$ is residuated, with residual the mapping $\Psi : E \to \mathrm{Res}\, E$ given by $\Psi(e) = \alpha_e$ where α_e is defined in Exercise 1.11.

1.17. Let S be a semigroup with a zero element 0. Let $\mathbb{P}_0(S)$ be the set of all subsets of S that contain 0. For each $A \in \mathbb{P}_0(S)$ let $\lambda_A : \mathbb{P}_0(S) \to \mathbb{P}_0(S)$ be given by
$$\lambda_A(X) = AX = \{ax \mid a \in A, x \in X\}.$$
Prove that λ_A is residuated and determine λ_A^+. Do likewise for the mapping $\rho_A : \mathbb{P}_0(S) \to \mathbb{P}_0(S)$ given by $\rho_A(X) = XA$.

1.18. Let **3** be the three-element chain $1 < 2 < 3$. Show that there are 10 isotone mappings $f : \mathbf{3} \to \mathbf{3}$, of which 6 are residuated. Obtain the Cayley tables for the semigroups $\operatorname{Res} E$ and $\operatorname{Res}^+ E$. Deduce that in general $\operatorname{Res} E \cup \operatorname{Res}^+ E$ is not a semigroup.

1.19. Let E be an ordered set and let $f \in \operatorname{Res} E$. Prove that the following statements are equivalent:

(1) there exist idempotents $a, b \in \operatorname{Res} E$ such that
$$\operatorname{Im} a = \operatorname{Im} f, \ \operatorname{Im} b^+ = \operatorname{Im} f^+;$$

(2) there exists $g \in \operatorname{Res} E$ such that $f = fgf$ and $g = gfg$.

1.20. Let E be a finite chain. If $f \in \operatorname{Res} E$ let $g : E \to E$ be given by
$$g(x) = \begin{cases} f^+(x) & \text{if } x \neq 0; \\ 0 & \text{if } x = 0. \end{cases}$$

Prove that $g \in \operatorname{Res} E$ with $f = fgf$. Deduce that for a finite chain E the semigroup $\operatorname{Res} E$ is regular.

1.4 Closures

We now consider an important type of isotone mapping that is intimately related to a residuated mapping.

Definition An isotone mapping $f : E \to E$ is a **closure** on E if it is such that $f = f^2 \geqslant \operatorname{id}_E$; and a **dual closure** if $f = f^2 \leqslant \operatorname{id}_E$.

Example 1.22 If $A \subseteq E$ then $\mu_A : \mathbb{P}(E) \to \mathbb{P}(E)$ given by $\mu_A(X) = A \cup X$ is a closure on $\mathbb{P}(E)$; and $\lambda_A : \mathbb{P}(E) \to \mathbb{P}(E)$ given by $\lambda_A(X) = A \cap X$ is a dual closure.

Theorem 1.7 *If E is an ordered set then $f : E \to E$ is a closure if and only if there is an ordered set F and a residuated mapping $g : E \to F$ such that $f = g^+ \circ g$.*

Proof $\Rightarrow$: Suppose that $f : E \to E$ is a closure. Let R be the kernel of f, i.e. the equivalence relation defined on E by
$$(x, y) \in R \iff f(x) = f(y).$$
Define the relation $\sqsubseteq$ on the quotient set E/R by
$$[x]_R \sqsubseteq [y]_R \iff f(x) \leqslant f(y).$$
It is readily seen that $\sqsubseteq$ is an order on E/R and, since f is isotone, the natural mapping $\natural_R : E \to E/R$ is isotone. Now since f is a closure every R-class has a top element, that in $[x]_R$ being $f(x)$. We can therefore define a mapping $g : E/R \to E$ by setting $g([x]_R) = f(x)$. We then have
$$\begin{cases} (g \circ \natural_R)(x) = g([x]_R) = f(x) \geqslant x; \\ (\natural_R \circ g)([x]_R) = \natural_R[f(x)] = [f(x)]_R = [x]_R. \end{cases}$$
It follows that $\natural_R$ is residuated with $\natural_R^+ = g$ and that $f = \natural_R^+ \circ \natural_R$.

$\Leftarrow$: Suppose conversely that there is an ordered set F and a residuated mapping $g : E \to F$ such that $f = g^+ \circ g$. Then on the one hand $g^+ \circ g \geqslant \mathrm{id}_E$; and on the other, by Theorem 1.5, $g = g \circ g^+ \circ g$, so that $g^+ \circ g = (g^+ \circ g)^2$. Since $g^+ \circ g$ is isotone it follows that $f = g^+ \circ g$ is a closure on E. $\square$

There is of course a dual result to Theorem 1.7, namely that $f : E \to E$ is a dual closure if and only if there is an ordered set F and a residuated mapping $g : E \to F$ such that $f = g \circ g^+$.

If $f : E \to E$ is a closure or a dual closure and if $x \in \mathrm{Im}\, f$ then $x = f(y)$ for some $y \in E$, whence we obtain $f(x) = f^2(y) = f(y) = x$. Consequently, we see that $\mathrm{Im}\, f = \{x \in E \mid f(x) = x\}$, the set of **fixed points** of f.

Definition A subset F of an ordered set E is called a **(dual) closure subset** if there is a (dual) closure $f : E \to E$ such that $F = \mathrm{Im}\, f$.

The concept of a closure is due to E. H. Moore [85] whereas that of a closure subset originated in the work of Riesz [94] in connection with topological spaces.

Example 1.23 In Example 1.22, for every $A \subseteq E$ the closure subset associated with $\mu_A : X \mapsto A \cup X$ is $\{X \in \mathbb{P}(E) \mid A \subseteq X\}$. The dual closure subset associated with $\lambda_A : X \mapsto A \cap X$ is $\mathbb{P}(A)$.

Closure subsets can be characterised as follows.

Theorem 1.8 *A subset F of an ordered set E is a closure subset of E if and only if for every $x \in E$ the set $x^\uparrow \cap F$ has a bottom element.*

Proof Suppose that F is a closure subset of E and let $f : E \to E$ be a closure such that $F = \mathrm{Im}\, f$. Then for every $x \in E$ the set $x^\uparrow \cap F$ is not empty since clearly it contains the element $f(x)$. Moreover, if $z \in x^\uparrow \cap F$ then $x \leqslant z$ and $f(x) \leqslant f(z) = z$. Consequently $x^\uparrow \cap F$ has a bottom element, namely $f(x)$.

Conversely, suppose that for every $x \in E$ the set $x^\uparrow \cap F$ has a bottom element, $x_\star$ say, and consider the mapping $f : E \to E$ given by $f(x) = x_\star$. If $x \leqslant y$ then $x^\uparrow \supseteq y^\uparrow$ gives $x^\uparrow \cap F \supseteq y^\uparrow \cap F$ whence it follows that $x_\star \leqslant y_\star$ and so f is isotone. Moreover, since $f(x) = x_\star \geqslant x$ for every $x \in E$ we also have $f \geqslant \mathrm{id}_E$. Now for any $y \in F$ we clearly have $y = y_\star = f(y) \in \mathrm{Im}\, f$. Applying this to $f(x) = x_\star \in F$ we obtain $f^2(x) = f(x)$. Hence $f^2 = f$ and so f is a closure with $\mathrm{Im}\, f = F$. $\square$

Definition A subset M of an ordered set E is said to be **bicomplete** if, for every $x \in E$, $x^\uparrow \cap M$ has a bottom element and $x^\downarrow \cap M$ has a top element. Equivalently, M is bicomplete if it is a closure subset of both E and E^d.

Theorem 1.9 *For an ordered set E there is a bijection between the set of residuated closure mappings on E and the set of bicomplete subsets of E, namely that given by $f \mapsto \mathrm{Im}\, f$.*

Proof Let f be a residuated closure on E. Then $f^+ \circ f = f^+ \circ f \circ f \geqslant \mathrm{id}_E \circ f = f$ and, by Theorem 1.5, $f = f \circ f^+ \circ f \geqslant \mathrm{id}_E \circ f^+ \circ f = f^+ \circ f$. Hence $f = f^+ \circ f$, and similarly $f^+ = f \circ f^+$. If now $x \in \mathrm{Im}\, f$ we have $x = f(x)$ and then $f^+(x) = (f^+ \circ f)(x) = f(x) = x$; and dually $x \in \mathrm{Im}\, f^+$ gives $x = f(x)$. Thus we see that $\mathrm{Im}\, f = \mathrm{Im}\, f^+$, whence it follows by Theorem 1.8 and its dual that $\mathrm{Im}\, f$ is bicomplete.

Suppose now that M is a bicomplete subset of E. Define $f, g : E \to E$ by setting, for every $x \in E$,

$$f(x) = \text{the bottom element of } x^\uparrow \cap M;$$
$$g(x) = \text{the top element of } x^\downarrow \cap M.$$

Then f and g are both isotone, $f \circ g = g \leqslant \mathrm{id}_E$ and $g \circ f = f \geqslant \mathrm{id}_E$. It follows that f is a residuated closure with $f^+ = g$. Clearly, we have $\mathrm{Im}\, f = M$. Since a (dual) closure mapping is completely determined by its image, the result follows.   □

EXERCISES

1.21. Let $X \mapsto \overline{X}$ be a closure on $\mathbb{P}(E)$. Prove that, for all $X, Y \in \mathbb{P}(E)$,
$$\overline{X \cup Y} = \overline{\overline{X} \cup Y} = \overline{X \cup \overline{Y}} = \overline{\overline{X} \cup \overline{Y}}.$$
Deduce that, for all $X, Y, Z \in \mathbb{P}(E)$, $\overline{X \cup Y \cup Z} = \overline{\overline{X} \cup Y} \cup Z$.

1.22. Let E be an ordered set. Prove that a mapping $f : E \to E$ is a closure if and only if
  (1) $(\forall x \in E) \quad x \leqslant f(x)$;
  (2) $(\forall x, y \in E) \quad x \leqslant f(y) \implies f(x) \leqslant f(y)$.

1.23. Consider the cartesian product set $E = \{0, 1\} \times \mathbb{N}$. Define an order on E by setting
$$\begin{cases} (0, m) \leqslant (0, n) \iff m \leqslant n; \\ (\forall n \in \mathbb{N}) \quad (0, n) < (1, n + 1). \end{cases}$$
Show that the only closure on E is the identity.

1.24. Let E be an ordered set and let $f, g : E \to E$ be closure mappings. Prove that the statements
$$f \leqslant g; \quad fg = g; \quad gf = g; \quad \mathrm{Im}\, g \subseteq \mathrm{Im}\, f$$
are equivalent.

1.25. Prove that the residuated mapping ξ_R of Theorem 1.4 is a closure if and only if R is reflexive and transitive. If R is an equivalence relation, prove that ξ_R is a **quantifier** in the sense that it is isotone and, for all $X, Y \in \mathbb{P}(E)$,
$$\xi_R[X \cap \xi_R(Y)] = \xi_R(X) \cap \xi_R(Y).$$

1.5 Isomorphisms of ordered sets

We now consider the notion of an *isomorphism* of ordered sets. Clearly, whatever properties of a bijection $f : E \to F$ that we require in order to define an isomorphism, we certainly want $f^{-1} : F \to E$ also to be an isomorphism. In this connection, we note that *simply choosing f to be an isotone bijection is not enough*. For example, consider the ordered sets with Hasse diagrams

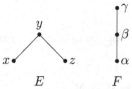

$$E \qquad F$$

The mapping $f : E \to F$ given by $f(x) = \beta$, $f(y) = \gamma$, $f(z) = \alpha$ is an isotone bijection, but f^{-1} is not isotone since $\alpha < \beta$ and $f^{-1}(\alpha) = z \parallel x = f^{-1}(\beta)$.

Definition By an **order isomorphism** from an ordered set E to an ordered set F we shall mean an isotone bijection $f : E \to F$ whose inverse $f^{-1} : F \to E$ is also isotone.

From the above results we can see that the notion of an order isomorphism is equivalent to that of a bijection f that is residuated, the residual of f being f^{-1}. If there is an order isomorphism $f : E \to F$ then we say that E, F are (order) **isomorphic**.

Example 1.24 The ordered sets with the following Hasse diagrams are isomorphic:

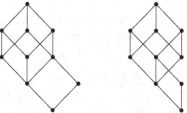

A useful criterion for an isomorphism of ordered sets is the following.

Theorem 1.10 *Ordered sets E and F are isomorphic if and only if there is a surjective mapping $f : E \to F$ such that*

$$x \leqslant y \iff f(x) \leqslant f(y).$$

Proof The necessity is clear. Suppose conversely that such a surjective mapping f exists. Then f is also injective; for if $f(x) = f(y)$ then from $f(x) \leqslant f(y)$ we obtain $x \leqslant y$, and from $f(x) \geqslant f(y)$ we obtain $x \geqslant y$, so that $x = y$. Hence f is a bijection. Clearly, f is isotone; and so also is f^{-1} since $x \leqslant y$ can be written $f[f^{-1}(x)] \leqslant f[f^{-1}(y)]$ which gives $f^{-1}(x) \leqslant f^{-1}(y)$. $\square$

We shall denote the fact that ordered sets E and F are isomorphic by writing $E \simeq F$. We shall say that E and F are **dually isomorphic** if $E \simeq F^d$ or, equivalently, $F \simeq E^d$. In the particular case where $E \simeq E^d$ we say that E is **self-dual**.

Example 1.25 Let $\mathrm{Sub}\,\mathbb{Z}$ be the set of subgroups of the additive abelian group $\mathbb{Z}$ and order $\mathrm{Sub}\,\mathbb{Z}$ by set inclusion. Then $(\mathbb{N}; |)$ is dually isomorphic to $(\mathrm{Sub}\,\mathbb{Z}; \subseteq)$ under the assignment $n \mapsto n\mathbb{Z}$. In fact, since every subgroup of $\mathbb{Z}$ is of the form $n\mathbb{Z}$ for some $n \in \mathbb{N}$, this assignment is surjective. Also, we have $n\mathbb{Z} \subseteq m\mathbb{Z}$ if and only if $m|n$. Note that we include 0 in $\mathbb{N}$. Then 0 is the top element of $(\mathbb{N}; |)$ and corresponds to the trivial subgroup $\{0\}$. The result therefore follows by Theorem 1.10.

EXERCISES

1.26. If E and F are ordered sets prove that, under the cartesian orders, $E \times F \simeq F \times E$.

1.27. Prove that $(E \times F)^d \simeq E^d \times F^d$.

1.28. Prove that $(\mathbb{P}(E); \subseteq)$ is self-dual.

1.29. Let E and F be finite ordered sets. If $f : E \to F$ is a bijection prove that f is an order isomorphism if and only if

$$(\forall x, y \in E) \quad x \prec y \iff f(x) \prec f(y).$$

1.30. Let **2** denote the two-element chain $0 < 1$. Prove that the mapping $f : \mathbb{P}(\{1, 2, \ldots, n\}) \to \mathbf{2}^n$ given by $f(X) = (x_1, \ldots, x_n)$ where

$$x_i = \begin{cases} 1 & \text{if } i \in X; \\ 0 & \text{otherwise,} \end{cases}$$

is an order isomorphism.

1.6 Galois connections

We now introduce a concept that is intimately related to that of a residuated mapping. In what follows it will often prove convenient to denote composites by juxtaposition, i.e. we shall write $f \circ g$ as simply fg.

Definition Given ordered sets E, F and *antitone* mappings $f : E \to F$ and $g : F \to E$, we say that the pair (f, g) establishes a **Galois connection** between E and F if $fg \geqslant \mathrm{id}_F$ and $gf \geqslant \mathrm{id}_E$.

Example 1.26 If F is a field and K is a separable normal extension of finite degree over F then there is a Galois connection between the set of subgroups of the Galois group $\mathrm{Gal}(K, F)$ and the set of subfields of K that contain F. This fundamental result of Galois in field theory is the model on which the notion of a Galois connection is founded.

Example 1.27 Let S be a semigroup with a zero element 0. For every $A \subseteq S$ define respectively the **left** and **right annihilators** of A by

$$L(A) = \{x \in S \mid (\forall a \in A) \ xa = 0\};$$
$$R(A) = \{x \in S \mid (\forall a \in A) \ ax = 0\}.$$

For convenience, when $A = \{x\}$ we shall write $L(x)$ for $L(\{x\})$, and similarly $R(x)$ for $R(\{x\})$. The mappings $L, R : \mathbb{P}(S) \to \mathbb{P}(S)$ so defined are both antitone and, as can readily be verified, we have $\mathrm{id}_{\mathbb{P}(S)} \leqslant LR$ and $\mathrm{id}_{\mathbb{P}(S)} \leqslant RL$. Consequently we see that (L, R) is a Galois connection.

Galois connections offer an alternative approach to residuated mappings. To see this, let (f, g) be a Galois connection between E and F. Then clearly we can regard f as an isotone mapping from E to F^d, and likewise g as an isotone mapping from F^d to E. In this way we see that $f : E \to F^d$ is residuated with residual g. As a consequence, by Theorem 1.5, we have $fgf = f$ and $gfg = g$. Moreover, by Theorem 1.7 and its dual, gf is a closure on E and fg

is a dual closure on F^d. Conversely, given a residuated mapping $f : E \to F$ we can consider f as an antitone mapping from E to F^d, and f^+ as an antitone mapping from F^d to E, whence (f, f^+) forms a Galois connection between E and F^d.

The algebraic benefit of residuated mappings lies in the fact that they can be composed to produce another residuated mapping, which is not so with Galois connections. In the next section we shall consider the semigroup Res E of residuated mappings on an ordered set E with the purpose of classifying it as a semigroup. As we shall see, for this the Galois connection of Example 1.27 is of particular significance.

1.7 Semigroups of residuated mappings

Definition Let S be a semigroup with an identity and a zero. Then we shall say that S is a **generalised Baer semigroup**[1] if the Galois connection (L, R) of Example 1.27 satisfies the property that for each $x \in S$ there exist $x_r, x_l \in S$ such that $LR(x) = L(x_r)$ and $RL(x) = R(x_l)$.

For every subset A of a semigroup S with zero let $\mathcal{L}(A) = \{L(x) \mid x \in A\}$ and $\mathcal{R}(A) = \{R(x) \mid x \in A\}$. Then we have the following characterisation of generalised Baer semigroups.

Theorem 1.11 *For a semigroup S with an identity and a zero the following statements are equivalent:*

    (1) *S is a generalised Baer semigroup;*
    (2) *$\mathcal{L}\mathcal{R}(S) = \mathcal{L}(S)$;*
    (3) *$\mathcal{R}\mathcal{L}(S) = \mathcal{R}(S)$.*

Proof (1) $\Rightarrow$ (2): If (1) holds then $LR(x) = L(x^r)$ and $L(x) = LRL(x) = LR(x^l)$, whence we obtain (2).

(2) $\Rightarrow$ (3): If (2) holds then on the one hand $L(x) = LR(x')$ for some $x' \in S$ whence $RL(x) = RLR(x') = R(x')$. On the other hand, $LR(x) = L(x')$ for some $x' \in S$ and so $R(x) = RLR(x) = RL(x')$. Thus we have (3).

(3) $\Rightarrow$ (1): If (3) holds then on the one hand $RL(x) = R(x')$ for some $x' \in S$. On the other hand, $R(x) = RL(x')$ for some $x' \in S$ whence $LR(x) = LRL(x') = L(x')$. Hence we have (1). $\square$

Given an ordered set E we now take a closer look at the semigroup Res E of residuated mappings on E. Clearly, Res E has an identity element, namely id_E. Observe now that if E has a bottom element 0 then the zero map $x \mapsto 0$ on E is residuated if and only if E has a top element 1. Thus, for a bounded ordered set E the semigroup Res E also has a zero element.

The connection with what has gone before is provided as follows.

Theorem 1.12 (Johnson [71]) *If E is a bounded ordered set then* Res E *is a generalised Baer semigroup and $E \simeq \mathcal{R}(\mathrm{Res}\, E)$.*

---

[1] The notion of a *Baer semigroup* will be defined in Chapter 2.

Proof For each $e \in E$ consider the mappings $\alpha_e, \beta_e : E \to E$ given by the prescriptions

$$\alpha_e(x) = \begin{cases} 0 & \text{if } x = 0; \\ e & \text{otherwise,} \end{cases} \qquad \beta_e(x) = \begin{cases} 0 & \text{if } x \leqslant e; \\ 1 & \text{otherwise.} \end{cases}$$

That α_e and β_e are isotone and idempotent is clear. They are also residuated; simple calculations show that

$$\alpha_e^+(x) = \begin{cases} 1 & \text{if } e \leqslant x; \\ 0 & \text{otherwise,} \end{cases} \qquad \beta_e^+(x) = \begin{cases} 1 & \text{if } x = 1; \\ e & \text{otherwise.} \end{cases}$$

Suppose now that $f \in \operatorname{Res} E$. Observe that $\beta_{f(1)} f = 0$ and $f \alpha_{f^+(0)} = 0$ so that $\beta_{f(1)} \in L(f)$ and $\alpha_{f^+(0)} \in R(f)$, whence $RL(f) \subseteq R(\beta_{f(1)})$ and $LR(f) \subseteq L(\alpha_{f^+(0)})$. To obtain the reverse inclusions, suppose that $g \in R(\beta_{f(1)})$ and $h \in L(f)$. Then $\beta_{f(1)} g = 0$ gives $g(1) \leqslant f(1)$, and $hf = 0$ gives $f(1) \leqslant h^+(0)$. Then $g(1) \leqslant h^+(0)$ and so $hg = 0$. Thus we have $g \in RL(f)$ and consequently $R(\beta_{f(1)}) \subseteq RL(f)$, whence we have $RL(f) = R(\beta_{f(1)})$. Dually we can see that $LR(f) = L(\alpha_{f^+(0)})$. Hence $\operatorname{Res} E$ is a generalised Baer semigroup. To prove that $E \simeq \mathcal{R}(\operatorname{Res} E)$, we observe that

$$RL(f) \subseteq RL(g) \;\Rightarrow\; L(g) \subseteq L(f) \;\Rightarrow\; \beta_{g(1)} \in L(f) \;\Rightarrow\; f(1) \leqslant g(1).$$

By Theorem 1.11 we may therefore define a mapping $\vartheta : \mathcal{R}(\operatorname{Res} E) \to E$ by setting $\vartheta\big(RL(\varphi)\big) = \varphi(1)$. Note that ϑ is surjective since for every $e \in E$ we have $e = \alpha_e(1)$. Note also that if $f(1) \leqslant g(1)$ then

$$h \in L(g) \;\Rightarrow\; hg(1) = 0 \;\Rightarrow\; hf(1) = 0 \;\Rightarrow\; h \in L(f),$$

so that if $f(1) \leqslant g(1)$ then $L(g) \subseteq L(f)$ and therefore $RL(f) \subseteq RL(g)$. We may now invoke Theorem 1.10 to conclude that ϑ is an isomorphism. $\qquad \square$

Definition If S is a semigroup with a zero then S is said to **coordinatise** an ordered set E if there is an order isomorphism $\mathcal{R}(S) \simeq E$.

Although the above coordinatisation of a bounded ordered set by a generalised Baer semigroup is not unique, we shall show that $\operatorname{Res} E$ is 'universal' in the sense that if S is a generalised Baer semigroup that coordinatises E then there is a morphism $\varphi : S \to \operatorname{Res} E$ with zero kernel such that $\operatorname{Im} \varphi$ is a generalised Baer semigroup that coordinatises E.

If S is a generalised Baer semigroup S then, using Theorem 1.11, we can define for each $z \in S$ a mapping $\varphi_z : \mathcal{R}(S) \to \mathcal{R}(S)$ by the prescription

$$\varphi_z[RL(x)] = RL(zx).$$

This mapping is well-defined and isotone, as can be seen from the implications

$$RL(x) \subseteq RL(y) \;\Rightarrow\; L(y) = LRL(y) \subseteq LRL(x) = L(x)$$
$$\Rightarrow\; L(zy) \subseteq L(zx)$$
$$\Rightarrow\; RL(zx) \subseteq RL(zy).$$

More importantly, we have the following result.

Theorem 1.13 *If S is a generalised Baer semigroup then $T = \{\varphi_z \mid z \in S\}$ is a subsemigroup of $\operatorname{Res} \mathcal{R}(S)$ and $\varphi : z \mapsto \varphi_z$ is a morphism with zero kernel.*

Proof Consider the mapping $\varphi_z^+ : \mathcal{R}(S) \to \mathcal{R}(S)$ given by the prescription $\varphi_z^+[R(x)] = R(xz)$. It is readily seen that this is well-defined and is isotone. Now $\varphi_z^+ \varphi_z[RL(x)] = \varphi_z^+[RL(zx)] = R\big((zx)_l z\big)$. But $zx \in RL(zx) = R\big((zx)_l\big)$ gives $(zx)_l zx = 0$ whence $x \in R\big((zx)_l z\big)$ and so $RL(x) \subseteq R\big((zx)_l z\big)$. It follows from these observations that $\varphi_z^+ \varphi_z \geqslant \mathrm{id}$. Dually, we have that $\varphi_z \varphi_z^+ \leqslant \mathrm{id}$. Consequently $\varphi_z \in \mathrm{Res}\,\mathcal{R}(S)$. Finally, observe that

$$\varphi_z \varphi_w[RL(x)] = \varphi_z[RL(wx)] = RL(zwx) = \varphi_{zw}[(RL(x)]$$

whence $\varphi_z \varphi_w = \varphi_{zw}$ and φ is a morphism. Finally, if $\varphi_z = 0$ then $RL(z) = \varphi_z[RL(1)] = \{0\}$ and consequently $z = 0$. $\square$

In order to simplify notation, we shall agree to identify E with $\mathcal{R}(S)$ whenever S coordinatises E. In this way the semigroup T of Theorem 1.13 becomes a subsemigroup of $\mathrm{Res}\,E$. The following result achieves our goal.

Theorem 1.14 *T is a generalised Baer semigroup that coordinatises E.*

Proof We observe first that the morphism φ of Theorem 1.13 is such that in T we have the equality $\varphi^\to[R(x)] = R(\varphi_x)$ for every $x \in S$. In fact, since φ has zero kernel,

$$y \in R(x) \iff xy = 0 \iff \varphi_x \varphi_y = \varphi_{xy} = 0 \iff \varphi_y \in R(\varphi_x).$$

It follows that for every subset A of S we have $\varphi^\to[R(A)] \subseteq \bigcap_{x \in A} R(\varphi_x)$. Taking $A = L(z)$ and noting that likewise $x \in L(z) \Leftrightarrow \varphi_x \in L(\varphi_z)$, we obtain

$$\varphi^\to[RL(z)] \subseteq \bigcap_{x \in L(z)} R(\varphi_x) = \bigcap_{\varphi_x \in L(\varphi_z)} R(\varphi_x) = RL(\varphi_z).$$

Since $\varphi^\to[RL(z)] = \varphi^\to[R(z_l)] = R(\varphi_{z_l})$, it follows that $R(\varphi_{z_l}) \subseteq RL(\varphi_z)$. But $z \in RL(z) = R(z_l)$ gives $\varphi_z \in R(\varphi_{z_l})$, and applying RL to this we obtain $RL(\varphi_z) \subseteq R(\varphi_{z_l})$. It therefore follows that $RL(\varphi_z) = R(\varphi_{z_l})$. Dually, we obtain $LR(\varphi_z) = L(\varphi_{z_r})$. Hence T is a generalised Baer semigroup. That it coordinatises E follows as in the proof of Theorem 1.12. $\square$

The coordinatisation of an ordered set by a generalised Baer semigroup provides a link between the theory of ordered sets on the one hand and the algebraic theory of semigroups on the other. The present general formulation has deep roots that originate in the work of Baer [2] and von Neumann [87]. As we proceed to consider special types of ordered set, we shall include a description of the associated type of semigroup.

EXERCISES

1.31. Let E be a bounded ordered set. If $f, g \in \mathrm{Res}\,E$ are such that $R(f) = g \circ \mathrm{Res}\,E$ with g idempotent, prove that $g(1) = f^+(0)$.

1.32. By an **involution** on an ordered set E we mean an antitone mapping $i : E \to E$ such that $i^2 = \mathrm{id}_E$. If E is an ordered set with an involution i and if $f \in \mathrm{Res}\,E$, define $f^* = i \circ f \circ i$. Prove that $f^* \in \mathrm{Res}\,E$ and determine its residual. Prove also that $(i \circ f, i \circ f^*)$ establishes a Galois connection.

1.33. If E is a bounded ordered set with an involution i and if, for every $e \in E$ the mappings α_e and β_e are defined as in Theorem 1.12, prove that $\alpha_e^\star = \beta_{i(e)}$ and $\beta_e^\star = \alpha_{i(e)}$.

1.34. By an **involution** on a semigroup S we mean a mapping $x \mapsto x^\star$ such that $(\forall x, y \in S)$ $(xy)^\star = y^\star x^\star$ and $x^{\star\star} = x$. If S is a generalised Baer semigroup with an involution * prove that $i : \mathcal{R}(S) \to \mathcal{R}(S)$ defined by $i[RL(x)] = L(x^\star)$ is an involution.

Lattices; lattice morphisms

2.1 Semilattices and lattices

If E is an ordered set and $x \in E$ then the canonical embedding of $x^\downarrow$ into E, i.e., the restriction to $x^\downarrow$ of the identity mapping on E, is clearly isotone. As we shall now see, consideration of when each such embedding is a residuated mapping has important consequences as far as the structure of E is concerned.

Theorem 2.1 *If E is an ordered set then the following are equivalent:*

(1) *for every $x \in E$ the canonical embedding of $x^\downarrow$ into E is residuated;*
(2) *the intersection of any two principal down-sets is a principal down-set.*

Proof For each $x \in E$ let $i_x : x^\downarrow \to E$ be the canonical embedding. Then (1) holds if and only if, for all $x, y \in E$, there exists $\alpha = \max\{z \in x^\downarrow \mid z = i_x(z) \leqslant y\}$. Clearly, this is equivalent to the existence of $\alpha \in E$ such that $x^\downarrow \cap y^\downarrow = \alpha^\downarrow$, which is (2). $\qquad\square$

Definition If E satisfies either of the equivalent conditions of Theorem 2.1 then we shall denote by $x \wedge y$ the element α such that $x^\downarrow \cap y^\downarrow = \alpha^\downarrow$, and call $x \wedge y$ the **meet** of x and y. In this situation we shall say that E is a **meet semilattice**. Equivalent terminology is a **$\wedge$-semilattice**.

Example 2.1 Every chain is a meet semilattice in which $x \wedge y = \min\{x, y\}$.

Example 2.2 $(\mathbb{N}; |)$ is a meet semilattice in which $m \wedge n = \mathrm{hcf}\{m, n\}$.

Example 2.3 The set $\mathrm{Equ}\, E$ of equivalence relations on E is a meet semilattice in which $\vartheta \wedge \varphi$ is given by

$$(x, y) \in \vartheta \wedge \varphi \iff \big((x, y) \in \vartheta \text{ and } (x, y) \in \varphi\big).$$

Meet semilattices can also be characterised in a purely algebraic way which we shall now describe. First we observe that in a meet semilattice E the assignment $(x, y) \mapsto x \wedge y$ defines a law of composition $\wedge$ on E. Now since $x^\downarrow \cap (y^\downarrow \cap z^\downarrow) = (x^\downarrow \cap y^\downarrow) \cap z^\downarrow$ we see that $\wedge$ is associative; and since $x^\downarrow \cap y^\downarrow = y^\downarrow \cap x^\downarrow$ it is commutative; and, moreover, since $x^\downarrow \cap x^\downarrow = x^\downarrow$ it is idempotent.

Hence $(E; \wedge)$ is a commutative idempotent semigroup. As the following result shows, the converse holds: every commutative idempotent semigroup gives rise in a natural way to a meet semilattice.

Theorem 2.2 *Every commutative idempotent semigroup can be ordered in such a way that it forms a meet semilattice.*

Proof Suppose that E is a commutative idempotent semigroup in which we denote the law of composition by multiplication. Define a relation R on E by $x R y \iff xy = x$. Then R is an order. In fact, since $x^2 = x$ for every $x \in E$ we have xRx, so that R is reflexive; if xRy and yRx then $x = xy = yx = y$, so that R is anti-symmetric; if xRy and yRz then $x = xy$ and $y = yz$ whence $x = xy = xyz = xz$ and therefore xRz, so that R is transitive. In what follows we write $\leqslant$ for R. If now $x, y \in E$ we have $xy = xxy = xyx$ and so $xy \leqslant x$. Inverting the roles of x, y we also have $xy \leqslant y$ and therefore $xy \in x^{\downarrow} \cap y^{\downarrow}$. Suppose now that $z \in x^{\downarrow} \cap y^{\downarrow}$. Then $z \leqslant x$ and $z \leqslant y$ give $z = zx$ and $z = zy$, whence $z = zy = zxy$ and therefore $z \leqslant xy$. It follows that $x^{\downarrow} \cap y^{\downarrow}$ has a top element, namely xy. Thus E is a meet semilattice in which $x \wedge y = xy$. $\square$

EXERCISES

2.1. Draw the Hasse diagrams for all possible meet semilattices with 4 elements.

2.2. If P and Q are meet semilattices prove that the set of isotone mappings from P to Q forms a meet semilattice with respect to the order described in Example 1.7.

Definition If E is an ordered set and F is a subset of E then $x \in E$ is said to be a **lower bound** of F if $(\forall y \in F) \; x \leqslant y$; and an **upper bound** of F if $(\forall y \in F) \; y \leqslant x$.

In what follows we shall denote the set of lower bounds of F in E by $F^{\downarrow}$, and the set of upper bounds of F by $F^{\uparrow}$.

Remark We note here that the notation $A^{\downarrow}$ is often used to denote the down-set generated by A, namely $\{x \in E \mid (\exists a \in A) \; x \leqslant a\}$, and $A^{\uparrow}$ to denote the up-set generated by A. Other commonly used notation for lower, upper bounds include A^{ℓ}, A^{u} and $A^{\blacktriangledown}$, $A^{\blacktriangle}$.

In particular, we have $\{x\}^{\downarrow} = x^{\downarrow}$ and $\{x\}^{\uparrow} = x^{\uparrow}$. Note that $F^{\downarrow}$ and $F^{\uparrow}$ may be empty, but not so when E is **bounded**, in the sense that it has both a top element 1 and a bottom element 0. If E has a top element 1 then $E^{\uparrow} = \{1\}$; otherwise $E^{\uparrow} = \emptyset$. Similarly, if E has a bottom element 0 then $E^{\downarrow} = \{0\}$; otherwise $E^{\downarrow} = \emptyset$. Note that if $F = \emptyset$ then every $x \in E$ satisfies (vacuously) the relation $y \leqslant x$ for every $y \in F$. Thus $\emptyset^{\uparrow} = E$; and similarly $\emptyset^{\downarrow} = E$.

Definition If E is an ordered set and F is a subset of E then by the **infimum**, or **greatest lower bound**, of F we mean the top element (when such exists) of the set $F^{\downarrow}$ of lower bounds of F. We denote this by $\inf_E F$ or simply $\inf F$ if there is no confusion.

Since $\emptyset^{\downarrow} = E$ we see that $\inf_E \emptyset$ exists if and only if E has a top element 1, in which case $\inf_E \emptyset = 1$.

It is immediate from what has gone before that a meet semilattice can be described as an ordered set in which every pair of elements x, y has a greatest lower bound; here we have $\inf\{x, y\} = x \wedge y$. A simple inductive argument shows that for every finite subset $\{x_1, \ldots, x_n\}$ of a meet semilattice we have that $\inf\{x_1, \ldots, x_n\}$ exists and is $x_1 \wedge \cdots \wedge x_n$.

We can of course develop the duals of the above, obtaining in this way the notion of a **join semilattice** which is characterised by the intersection of any two principal up-sets being a principal up-set, the element β such that $x^{\uparrow} \cap y^{\uparrow} = \beta^{\uparrow}$ being denoted by $x \vee y$ and called the **join** of x and y. Equivalent terminology for this is a $\vee$-**semilattice**. Then Theorem 2.2 has an analogue for join semilattices in which the order is defined by $x \, R \, y \iff xy = y$. Likewise by duality we can define the notion of that of **supremum** or **least upper bound** of a subset F, denoted by $\sup_E F$. In particular, we see that $\sup_E \emptyset$ exists if and only if $\emptyset^{\uparrow} = E$ has a bottom element 0, in which case $\sup_E \emptyset = 0$. In a join semilattice we have $\sup\{x, y\} = x \vee y$ and, by induction, $\sup\{x_1, \ldots, x_n\} = x_1 \vee \cdots \vee x_n$.

Definition A **lattice** is an ordered set $(E; \leqslant)$ which, with respect to its order, is both a meet semilattice and a join semilattice.

Thus a lattice is an ordered set in which every pair of elements (and hence every finite subset) has an infimum and a supremum. We often denote a lattice by $(E; \wedge, \vee, \leqslant)$.

Theorem 2.3 *A set E can be given the structure of a lattice if and only if it can be endowed with two laws of composition $(x, y) \mapsto x \, ⋒ \, y$ and $(x, y) \mapsto x \, ⋓ \, y$ such that*

(1) *$(E; ⋒)$ and $(E; ⋓)$ are commutative semigroups;*

(2) *the following **absorption laws** hold:*

$$(\forall x, y \in E) \quad x \, ⋒ \, (x \, ⋓ \, y) = x = x \, ⋓ \, (x \, ⋒ \, y).$$

Proof $\Rightarrow$: If E is a lattice then E has two laws of composition that satisfy (1), namely $(x, y) \mapsto x \wedge y$ and $(x, y) \mapsto x \vee y$. To show that (2) holds, we observe that $x \leqslant \sup\{x, y\} = x \vee y$ and so $x \wedge (x \vee y) = \inf\{x, x \vee y\} = x$; and similarly $x \wedge y = \inf\{x, y\} \leqslant x$ gives $x \vee (x \wedge y) = \sup\{x, x \wedge y\} = x$.

$\Leftarrow$: Suppose now that E has two laws of composition $⋒$ and $⋓$ that satisfy (1) and (2). Using (2) twice, we see that $x \, ⋓ \, x = x \, ⋓ \, [x \, ⋒ \, (x \, ⋓ \, x)] = x$, and similarly that $x \, ⋒ \, x = x$. This, together with Theorem 2.2 and its dual shows that $(E; ⋒)$ and $(E; ⋓)$ are semilattices. In order to show that $(E; ⋓, ⋒)$ is a lattice with (for example) $⋒$ as $\wedge$, and $⋓$ as $\vee$, we must show that the orders defined by $⋒$ and $⋓$ coincide. In other words, we must show that $x \, ⋒ \, y = x$ is equivalent to $x \, ⋓ \, y = y$. But if $x \, ⋒ \, y = x$ then, using the absorption laws, we have $y = (x \, ⋒ \, y) \, ⋓ \, y = x \, ⋓ \, y$; and if $x \, ⋓ \, y = y$ then $x = x \, ⋒ \, (x \, ⋓ \, y) = x \, ⋒ \, y$. Thus we see that E is a lattice in which $x \leqslant y$ is described equivalently by $x \, ⋒ \, y = x$ or by $x \, ⋓ \, y = y$. $\square$

Example 2.4 Every chain is a lattice; here we have $\inf\{x, y\} = \min\{x, y\}$ and $\sup\{x, y\} = \max\{x, y\}$.

Example 2.5 For every set E, $(\mathbb{P}(E); \cap, \cup, \subseteq)$ is a bounded lattice.

Example 2.6 For every infinite set E let $\mathbb{P}_f(E)$ be the set of finite subsets of E. Then $(\mathbb{P}_f(E); \cap, \cup, \subseteq)$ is a lattice with no top element.

Example 2.7 $(\mathbb{N}; |)$ is a bounded lattice. The bottom element is 1 and the top element is 0. Here we have $\inf\{m, n\} = \mathrm{hcf}\{m, n\}$ and $\sup\{m, n\} = \mathrm{lcm}\{m, n\}$.

Example 2.8 If V is a vector space and if $\mathrm{Sub}\,V$ denotes the set of subspaces of V then in the ordered set $(\mathrm{Sub}\,V; \subseteq)$ we have $\inf\{A, B\} = A \cap B$ since $A \cap B$ is the biggest subspace that is contained in both A and B. Also, $\sup\{A, B\}$ exists and is the smallest subspace to contain both A and B, namely the subspace $A + B = \{a + b \mid a \in A, b \in B\}$. Thus $(\mathrm{Sub}\,V; \cap, +, \subseteq)$ is a lattice.

Example 2.9 If L, M are lattices then the set of isotone mappings $f : L \to M$ forms a lattice in which $f \wedge g$ and $f \vee g$ are given by the prescriptions

$$(f \wedge g)(x) = f(x) \wedge g(x), \quad (f \vee g)(x) = f(x) \vee g(x).$$

The concept of a lattice was introduced by Peirce [91] and Schröder [101] towards the end of the nineteenth century. It derives from pioneering work by Boole [35], [36] on the formalisation of propositional logic. The terms idempotent, commutative, associative, and absorption are mostly due to Boole. The study of lattices became systematic with Birkhoff's first paper [8] in 1933 and his book [13] the first edition of which appeared in 1940 and was for several decades the bible of lattice theorists. Over the years the theory of lattices and its many applications has grown considerably. Notable reference works include books by Abbott [1], Balbes and Dwinger [3], Crawley and Dilworth [40], Davey and Priestley [42], Dubreil-Jacotin, Lesieur and Croisot [46], Freese, Ježek and Nation [50], Ganter and Wille [54], Hermes [63], Maeda and Maeda [83], Rutherford [96], Salii [100], Sikorski [102], and Szász [107]. In recent times the Birkhoff bible has been replaced by that of Grätzer [58].

EXERCISES

2.3. If L is a lattice and $x, y, z \in L$ prove that
$$[(x \wedge y) \vee (x \wedge z)] \wedge [(x \wedge y) \vee (y \wedge z)] = x \wedge y.$$

2.4. If x_{ij} $(i = 1, \ldots, m;\ j = 1, \ldots, n)$ are mn elements of a lattice L, establish the **minimax inequality**

$$\bigvee_{j=1}^{n} \bigwedge_{i=1}^{m} x_{ij} \leqslant \bigwedge_{i=1}^{m} \bigvee_{j=1}^{n} x_{ij}.$$

A regiment of soldiers, each of a different height, stands at attention in a rectangular array. Of the soldiers who are the tallest in their row, the smallest is Sergeant Mintall; and of the soldiers who are the smallest in their column, the tallest is Corporal Max Small. Which of these two soldiers is the taller?

2.5. If L is a lattice and $a, b \in L$ define $f_{a,b} : L \to L$ by the prescription
$$f_{a,b}(x) = [(a \wedge b) \vee x] \wedge (a \vee b).$$
Prove that $f_{a,b}$ is isotone and idempotent. What is $\operatorname{Im} f_{a,b}$?

2.6. For $p \leqslant q$ in a lattice L let $[p, q] = \{x \in L \mid p \leqslant x \leqslant q\}$. Given any $a, b \in L$, prove that the mapping $f : [a \wedge b, b] \to [a, a \vee b]$ defined by $f(x) = x \vee a$ is residuated and determine f^{+}.

2.7. Prove that the set $N(G)$ of normal subgroups of a group G forms a lattice in which $\sup\{H, K\} = \{hk \mid h \in H, k \in K\}$.

2.8. Draw the Hasse diagram of the lattice of subgroups of the alternating group $\mathcal{A}_4$.

2.9. Let L, M be lattices and let $\operatorname{Res}(L, M)$ be the set of residuated mappings from L to M. Prove that if $f, g \in \operatorname{Res}(L, M)$ then $f \vee g \in \operatorname{Res}(L, M)$.

2.10. Consider the lattice L described by the following Hasse diagram:

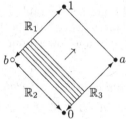

in which each $\mathbb{R}_i$ is a copy of the chain of real numbers. Let L^* be the lattice $L \backslash \{b\}$. Prove that the mapping $f_a : L^* \to L^*$ given by
$$f_a(x) = \begin{cases} x & \text{if } x \leqslant a; \\ a & \text{otherwise,} \end{cases}$$
is residuated. If $g \in \operatorname{Res} L^*$ is such that $g \leqslant f_a$ and $g \leqslant \operatorname{id}_{L^*}$ prove that there exists $c \in \mathbb{R}_1$ such that $c < g^{+}(0)$. Show further that if $h_c : L^* \to L^*$ is given by
$$h_c(x) = \begin{cases} 0 & \text{if } x \leqslant c; \\ x \wedge a & \text{otherwise,} \end{cases}$$
then $h_c \in \operatorname{Res} L^*$ with $h_c \leqslant f_a$ and $h_c \leqslant \operatorname{id}_{L^*}$. Moreover, show that $g < h_c$. Conclude from this that $\operatorname{Res} L^*$ is not a $\wedge$-semilattice.

2.11. Given a lattice L, let f be a residuated closure on L and let g be a residuated dual closure on $\operatorname{Im} f$. Prove that if $\alpha : L \to L$ is given by the prescription
$$(\forall x \in L) \quad \alpha(x) = g[f(x)]$$
then α is an idempotent element of the semigroup $\operatorname{Res} L$. Prove further that every idempotent of $\operatorname{Res} L$ arises in this way.

2.2 Down-set lattices

If E is an ordered set and A, B are down-sets of E then clearly so also are $A \cap B$ and $A \cup B$. Thus the set of down-sets of E is a lattice in which $\inf\{A, B\} = A \cap B$ and $\sup\{A, B\} = A \cup B$. We shall denote this lattice by $\mathcal{O}(E)$.

We recall from the definition of a down-set that we include the empty subset as such. Thus the lattice $\mathcal{O}(E)$ is bounded with top element E and bottom element $\emptyset$.

Example 2.10

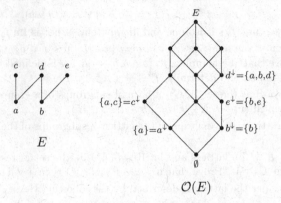

$\mathcal{O}(E)$

Down-set lattices will be of considerable interest to us later. For the moment we shall consider how to compute the cardinality of $\mathcal{O}(E)$ when the ordered set E is finite. Upper and lower bounds for this are provided by the following result.

Theorem 2.4 *If E is a finite ordered set with $|E| = n$ then*

$$n + 1 \leqslant |\mathcal{O}(E)| \leqslant 2^n.$$

Proof Clearly, E has the least number of down-sets when it is a chain, in which case $\mathcal{O}(E)$ is also a chain, of cardinality $n + 1$. Correspondingly, E has the greatest number of down-sets when it is an anti-chain, in which case $\mathcal{O}(E) = \mathbb{P}(E)$ which is of cardinality 2^n. $\qquad\square$

In certain cases $|\mathcal{O}(E)|$ can be calculated using an ingenious algorithm that we shall describe. For this purpose, we shall denote by $E \backslash x$ the ordered set obtained from E by deleting the element x and related comparabilities whilst retaining all comparabilities resulting from transitivity through x.

Example 2.11

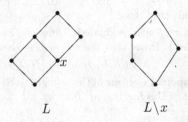

L $L \backslash x$

We shall also use the notation $x^{\updownarrow}$ to denote the **cone** through x, namely the set of elements that are comparable to x; formally,

$$x^{\updownarrow} = x^{\downarrow} \cup x^{\uparrow} = \{ y \in E \mid y \nparallel x \}.$$

Example 2.12 If L is as in Example 2.11 then $L \backslash x^{\updownarrow}$ is a singleton.

Finally, we shall say that $x \in E$ is **maximal** if there is no $y \in E$ such that $y > x$. The dual notion is that of a **minimal** element. Clearly, a top (bottom) element can be characterised as a unique maximal (minimal) element.

Theorem 2.5 (Berman–Köhler [4]) *If E is a finite ordered set then*
$$|\mathcal{O}(E)| = |\mathcal{O}(E \setminus x)| + |\mathcal{O}(E \setminus x^{\updownarrow})|.$$

Proof Observe first that every non-empty down-set X of E is determined by a unique antichain in E, namely the set of maximal elements of X. Counting $\emptyset$ as an antichain, we thus see that $|\mathcal{O}(E)|$ is the number of antichains in E. For any given element x of E this can be expressed as the number of antichains that contain x plus the number that do not contain x.

Now if an antichain A contains a particular element x of E then A contains no other elements of the cone $x^{\updownarrow}$. Thus every antichain that contains x determines a down-set of $\mathcal{O}(E \setminus x^{\updownarrow})$, and conversely. Hence we see that the number of antichains that contain x is precisely $|\mathcal{O}(E \setminus x^{\updownarrow})|$. Since likewise the number of antichains that do not contain x is precisely $|\mathcal{O}(E \setminus x)|$, the result follows. $\qquad\square$

Example 2.13 By an **even fence** we shall mean an ordered set F_{2n} of the form

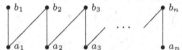

it being assumed that all the elements are distinct.

We can also define two non-isomorphic **odd fences**, namely by setting $F_{2n+1} = F_{2n} \cup \{b_{n+1}\}$ with the single extra relation $a_n < b_{n+1}$; and its dual $F^d_{2n+1} = F_{2n} \cup \{a_0\}$ with the single extra relation $a_0 < b_1$.

If we apply Theorem 2.5 to F_{2n} with $x = a_n$, we obtain
$$|\mathcal{O}(F_{2n})| = |\mathcal{O}(F_{2n} \setminus a_n)| + |\mathcal{O}(F_{2n-2})| = |\mathcal{O}(F_{2n-1})| + |\mathcal{O}(F_{2n-2})|;$$
and then to F_{2n-1} with $x = b_n$, we obtain
$$|\mathcal{O}(F_{2n-1})| = |\mathcal{O}(F_{2n-2})| + |\mathcal{O}(F_{2n-3})|.$$
Writing $\alpha_k = |\mathcal{O}(F_k)|$, we thus see that α_k satisfies the recurrence relation
$$\alpha_k = \alpha_{k-1} + \alpha_{k-2}.$$

Now in recognising this recurrence relation the reader will recall that the **Fibonacci sequence** $(f_n)_{n \geq 0}$ is defined by
$$f_0 = 0, \quad f_1 = 1, \quad (n \geq 2)\ f_n = f_{n-1} + f_{n-2}.$$

Furthermore, as is readily computed, we have $\alpha_2 = |\mathcal{O}(F_2)| = 3 = f_4$ and $\alpha_3 = |\mathcal{O}(F_3)| = 5 = f_5$. We therefore conclude from the above that α_k, the cardinality of $\mathcal{O}(F_k)$, is the Fibonacci number f_{k+2}.

EXERCISES

2.12. If E and F are finite ordered sets prove that
$$\mathcal{O}(E \,\dot\cup\, F) \simeq \mathcal{O}(E) \times \mathcal{O}(F).$$

2.13. Let **2** denote the 2-element chain $0 < 1$ and for every ordered set E let Isomap$(E, \mathbf{2})$ be the set of isotone mappings $f : E \to \mathbf{2}$. Prove that the ordered sets $\mathcal{O}(E)$ and Isomap$(E, \mathbf{2})$ are dually isomorphic.
[*Hint.* Consider $\alpha : \text{Isomap}(E, \mathbf{2}) \to \mathcal{O}(E)$ given by $\alpha(f) = f^{\leftarrow}\{0\}$.]

2.14. Draw the Hasse diagram of the lattice of down-sets of each of the following ordered sets:

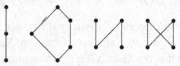

2.15. If P_1 and P_2 are the ordered sets

draw the Hasse diagram of the lattice of down-sets of $P_1 \cup P_2$.

2.16. The **Lucas sequence** $(\ell_n)_{n \geqslant 0}$ is defined by

$$\ell_0 = 1, \quad \ell_1 = 1, \quad (n \geqslant 2) \, \ell_n = \ell_{n-1} + \ell_{n-2}.$$

If f_i denotes the i-th Fibonacci number, establish the identity

$$\ell_{2n} = f_{2n+2} - f_{2n-2}.$$

By a **crown** we mean an ordered set C_{2n} of the form

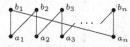

it being assumed that all the elements are distinct. Prove that $|\mathcal{O}(C_{2n})| = \ell_{2n}$.

2.17. Let E_{2n} be the ordered set obtained from the crown C_{2n} by adjoining comparabilities in such a way that $a_i < b_j$ for all i, j. Determine $|\mathcal{O}(E_{2n})|$.

2.3 Sublattices

As we have seen, important substructures of an ordered set are the down-sets and the principal down-sets. We now consider substructures of (semi)lattices.

Definition By a $\wedge$-**subsemilattice** of a $\wedge$-semilattice L we mean a non-empty subset E of L that is closed under the meet operation, in the sense that if $x, y \in E$ then $x \wedge y \in E$. A $\vee$-**subsemilattice** of a $\vee$-semilattice is defined dually. By a **sublattice** of a lattice we mean a subset that is both a $\wedge$-subsemilattice and a $\vee$-subsemilattice.

Example 2.14 If V is a vector space then, by Example 2.8, the set Sub V of subspaces of V is a $\cap$-subsemilattice of the lattice $\mathbb{P}(V)$.

Example 2.15 For every ordered set E the lattice $\mathcal{O}(E)$ of down-sets of E is a sublattice of the lattice $\mathbb{P}(E)$.

Particularly important sublattices of a lattice are the following.

Definition By an **ideal** of a lattice L we shall mean a sublattice of L that is also a down-set; dually, by a **filter** of L we mean a sublattice that is also an up-set.

Theorem 2.6 *If L is a lattice then, ordered by set inclusion, the set $\mathcal{I}(L)$ of ideals of L forms a lattice in which the lattice operations are given by*

$$\begin{cases} \inf\{J, K\} = J \cap K; \\ \sup\{J, K\} = \{x \in L \mid (\exists j \in J)(\exists k \in K) \; x \leqslant j \vee k\}. \end{cases}$$

Proof It is clear that if J and K are ideals of L then so is $J \cap K$, and that this is the biggest ideal of L that is contained in both J and K. Hence $\inf\{J, K\}$ exists in $\mathcal{I}(L)$ and is $J \cap K$.

Now any ideal that contains both J and K must clearly contain all the elements x such that $x \leqslant j \vee k$ where $j \in J$ and $k \in K$. Conversely, the set of all such x clearly contains both J and K, and is contained in every ideal of L that contains both J and K. Moreover, this set is also an ideal of L. Thus we see that $\sup\{J, K\}$ exists in $\mathcal{I}(L)$ and is as described above. $\qquad\square$

Note from Theorem 2.6 that although $\mathcal{I}(L)$ is a $\cap$-subsemilattice of $\mathcal{O}(L)$ it is not a sublattice since suprema are not the same. This situation, in which a subsemilattice of a given lattice L that is not a sublattice of L can also form a lattice with respect to the same order as L, is quite common in lattice theory. Another instance of this has been seen before in Example 2.8 where the set $\mathrm{Sub}\,V$ of subspaces of a vector space V forms a lattice in which $\inf\{A, B\} = A \cap B$ and $\sup\{A, B\} = A + B$, so that $(\mathrm{Sub}\,V; \subseteq)$ forms a lattice that is a $\cap$-subsemilattice, but not a sublattice, of $(\mathbb{P}(V); \subseteq)$. As we shall now see, a further instance is provided by a closure mapping on a lattice.

Theorem 2.7 *Let L be a lattice and let $f : L \to L$ be a closure. Then $\mathrm{Im}\,f$ is a lattice in which the lattice operations are given by*

$$\inf\{a, b\} = a \wedge b, \quad \sup\{a, b\} = f(a \vee b).$$

Proof Recall that for a closure f on L we have $\mathrm{Im}\,f = \{x \in L \mid x = f(x)\}$. If then $a, b \in \mathrm{Im}\,f$ we have, since f is isotone with $f \geqslant \mathrm{id}_L$, $f(a) \wedge f(b) = a \wedge b \leqslant f(a \wedge b) \leqslant f(a) \wedge f(b)$ and the resulting equality gives $a \wedge b \in \mathrm{Im}\,f$. It follows that $\mathrm{Im}\,f$ is a $\wedge$-subsemilattice of L.

As for the supremum in $\mathrm{Im}\,f$ of $a, b \in \mathrm{Im}\,f$, we observe first that $a \vee b \leqslant f(a \vee b)$ and so $f(a \vee b) \in \mathrm{Im}\,f$ is an upper bound of $\{a, b\}$. Suppose now that $c = f(c) \in \mathrm{Im}\,f$ is any upper bound of $\{a, b\}$ in $\mathrm{Im}\,f$. Then from $a \vee b \leqslant c$ we obtain $f(a \vee b) \leqslant f(c) = c$. Thus, in the subset $\mathrm{Im}\,f$, the upper bound $f(a \vee b)$ is less than or equal to every upper bound of $\{a, b\}$. Consequently, $\sup\{a, b\}$ exists in $\mathrm{Im}\,f$ and is $f(a \vee b)$. $\qquad\square$

Example 2.16 Consider the lattice L with Hasse diagram

Let $f : L \to L$ be given by

$$f(t) = \begin{cases} 1 & \text{if } t = z; \\ t & \text{otherwise.} \end{cases}$$

It is readily seen that f is a closure with $\operatorname{Im} f = \{0, x, y, 1\}$. In the corresponding lattice (the elements of which are denoted by $\bullet$) we have $\sup\{x, y\} = f(x \vee y) = f(z) = 1$.

EXERCISES

2.18. If L is a lattice prove that, ordered by set inclusion, the set $\mathcal{F}(L)$ of filters of L forms a lattice and determine the lattice operations.

2.19. Let L, M be lattices and let $f : L \to M$ be residuated. Prove that $\operatorname{Im} f$ is a lattice in which $\sup\{x, y\} = x \vee y$ and $\inf\{x, y\} = f f^+(x \wedge y)$.

2.20. Prove that in the ordered set Equ E of equivalence relations on E the supremum of ϑ and φ is the relation ψ given by $(x, y) \in \psi$ if and only if there exist $a_1, \ldots, a_n \in E$ such that

$$x \equiv a_1 \equiv a_2 \equiv \cdots \equiv a_{n-1} \equiv a_n \equiv y$$

where each $\equiv$ denotes ϑ or φ.

2.4 Lattice morphisms

We now consider isotone mappings that preserve lattice operations.

Definition If L and M are $\vee$-semilattices then $f : L \to M$ is said to be a $\vee$-**morphism** if $f(x \vee y) = f(x) \vee f(y)$ for all $x, y \in L$. The notion of a $\wedge$-**morphism** is defined dually. If L and M are lattices then $f : L \to M$ is a **lattice morphism** if it is both a $\vee$-morphism and a $\wedge$-morphism. If L and M are $\vee$-semilattices then a mapping $f : L \to M$ is said to be a **complete** $\vee$-**morphism** if, for every family $(x_\alpha)_{\alpha \in I}$ of elements of L such that $\bigvee\limits_{\alpha \in I} x_\alpha$ exists in L, $\bigvee\limits_{\alpha \in I} f(x_\alpha)$ exists in M and $f\left(\bigvee\limits_{\alpha \in I} x_\alpha \right) = \bigvee\limits_{\alpha \in I} f(x_\alpha)$. The notion of a **complete** $\wedge$-**morphism** is defined dually.

Theorem 2.8 *If L and M are $\vee$-semilattices then every residuated mapping $f : L \to M$ is a complete $\vee$-morphism.*

Proof Suppose that $(x_\alpha)_{\alpha \in I}$ is a family of elements of L such that $x = \bigvee\limits_{\alpha \in I} x_\alpha$ exists in L. Clearly, for each $\alpha \in I$ we have $f(x) \geqslant f(x_\alpha)$. Now if $y \geqslant f(x_\alpha)$ for each $\alpha \in I$ then $f^+(y) \geqslant f^+[f(x_\alpha)] \geqslant x_\alpha$ and so $f^+(y) \geqslant \bigvee\limits_{\alpha \in I} x_\alpha = x$. But then $y \geqslant f[f^+(y)] \geqslant f(x)$. Thus we see that $\bigvee\limits_{\alpha \in I} f(x_\alpha)$ exists and is $f(x)$. $\square$

Definition We shall say that lattices L and M are **isomorphic** if they are isomorphic as ordered sets.

Theorem 2.9 *Lattices L, M are isomorphic if and only if there is a bijection $f : L \to M$ that is a $\vee$-morphism.*

Proof $\Rightarrow$: If $L \simeq M$ then there is a residuated bijection $f : L \to M$, and by Theorem 2.8 this is a $\vee$-morphism.

$\Leftarrow$: If $f : L \to M$ is a bijection and a $\vee$-morphism then we have

$$x \leqslant y \iff y = x \vee y \iff f(y) = f(x \vee y) = f(x) \vee f(y)$$
$$\iff f(x) \leqslant f(y),$$

whence, by Theorem 1.10, $L \simeq M$. $\qquad\qquad\qquad\qquad\qquad\qquad\qquad\square$

EXERCISES

2.21. Let L be a lattice. Prove that every isotone mapping from L to an arbitrary lattice M is a lattice morphism if and only if L is a chain.
[*Hint.* If L is not a chain then there exist $a, b \in L$ with $a \parallel b$. Construct a lattice M by substituting a chain $a \wedge b < \alpha < \beta < a \vee b$ for the sublattice $[a, b] = \{x \in L \mid a \leqslant x \leqslant b\}$. Consider the mapping $f : L \to M$ given by

$$f(x) = \begin{cases} \alpha & \text{if } a \wedge b < x < a; \\ \beta & \text{if } a \wedge b < x < a \vee b \text{ and } x \nleqslant a; \\ x & \text{otherwise.} \end{cases}$$

2.22. If L is a lattice and $a, b \in L$ let

$$X_{a,b} = \{x \in L \mid x = (x \vee b) \wedge a\}, \quad Y_{a,b} = \{y \in L \mid y = (y \wedge a) \vee b\}.$$

Prove that $X_{a,b}$ and $Y_{a,b}$ are isomorphic lattices.

2.5 Complete lattices

We have seen that in a meet semilattice the infimum of every finite subset exists. We now extend this concept to arbitrary subsets.

Definition A $\wedge$-semilattice L is said to be $\wedge$**-complete** if every subset $E = \{x_\alpha \mid \alpha \in A\}$ of L has an infimum which we denote by $\inf_L E$ or by $\bigwedge_{\alpha \in A} x_\alpha$. In a dual manner we define the notion of a $\vee$**-complete** $\vee$**-semilattice**, in which we use the notation $\sup_L E$ or $\bigvee_{\alpha \in A} x_\alpha$. A lattice is said to be **complete** if it is both $\wedge$-complete and $\vee$-complete.

Theorem 2.10 *Every complete lattice has a top and a bottom element.*

Proof Clearly, if L is complete then $\sup_L L$ is the top element of L, and $\inf_L L$ is the bottom element. $\qquad\qquad\qquad\qquad\qquad\qquad\qquad\qquad\qquad\qquad\square$

Example 2.17 For every non-empty set E the power set lattice $\mathbb{P}(E)$ is complete. The top element is E and the bottom element is $\emptyset$.

Example 2.18 Let L be the lattice that is formed by adding to the chain $\mathbb{Q}$ of rationals a top element ∞ and a bottom element $-\infty$. Then L is bounded but is not complete; for example $\sup_L \{x \in \mathbb{Q} \mid x^2 \leqslant 2\}$ does not exist.

Example 2.19 For every non-empty set E the set $\operatorname{Equ} E$ of equivalence relations on E is a complete lattice. In fact, if $F = (R_\alpha)_{\alpha \in A}$ is a family of equivalence relations on E, then $\inf_{\alpha \in A} R_\alpha$ clearly exists in $\operatorname{Equ} E$ and is the relation $\bigwedge_{\alpha \in A} R_\alpha$ given by

$$(x, y) \in \bigwedge_{\alpha \in A} R_\alpha \iff (\forall \alpha \in A) \ (x, y) \in R_\alpha.$$

As for the supremum of this family, consider the relation ϑ defined by $(x, y) \in \vartheta$ if and only if there exist $z_1, \ldots, z_n$ and $R_{\alpha_1}, \ldots, R_{\alpha_{n+1}}$ such that

$$x \overset{R_{\alpha_1}}{\equiv} z_1 \overset{R_{\alpha_2}}{\equiv} z_2 \overset{R_{\alpha_3}}{\equiv} \cdots \overset{R_{\alpha_n}}{\equiv} z_n \overset{R_{\alpha_{n+1}}}{\equiv} y.$$

It is clear that $\vartheta \in \operatorname{Equ} E$. If $x \overset{R_\alpha}{\equiv} y$ for any $R_\alpha \in F$ then since this is a trivial example of such a display it is clear that $R_\alpha \subseteq \vartheta$. Thus ϑ is an upper bound of F. Observe now that, by the transitivity of ϑ, every relation on E that is implied by every R_{α_i} (i.e. every upper bound of F) is also implied by ϑ. We therefore conclude that $\vartheta = \sup_{\alpha \in A} R_\alpha = \bigvee_{\alpha \in A} R_\alpha$. Hence $\operatorname{Equ} E$ forms a complete lattice. The relation ϑ so described is called the **transitive product** of the family $(R_\alpha)_{\alpha \in A}$.

The relationship between complete semilattices and complete lattices is highlighted by the following useful result (and its dual).

Theorem 2.11 *A $\wedge$-complete $\wedge$-semilattice is a complete lattice if and only if it has a top element.*

Proof The condition is clearly necessary. To show that it is also sufficient, let L be a $\wedge$-complete $\wedge$-semilattice with top element 1. Let $X = \{x_\alpha \mid \alpha \in A\}$ be a non-empty subset of L. We show as follows that $\sup_L X$ exists.

Observe first that the set $X^\uparrow$ of upper bounds of X is not empty since it contains the top element 1. Let $X^\uparrow = \{m_\beta \mid \beta \in B\}$. Then, since L is $\wedge$-complete, $\bigwedge_{\beta \in B} m_\beta$ exists. Now clearly we have $x_\alpha \leqslant m_\beta$ for all α and β. It follows that $x_\alpha \leqslant \bigwedge_{\beta \in B} m_\beta$ for every $x_\alpha \in X$, whence $\bigwedge_{\beta \in B} m_\beta \in X^\uparrow$. By its very definition, $\bigwedge_{\beta \in B} m_\beta$ is then the supremum X in L. Hence L is a complete lattice. $\qquad \square$

Example 2.20 Let E be an infinite set and let $\mathbb{P}_f(E)$ be the set of all finite subsets of E. Ordered by set inclusion, $\mathbb{P}_f(E)$ is a lattice which is clearly $\cap$-complete. By Theorem 2.11, $\mathbb{P}_f(E) \cup \{E\}$ is then a complete lattice.

Example 2.21 If G is a group let $\operatorname{Sub} G$ be the set of all subgroups of G. Ordered by set inclusion, $\operatorname{Sub} G$ is clearly a $\cap$-semilattice that is $\cap$-complete. By Theorem 2.11, $\operatorname{Sub} G$ is a complete lattice. In this, joins are given as follows. If X is any subset of G then the subgroup generated by X (i.e. the smallest subgroup of G that contains X) is the set

$$\langle X \rangle = \Big\{ \prod_{i=1}^{n} a_i \mid a_i \in X \text{ or } a_i^{-1} \in X \Big\}.$$

Thus, if $(H_i)_{i \in I}$ is a family of subgroups of G then the subgroup $\bigvee_{i \in I} H_i$ generated by $\bigcup_{i \in I} H_i$ is the set of all finite products $\prod_{i=1}^{n} h_i$ where each $h_i \in \bigcup_{i \in I} H_i$. Consequently $(\operatorname{Sub} G; \subseteq, \cap, \vee)$ is a complete lattice.

Example 2.22 Consider the lattice $(\mathbb{N}; |)$. This is bounded above by 0 and bounded below by 1. If X is any non-empty subset of $\mathbb{N}$ then $\inf_{\mathbb{N}} X$ exists, being the greatest common divisor of the elements of X. It follows by Theorem 2.11 that $(\mathbb{N}; |)$ is a complete lattice. For X finite $\sup_{\mathbb{N}} X$ is the least common multiple of the elements of X; and for X infinite the supremum is the top element 0, this following from the observation that an infinite subset of $\mathbb{N}$ contains integers that are greater than any fixed positive integer.

Concerning complete lattices we have the following remarkable result.

Theorem 2.12 (Knaster [74]) *If L is a complete lattice and if $f : L \to L$ is an isotone mapping then f has a fixed point.*

Proof Consider the set $A = \{x \in L \mid x \leqslant f(x)\}$. Observe that $A \neq \emptyset$ since L has a bottom element 0, and $0 \in A$. By completeness, there exists $\alpha = \sup_L A$. Now for every $x \in A$ we have $x \leqslant \alpha$ and therefore $x \leqslant f(x) \leqslant f(\alpha)$. It follows from this that $\alpha = \sup_L A \leqslant f(\alpha)$. Consequently $f(\alpha) \leqslant f[f(\alpha)]$, which gives $f(\alpha) \in A$ and therefore $f(\alpha) \leqslant \sup_L A = \alpha$. Thus we have $f(\alpha) = \alpha$. □

An interesting application of Theorem 2.12 is to a proof of the following important set-theoretic result.

Theorem 2.13 (Bernstein [5]) *If E and F are sets and if there are injections $f : E \to F$ and $g : F \to E$ then E and F are equipotent.*

Proof We use the notation $i_X : \mathbb{P}(X) \to \mathbb{P}(X)$ to denote the antitone mapping that sends every subset of X to its complement in X. Consider the mapping $\zeta : \mathbb{P}(E) \to \mathbb{P}(E)$ given by $\zeta = i_E \circ g^{\rightarrow} \circ i_F \circ f^{\rightarrow}$. Since $f^{\rightarrow}$ and $g^{\rightarrow}$ are isotone, so also is ζ. By Theorem 2.12, there exists $G \subseteq E$ such that $\zeta(G) = G$, and therefore $i_E(G) = (g^{\rightarrow} \circ i_F \circ f^{\rightarrow})(G)$. The situation may be summarised pictorially:

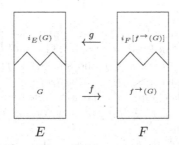

Now since f and g are injective by hypothesis this configuration shows that we can define a bijection $h : E \to F$ by the prescription

$$h(x) = \begin{cases} f(x) & \text{if } x \in G; \\ \text{the unique element of } g^{\leftarrow}\{x\} & \text{if } x \notin G. \end{cases}$$

Hence E and F are equipotent. □

EXERCISES

2.23. If L is a lattice prove that the ideal lattice $\mathcal{I}(L)$ is complete if and only if L has a bottom element.

2.24. If V is a vector space prove that the lattice $\operatorname{Sub} V$ of subspaces of V is complete.

2.25. If E is an ordered set prove that the set of closure mappings on E is a complete lattice.

2.26. Let T be the subset of $\operatorname{Rel} E$ consisting of the transitive relations on E. Prove that T is a $\cap$-complete $\cap$-semilattice. Given $R \in \operatorname{Rel} E$ let $T(R)$ be the set of transitive relations on E that contain R, and let $\overline{R} = \inf T(R)$. Show that

$$(x, y) \in \overline{R} \iff (\exists a_0, \ldots, a_n \in E)\ x = a_0 \overset{R}{\equiv} a_1 \overset{R}{\equiv} \cdots \overset{R}{\equiv} a_n = y.$$

In the complete lattice $\operatorname{Equ} E$ prove that $\sup_{\alpha \in A} R_\alpha = \overline{\bigcup_{\alpha \in A} R_\alpha}$.

2.27. Let L be a complete lattice and let $f : L \to L$ be an isotone mapping. If ω is a fixed point of f and $a = \bigvee_{n \geq 0} f^n(0)$ prove that $a \leq \omega$. Hence show that f has a smallest fixed point.

2.28. Let L be a complete lattice with top element 1 and bottom element 0. If $f : L \to L$ is a closure mapping prove that f is residuated if and only if $\operatorname{Im} f$ is a complete sublattice of L containing 0 and 1.

2.29. Prove that if L and M are complete lattices then a mapping $f : L \to M$ is residuated if and only if it is a complete $\vee$-morphism and $f(0_L) = 0_M$.

Using Theorem 2.11 we can extend as follows the result of Theorem 2.7 to complete lattices.

Theorem 2.14 (Ward [111]) *Let L be a complete lattice. If f is a closure on L then $\operatorname{Im} f$ is a complete lattice. Moreover, for every non-empty subset A of $\operatorname{Im} f$,*

$$\inf\nolimits_{\operatorname{Im} f} A = \inf\nolimits_L A \quad and \quad \sup\nolimits_{\operatorname{Im} f} A = f(\sup\nolimits_L A).$$

Proof First we observe that $\operatorname{Im} f$ is a $\wedge$-complete $\wedge$-semilattice. To see this, recall that $\operatorname{Im} f$ is the set of fixed points of f. Given $C \subseteq \operatorname{Im} f$ let $a = \inf_L C$. Then for every $x \in C$ we have $a \leq x$ and so $f(a) \leq f(x) = x$. Thus $f(a) \leq \inf_L C = a$ and consequently $f(a) = a$, whence $a \in \operatorname{Im} f$ and $\operatorname{Im} f$ is $\wedge$-complete. Now since L is complete it has a top element 1; and since $f \geq \operatorname{id}_L$ we have necessarily $1 = f(1) \in \operatorname{Im} f$. It now follows by Theorem 2.11 that $\operatorname{Im} f$ is a complete lattice.

Suppose now that $A \subseteq \operatorname{Im} f$. If $a = \inf_L A$ then, from the above, we have $a = f(a) \in \operatorname{Im} f$. If now $y \in \operatorname{Im} f$ is such that $y \leq x$ for every $x \in A$ then $y \leq a$. Consequently we have $a = \inf_{\operatorname{Im} f} A$.

Now let $b = \sup_L A$ and $b^\star = \sup_{\operatorname{Im} f} A$. Since $\operatorname{Im} f$ is complete we have $b^\star \in \operatorname{Im} f$; and since $b^\star \geq x$ for every $x \in A$ we have $b^\star \geq \sup_L A = b$. Thus $b^\star = f(b^\star) \geq f(b)$. But $f(b) \geq f(x) = x$ for every $x \in A$, and so we also have $f(b) \geq \sup_{\operatorname{Im} f} A = b^\star$. Thus $b^\star = f(b)$ as asserted. $\qquad\square$

We now proceed to describe an important application of Theorem 2.14. For this purpose, given an ordered set E, consider the mapping $\vartheta : \mathbb{P}(E) \to \mathbb{P}(E)$ given by $\vartheta(A) = A^{\downarrow}$ and the mapping $\varphi : \mathbb{P}(E) \to \mathbb{P}(E)$ given by $\varphi(A) = A^{\uparrow}$. If $A \subseteq B$ then clearly every lower bound of B is a lower bound of A, whence $B^{\downarrow} \subseteq A^{\downarrow}$. Hence ϑ is antitone. Dually, so is φ. Now every element of A is clearly a lower bound of the set of upper bounds of A, whence $A \subseteq A^{\uparrow\downarrow}$ and therefore $\mathrm{id}_{\mathbb{P}(E)} \leqslant \vartheta\varphi$. Dually, every element of A is an upper bound of the set of lower bounds of A, so $A \subseteq A^{\downarrow\uparrow}$ and therefore $\mathrm{id}_{\mathbb{P}(E)} \leqslant \varphi\vartheta$. Consequently we see that (ϑ, φ) establish a Galois connection on $\mathbb{P}(E)$. We shall focus on the associated closure $A \mapsto A^{\uparrow\downarrow}$. For this purpose we shall also require the following facts.

Theorem 2.15 *Let E be an ordered set. If $(A_\alpha)_{\alpha \in I}$ is a family of subsets of E then*

$$\left(\bigcup_{\alpha \in I} A_\alpha\right)^{\uparrow} = \bigcap_{\alpha \in I} A_\alpha^{\uparrow} \quad and \quad \left(\bigcup_{\alpha \in I} A_\alpha\right)^{\downarrow} = \bigcap_{\alpha \in I} A_\alpha^{\downarrow}.$$

Proof Since each A_α is contained in $\bigcup_{\alpha \in I} A_\alpha$ and $A \mapsto A^{\uparrow}$ is antitone, we have that $\left(\bigcup_{\alpha \in I} A_\alpha\right)^{\uparrow} \subseteq \bigcap_{\alpha \in I} A_\alpha^{\uparrow}$. To obtain the reverse inclusion, observe that if $x \in \bigcap_{\alpha \in I} A_\alpha^{\uparrow}$ then x is an upper bound of A_α for every $\alpha \in I$, whence x is an upper bound of $\bigcup_{\alpha \in I} A_\alpha$ and therefore belongs to $\left(\bigcup_{\alpha \in I} A_\alpha\right)^{\uparrow}$. The second statement is proved similarly. $\square$

Definition By an **embedding** of an ordered set E into a lattice L we mean a mapping $f : E \to L$ such that, for all $x, y \in E$,

$$x \leqslant y \iff f(x) \leqslant f(y).$$

Theorem 2.16 (Dedekind–MacNeille [79]) *Every ordered set E can be embedded in a complete lattice L in such a way that meets and joins that exist in E are preserved in L.*

Proof If E does not have a top element or a bottom element we begin by adjoining whichever of these bounds is missing. Then, without loss of generality we may asume that E is a bounded ordered set.

Let $f : \mathbb{P}(E) \to \mathbb{P}(E)$ be the closure mapping given by $f(A) = A^{\uparrow\downarrow}$. Then, by Theorem 2.14, $L = \mathrm{Im}\, f$ is a complete lattice. Observe that $f(\{x\}) = \{x\}^{\uparrow\downarrow} = x^{\downarrow}$ for all $x \in E$ and hence that $x \leqslant y \iff f(\{x\}) \subseteq f(\{y\})$. It follows that f induces an embedding $f^{\star} : E \to L$, namely that given by $f^{\star}(x) = f(\{x\}) = \{x\}^{\uparrow\downarrow} = x^{\downarrow}$.   Suppose now that $A = \{x_\alpha \mid \alpha \in I\} \subseteq E$. If $a = \bigwedge_{\alpha \in I} x_\alpha$ exists then clearly $a^{\downarrow} = \bigcap_{\alpha \in I} x_\alpha^{\downarrow}$ so that $f^{\star}(a) = \bigcap_{\alpha \in I} f^{\star}(x_\alpha)$, i.e. existing infima are preserved.

Suppose now that $b = \bigvee_{\alpha \in I} x_\alpha$ exists. Since

$$y \geqslant b \iff (\forall \alpha \in I) \; y \in x_\alpha^\uparrow = \{x_\alpha\}^{\uparrow\downarrow\uparrow}$$

$$\iff y \in \bigcap_{\alpha \in I} \{x_\alpha\}^{\uparrow\downarrow\uparrow} = \left(\bigcup_{\alpha \in I} \{x_\alpha\}^{\uparrow\downarrow}\right)^\uparrow \quad \text{(Theorem 2.15)},$$

we see that $b^\uparrow = \left(\bigcup_{\alpha \in I} \{x_\alpha\}^{\uparrow\downarrow}\right)^\uparrow$. Consequently,

$$f^\star(b) = \{b\}^{\uparrow\downarrow} = \left(\bigcup_{\alpha \in I} \{x_\alpha\}^{\uparrow\downarrow}\right)^{\uparrow\downarrow} = f\big(\sup_{\mathbb{P}(E)}\{\{x_\alpha\}^{\uparrow\downarrow} \mid \alpha \in I\}\big)$$

$$= \sup_{\mathrm{Im}\, f}\{\{x_\alpha\}^{\uparrow\downarrow} \mid \alpha \in I\} \quad \text{(Theorem 2.14)}$$

$$= \sup_{\mathrm{Im}\, f}\{f^\star(x_\alpha) \mid \alpha \in I\},$$

so that existing suprema are also preserved. □

Definition The complete lattice $L = \mathrm{Im}\, f = \{A^{\uparrow\downarrow} \mid A \in \mathbb{P}(E)\}$ in the above is called the **Dedekind–MacNeille completion** of E.

This construction is also known as the **completion by cuts** of E since it generalises the method of constructing $\mathbb{R}$ from $\mathbb{Q}$ by Dedekind sections.

The Dedekind–MacNeille completion of E has the following property.

Theorem 2.17 *Let E be an ordered set and let $\mathrm{DMac}\, E$ together with the embedding $f^\star : E \to \mathrm{DMac}\, E$ be the Dedekind–MacNeille completion of E. If $g : E \to M$ is any embedding of E into a complete lattice M then there is an embedding $\zeta : \mathrm{DMac}\, E \to M$ such that $\zeta \circ f^\star = g$.*

Proof We have the situation

$$
\begin{array}{ccc}
E & \xrightarrow{\;\;g\;\;} & M \\
{\scriptstyle f^\star}\big\downarrow & & \\
\mathrm{DMac}\, E & &
\end{array}
$$

in which $f^\star : x \mapsto x^\downarrow$, and the requirement is to produce an embedding $\zeta : \mathrm{DMac}\, E \to M$ such that $\zeta \circ f^\star = g$. For this purpose consider the mapping $\zeta : \mathrm{DMac}\, E \to M$ defined by the prescription

$$\zeta(X) = \sup_M\{g(x) \mid x \in X\}.$$

It is clear that ζ is isotone. Suppose now that $\zeta(X) \leqslant \zeta(Y)$. Then for every $x \in X$ we have

$$g(x) \leqslant \zeta(X) \leqslant \zeta(Y) = \sup_M\{g(y) \mid y \in Y\}$$

and so $g(x) \leqslant g(z)$ for all $z \in Y^\uparrow$. Since g is an embedding we deduce that $x \leqslant z$ for all $z \in Y^\uparrow$. Hence $x \in Y^{\uparrow\downarrow} = Y$ and so $X \subseteq Y$. Hence ζ is an embedding.

Now for every $x \in E$ we have

$$\zeta[f^\star(x)] = \zeta(x^\downarrow) = \sup_M\{g(t) \mid t \in x^\downarrow\} = g(x).$$

Consequently we have $\zeta \circ f^\star = g$. □

EXERCISE

2.30. Construct the Dedekind–MacNeille completion of each of the following:
   (1) a finite chain;
   (2) a finite antichain;
   (3) the 4-element fence;
   (4) the 4-element crown.

2.6 Baer semigroups

We now show how the coordinatisation of a bounded ordered set can be extended to that of a bounded lattice. For this purpose we require the notion of a Baer semigroup, which pre-dates that of a generalised Baer semigroup (hence the terminology for the latter).

Definition Let S be a semigroup with a zero element. Then we say that S is a **Baer semigroup** if the Galois connection (L, R) of Example 1.27 has the property that for each $x \in S$ there are idempotents $e, f \in S$ such that $R(x) = eS$ and $L(x) = Sf$.

Thus a semigroup S with a 0 is a Baer semigroup if and only if the right annihilator of every $x \in S$ is an idempotent-generated principal right ideal, and the left annihilator of every $x \in S$ is an idempotent-generated principal left ideal. In particular, we note that since $S = R(0) = L(0)$ there exist idempotents e, f such that $S = eS = Sf$. Then e is a left identity and f is a right identity, whence $e = f$ and so S has an identity element.

Example 2.23 Let V be a vector space. Consider the ring A of endomorphisms on V as a semigroup under composition. Given $\vartheta \in A$ let e be the projection onto $\operatorname{Ker} \vartheta$, and let f be the projection onto $\operatorname{Im} \vartheta$. Then $R(\vartheta) = eA$ and $L(\vartheta) = A(\mathrm{id}_V - f)$. Hence A is a Baer semigroup.

Example 2.24 The set of square matrices over a field is a Baer semigroup.

Example 2.25 If S is a Baer semigroup and if T is a full subsemigroup of S (i.e., T contains all the idempotents of S) then T is a Baer semigroup.

Example 2.26 Let X be a non-empty set and let $\operatorname{Rel} X$ be the set of binary relations on X. If $S \in \operatorname{Rel} X$ and $M \subseteq X$ define

$$S(M) = \{y \in X \mid (\exists x \in M) \ (x, y) \in S\}.$$

Then the **image** of S is the set $\operatorname{Im} S = S(X)$, and the **domain** of S is the set $\operatorname{Dom} S = S^d(X)$ where S^d denotes the dual of S. Given $S, T \in \operatorname{Rel} X$, define the composite ST by

$$(x, y) \in ST \iff (\exists z \in X) \ (x, z) \in T \text{ and } (z, y) \in S.$$

Then, with respect to this law of composition, $\operatorname{Rel} X$ becomes a semigroup in which the empty relation $\emptyset$ acts as a zero element. We show as follows that $\operatorname{Rel} X$ is in fact a Baer semigroup.

For this purpose, suppose first that $ST = \emptyset$. If $z \in \operatorname{Im} T$ then we cannot have $z \in \operatorname{Dom} S$ and so $\operatorname{Im} T \subseteq [\operatorname{Dom} S]'$, the complement of $\operatorname{Dom} S$. On the other hand, if $\operatorname{Im} T \subseteq [\operatorname{Dom} S]'$ then clearly there can be no elements $x, y \in X$ such that $(x, y) \in ST$, and therefore $ST = \emptyset$. Thus we see that

$$ST = \emptyset \iff \operatorname{Im} T \subseteq (\operatorname{Dom} S)'.$$

For each subset M of X define the relation I_M by

$$(x, y) \in I_M \iff x = y \in M.$$

Observe that for every subset M of X the relation I_M is an idempotent of $\operatorname{Rel} X$. If now $T \in \operatorname{Rel} X$ is such that $T = I_A T$ then $\operatorname{Im} T = T(X) = I_A[T(X)] \subseteq I_A(X) = A$; and conversely, if $\operatorname{Im} T \subseteq A$ then $T = I_A T$. Thus we have

$$T \in I_A \cdot \operatorname{Rel} X \iff \operatorname{Im} T \subseteq A.$$

It follows from these observations that, for every $S \in \operatorname{Rel} X$,

$$R(S) = I_A \cdot \operatorname{Rel} X \text{ where } A = (\operatorname{Dom} S)'.$$

Since $TS = \emptyset \iff S^d T^d = \emptyset$, a dual argument produces the fact that

$$L(S) = \operatorname{Rel} X \cdot I_B \text{ where } B = (\operatorname{Dom} S^d)' = (\operatorname{Im} S)'.$$

Hence $\operatorname{Rel} X$ is a Baer semigroup.

We observe that if S is a Baer semigroup then since S has an identity element we have $R(x) = R(Sx)$ and $L(x) = L(xS)$ for every $x \in S$. Thus, if $R(x) = eS$ and $L(x) = Sf$ then $LR(x) = L(eS) = L(e)$ and $RL(x) = R(Sf) = R(f)$. Consequently, every Baer semigroup is a generalised Baer semigroup.

Theorem 2.18 *If S is a Baer semigroup then $\mathcal{R}(S)$ and $\mathcal{L}(S)$ are dually isomorphic bounded lattices.*

Proof Since the restriction to $\mathcal{R}(S)$ of RL is the identity and since the restriction to $\mathcal{L}(S)$ of LR is the identity, the ordered sets $\mathcal{R}(S)$ and $\mathcal{L}(S)$ are dually isomorphic. Note that

$$xS \in \mathcal{R}(S) \iff xS = RL(x).$$

In fact, if $xS \in \mathcal{R}(S)$ then $xS = R(y)$ for some $y \in S$ whence $RL(x) = RL(xS) = RLR(y) = R(y) = xS$. Conversely, if $RL(x) = xS$ let $L(x) = Sf$. Then $xS = R(Sf) = R(f) \in \mathcal{R}(S)$.

Suppose then that $eS, fS \in \mathcal{R}(S)$. Then $eS = RL(eS) = RL(e) = R(e_l)$ and $fS = RL(fS) = RL(f) = R(f_l)$. Let $R(f_l e) = gS$. Then we have $eg \in R(f_l e) = gS$ whence $eg = geg$ and so eg is idempotent. Observe that

(1) $egS = R\{e_l, f_l\}$.

In fact, if $x \in egS$ then $x = egx$ whence on the one hand $e_l x = e_l egx = 0gx = 0$, and on the other hand $f_l x = f_l egx = 0x = 0$. Thus $x \in R\{e_l, f_l\}$. Conversely, if $x \in R\{e_l, f_l\}$ then $x \in R(e_l) = eS$ and so $x = ex$ whence $f_l ex = f_l x = 0$. Thus $x \in R(f_l e) = gS$ and therefore $x = gx = egx \in egS$.

It follows from (1) that we have $RL(eg) = RL(egS) = egS$ and consequently $egS \in \mathcal{R}(S)$. We now see that

$$eS \cap fS = R(e_l) \cap R(f_l) = R\{e_l, f_l\} = egS.$$

We deduce from this that $R(S)$ is a $\cap$-semilattice with bottom element 0. In a dual manner we have that $\mathcal{L}(S)$ is a $\cap$-semilattice with bottom element 0. As $\mathcal{R}(S)$ and $\mathcal{L}(S)$ are dually isomorphic, each is then a bounded lattice.   □

The coordinatisation theorem for bounded lattices is the following.

Theorem 2.19 (Janowitz [68],[69]) *For a bounded ordered set E the follow-ing statements are equivalent:*

(1) *E is a lattice;*
(2) *$\operatorname{Res} E$ is a Baer semigroup;*
(3) *E can be coordinatised by a Baer semigroup.*

Proof (1) $\Rightarrow$ (2): For each $e \in E$ consider the mappings $\vartheta_e, \psi_e : E \to E$ given by the prescriptions

$$\vartheta_e(x) = \begin{cases} x & \text{if } x \leqslant e; \\ e & \text{otherwise,} \end{cases} \qquad \psi_e(x) = \begin{cases} 0 & \text{if } x \leqslant e; \\ x \vee e & \text{otherwise.} \end{cases}$$

It is clear that ϑ_e and ψ_e are isotone and idempotent. They are also residuated; simple calculations show that

$$\vartheta_e^+(x) = \begin{cases} 1 & \text{if } x \geqslant e; \\ x \wedge e & \text{otherwise,} \end{cases} \qquad \psi_e^+(x) = \begin{cases} x & \text{if } x \geqslant e; \\ e & \text{otherwise.} \end{cases}$$

Given $f \in \operatorname{Res} E$, observe that $g \in R(f)$ if and only if $g(1) \leqslant f^+(0)$. Thus, if $g \in R(f)$ we have $g = \vartheta_{f^+(0)}g \in \vartheta_{f^+(0)} \circ \operatorname{Res} E$. Conversely, if $g = \vartheta_{f^+(0)}g$ then $g(1) \leqslant \vartheta_{f^+(0)}(1) = f^+(0)$. Thus we see that $R(f) = \vartheta_{f^+(0)} \circ \operatorname{Res} E$. Now a dual argument in the semigroup $\operatorname{Res}^+ E$ shows that $R(f^+) = \psi_{f(1)}^+ \circ \operatorname{Res}^+ E$, so in $\operatorname{Res} E$ we have $L(f) = \operatorname{Res} E \circ \psi_{f(1)}$. Hence $\operatorname{Res} E$ is a Baer semigroup.

(2) $\Rightarrow$ (3): This follows exactly as in the proof of Theorem 1.12 with $\vartheta : \mathcal{R}(\operatorname{Res} E) \to E$ given by $\vartheta(\varphi S) = \varphi(1)$.

(3) $\Rightarrow$ (1): This is immediate from Theorem 2.18.   □

We note here that a more general definition of a Baer semigroup was de-veloped by Blyth and Janowitz [24] in which the existence of a zero element is replaced by that of a principal ideal K that is generated by a central idem-potent k.

EXERCISES

2.31. If S is a Baer semigroup prove that the join operation in the lattice $\mathcal{R}(S)$ is given by
$$eS \vee fS = R(Se_l \cap Sf_l).$$

2.32. If S is a Baer semigroup and $e \in S$ is idempotent prove that eSe is a Baer semigroup. Show also that $\mathcal{R}(eSe)$ is isomorphic to the set of fixed points of $\varphi_e \in \operatorname{Res} \mathcal{R}(E)$.

2.33. A Baer semigroup S is said to be **complete** if the right annihilator of every *subset* of S is a principal right ideal generated by an idempotent. Show that if S is a complete Baer semigroup then the left annihilator of every subset of S is a principal left ideal generated by an idempotent. Show further that the following statements are equivalent:

(1) E is a complete lattice;

(2) Res E is a complete Baer semigroup;

(3) E can be coordinatised by a complete Baer semigroup.

3

Regular equivalences

3.1 Ordering quotient sets

In an algebraic structure A (such as a group, a ring, a module, ...) the notion of a **congruence**, i.e. an equivalence relation ϑ that is compatible with the operations of A, plays an important role. It is well known that an equivalence relation ϑ on A is a congruence if and only if, on the quotient set A/ϑ (consisting of the ϑ-classes $[x]_\vartheta$), there can be defined (unique) laws of composition such that A/ϑ is the same type of algebra as A and the **natural mapping** $\natural_\vartheta : A \to A/\vartheta$ given by $\natural_\vartheta(x) = [x]_\vartheta$ is a morphism. Here we shall consider the corresponding situation for ordered sets.

Definition If E is an ordered set and ϑ is an equivalence relation on E then we say that ϑ is **regular** if the quotient set E/ϑ can be ordered in such a way that the natural map $\natural_\vartheta : E \to E/\vartheta$ is isotone.

If $f : E \to F$ then the **kernel** of f is the equivalence relation ϑ_f (also written as $\ker f$) defined on E by

$$(x, y) \in \vartheta_f \iff f(x) = f(y).$$

Example 3.1 If $f : E \to F$ is isotone then the relation

$$[x]_{\vartheta_f} \leqslant [y]_{\vartheta_f} \iff f(x) \leqslant f(y)$$

is an order on the quotient set E/ϑ_f. Moreover, $\natural_{\vartheta_f} : E \to E/\vartheta_f$ is isotone since f is isotone. Thus ϑ_f is regular on E.

It is useful to have a criterion for regularity in terms only of the given equivalence relation and the order on E and not on that of the quotient set.

Definition By a ϑ-**fence** we shall mean an ordered set of the form

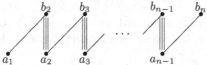

in which $a_i \leqslant b_{i+1}$ and three vertical lines indicate equivalence modulo ϑ.

This terminology, though convenient, should not be confused with that of a *fence* (Example 2.13) since in a ϑ-fence the elements a_i and b_i need not be comparable.

Similarly, we define a ϑ-**crown** to be an ordered set of the form

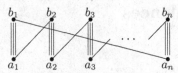

We shall often write a a ϑ-fence as $\langle a_1, b_n \rangle_\vartheta$ and a ϑ-crown as $\langle\langle a_i, b_i \rangle\rangle_\vartheta$. We shall also say that a ϑ-fence $\langle a_1, b_n \rangle_\vartheta$ *joins* a_1 to b_n.

A ϑ-crown $\langle\langle a_i, b_i \rangle\rangle_\vartheta$ will be called ϑ-**closed** when all the elements a_i, b_j are equivalent modulo ϑ.

If ϑ is an equivalence relation on an ordered set E, consider the relation $\leqslant_\vartheta$ defined on E/ϑ by

$$[x]_\vartheta \leqslant_\vartheta [y]_\vartheta \iff \text{there is a } \vartheta\text{-fence that joins } x \text{ to } y.$$

Clearly, $\leqslant_\vartheta$ is reflexive. It is also transitive since we can join together a ϑ-fence that joins x to y and a ϑ-fence that joins y to z to obtain a ϑ-fence that joins x to z. Precisely when $\leqslant_\vartheta$ is an order is the substance of the following result, which also gives a characterisation of regular equivalences.

Theorem 3.1 *If ϑ is an equivalence relation on an ordered set E then the following statements are equivalent*:

(1) *$\leqslant_\vartheta$ is an order on E/ϑ;*

(2) *ϑ is regular;*

(3) *every ϑ-crown in E is ϑ-closed.*

Proof (1)$\Rightarrow$(2): Suppose that $\leqslant_\vartheta$ is an order. If $x < y$ in E then the trivial 2-element fence consisting of x and y gives $[x]_\vartheta \leqslant_\vartheta [y]_\vartheta$. Consequently $\natural_\vartheta$ is isotone and so ϑ is regular.

(2)$\Rightarrow$(3): If ϑ is regular then there is an order $\leqslant$ on E/ϑ such that $\natural_\vartheta$ is isotone. If $\langle\langle a_i, b_i \rangle\rangle_\vartheta$ is a ϑ-crown in E then we have

$$[b_1]_\vartheta = [a_1]_\vartheta \leqslant [b_2]_\vartheta = [a_2]_\vartheta \leqslant \cdots \leqslant [b_n]_\vartheta = [a_n]_\vartheta \leqslant [b_1]_\vartheta,$$

whence we have equality throughout and hence $\langle\langle a_i, b_i \rangle\rangle_\vartheta$ is ϑ-closed.

(3)$\Rightarrow$(1): Observe that a ϑ-fence that joins x to y and a ϑ-fence that joins y to x can be joined together to form a ϑ-crown that contains both x and y. Since, by hypothesis (3), this ϑ-crown is ϑ-closed we have $[x]_\vartheta = [y]_\vartheta$. Consequently $\leqslant_\vartheta$ is also anti-symmetric and so is an order. $\square$

If E is an ordered set and $x, y \in E$ are such that $x \leqslant y$ then the **interval** $[x, y]$ is defined by

$$[x, y] = x^\uparrow \cap y^\downarrow = \{z \in L \mid x \leqslant z \leqslant y\}.$$

A non-empty subset A of an ordered set E is said to be **convex** if $[a, b] \subseteq A$ for all $a, b \in A$ with $a \leqslant b$. Clearly, every interval of an ordered set is a convex subset.

With this terminology we have the following important result.

Theorem 3.2 *If ϑ is a regular equivalence on an ordered set E then the ϑ-classes are convex subsets of E.*

Proof If $x < y < z$ with $(x, z) \in \vartheta$ then $\natural_\vartheta(x) \leqslant_\vartheta \natural_\vartheta(y) \leqslant_\vartheta \natural_\vartheta(z) = \natural_\vartheta(x)$, whence we have equality and the result follows. $\square$

EXERCISES

3.1. Prove that the set $\operatorname{Reg} E$ of regular equivalences on an ordered set E is a complete lattice which is a meet subsemilattice of the lattice of equivalence relations on E.

3.2. Consider the ordered sets

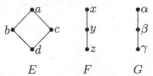

$$E \qquad F \qquad G$$

and the isotone mappings $f : E \to F$ and $g : E \to G$ given by
$$f(a) = f(b) = x, \ f(c) = y, \ f(d) = z;$$
$$g(a) = \alpha, \ g(c) = \beta, \ g(b) = g(d) = \gamma.$$

Show that the transitive product $\vartheta_f \vee \vartheta_g$ is not regular. Deduce that the complete lattice $\operatorname{Reg} E$ is not a join subsemilattice of the lattice of equivalence relations on E.

3.3. Let E be an ordered set and ϑ an equivalence relation on E. Prove that the smallest regular equivalence that contains ϑ (the **regular closure** of ϑ) is the relation ϑ^* given by

$$(x, y) \in \vartheta^* \iff \text{there is a } \vartheta\text{-crown containing } x \text{ and } y.$$

Deduce that joins in the lattice $\operatorname{Reg} E$ are given by regular closures of transitive products.

3.4. Prove that a non-empty subset of an ordered set E is an equivalence class of a regular equivalence on E if and only if it is convex.

3.2 Strongly upper regular equivalences

Definition We shall say that an equivalence relation ϑ on an ordered set E satisfies the **link property** if, whenever $(a^\star, a) \in \vartheta$ with $a \leqslant b$, there exists $b^\star \in E$ such that $(b^\star, b) \in \vartheta$ with $a^\star \leqslant b^\star$; in other words, if

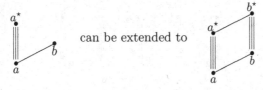

can be extended to

By a **strongly upper regular** equivalence on E we shall mean a regular equivalence that satisfies the link property.

Theorem 3.3 *If ϑ is a strongly upper regular equivalence on the ordered set E then the order $\leqslant_\vartheta$ on E/ϑ can be described by*

$$[x]_\vartheta \leqslant_\vartheta [y]_\vartheta \iff (\forall x^\star \in [x]_\vartheta)(\exists y^\star \in [y]_\vartheta)\ x^\star \leqslant y^\star.$$

Proof If $[x]_\vartheta \leqslant_\vartheta [y]_\vartheta$ then there is a ϑ-fence $\langle x, y \rangle_\vartheta$ that joins x to y. By applying the link property repeatedly to such a fence, we can reduce it to

A further application of the link property shows that for every $x^\star \in [x]_\vartheta$ there exists $y^\star \in [y]_\vartheta$ such that $x^\star \leqslant y^\star$. Conversely, taking $x^\star = x$ we obtain the fence

whence $[x]_\vartheta \leqslant_\vartheta [y]_\vartheta$. □

Strongly upper regular equivalences can be characterised as follows.

Theorem 3.4 *An equivalence relation on an ordered set is strongly upper regular if and only if it has convex classes and satisfies the link property.*

Proof The necessity of the conditions is immediate from Theorem 3.2. To show that they are sufficient, suppose that we have a ϑ-crown $\langle\!\langle a_i, b_i \rangle\!\rangle_\vartheta$. Starting at any index j, apply the link property repeatedly to construct the picture

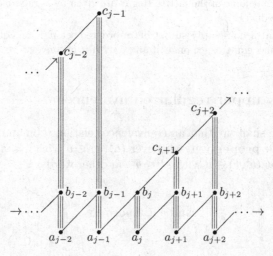

This process yields $a_{j-1} \leqslant b_j \leqslant c_{j-1}$ with $(a_{j-1}, c_{j-1}) \in \vartheta$. Since this holds for each j, it follows by convexity that the ϑ-crown $\langle\!\langle a_i, b_i \rangle\!\rangle_\vartheta$ is ϑ-closed and so, by Theorem 3.1, ϑ is regular. Hence ϑ is strongly upper regular. □

Definition By a **closure equivalence** on an ordered set E we mean an equivalence ϑ such that $\vartheta = \ker f$ for some closure mapping $f : E \to E$.

Theorem 3.5 *An equivalence relation ϑ on an ordered set E is a closure equivalence if and only if every ϑ-class has a top element and ϑ satisfies the link property.*

Proof $\Rightarrow$: Suppose that ϑ is a closure equivalence and let $f : E \to E$ be a closure mapping with kernel ϑ. Then for every $x \in E$ the ϑ-class $[x]_\vartheta$ is convex with top element $f(x)$. If now $(x, y) \in \vartheta$ and $x \leqslant z$ then we have $y \leqslant f(y) = f(x) \leqslant f(z) \in [z]_\vartheta$. Consequently ϑ satisfies the link property.

$\Leftarrow$: Suppose conversely that every ϑ-class has a top element and that ϑ satisfies the link property. For every $x \in E$ let the top element of $[x]_\vartheta$ be $x^\star$ and consider the mapping $f : E \to E$ given by $f(x) = x^\star$. Clearly, we have $f = f^2 \geqslant \mathrm{id}_E$ and ϑ is the kernel of f. Now the link property applied to $x^\star \overset{\vartheta}{\equiv} x \leqslant y$ gives the existence of y' such that $x^\star \leqslant y' \overset{\vartheta}{\equiv} y$. Since necessarily $y' \leqslant y^\star$, we see that $x \leqslant y$ implies that $f(x) = x^\star \leqslant y^\star = f(y)$. Thus f is also isotone and is therefore a closure mapping. $\square$

Corollary *If f is a closure mapping then its kernel ϑ_f is strongly upper regular and the order $\leqslant_{\vartheta_f}$ is given by*

$$[x]_{\vartheta_f} \leqslant_{\vartheta_f} [y]_{\vartheta_f} \iff x \leqslant f(y).$$

Moreover, $\natural_{\vartheta_f}$ is residuated.

Proof That ϑ_f is strongly upper regular is immediate from the above and Theorem 3.4. The description of $\leqslant_{\vartheta_f}$ follows from Theorem 3.3, and it is clear from this description that $\natural_{\vartheta_f}$ is residuated. $\square$

The above results relate to residuated mappings as follows.

Theorem 3.6 *If $f : E \to F$ is residuated then its kernel ϑ_f is a closure equivalence on E and can be characterised as an equivalence relation with convex classes such that each ϑ_f-class contains a unique element of $\mathrm{Im}\, f^+$ which is the top element in its class.*

Proof Since $f \circ f^+ \circ f = f$ we have $f^+ f(x) = f^+ f(y) \iff f(x) = f(y)$ and so the closure $f^+ f$ has the same kernel as f. Hence ϑ_f is a closure equivalence.

Clearly, $f^+ f(x)$ is the top element of $[x]_{\vartheta_f}$. Moreover, if $f^+(y) \overset{\vartheta_f}{\equiv} f^+(z)$ then $f f^+(y) = f f^+(z)$ whence $f^+(y) = f^+ f f^+(y) = f^+ f f^+(z) = f^+(z)$. Thus each ϑ_f-class contains a unique element of $\mathrm{Im}\, f^+$ which is the top of its class.

Conversely, suppose that ϑ is an equivalence relation on E such that ϑ has convex classes and each ϑ-class contains a unique element of $\mathrm{Im}\, f^+$ which is the top element in its class. Given any $a \in E$ there then exists a unique $f^+(y)$ such that $a \overset{\vartheta}{\equiv} f^+(y)$ and $a \leqslant f^+(y)$. Now $a \leqslant f^+ f(a) \leqslant f^+ f f^+(y) = f^+(y)$ and so the convexity of the classes gives $f^+ f(a) \overset{\vartheta}{\equiv} f^+(y)$. But since each ϑ-class contains a unique element of $\mathrm{Im}\, f^+$ it follows that we must have $f^+ f(a) = f^+(y)$. Consequently,

$$(a,b) \in \vartheta \;\Rightarrow\; f^+ f(a) = f^+(y) = f^+ f(b) \;\Rightarrow\; (a,b) \in \ker f^+ f = \vartheta_f$$

and therefore $\vartheta \subseteq \vartheta_f$. Since each ϑ_f-class contains a unique element of $\operatorname{Im} f^+$ it follows that we must have $\vartheta = \vartheta_f$. $\square$

EXERCISES

3.5. Show that on the 4-element crown there is an equivalence relation that has convex classes but does not satisfy the link property.

3.6. Consider the equivalence relation R on $\mathbb{Z}$ that partitions $\mathbb{Z}$ into two classes, one consisting of the even integers and the other consisting of the odd integers. Show that R is not regular but satisfies the link property.

3.7. Prove that if $(R_\alpha)_{\alpha \in A}$ is a family of equivalence relations on an ordered set E each of which satisfies the link property then so does their transitive product. Deduce that the transitive product of a family of strongly upper regular equivalences is strongly upper regular if and only if its classes are convex.

3.8. Prove that if $(R_\alpha)_{\alpha \in A}$ is a family of strongly upper regular equivalences on an ordered set E then the transitive closure of their transitive product satisfies the link property. Deduce that the strongly upper regular equivalences on E form a join subsemilattice of the complete lattice of regular equivalences on E.

3.9. Consider the ordered sets

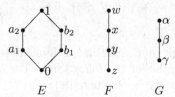

$$\qquad\qquad E \qquad\qquad F \qquad G$$

and the mappings $f : E \to F$ and $g : E \to G$ given by
$$f(1) = w,\; f(a_2) = f(b_2) = f(b_1) = x,\; f(a_1) = y,\; f(0) = z;$$
$$g(1) = g(b_2) = \alpha,\; g(a_1) = g(a_2) = g(b_1) = \beta,\; g(0) = \gamma.$$
Show that ϑ_f and ϑ_g are strongly upper regular whereas $\vartheta_f \cap \vartheta_g$ is not. Deduce that the lattice of strongly upper regular equivalences on E is not a meet subsemilattice of the lattice of regular equivalences on E.

3.10. If ϑ is a closure equivalence on E prove that $\natural_\vartheta$ is residuated and determine its residual.

3.11. An equivalence relation ϑ on an ordered set E is said to satisfy the **interpolation property** if the ϑ-fence

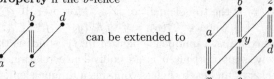

can be extended to

Prove that if ϑ satisfies the link property or its dual then ϑ satisfies the interpolation property.

3.12. An equivalence relation ϑ on an ordered set E is said to be **inter-regular** if it is regular and satisfies the interpolation property. Prove that ϑ is inter-regular if and only if it has convex classes and satisfies the interpolation property.

3.13. If ϑ is inter-regular prove that the order $\leqslant_\vartheta$ is given by
$$[x]_\vartheta \leqslant_\vartheta [y]_\vartheta \iff (\exists x^\star \in [x]_\vartheta)(\exists y^\star \in [y]_\vartheta)\ x^\star \leqslant y^\star.$$

3.14. Let E and F be ordered sets and let $f : E \to F$ be isotone and such that there exists an isotone $g : F \to E$ such that $f = fgf$. Prove that $\ker f$ is inter-regular on E.

3.3 Lattice congruences

Definition If L is a $\vee$-semilattice then an equivalence relation ϑ on E is said to be $\vee$-**compatible** if
$$(x, y) \in \vartheta \ \Rightarrow\ (\forall z \in L)\ (x \vee z, y \vee z) \in \vartheta.$$

When ϑ is $\vee$-compatible, the quotient set L/ϑ becomes a $\vee$-semilattice under the induced law $[x]_\vartheta \sqcup [y]_\vartheta = [x \vee y]_\vartheta$, the semilattice order $\leqslant$ being given by
$$[x]_\vartheta \leqslant [y]_\vartheta \iff [y]_\vartheta = [x]_\vartheta \sqcup [y]_\vartheta \iff (y, x \vee y) \in \vartheta.$$

Clearly, if $x \leqslant y$ then $[x]_\vartheta \leqslant [y]_\vartheta$. Thus every $\vee$-compatible equivalence is regular. The connection between $\vee$-compatibility and what has gone before is the substance of the following result.

Theorem 3.7 *Let L be a $\vee$-semilattice. If ϑ is a $\vee$-compatible equivalence on L then ϑ is strongly upper regular and the orders $\leqslant$ and $\leqslant_\vartheta$ on L/ϑ coincide.*

Proof We note first that the ϑ-classes are convex; for if $x \leqslant y \leqslant z$ with $(x, z) \in \vartheta$ then $(y, z) = (x \vee y, z \vee y) \in \vartheta$. Moreover, ϑ satisfies the link property; for if $(x^\star, x) \in \vartheta$ with $x \leqslant y$ then $(x^\star \vee y, y) = (x^\star \vee y, x \vee y) \in \vartheta$. Thus ϑ is strongly upper regular.

Suppose now that $[x]_\vartheta \leqslant_\vartheta [y]_\vartheta$. Then by Theorem 3.3 there exists $y^\star \in [y]_\vartheta$ such that $x \leqslant y^\star$. This gives $y^\star = x \vee y^\star$ and so $[x]_\vartheta \leqslant [y^\star]_\vartheta = [y]_\vartheta$. Conversely, if $[x]_\vartheta \leqslant [y]_\vartheta$ then $[x \vee y]_\vartheta = [y]_\vartheta$ and therefore $(x \vee y, y) \in \vartheta$. If then $(x^\star, x) \in \vartheta$ we have $(x^\star \vee y, y) \in \vartheta$. Thus, for every $x^\star \in [x]_\vartheta$ there exists $y^\star = x^\star \vee y \in [y]_\vartheta$ such that $x^\star \leqslant y^\star$. Hence we have $[x]_\vartheta \leqslant_\vartheta [y]_\vartheta$. $\square$

In a dual manner to the above, we can formulate the notion of a **strongly lower regular** equivalence. An equivalence relation that is both strongly upper regular and strongly lower regular will be called **strongly regular**. Likewise we may define the dual notion of a $\wedge$-**compatible** equivalence.

An equivalence relation on a lattice that is both $\vee$-compatible and $\wedge$-compatible is called a **congruence**. By Theorem 3.7 and its dual, every congruence on a lattice is strongly regular.

Given a congruence ϑ on a lattice L we can define laws of composition on the quotient set L/ϑ by $[x]_\vartheta \sqcap [y]_\vartheta = [x \wedge y]_\vartheta$ and $[x]_\vartheta \sqcup [y]_\vartheta = [x \vee y]_\vartheta$. Then, by Theorem 2.3, $(L/\vartheta; \sqcap, \sqcup, \leqslant_\vartheta)$ is a lattice.

The following isomorphism theorem is fundamental.

Theorem 3.8 *If $f : L \to M$ is a lattice morphism then there is a lattice isomorphism $\operatorname{Im} f \simeq L/\ker f$.*

Proof Writing ker f as ϑ_f we have

$$[x]_{\vartheta_f} = [y]_{\vartheta_f} \iff f(x) = f(y).$$

The assignment $[x]_{\vartheta_f} \mapsto f(x)$ therefore defines an injective mapping ϑ from $L/\ker f$ to the sublattice $\operatorname{Im} f$ of M. Clearly, ϑ is also surjective. Finally, ϑ is a lattice morphism. Indeed, for all $x, y \in L$ we have

$$\begin{aligned} \vartheta([x]_{\vartheta_f} \sqcap [y]_{\vartheta_f}) = \vartheta([x \wedge y]_{\vartheta_f}) &= f(x \wedge y) \\ &= f(x) \wedge f(y) \\ &= \vartheta([x]_{\vartheta_f}) \sqcap \vartheta([y]_{\vartheta_f}), \end{aligned}$$

and similarly for joins. $\qquad\square$

If L is a lattice then in what follows we shall denote by $\operatorname{Con} L$ the set of congruences on L.

Theorem 3.9 *If L is a lattice then $\operatorname{Con} L$ is a complete lattice and is a sublattice of the complete lattice of equivalence relations on L.*

Proof We have seen in Example 2.19 that the set of equivalence relations on L is a complete lattice and have described there both $\bigwedge\limits_{\alpha \in I} R_\alpha$ and $\bigvee\limits_{\alpha \in I} R_\alpha$ for any family $\{R_\alpha \; ; \; \alpha \in I\}$ of equivalence relations on L.

Suppose now that $F = \{\vartheta_\alpha \; ; \; \alpha \in I\}$ is a family of congruences on L. It is clear that $\bigwedge\limits_{\alpha \in I} \vartheta_\alpha$ is also a congruence on L and so this is the infimum of F in $\operatorname{Con} L$. It therefore suffices to show that $\bigvee\limits_{\alpha \in I} \vartheta_\alpha$ is also a congruence on L, whence this will be the supremum of F in $\operatorname{Con} L$. Now if $(x, y) \in \bigvee\limits_{\alpha \in I} \vartheta_\alpha$ then there is a finite sequence of equivalences of the form

$$(x, a_1) \in \vartheta_{\alpha_1}, \; (a_1, a_2) \in \vartheta_{\alpha_2}, \; \ldots, \; (a_n, y) \in \vartheta_{\alpha_{n+1}}$$

where each $\vartheta_{\alpha_i} \in F$. Then for all $z \in L$ we have

$$(x \vee z, a_1 \vee z) \in \vartheta_{\alpha_1}, \; (a_1 \vee z, a_2 \vee z) \in \vartheta_{\alpha_2}, \; \ldots, \; (a_n \vee z, y \vee z) \in \vartheta_{\alpha_{n+1}}$$

whence $(x \vee y, y \vee z) \in \bigvee\limits_{\alpha \in I} \vartheta_\alpha$. Similarly $(x \wedge z, y \wedge z) \in \bigvee\limits_{\alpha \in I} \vartheta_\alpha$ and consequently $\bigvee\limits_{\alpha \in I} \vartheta_\alpha$ is a congruence on L, as required. $\qquad\square$

EXERCISES

3.15. Give an example of a lattice L that contains a convex sublattice C that is not a ϑ-class for any congruence ϑ on L.

3.16. If L is a 4-element chain, show that $|\operatorname{Con} L| = 8$ and describe the Hasse diagram.

Lattice congruences that have bounded classes arise from residuated dual closures. This is the substance of the following result.

Theorem 3.10 *If L is a lattice and f is a residuated dual closure on L then* $\ker f$ *has bounded classes. Conversely, if ϑ is a congruence on L that has bounded classes then $x \mapsto \inf_L \{ y \in L \mid y \in [x]_\vartheta \}$ describes a residuated dual closure on L. Moreover, the assignment $f \mapsto \ker f$ is a bijection between the set of residuated dual closures on L and the set of congruences on L that have bounded classes.*

Proof If f is a residuated dual closure then, as in the proof of Theorem 1.9, we have $f = f \circ f^+$ and $f^+ = f^+ \circ f$. It follows immediately that $f(x) = f(y)$ if and only if $f^+(x) = f^+(y)$. Suppose now that $(x, y) \in \vartheta_f = \ker f$. Then by Theorem 2.8, for every $t \in L$, $f(x \vee t) = f(x) \vee f(t) = f(y) \vee f(t) = f(y \vee t)$ and so $(x \vee t, y \vee t) \in \vartheta_f$. Dually, $f^+(x \wedge t) = f^+(y \wedge t)$ gives $(x \wedge t, y \wedge t) \in \vartheta_f$. Hence ϑ_f is a congruence. Now if $(x, y) \in \vartheta_f$ we have $f(x) = f(y)$ whence $f^+(x) = f^+(y)$ and therefore $y \in [f(x), f^+(x)]$. Conversely, $y \in [f(x), f^+(x)]$ gives $f(x) = f^2(x) \leqslant f(y) \leqslant ff^+(x) = f(x)$ whence $f(x) = f(y)$. Consequently ϑ_f has bounded classes. Furthermore, f may be recaptured from ϑ_f since for every $x \in L$ we have $f(x) = \inf[x]_{\vartheta_f}$.

Suppose now that ϑ is a congruence on L that has bounded classes, say $[x]_\vartheta = [x_\vartheta, x^\vartheta]$ for every $x \in L$. Define $f, g : L \to L$ by the prescriptions $f(x) = x_\vartheta$ and $g(x) = x^\vartheta$. Then if $x \leqslant y$ we have $x_\vartheta = x_\vartheta \wedge y \overset{\vartheta}{\equiv} x_\vartheta \wedge y_\vartheta$ whence $x_\vartheta \leqslant y_\vartheta$. Thus f is isotone; and dually so is g. Observing that $f \circ g = f \leqslant \mathrm{id}_L$ and $g \circ f = g \geqslant \mathrm{id}_L$, we conclude that f is a residuated dual closure with $f^+ = g$. Finally, ϑ can be recaptured from f since $(x, y) \in \vartheta$ if and only if $f(x) = f(y)$. $\qquad\square$

We shall return later to congruences on specific types of lattice. For the present we give the following description of congruences on cartesian products.

Theorem 3.11 *Let L and M be lattices. If $\vartheta \in \mathrm{Con}\, L$ and $\varphi \in \mathrm{Con}\, M$ then the relation $\vartheta \times \varphi$ defined on $L \times M$ by*

$$\big((x, a), (y, b)\big) \in \vartheta \times \varphi \iff (x, y) \in \vartheta \quad \text{and} \quad (a, b) \in \varphi$$

is a congruence. Moreover, every congruence ξ on $L \times M$ is of the form $\xi_L \times \xi_M$ for some $\xi_L \in \mathrm{Con}\, L$ and $\xi_M \in \mathrm{Con}\, M$.

Proof It is clear that $\vartheta \times \varphi \in \mathrm{Con}(L \times M)$. Suppose now that ξ is a congruence on $L \times M$ and define ξ_L on L by

$$(x, y) \in \xi_L \iff (\exists c \in M) \ \big((x, c), (y, c)\big) \in \xi.$$

Then clearly $\xi_L \in \mathrm{Con}\, L$. We can in fact also describe ξ_L by

$$(x, y) \in \xi_L \iff (\forall d \in M) \ \big((x, d), (y, d)\big) \in \xi.$$

To justify this change of quantifiers, observe that for any $d \in M$ we have

$$[(x, c) \vee (x \wedge y, d)] \wedge (x \vee y, d) = (x, d).$$

Similarly we can define $\xi_M \in \mathrm{Con}\, M$ by

$$(a, b) \in \xi_M \iff (\exists z \in L) \ \big((z, a), (z, b)\big) \in \xi,$$

and here we can likewise replace the existential quantifier $\exists$ by the universal quantifier $\forall$.

Now if $\big((x,a),(y,b)\big) \in \xi_L \times \xi_M$ then $(x,y) \in \xi_L$ and $(a,b) \in \xi_M$ give $\big((x,d),(y,d)\big) \in \xi$ and $\big((z,a),(z,b)\big) \in \xi$. Taking $z = x \wedge y$ and $d = a \wedge b$ we then have $\big((x, a\wedge b)\vee(x\wedge y, a),(y, a\wedge b)\vee(x\wedge y, b)\big) \in \xi$, i.e. $\big((x,a),(y,b)\big) \in \xi$.

Conversely, if $\big((x,a),(y,b)\big) \in \xi$ then taking infima with $(x \vee y, a \wedge b)$ we obtain $\big((x, a \wedge b),(y, a \wedge b)\big) \in \xi$ so that $(x,y) \in \xi_L$; and similarly $(a,b) \in \xi_M$. Consequently $\big((x,a),(y,b)\big) \in \xi_L \times \xi_M$.

It follows from these observations that $\xi = \xi_L \times \xi_M$. $\qquad\qquad$ $\square$

EXERCISES

3.17. A congruence relation ϑ on a complete lattice L is said to be **complete** if

$$(\forall \alpha \in I)\ x_\alpha \stackrel{\vartheta}{\equiv} x \implies \bigwedge_{\alpha \in I} x_\alpha \stackrel{\vartheta}{\equiv} x \stackrel{\vartheta}{\equiv} \bigvee_{\alpha \in I} x_\alpha.$$

Prove that there is a bijection between the set of complete congruences on a complete lattice L and the set of residuated dual closure mappings on L.

3.18. Let L be the 3-element chain $x < y < z$ and let M be the 4-element chain $a < b < c < d$. If ξ_L is the congruence on L with partition $\{\{x,y\},\{z\}\}$ and if ξ_M is the congruence on M with partition $\{\{a,b\},\{c,d\}\}$, indicate on a Hasse diagram the classes of the congruence $\xi_L \times \xi_M$ on $L \times M$.

3.19. If L is a 4-element chain determine $|\text{Con}\,L^2|$.

3.20. Generalise Theorem 3.11 to finitely many lattices.

3.21. Consider the chain

$$\overline{A} \equiv \underbrace{a_1 < a_2 < a_3 \cdots \to}_{A} \overbrace{\leftarrow \cdots < a_3^* < a_2^* < a_1^*}^{A^*}.$$

Describe the classes of the smallest congruence that identifies the elements x and y when

(1) $x,y \in A$;
(2) $x,y \in A^*$;
(3) $x \in A, y \in A^*$.

Modular lattices

4.1 Modular pairs; Dedekind's modularity criterion

We now proceed to investigate particular classes of lattices. For this purpose we consider the following type of residuated mapping.

Definition If E, F are ordered sets and $f : E \to F$ is residuated then we shall say that f is **range-closed** if $\operatorname{Im} f$ is a down-set of F.

Example 4.1 Let L be a lattice with a top element 1. Given $e \in L$, consider the mapping $\vartheta_e : L \to L$ given by

$$\vartheta_e(x) = \begin{cases} x & \text{if } x \leqslant e; \\ e & \text{otherwise.} \end{cases}$$

As observed in the proof of Theorem 2.19, ϑ_e is residuated. Clearly, $\operatorname{Im} \vartheta_e$ is the down-set $e^{\downarrow}$ of L, so ϑ_e is range-closed.

Range-closed residuated mappings can be characterised as follows.

Theorem 4.1 *Let E and F be ordered sets and suppose that E has a top element 1_E. If $f : E \to F$ is residuated then the following statements are equivalent:*

(1) *f is range-closed;*
(2) *$(\forall x \in F)$ $x \wedge f(1_E)$ exists in F and is $ff^+(x)$.*

Proof (1) $\Rightarrow$ (2): Clearly, $ff^+(x) \leqslant x$ and $ff^+(x) \leqslant f(1_E)$ for every $x \in F$. Suppose then that $y \in F$ is such that $y \leqslant x$ and $y \leqslant f(1_E)$. By (1) we have $y = f(z)$ for some $z \in E$. Then $f(z) \leqslant x$ gives $z \leqslant f^+(x)$ whence $y = f(z) \leqslant ff^+(x)$. Thus we see that $x \wedge f(1_E)$ exists in F and is $ff^+(x)$.

(2) $\Rightarrow$ (1): If $x \leqslant f(1_E)$ then, by (2), $x = x \wedge f(1_E) = ff^+(x) \in \operatorname{Im} f$ and consequently f is range-closed. $\square$

There is of course a dual characterisation of residuated mappings that are **dually range-closed**, i.e. those for which $\operatorname{Im} f^+$ is an up-set. If F has a bottom element 0_F then $f : E \to F$ is dually range-closed if and only if

$$(\forall x \in E) \quad x \vee f^+(0_F) \text{ exists in } E \text{ and is } f^+f(x).$$

Definition If E, F are ordered sets then a residuated mapping $f : E \to F$ is said to be **weakly regular** if it is both range-closed and dually range-closed.

Example 4.2 Let V and W be vector spaces over a field F and let $f : V \to W$ be linear. Then f induces a mapping $f^\to : \mathrm{Sub}\,V \to \mathrm{Sub}\,W$. This mapping is residuated, its residual being the induced inverse image mapping $f^\leftarrow$. Now, as the reader will easily verify, we have the identities

$$(\forall B \in \mathrm{Sub}\,W) \quad (f^\to \circ f^\leftarrow)(B) = B \cap \mathrm{Im}\,f;$$
$$(\forall A \in \mathrm{Sub}\,V) \quad (f^\leftarrow \circ f^\to)(A) = A + \mathrm{Ker}\,f.$$

Consequently, by Theorem 4.1 and its dual, $f^\to$ is weakly regular.

EXERCISES

4.1. Let E and F be ordered sets and let $f : E \to F$ be residuated. Let I be the down-set of F that is generated by $\mathrm{Im}\,f$. Prove that f is range-closed if and only if the restriction of f^+ to I is injective.

4.2. If $A \subseteq E$ prove that the dual closure $\lambda_A : \mathbb{P}(E) \to \mathbb{P}(E)$ given by $\lambda_A(X) = A \cap X$ is a weakly regular residuated mapping.

4.3. Let L be a bounded lattice and let $f \in \mathrm{Res}\,L$ be weakly regular and idempotent. Prove that $(\forall x \in L) \quad f(x) = [x \vee f^+(0)] \wedge f(1)$.

If L is a lattice and $a, b \in L$ with $a \neq b$, consider the mappings

$$\alpha_a : b^\downarrow \to a^\uparrow \text{ given by } \alpha_a(x) = a \vee x;$$
$$\beta_b : a^\uparrow \to b^\downarrow \text{ given by } \beta_b(y) = b \wedge y.$$

For every $x \in b^\downarrow$ we have

$$\beta_b \alpha_a(x) = b \wedge (a \vee x) \geqslant b \wedge x = x$$

so that $\beta_b \alpha_a \geqslant \mathrm{id}_{b^\downarrow}$, and for every $y \in a^\uparrow$ we have

$$\alpha_a \beta_b(y) = a \vee (b \wedge y) \leqslant a \vee y = y$$

so that $\alpha_a \beta_b \leqslant \mathrm{id}_{a^\uparrow}$. Since α_a and β_b are clearly isotone it follows that α_a is residuated with residual $\alpha_a^+ = \beta_b$. Precisely when α_a is weakly regular is the substance of the following result.

Theorem 4.2 *If L is a lattice and $a, b \in L$ with $a \neq b$ then the following statements are equivalent:*

(1) α_a *is weakly regular;*

(2) a *and b are such that*

$$M(a,b) : \quad (\forall x \in L) \quad x \leqslant b \Rightarrow x \vee (a \wedge b) = (x \vee a) \wedge b,$$
$$M^\star(b,a) : \quad (\forall y \in L) \quad y \geqslant a \Rightarrow y \wedge (b \vee a) = (y \wedge b) \vee a;$$

(3) $\alpha_a|_{[a \wedge b, b]}$ *and $\beta_b|_{[a, a \vee b]}$ are mutually inverse bijections.*

Proof (1) $\Leftrightarrow$ (2): Note that for every $x \in b^\downarrow$ we have $\alpha_a^+ \alpha_a'(x) = b \wedge (a \vee x)$ and $x \vee \alpha_a^+(a) = x \vee \beta_b(a) = x \vee (b \wedge a)$. Then α_a being dually range-closed translates into the property $M(a, b)$; and, dually, α_a being range-closed translates into the property $M^\star(b, a)$.

$(2) \Rightarrow (3)$: If (2) holds and $x \in [a \wedge b, b]$ then $M(a, b)$ gives $x = (x \vee a) \wedge b = \beta_b \alpha_a(x)$; and if $y \in [a, a \vee b]$ then $M^*(b, a)$ gives $y = (y \vee b) \wedge a = \alpha_a \beta_b(y)$. Statement (3) is now immediate.

$(3) \Rightarrow (2)$: Suppose that (3) holds and let $x \leqslant b$. Then $(a \wedge b) \vee x \in [a \wedge b, b]$ and so there exists $y \in [a, a \vee b]$ such that $(a \wedge b) \vee x = \beta_b(y) = y \wedge b$. Since $y \geqslant a, x$ we have $y \geqslant a \vee x$ and therefore $(a \wedge b) \vee x \geqslant (a \vee x) \wedge b$. The reverse inequality holds since $x \leqslant b$ gives, by the isotonicity of $z \mapsto z \vee x$,

$$(a \wedge b) \vee x \leqslant (a \vee x) \wedge (b \vee x) = (a \vee x) \wedge b.$$

Thus we have $M(a, b)$ and, dually, $M^*(b, a)$.   □

Corollary *If L is a lattice and $a, b \in L$ are such that $M(a, b)$ and $M^*(b, a)$ hold then the intervals $[a \wedge b, b]$ and $[a, a \vee b]$ are isomorphic.*

Proof This is immediate from (3) above and Theorem 2.9.   □

The property $M(a, b)$ that appears in Theorem 4.2 was first considered by Wilcox [113] who introduced the following notion.

Definition If L is a lattice and $a, b \in L$ satisfy the property $M(a, b)$ then the ordered pair (a, b) is said to be **modular**.

Example 4.3 It is readily seen that in every lattice an ordered pair (a, b) is modular whenever $a \nparallel b$. In the lattice

all pairs (x, y) are modular except the pair (a, c).

Modular pairs can be characterised in the following useful way.

Theorem 4.3 *$M(a, b)$ holds in a lattice L if and only if L does not contain a sublattice of the form*

in which $y \leqslant b$.

Proof If $M(a, b)$ holds and there exists a sublattice of the form indicated then $x < y \leqslant b$ and so $(x \vee a) \wedge b = x \vee (a \wedge b) = x$ which gives the contradiction $y = (x \vee a) \wedge y = (x \vee a) \wedge b \wedge y = x \wedge y = x$. If, on the other hand, $M(a, b)$ failed then we could find an element $c < b$ such that $c \vee (a \wedge b) < (c \vee a) \wedge b$. Let $x = c \vee (a \wedge b)$ and $y = (c \vee a) \wedge b$. Note that $a \wedge b \leqslant a \wedge y = a \wedge (c \vee a) \wedge b = a \wedge b$ and so $a \wedge b = a \wedge y = a \wedge x$. Dually, $a \vee x = a \vee c \vee (a \wedge b) = a \vee c \geqslant a \vee y \geqslant a \vee x$ shows that $a \vee x = a \vee y = a \vee c$. This produces a sublattice of the stated form.   □

Corollary *$M(a, b)$ holds in L if and only if it holds in $[a \wedge b, a \vee b]$.*   □

EXERCISES

4.4. If $M(a,b)$ holds in a lattice L prove that so does $M(x,y)$ for all $x \in [a \wedge b, a]$ and $y \in [a \wedge b, b]$.

4.5. If $M(a,b)$ holds in an interval $[c,d]$ of a lattice L prove that $M(a,b)$ holds in L.

4.6. A lattice L is M-**symmetric** when $M(a,b)$ implies $M(b,a)$. Prove that if $M(a,b)$ implies $M^*(b,a)$ then L is M-symmetric.

Definition A lattice L is said to be **modular** if $M(a,b)$ holds for all $a,b \in L$.

A useful working test for modularity is the following, often referred to as **Dedekind's modularity criterion**.

Theorem 4.4 *A lattice L is modular if and only if it has no sublattice of the form*

Equivalently, L is modular if and only if $x \leqslant y$, $a \wedge x = a \wedge y$, $a \vee x = a \vee y$ imply that $x = y$.

Proof This is immediate from Theorem 4.3.   □

Observe that in every lattice the inequality $x \vee (a \wedge b) \leqslant (x \vee a) \wedge (x \vee b)$ holds for all x, a, b, so that the implication $x \leqslant b \Rightarrow x \vee (a \wedge b) \leqslant (x \vee a) \wedge b$ always holds. Thus, to establish modularity it suffices to establish the implication

$$x \leqslant b \Rightarrow x \vee (a \wedge b) \geqslant (x \vee a) \wedge b.$$

Example 4.4 Let G be a group and denote by $N(G)$ the set of all normal subgroups of G. Then $(N(G); \cap, \vee, \subseteq)$ is a lattice in which $H \vee K = HK = \{hk \; ; \; h \in H, k \in K\}$. To see that this lattice is modular, all we need do is to show that $H \subseteq K$ implies that $H(J \cap K) \supseteq HJ \cap K$. Suppose then that $H \subseteq K$, and let $x \in HJ \cap K$. Then $x = yz$ where $y \in H, z \in J$ and $x \in K$, so that $z = y^{-1}x \in HK \subseteq KK = K$ whence $z \in J \cap K$ and therefore $x = yz \in H(J \cap K)$.

Example 4.5 Every sublattice M of a modular lattice L is modular, for clearly every pair of elements of M is a modular pair. In particular, therefore, the lattice $L(V)$ of subspaces of a vector space V is modular since it is a sublattice of the modular lattice of (normal) subgroups of the additive group of V.

EXERCISES

4.7. Prove that if a lattice is modular then so is its dual.

4.8. If $f : L \to M$ is a lattice morphism and L is modular, prove that $\operatorname{Im} f$ is a modular sublattice of M.

4.9. Given a family $(L_i)_{i \in I}$ of lattices, prove that $\bigtimes_{i \in I} L_i$ is modular if and only if each L_i is modular.

4.10. Prove that a lattice L is modular if and only if

$$(\forall x, y, z \in L) \quad x \vee [y \wedge (x \vee z)] = (x \vee y) \wedge (x \vee z).$$

4.11. Prove that a lattice L is modular if and only if

$$(\forall x, y, z \in L) \quad \{x \vee [y \wedge (x \vee z)]\} \wedge z = (x \vee y) \wedge z.$$

4.12. Prove that a lattice L is modular if and only if, for all $x, y \in L$, the mapping $f : [x \wedge y, x] \to [y, x \vee y]$ given by $f(a) = a \vee y$ is injective.

4.13. Prove that a lattice L is modular if and only if, for all $x, y \in L$, the mapping

$$f : [x \wedge y, x] \times [x \wedge y, y] \to [x \wedge y, x \vee y]$$

given by $f(a, b) = a \vee b$ is injective.

4.14. If, in a modular lattice with bottom element 0, elements $a, x_1, \ldots, x_n$ are such that $a \wedge (x_1 \vee \cdots \vee x_n) = 0$ prove that

$$a \vee (x_1 \wedge \cdots \wedge x_n) = (a \vee x_1) \wedge \cdots \wedge (a \vee x_n).$$

4.15. If L is a lattice prove that the ideal lattice $I(L)$ is modular if and only if L is modular. Deduce that every modular lattice with a bottom element can be embedded in a complete modular lattice.

4.16. Intervals $[a, b]$ and $[c, d]$ of a lattice L are said to be **perspective** if $a = b \wedge c$ and $d = b \vee c$, in which case we write $[a, b] \approx [c, d]$. We say that $[a, b]$ and $[c, d]$ are **projective** if there is a finite sequence of intervals $[x_i, y_i]$ such that

$$[a, b] \approx [x_1, y_1] \approx \cdots \approx [x_n, y_n] \approx [c, d],$$

in which case we write $[a, b] \sim [c, d]$.

If L is a modular lattice and $a, b, c, d \in L$ are such that $a < b$ and $c < d$, establish the diagram

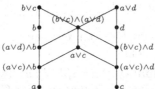

and hence deduce the **Zassenhaus Lemma** that

$$[(a \vee c) \wedge b, (a \vee d) \wedge b] \sim [(a \vee c) \wedge d, (b \vee c) \wedge d].$$

4.17. Establish the converse of the Zassenhaus Lemma, namely that if L is a lattice and if $a, b, c, d \in L$ with $a \leqslant b$, $c \leqslant d$ are such that the intervals $[(a \vee c) \wedge b, (a \vee d) \wedge b]$ and $[a \vee c, (b \vee c) \wedge (a \vee d)]$ are perspective then L is modular.

4.18. Let $x = a_1 \geqslant a_2 \geqslant \cdots \geqslant a_n = y$ and $x = b_1 \geqslant b_2 \geqslant \cdots \geqslant b_p = y$ be chains joining elements x and y in a modular lattice. Define $a_{ij} = (a_i \wedge b_j) \vee a_{i+1}$ and $b_{ji} = (a_i \wedge b_j) \vee b_{j+1}$. Establish the chains

$$x = a_{11} \geqslant \cdots \geqslant a_{1p} = a_{21} \geqslant \cdots \geqslant a_{2p} = a_{31} \geqslant \cdots \geqslant a_{n-1,p} = y;$$
$$x = b_{11} \geqslant \cdots \geqslant b_{1n} = b_{21} \geqslant \cdots \geqslant b_{2n} = b_{31} \geqslant \cdots \geqslant b_{p-1,n} = y,$$

in which $[a_{i,j+1}, a_{ij}] \sim [b_{j,i+1}, b_{ji}]$.

4.2 Chain conditions

We now examine chains of elements in ordered sets with particular reference to modular lattices.

Definition We define the **length** of a chain of $n \geqslant 1$ elements to be $n - 1$; and the **length** of an ordered set E to be the supremum of the lengths of the chains in E.

Example 4.6 Every finite lattice is of finite length.

Example 4.7 The lattice $L(\mathbb{R}^3)$ of subspaces of the real vector space $\mathbb{R}^3$ is infinite, but is of finite length. In fact, it is of length 3 since the longest chains in $L(\mathbb{R}^3)$ are of the form $\{0\} \prec A \prec B \prec \mathbb{R}^3$ where $\dim A = 1$ and $\dim B = 2$.

Definition A lattice L is said to be of **locally finite length** if every interval of L is of finite length.

Example 4.8 $(\mathbb{N}, |)$ is of locally finite length but is not of finite length.

Example 4.9 A bounded lattice is of locally finite length if and only if it is of finite length. For each positive integer n let $\mathbf{n}$ denote the chain $1 < 2 < \cdots < n$. If we add top and bottom elements to the disjoint union $\underset{n \geqslant 1}{\bigcup} \mathbf{n}$ then we obtain a bounded lattice that is not of locally finite length.

Definition By a **maximal chain** in an ordered set E we mean a chain that is not a proper subchain of any other chain. Equivalently, a chain C is maximal in E if the only elements of E that are comparable to every element of C are the elements of C.

In the theory of ordered sets an important result, which can be shown to be equivalent to the Axiom of Choice, is the

Hausdorff Maximality Principle   *For any chain C in an ordered set E there is a maximal chain in E that contains C.*

Now in general not all maximal chains with the same end points have the same length. The 5-element non-modular lattice of Theorem 4.4 is an immediate example. For an infinite example, consider the following.

Example 4.10 In the cartesian product lattice $\mathbb{R} \times \mathbb{Q}$ the set C of elements of the form (x, x) with $0 \leqslant x \leqslant 1$ forms a maximal chain from $(0, 0)$ to $(1, 1)$. In fact, if $(p, q) \in \mathbb{R} \times \mathbb{Q}$ with $0 \leqslant p \leqslant 1$, $0 \leqslant q \leqslant 1$ and $p \neq q$ then there exists $z \in \mathbb{Q}$ lying between p and q and we have $(p, q) \parallel (z, z)$. Thus (p, q) does not belong to a chain that properly contains C. The chain C has the same cardinality as $\mathbb{Q}$ and so is countable. On the other hand, the elements of the form $(0, x)$ and $(y, 1)$ where $0 \leqslant x \leqslant 1$ and $0 \leqslant y \leqslant 1$ also form a maximal chain from $(0, 0)$ to $(1, 1)$, and this chain has cardinality that of $\mathbb{R}$, so is uncountable.

Definition An ordered set is said to satisfy the **Jordan–Dedekind chain condition** if all maximal chains between the same end points have the same length.

Theorem 4.5 *If L is a lattice of locally finite length that satisfies the Jordan–Dedekind chain condition then there exists a mapping $d : L \to \mathbb{Z}$ such that*

$$x \prec y \iff (x < y \text{ and } d(y) = d(x) + 1).$$

Proof We begin by choosing any $t \in L$ and defining $d(t)$ to be any integer d_0. Given $x, y \in L$ let the length of any maximal chain from $t \wedge x$ to y be $\ell_{t \wedge x, y}$. Now define $d : L \to \mathbb{Z}$ by the prescription $d(x) = d_0 - \ell_{t \wedge x, t} + \ell_{t \wedge x, x}$. Then we have

$$d(y) - d(x) = -\ell_{t \wedge y, t} + \ell_{t \wedge y, y} + \ell_{t \wedge x, t} - \ell_{t \wedge x, x}. \tag{4.1}$$

Now let β be the length of all maximal chains from x to y. Then from the diagram

we see that

$$\ell_{t \wedge x, t} - \ell_{t \wedge y, t} + \ell_{t \wedge y, y} = \ell_{t \wedge x, x} + \beta.$$

It follows from this and (4.1) that $d(y) - d(x) = \beta$. Since $x \prec y$ is equivalent to $x < y$ and $\beta = 1$, the result follows. $\qquad\square$

Definition A mapping $d : L \to \mathbb{Z}$ that satisfies the property stated in Theorem 4.5 is called a **dimension function** on L.

Observe that if we suppose in Theorem 4.5 that L has a bottom element 0_L then, by taking $t = 0_L$ and $d(t) = 0$, we see that $d(x)$ becomes the length of the interval $[0_L, x]$. This is often called the **height** of x.

Definition A lattice L is said to be **upper semi-modular** if

$$x \wedge y \prec x \implies y \prec x \vee y,$$

and **lower semi-modular** if the dual property (which is given by the converse implication) holds.

It is immediate from the Corollary to Theorem 4.2 that every modular lattice is both upper semi-modular and lower semi-modular. As we shall now show, for a lattice of locally finite length the converse is true.

Theorem 4.6 *For a lattice L of locally finite length the following statements are equivalent:*

(1) *L is modular;*

(2) *L is both upper and lower semi-modular;*

(3) *L satisfies the Jordan–Dedekind chain condition and every dimension function $d : L \to \mathbb{Z}$ is such that*

$$(\forall x, y \in L) \qquad d(x) + d(y) = d(x \wedge y) + d(x \vee y).$$

Proof (1) $\Rightarrow$ (2): This has been observed above.

(2) $\Rightarrow$ (3): Suppose that (2) holds and let $a, b \in L$. Let

$$a = a_0 \prec a_1 \prec \cdots \prec a_n = b$$

be a maximal chain joining a and b. We shall show that the length of any chain joining a and b is at most n, whence the Jordan–Dedekind chain condition will follow.

The proof is by induction on n. The case $n = 1$ being trivial, let $n \geqslant 2$ and suppose that the above proposition is true for every interval containing a maximal chain of length at most $n - 1$. Let $a = x_0 < x_1 < \cdots < x_m = b$ be any chain from a to b and consider the chain

$$a_1 = a_1 \vee a = a_1 \vee x_0 \leqslant a_1 \vee x_1 \leqslant \cdots \leqslant a_1 \vee x_m = b. \qquad (4.2)$$

Since a maximal chain of length $n - 1$ exists between a_1 and b, the induction hypothesis implies that any chain from a_1 to b is of length at most $n - 1$. Let the smallest x_i that is greater than a_1 be x_t. Then for $j \geqslant t$ we have $a_1 \vee x_j = x_j < x_{j+1} = a_1 \vee x_{j+1}$; for $j = t - 1$ we have $a_1 \vee x_j \leqslant a_1 \vee x_{j+1} = x_{j+1}$; and for $j \leqslant t - 2$ we have $a \leqslant a_1 \wedge x_j < a_1$ and $a \leqslant a_1 \wedge x_{j+1} < a_1$ whence, since $a \prec a_1$, we see that $a = a_1 \wedge x_j = a_1 \wedge x_{j+1} \prec a_1$. Upper semi-modularity now gives

$$x_j \prec a_1 \vee x_j \leqslant a_1 \vee x_{j+1}, \quad x_j < x_{j+1} \prec a_1 \vee x_{j+1}$$

which imply that $a_1 \vee x_j < a_1 \vee x_{j+1}$. These observations show that the length of the chain (4.2) is at least $m - 1$. Thus we have $m - 1 \leqslant n - 1$ whence $m \leqslant n$. This completes the induction and the Jordan–Dedekind chain condition follows.

As for the statement concerning dimension functions, suppose that $x \parallel y$ and consider a maximal chain $x \wedge y = b_0 \prec b_1 \prec \cdots \prec b_r = y$. There follows the chain

$$x \preceq x \vee b_1 \preceq x \vee b_2 \preceq \cdots \preceq x \vee b_r = x \vee y.$$

For example, we cannot have $x \vee b_1 \leqslant b_2$ (otherwise $x \leqslant y$) so either $x \vee b_1 > b_2$ or $x \vee b_1 \parallel b_2$. In the former case we have $x \vee b_1 = x \vee b_1 \vee b_2 = x \vee b_2$, and in the latter case we have $b_1 = (x \vee b_1) \wedge b_2 \prec b_2$ whence upper semi-modularity gives $x \vee b_1 \prec (x \vee b_1) \vee b_2 = x \vee b_2$.

Consequently, for any dimension function d on L, we have

$$d(x \vee b_{i+1}) - d(x \vee b_i) \in \{0, 1\}$$

and therefore

$$\begin{aligned}
d(x \vee y) - d(x) &= d(x \vee b_r) - d(x \vee b_0) \\
&= \sum_{i=0}^{r-1} \big(d(x \vee b_{i+1}) - d(x \vee b_i)\big) \\
&\leqslant r \\
&= d(y) - d(x \wedge y).
\end{aligned}$$

Thus we see that $d(x) + d(y) \geqslant d(x \wedge y) + d(x \vee y)$. A dual argument using lower semi-modularity gives the reverse inequality.

(3) $\Rightarrow$ (1): Suppose now that (3) holds and that L has the sublattice

Then we have

$$d(x) + d(y) = d(x \wedge y) + d(x \vee y) = d(x \wedge z) + d(x \vee z) = d(x) + d(z),$$

whence we obtain the contradiction $d(y) = d(z)$. Thus L has no such sublattice and so, by Theorem 4.4, is modular.   □

An example of an upper semi-modular lattice that is not of finite length is given in the following result.

Theorem 4.7 *For every non-empty set E the lattice of equivalence relations on E is upper semi-modular, and is of finite length if and only if E is finite.*

Proof If ϑ and φ are equivalence relations on E then φ covers ϑ if and only if the φ-classes coincide with the ϑ-classes except for one which is the union of two ϑ-classes. Suppose then that $\vartheta \wedge \varphi \prec \varphi$. Then the φ-classes coincide with the $(\vartheta \wedge \varphi)$-classes except for one φ-class A which is the union of two $(\vartheta \wedge \varphi)$-classes B and C. Let B_ϑ and C_ϑ be the ϑ-classes containing B and C respectively. Then given $x \in B_\vartheta$ and $y \in C_\vartheta$ we have, for all $b \in B$ and all $c \in C$, the relations $(x, b) \in \vartheta, (b, c) \in \varphi, (c, y) \in \vartheta$. Consequently we see that $(x, y) \in \vartheta \vee \varphi$ and it follows that $B_\vartheta \cup C_\vartheta$ is a single $(\vartheta \vee \varphi)$-class. Since $\vartheta \wedge \varphi$ coincides with $\vartheta \vee \varphi$ on $E \backslash (B_\vartheta \cup C_\vartheta)$ we deduce that $\vartheta \prec \vartheta \vee \varphi$. Thus the lattice is upper semi-modular. Clearly, it is of finite length if and only if E is finite.   □

EXERCISES

4.19. If L is a modular lattice with bottom element 0 prove that the elements of finite height in L form an ideal of L.

4.20. If L is a modular lattice prove that the relation $\equiv$ defined on L by

$$x \equiv y \iff [x \wedge y, x \vee y] \text{ is of finite length}$$

is an equivalence relation.
[*Hint.* For transitivity consider the interval $[(x \wedge y) \vee z, x \vee y \vee z]$.]

4.21. Prove that a lattice L is upper semi-modular if and only if

$$a \prec b \Rightarrow (\forall x \in L) \; a \vee x \preceq b \vee x.$$

4.22. Prove that every M-symmetric lattice (Exercise 4.6) is upper semi-modular.

4.23. If L is an upper semi-modular lattice of finite length and $a, b \in L$, prove that $M(a, b)$ holds if and only if $h(a) + h(b) = h(a \wedge b) + h(a \vee b)$. Deduce that L is M-symmetric.

4.24. Let $\mathbb{N}^\infty = \mathbb{N} \cup \{\infty\}$ denote the chain of natural numbers with a top element ∞ adjoined. If **2** denotes the chain $0 < 1$, consider the cartesian ordered cartesian product lattice $(\mathbb{N}^\infty \times \mathbb{N}^\infty \times \mathbf{2}) \backslash \{(\infty, \infty, 0)\}$. Use this lattice to show that a lattice that is both upper and lower semi-modular is not necessarily modular.

4.25. If L is a lattice then a **valuation** on L is a mapping $f : L \to \mathbb{R}$ such that
$$(\forall x, y \in L) \quad f(x) + f(y) = f(x \wedge y) + f(x \vee y).$$
A **positive valuation** is a valuation f such that $x < y$ implies $f(x) < f(y)$. A **metric lattice** is defined to be a lattice with a positive valuation. Prove that every metric lattice is modular and is a metric space with respect to the distance function d defined by
$$d(x, y) = f(x \vee y) - f(x \wedge y).$$

4.3 Join-irreducibles

We shall now consider a special type of element which, as we shall see, is of particular importance.

Definition If L is a lattice then $a \in L$ (with $a \neq 0$ if L has a bottom element 0) is said to be **join-irreducible** if
$$x \vee y = a \Rightarrow x = a \ \text{or} \ y = a.$$

Thus, $a \in L$ is join-irreducible if it cannot be expressed as the join of two elements that are strictly less than a. We denote the set of join-irreducible elements of a lattice L by $\mathcal{J}(L)$. Dually, we can define the set $\mathcal{M}(L)$ of **meet-irreducible** elements.

Example 4.11 In the lattices

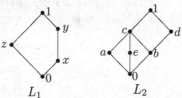

we have $\mathcal{J}(L_1) = \{x, y, z\} = \mathcal{M}(L_1)$ and $\mathcal{J}(L_2) = \{a, b, d, e\}$, $\mathcal{M}(L_2) = \{a, c, d, e\}$.

Example 4.12 In a chain every element (other than a bottom element if it exists) is join-irreducible. In contrast, in the cartesian lattice $\mathbb{Z} \times \mathbb{Z}$ there are no elements that are join-irreducible.

EXERCISES

4.26. For any set E identify $\mathcal{J}\big(\mathbb{P}(E); \cap, \cup, \subseteq\big)$.

4.27. For any ordered set E identify $\mathcal{J}\big(\mathcal{O}(E); \cap, \cup, \subseteq\big)$.

4.28. If L_1 and L_2 are lattices each having a bottom element, prove that
$$\mathcal{J}(L_1 \times L_2) \simeq \mathcal{J}(L_1) \,\dot\cup\, \mathcal{J}(L_2).$$

Definition We say that an ordered set E satisfies the **descending chain condition** if all descending chains in E are finite.

Theorem 4.8 *Let L be a lattice satisfying the descending chain condition. Then L has a bottom element 0 and every $x \in L \setminus \{0\}$ can be represented as a join of a finite number of join-irreducibles.*

Proof It is clear that L has a bottom element 0. If every $x \in L \setminus \{0\}$ is join-irreducible then there is nothing more to do. Otherwise there exists $x \in L \setminus \{0\}$ such that $x = y \vee z$ where $y < x$ and $z < x$. If both y and z are join-irreducible then we are done; otherwise we can write y or z as a join $a \vee b$ and repeat the argument. Since L has no infinite descending chains, this process ends after a finite number of steps and the result follows. $\qquad\square$

We now investigate the uniqueness of such representations.

Theorem 4.9 *Let L be a modular lattice satisfying the descending chain condition and let $x \in L \setminus \{0\}$ have two representations*

$$x = \bigvee_{i=1}^{m} x_i = \bigvee_{j=1}^{n} y_j$$

as joins of join-irreducibles. Then for every x_i there is a y_j such that

$$x = x_1 \vee \cdots \vee x_{i-1} \vee y_j \vee x_{i+1} \vee \cdots \vee x_m.$$

Proof Let $a = \bigvee_{k \neq i} x_k$ and for $t = 1, \ldots, n$ let $z_t = y_t \vee a$. Then

$$\bigvee_{t=1}^{n} z_t = \bigvee_{t=1}^{n} y_t \vee a = x \vee a = x.$$

Now since L is modular it follows by the Corollary to Theorem 4.2 that $[x_i \wedge a, x_i] \simeq [a, x_i \vee a]$. Since x_i is join-irreducible in L it is join-irreducible in $[x_i \wedge a, x_i]$ and hence $x = x_i \vee a$ is join-irreducible in $[a, x_i \vee a] = [a, x]$. Then since $x = \bigvee_{t=1}^{n} z_t \geq z_t \geq a$ for each t we deduce that $x = z_j$ for some j. Consequently

$$x = z_j = y_j \vee a = y_j \vee \bigvee_{k \neq i} x_k,$$

as required. $\qquad\square$

Definition We shall say that a join $\bigvee_{i=1}^{m} x_i$ is **irredundant** if it is such that, for every k,

$$\bigvee_{i=1}^{m} x_i > x_1 \vee \cdots \vee x_{k-1} \vee x_{k+1} \vee \cdots \vee x_m = \bigvee_{j \neq k} x_j.$$

Roughly speaking, a join is irredundant if the removal of any term results in something smaller. If

$$\bigvee_{i=1}^{m} x_i = \bigvee_{i \neq k} x_i$$

then we say that the element x_k is **redundant**.

Theorem 4.10 (Kurosh–Ore [75], [89]) *If L is a modular lattice that satisfies the descending chain condition then all irredundant representations of an element as finite joins of join-irreducibles have the same number of components.*

Proof Suppose that we have two irredundant join-irreducible representations
$$x = \bigvee_{i=1}^{m} x_i = \bigvee_{j=1}^{n} y_j$$
and that m is the least possible number of components that can arise. Then clearly $m \leqslant n$. Now by Theorem 4.9 we can replace x_1 by some y_{i_1} to obtain $x = y_{i_1} \vee x_2 \vee \cdots \vee x_m$. This representation is also irredundant, by the minimality of m. So we can apply Theorem 4.9 again to replace x_2 by some y_2, and so on. In this way we can replace each x_j by some y_{i_j}. Since the representation so obtained is irredundant, it must contain all of the y_i (otherwise $x = \bigvee_{j=1}^{n} y_j$ would not be irredundant). Hence $n \leqslant m$ and therefore $m = n$. $\qquad\square$

Note that in the Kurosh–Ore Theorem the condition that L be modular cannot be relaxed. To see this, consider the upper semi-modular lattice with Hasse diagram

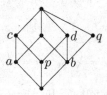

Here the join-irreducibles are a, b, p, q and we have the following irredundant representations of the top element
$$1 = a \vee b \vee p = q \vee p.$$

Note also that the representations in the Kurosh–Ore Theorem are not unique in general. To see this, consider the modular lattice

Here we have $1 = a \vee b = a \vee c$.

In the next chapter we shall consider a subclass of the class of modular lattices in which such representations are unique.

EXERCISES

4.29. Prove that a lattice L satisfies the descending chain condition if and only if every filter of L is principal.

4.30. Let L be a lattice satisfying the descending chain condition and let $x, y \in L$ be such that $x \nleqslant y$. Prove that there exists $a \in \mathcal{J}(L)$ such that $a \leqslant x$ and $a \nleqslant y$.

4.31. Let L be an upper semi-modular lattice that satisfies the descending chain condition. By an **atom** of L we mean an element p such that $0 \prec p$. Prove that if $p \in L$ is an atom then $M^\star(p, a)$ holds for all $a \in L$.

4.32. Let L be a modular lattice with bottom element 0. If $a, b, c \in L \setminus \{0\}$ prove that
$$(a \vee b) \wedge c = 0 \implies a \wedge (b \vee c) = a \wedge b.$$

4.33. Let L be a modular lattice with bottom element 0. If $a_1, \ldots, a_n \in L \setminus \{0\}$ then $a_1, \ldots, a_n$ are said to be **independent** if
$$(i = 1, \ldots, n) \quad a_i \wedge \bigvee_{j \neq i} a_j = 0.$$

Prove that if $a_1, \ldots, a_n$ are independent then the sublattice generated by the intervals $[0, a_i]$ (i.e. the smallest sublattice of L that contains each of these intervals) is isomorphic to the cartesian lattice $\underset{i=1}{\overset{n}{\times}} [0, a_i]$.

4.4 Baer semigroups and modularity

For the purpose of investigating the type of Baer semigroup involved in the co-ordinatisation of modular lattices, we introduce the following mappings which provide an interesting further characterisation of these lattices.

Let L be a bounded lattice and let $a, b \in L$ be such that $a \parallel b$. Define the mappings $\vartheta_{a,b}, \zeta_{a,b} : L \to L$ by

$$\vartheta_{a,b}(x) = \begin{cases} x & \text{if } x \leqslant a; \\ a \wedge (b \vee x) & \text{if } x \not\leqslant a, \ x \leqslant a \vee b; \\ a & \text{if } x \not\leqslant a \vee b, \end{cases}$$

$$\zeta_{a,b}(x) = \begin{cases} 1 & \text{if } x \geqslant a; \\ b \vee (a \wedge x) & \text{if } x \not\geqslant a, \ x \geqslant a \wedge b; \\ a \wedge x & \text{if } x \not\geqslant a \wedge b. \end{cases}$$

Routine verification reveals that both $\vartheta_{a,b}$ and $\zeta_{a,b}$ are isotone.

Theorem 4.11 *For a bounded lattice L and $a, b \in L$ with $a \parallel b$ the following statements are equivalent*:

(1) *$\vartheta_{a,b}$ is residuated with residual $\zeta_{a,b}$*;
(2) *$M(b, a)$ and $M^\star(a, b)$ hold.*

Proof (1) $\Rightarrow$ (2): Suppose that $\vartheta_{a,b}$ is residuated and, by way of obtaining a contradiction, suppose that $M(b, a)$ does not hold. Then, by Theorem 4.3, L contains a sublattice of the form in which $y \leqslant a$. Since on the one hand $\vartheta_{a,b}(b \vee x) = a \wedge (b \vee x) \geqslant y$, and on the other $\vartheta_{a,b}(b) \vee \vartheta_{a,b}(x) = (a \wedge b) \vee x = x$ we see that $\vartheta_{a,b}$ is not a $\vee$-morphism and so, by Theorem 2.8, fails to be residuated. Thus we have $M(b, a)$.

To show that $M^\star(a, b)$ holds, let $c \in L$ be such that $c \geqslant b$. There are two cases to consider:

(α) $c \geqslant a \vee b$: Clearly, we have $c \wedge (a \vee b) = a \vee b = (c \wedge a) \vee b$.

(β) $c \not\geqslant a \vee b$: In this case we have $a \vee b > c \wedge (a \vee b) \geqslant b$. But if $a \vee b > x \geqslant b$ we have $x \leqslant \zeta_{a,b}\vartheta_{a,b}(x) = b \vee [a \wedge (b \vee x)] = b \vee (a \wedge x)$. Taking $x = c \wedge (a \vee b)$ we obtain $c \wedge (a \vee b) \leqslant b \vee [a \wedge c \wedge (a \vee b)] = b \vee (a \wedge c)$. But since $c \geqslant b$ we also have $b \vee (a \wedge c) \leqslant (b \vee a) \wedge (b \vee c) = (b \vee a) \wedge c$. Thus in this case also we have $c \wedge (a \vee b) = (c \wedge a) \vee b$ and hence $M^\star(a, b)$ holds.

(2) $\Rightarrow$ (1): Conversely, suppose that $M(b, a)$ and $M^\star(a, b)$ hold. Then a routine verification (which we leave to the reader) shows that $\vartheta_{a,b}\zeta_{a,b} \leqslant \mathrm{id}_L$ and $\zeta_{a,b}\vartheta_{a,b} \geqslant \mathrm{id}_L$, so that $\vartheta_{a,b}$ is residuated with residual $\zeta_{a,b}$. $\qquad\square$

Corollary *The following conditions on a bounded lattice L are equivalent*:

(1) *L is modular*;

(2) *for all $a, b \in L$ with $a \parallel b$, the mapping $\vartheta_{a,b}$ is residuated.* $\qquad\square$

If L is modular then the mappings $\vartheta_{a,b}$ described above are clearly range-closed idempotents of $\mathrm{Res}\, L$. Our objective now is to show how they may be used in order to obtain a Baer semigroup coordinatisation of bounded modular lattices. For this purpose, we recall the residuated mappings ϑ_e as used in Theorem 2.19:

$$\vartheta_e(x) = \begin{cases} x & \text{if } x \leqslant e; \\ e & \text{otherwise,} \end{cases} \qquad \vartheta_e^+(x) = \begin{cases} 1 & \text{if } x \geqslant e; \\ x \wedge e & \text{otherwise.} \end{cases}$$

In relation to these, we have the following properties.

Theorem 4.12 *Let L be a bounded lattice and let $a, b \in L$.*

(1) *If $a \not\parallel b$ then $\vartheta_a \vartheta_b = \vartheta_{a \wedge b} = \vartheta_b \vartheta_a$.*

(2) *If $a \parallel b$ then $\vartheta_{a,b}\vartheta_a = \vartheta_a$, $\vartheta_a \vartheta_{a,b} = \vartheta_{a,b}$, $\vartheta_{a,b}\vartheta_b = \vartheta_{a,b}\vartheta_{b,a} = \vartheta_{a \wedge b}$.*

Proof (1) If, for example, $a \leqslant b$ then

$$\vartheta_a[\vartheta_b(x)] = \vartheta_a \begin{cases} x & \text{if } x \leqslant b; \\ b & \text{otherwise,} \end{cases} = \begin{cases} x & \text{if } x \leqslant a; \\ a & \text{otherwise,} \end{cases} = \vartheta_a(x),$$

so that $\vartheta_a \circ \vartheta_b = \vartheta_a$; and since $\vartheta_a(x) \leqslant a \leqslant b$ for all $x \in L$ we have $\vartheta_b[\vartheta_a(x)] = \vartheta_a(x)$ so that $\vartheta_b \circ \vartheta_a = \vartheta_a$.

(2) Suppose now that $a \parallel b$. Then we note that

$$\vartheta_{a,b}[\vartheta_a(x)] = \vartheta_{a,b} \begin{cases} x & \text{if } x \leqslant a; \\ a & \text{otherwise,} \end{cases} = \begin{cases} x & \text{if } x \leqslant a; \\ a & \text{otherwise,} \end{cases} = \vartheta_a(x),$$

so that $\vartheta_{a,b} \circ \vartheta_a = \vartheta_a$. Also, since $\vartheta_{a,b}(x) \leqslant a$ for every $x \in L$ it is clear that $\vartheta_a \circ \vartheta_{a,b} = \vartheta_{a,b}$. Finally,

$$\vartheta_{a,b}[\vartheta_{b,a}(x)] = \vartheta_{a,b} \begin{cases} x & \text{if } x \leqslant b; \\ b \wedge (a \vee x) & \text{if } x \nleqslant b, \ x \leqslant b \vee a; \\ b & \text{if } x \nleqslant b \vee a, \end{cases}$$

$$= \begin{cases} x & \text{if } x \leqslant a \wedge b; \\ a \wedge b & \text{otherwise}, \end{cases}$$

$$= \vartheta_{a \wedge b}(x)$$

Thus we have $\vartheta_{a,b} \circ \vartheta_{b,a} = \vartheta_{a \wedge b}$.

The remaining equalities follow similarly. $\qquad\square$

Corollary $\vartheta_{a,b}$ *and* $\vartheta_{b,a}$ *commute.* $\qquad\square$

We recall from Theorem 2.19 that if L is a bounded lattice then in the coordinatising Baer semigroup $\operatorname{Res} L$ we have

$$R(f) = \vartheta_{f^+(0)} \circ \operatorname{Res} L, \qquad L(f) = \operatorname{Res} L \circ \psi_{f(1)}.$$

Moreover, writing $S = \operatorname{Res} L$, if $fS \in \mathcal{R}(S)$ then

$$fS = RL(f) = R(\psi_{f(1)}) = \vartheta_{\psi_{f(1)}^+(0)} S = \vartheta_{f(1)} S, \qquad (4.3)$$

and the fundamental coordinatisation $\mathcal{R}(S) \simeq L$ is described by $\overline{f} S \mapsto \overline{f}(1)$ where $\overline{f}$ is an idempotent generator of fS. On taking $\overline{f} = \vartheta_{f(1)}$, we see that

$$fS \cap gS \mapsto \vartheta_{f(1)}(1) \wedge \vartheta_{g(1)}(1) = f(1) \wedge g(1)$$

and likewise

$$\vartheta_{f(1) \wedge g(1)} S \mapsto \vartheta_{f(1) \wedge g(1)}(1) = f(1) \wedge g(1),$$

so that we have

$$fS \cap gS = \vartheta_{f(1) \wedge g(1)} S.$$

Moreover, by the same lattice isomorphism,

$$fS \| gS \iff f(1) \| g(1).$$

In view of the above observations, a coordinatisation of bounded modular lattices can now be formulated as follows.

Theorem 4.13 (Blyth [18]) *For a bounded lattice L the following statements are equivalent*:

(1) L *is modular*;

(2) L *can be coordinatised by a Baer semigroup S such that if $eS, fS \in \mathcal{R}(S)$ then there exist idempotents $\overline{e}, \overline{f} \in S$ such that $\overline{e}S = eS$, $\overline{f}S = fS$, and $\overline{e}, \overline{f}$ commute*;

(3) L *can be coordinatised by a Baer semigroup S such that if $eS, fS \in \mathcal{R}(S)$ with e idempotent then there is an idempotent $\overline{f} \in S$ such that $\overline{f}S = fS$ and $\overline{f}e = e\overline{f}e$.*

Proof (1) $\Rightarrow$ (2): Suppose that L is modular and that $S = \operatorname{Res} L$ with $eS, fS \in \mathcal{R}(S)$. Supppose first that $eS \subseteq fS$. From the fundamental isomorphism $\mathcal{R}(S) \simeq L$ of Theorem 2.19 we have $e(1) \leqslant f(1)$ and therefore, by Theorem 4.12(1), $\vartheta_{e(1)}\vartheta_{f(1)} = \vartheta_{e(1)} = \vartheta_{f(1)}\vartheta_{e(1)}$. Taking $\overline{e} = \vartheta_{e(1)}$ and $\overline{f} = \vartheta_{f(1)}$

we see that $\bar{e}, \bar{f}$ commute. Since, by (4.3), $eS = \vartheta_{e(1)}S$ and $fS = \vartheta_{f(1)}S$, it follows that (2) holds in this case. Clearly, a similar result holds if $fS \subseteq eS$.

Suppose now that $eS \parallel fS$. Then $e(1) \parallel f(1)$. Since L is modular by hypothesis, it follows by the Corollary to Theorem 4.11 that $\vartheta_{e(1),f(1)} \in S$ and $\vartheta_{f(1),e(1)} \in S$. Moreover, by Theorem 4.12(2), we have

$$\vartheta_{e(1),f(1)}S = \vartheta_{e(1)}S = eS, \qquad \vartheta_{f(1),e(1)}S = \vartheta_{f(1)}S = fS.$$

In view of the Corollary to Theorem 4.12, the result holds in this case on taking $\bar{e} = \vartheta_{e(1),f(1)}$ and $\bar{f} = \vartheta_{f(1),e(1)}$.

(2) $\Rightarrow$ (3): Suppose now that (2) holds. Since $\bar{e}$ and $\bar{f}$ commute we have

$$eS \cap \bar{f}S = \bar{e}S \cap \bar{f}S = \bar{e}\bar{f}S = \bar{f}\bar{e}S = \bar{f}eS,$$

whence $\bar{f}e = e\bar{f}e$ and (3) follows.

(3) $\Rightarrow$ (1): If (3) holds then from $\bar{f}e = e\bar{f}e$ we obtain $\bar{f}e \in eS$ so that $\bar{f}eS \subseteq eS \cap \bar{f}S$. Conversely, if $x \in eS \cap \bar{f}S$ then $x = ex = \bar{f}x = \bar{f}ex$ and so $x \in \bar{f}eS$. Hence $\bar{f}eS = eS \cap \bar{f}S$. To show that $M(eS, \bar{f}S)$ it suffices, by Theorem 4.3 to show that $\mathcal{R}(S)$ has no sublattice of the form

in which $gS \subseteq \bar{f}S$. Now since $L : \mathcal{R}(S) \to \mathcal{R}(S)$ given by $L(eS) = L(e) = Se^*$ is a dual isomorphism it suffices to show that $\mathcal{L}(S)$ has no sublattice of the form

$$L(P)$$
$$L(e) = Se^* \bullet \qquad \bullet Sh^* = L(h)$$
$$\bullet Sg^* = L(g)$$
$$L(Q)$$

in which $S\bar{f}^* \subseteq Sg^*$. By way of obtaining a contradiction, suppose that $\mathcal{L}(S)$ has such a sublattice. Let $x \in Sh^*$. Then $x = xh^*$ so $xh = xh^*h = 0$. But $hS \subseteq \bar{f}S$ so $h = \bar{f}h$ and consequently

$$0 = xh = x\bar{f}h. \tag{4.4}$$

Now $P = eS \cap fS == eS \cap \bar{f}S = \bar{f}eS$ so $L(P) = L(\bar{f}e)$. Since $x \in Sh^* \subseteq L(P)$ we thus have

$$0 = x\bar{f}e. \tag{4.5}$$

By (4.4) and (4.5) we obtain

$$x\bar{f} \in L(e) \cap L(h) = Se^* \cap Sh^* = L(Q) \subseteq Sg^*$$

so that $x\bar{f}g = 0$. But $gS \subseteq \bar{f}S$ so $g = \bar{f}g$. Thus we have $0 = x\bar{f}g = xg$ and therefore $x \in L(g) = Sg^*$. This gives the contradiction $Sh^* \subseteq Sg^*$ and we conclude that $M(eS, fS)$ holds. $\square$

5

Distributive lattices

5.1 Birkhoff's distributivity criterion

In this chapter we introduce an important subclass of the class of modular lattices, namely that of the distributive lattices. These were the first to be considered in the earliest of investigations. For this purpose, consider for any element x of a lattice L the mapping $\mu_x : L \to L$ given by $\mu_x(y) = x \vee y$. Clearly, μ_x is a $\vee$-morphism. It is natural to investigate the situation when every μ_x is a lattice morphism, which is equivalent to the property

$$(\forall x, a, b \in L) \qquad x \vee (a \wedge b) = (x \vee a) \wedge (x \vee b).$$

If this property holds then it is clear that $M(a, b)$ holds for all $a, b \in L$. Hence every lattice that satisfies the above property is modular.

Definition A lattice L is said to be **distributive** if, for every $x \in L$, the mapping μ_x is a lattice morphism.

We begin by noting that the above property is self-dual. Put another way, if each μ_x is a morphism then so also is each $\lambda_x : L \to L$ given by $\lambda_x(y) = x \wedge y$. That this is so can be seen by two applications of the property and the absorption law. In fact, if each μ_x is a morphism then, for all $x, y, z \in L$,

$$
\begin{aligned}
(x \wedge y) \vee (x \wedge z) &= [(x \wedge y) \vee x] \wedge [(x \wedge y) \vee z] \\
&= x \wedge [(x \wedge y) \vee z] \\
&= x \wedge (x \vee z) \wedge (y \vee z) \\
&= x \wedge (y \vee z),
\end{aligned}
$$

and so λ_x is also a morphism.

It is a curious historical fact that it was originally thought that every lattice was distributive! That this is not so is shown by the following example of a modular lattice that is not distributive:

Here we clearly have $x \vee (y \wedge z) = x$ whereas $(x \vee y) \wedge (x \vee z) = 1$.

Example 5.1 Every sublattice of a distributive lattice is distributive.

Example 5.2 $(\mathbb{P}(E); \cap, \cup, \subseteq)$ is a distributive lattice.

Example 5.3 If E is an ordered set then the lattice $\mathcal{O}(E)$ of down-sets of E is a sublattice of $\mathbb{P}(E)$ and so is distributive.

Example 5.4 If $\mathbb{P}_f(E)$ is the set of all finite subsets of an infinite set E then $\mathbb{P}_f(E)$ is a distributive lattice with bottom element $\emptyset$ and no top element.

Example 5.5 If D is a distributive lattice and $f : D \to L$ is a lattice morphism then $\operatorname{Im} f$ is a distributive sublattice of L.

Example 5.6 If D is a distributive lattice and E is a non-empty set then the set D^E of mappings $f : E \to D$ can be ordered by

$$f \sqsubseteq g \iff (\forall x \in E) \; f(x) \leqslant g(x).$$

Then $(D^E; \sqsubseteq)$ is also a distributive lattice. Here $\sup\{f, g\} = f \sqcup g$ is given by $(f \sqcup g)(x) = f(x) \vee g(x)$, and $\inf\{f, g\} = f \sqcap g$ is given by $(f \sqcap g)(x) = f(x) \wedge g(x)$. By the distributivity of D, we have $f \sqcap (g \sqcup h) = (f \sqcap g) \sqcup (f \sqcap h)$.

Example 5.7 The set $P = \{1, 2, 3, \dots\}$, ordered by divisibility, is a distributive lattice. In fact, by the fundamental theorem of arithmetic we can represent any given positive integer m in the form

$$m = 2^{\alpha_m(1)} 3^{\alpha_m(2)} 5^{\alpha_m(3)} \cdots p_k^{\alpha_m(k)}$$

where p_k is the k-th prime. For all $n > k$ define $\alpha_m(n) = 0$. Then in this way we can assign to each positive integer m a mapping $\alpha_m : P \to \mathbb{N}$. Under this assignment $\operatorname{hcf}\{m, n\} = m \wedge n$ maps to $\alpha_m \sqcap \alpha_n$, and $\operatorname{lcm}\{m, n\} = m \vee n$ maps to $\alpha_m \sqcup \alpha_n$. Thus $\alpha : m \mapsto \alpha_m$ is a lattice morphism. Since clearly

$$m|n \iff \alpha_m \sqsubseteq \alpha_n,$$

it follows that $(P; |)$ is isomorphic to the sublattice $\operatorname{Im} \alpha$ of the distributive lattice $\mathbb{N}^P$ and so is distributive.

Example 5.8 A **ring of sets** is a family $\mathcal{F} = (E_\alpha)_{\alpha \in I}$ of subsets of some set E such that $E_\alpha \cap E_\beta \in \mathcal{F}$ and $E_\alpha \cup E_\beta \in \mathcal{F}$ for all $\alpha, \beta \in I$. Clearly, a ring of sets is a distributive lattice. For example, let E be an infinite set and define a subset of E to be **cofinite** if its complement in E is finite. Then it is easy to see that the family $\mathcal{F}$ consisting of those subsets of E that are finite or cofinite is a ring of sets and is therefore a distributive lattice.

Example 5.9 A group G is said to be **generalised cyclic** if the subgroup generated by every finite subset of G is cyclic. Such a group is necessarily abelian since the subgroup generated by $\{x, y\}$, being cyclic, is abelian. It can be shown that a group G has a distributive lattice of subgroups if and only if G is generalised cyclic. This result is due to Ore [90]; see also Suzuki [106].

EXERCISES

5.1. Let R be a commutative ring. Show that the set E of idempotents of R can be ordered by
$$e \leq f \iff e = ef.$$
Prove that with respect to this order E forms a distributive lattice in which $\inf\{e, f\} = ef$ and $\sup\{e, f\} = e + f - ef$.

5.2. Let L be a modular lattice. If $a, b, c \in L$ are such that the distributive equality $a \wedge (b \vee c) = (a \wedge b) \vee (a \wedge c)$ holds, prove that so also do the five others involving a, b, c.
[*Hint.* Let $x = (a \wedge b) \vee c$ and $y = (a \vee c) \wedge (b \vee c)$. Prove that $a \vee x = a \vee y$ and $a \wedge x = a \wedge y$. Now use $x \leq y$ and modularity to show that $x = y$, so that the distributive equality $c \vee (a \wedge b) = (c \vee a) \wedge (c \vee b)$ holds. Now use the dual result to deduce that the distributive equality for one arrangement α, β, γ of a, b, c implies its validity for the cyclically permuted arrangement β, γ, α.]

5.3. Prove that a lattice L is distributive if and only if
$$(\forall x, y, z \in L) \quad (x \wedge y) \vee (y \wedge z) \vee (z \wedge x) = (x \vee y) \wedge (y \vee z) \wedge (z \vee x).$$

5.4. Let L be a lattice. Prove that the ideal lattice $I(L)$ is distributive if and only if L is distributive.

5.5. If $L = \mathbb{Z} \times \mathbb{Z}$ under the cartesian order, describe the elements of the ideal lattice $I(L)$ and sketch its Hasse diagram.

For distributive lattices we have the following **Birkhoff distributivity criterion** which is the analogue of Theorem 4.4.

Theorem 5.1 (Birkhoff [9])   *A lattice L is distributive if and only if it has no sublattice of either of the forms*

Equivalently, L is distributive if and only if $z \wedge x = z \wedge y$ and $z \vee x = z \vee y$ imply that $x = y$.

Proof Observe first that the two statements are equivalent. In fact, if $x \wedge z = y \wedge z$ and $x \vee z = y \vee z$ with $x \neq y$ then the two lattices in question arise from the cases $x \nparallel y$ and $x \parallel y$.

$\Rightarrow$: Suppose that L is distributive and that there exist $x, y, z \in L$ such that $x \wedge z = y \wedge z$ and $x \vee z = y \vee z$. Then we have
$$\begin{aligned}
x = x \wedge (x \vee z) = x \wedge (y \vee z) &= (x \wedge y) \vee (x \wedge z) \\
&= (x \wedge y) \vee (y \wedge z) \\
&= y \wedge (x \vee z) = y \wedge (y \vee z) = y.
\end{aligned}$$
Consequently L has no sublattice of either of the above forms.

$\Leftarrow$: If L has no sublattice of either of the above forms then by Theorem 4.4 we see that L must be modular. Given $a, b, c \in L$, define

$$a^\star = (b \vee c) \wedge a, \quad b^\star = (c \vee a) \wedge b, \quad c^\star = (a \vee b) \wedge c.$$

Then clearly $a^\star \wedge c^\star = a \wedge c$, $b^\star \wedge c^\star = b \wedge c$, and $a^\star \wedge b^\star = a \wedge b$. Now let

$$d = (a \vee b) \wedge (b \vee c) \wedge (c \vee a).$$

Then, using modularity twice, we have

$$
\begin{aligned}
a^\star \vee c^\star &= a^\star \vee [(a \vee b) \wedge c] \\
&= (a^\star \vee c) \wedge (a \vee b) && \text{since } a^\star \leqslant a \vee b \\
&= ([(b \vee c) \wedge a] \vee c) \wedge (a \vee b) \\
&= (b \vee c) \wedge (a \vee c) \wedge (a \vee b) && \text{since } c \leqslant b \vee c \\
&= d.
\end{aligned}
$$

By symmetry we deduce that

$$a^\star \vee c^\star = a^\star \vee b^\star = b^\star \vee c^\star = d.$$

We now observe that

$$
\begin{cases}
c^\star \vee a^\star \vee (b \wedge c) = d; \\
c^\star \wedge [a^\star \vee (b \wedge c)] = (c^\star \wedge a^\star) \vee (b \wedge c) = (a \wedge c) \vee (b \wedge c),
\end{cases}
$$

and by symmetry that

$$
\begin{cases}
c^\star \vee b^\star \vee (a \wedge c) = d; \\
c^\star \wedge [b^\star \vee (a \wedge c)] = (a \wedge c) \vee (b \wedge c).
\end{cases}
$$

By the hypothesis, therefore, we deduce that $a^\star \vee (b \wedge c) = b^\star \vee (a \wedge c)$ whence

$$a^\star \vee (b \wedge c) = a^\star \vee (b \wedge c) \vee b^\star \vee (a \wedge c) = a^\star \vee b^\star = d.$$

It follows from this that

$$(a \vee b) \wedge c = c^\star = c^\star \wedge d = c^\star \wedge [a^\star \vee (b \wedge c)] = (a \wedge c) \vee (b \wedge c)$$

and so L is distributive. $\square$

EXERCISES

5.6. Use Theorem 5.1 and the identity $mn = \mathrm{hcf}\{m,n\}\,\mathrm{lcm}\{m,n\}$ to show that $(P; \,|\,)$ is a distributive lattice (cf. Example 5.7).

5.7. Prove that a lattice L is distributive if and only if, for every $x \in L$, the mapping $\alpha_x : L \to x^\uparrow \times x^\downarrow$ given by $\alpha_x(a) = (x \vee a, x \wedge a)$ is an injective lattice morphism.

5.8. Prove that a lattice is distributive if and only if, for all ideals I, J of L, $I \vee J = \{i \vee j \mid i \in I, j \in J\}$.

In a complete lattice there are two **infinite distributive laws** to consider, namely

$$x \wedge \bigvee_{\alpha \in I} y_\alpha = \bigvee_{\alpha \in I} (x \wedge y_\alpha); \qquad x \vee \bigwedge_{\alpha \in I} y_\alpha = \bigwedge_{\alpha \in I} (x \vee y_\alpha).$$

Unlike ordinary distributivity which is self-dual, these laws do not imply each other in general.

Example 5.10 Let E be an infinite set and let $\mathbb{P}_f(E)$ be the set of all finite subsets of E. Then $L = \mathbb{P}_f(E) \cup \{E\}$ is a complete lattice. The infinite distributive law $X \vee \bigwedge_{\alpha \in I} Y_\alpha = \bigwedge_{\alpha \in I} (X \vee Y_\alpha)$ holds in L. To see this, there are three cases to consider:

(1) $\bigwedge\limits_{\alpha \in I} Y_\alpha = E$: in this case every $Y_\alpha = E$ and the identity holds trivially;

(2) $X = E$: again the identity holds trivially;

(3) $\bigwedge\limits_{\alpha \in I} Y_\alpha \neq E$ and $X \neq E$: in this case $Y_\alpha \neq E$ for all α and the result

follows by the usual $\in$-argument.

In contrast, the infinite distributive law $X \wedge \bigvee\limits_{\alpha \in I} Y_\alpha = \bigvee\limits_{\alpha \in I} (X \wedge Y_\alpha)$ does

not hold in the lattice L. To see this, take $X = \{x\}$ and let $\{Y_\alpha \; ; \; \alpha \in I\}$ consist of all the subsets of E that do not contain x. Then $X \wedge Y_\alpha = \emptyset$ for every α so the right-hand side is $\emptyset$; and since $\bigvee\limits_{\alpha \in I} Y_\alpha = E$, the left-hand side is X.

EXERCISES

5.9. If L is a complete distributive lattice prove that the infinite distributive law $x \wedge \bigvee\limits_{\alpha \in I} y_\alpha = \bigvee\limits_{\alpha \in I} (x \wedge y_\alpha)$ holds if and only if, for every $x \in L$, the mapping $\lambda_x : y \mapsto x \wedge y$ is residuated.

5.10. Let L be a distributive lattice. If L has a bottom element prove that the ideal lattice $I(L)$ is complete and satisfies the infinite distributive law

$$J \wedge \bigvee\limits_{\alpha \in I} K_\alpha = \bigvee\limits_{\alpha \in I} (J \wedge K_\alpha).$$

Does L satisfy the other infinite distributive law?

5.11. Prove that in $(\mathbb{N}; |)$ the infinite distributive law

$$x \vee \bigwedge\limits_{\alpha \in I} y_\alpha = \bigwedge\limits_{\alpha \in I} (x \vee y_\alpha)$$

holds. Does L satisfy the other infinite distributive law?

5.2 More on join-irreducibles

In distributive lattices with the descending chain condition we have the following strengthening of the Kurosh–Ore Theorem.

Theorem 5.2 *If L is a distributive lattice that satisfies the descending chain condition then every element of $L \backslash \{0\}$ can be expressed uniquely as an irredundant join of join-irreducibles.*

Proof Suppose that $x \in L \backslash \{0\}$ can be expressed in two ways as irredundant joins of join-irreducibles, say $x = \bigvee\limits_{i=1}^{m} x_i = \bigvee\limits_{j=1}^{n} y_j$. Then by Theorem 4.10 we have $m = n$. Suppose now that $t = \bigvee\limits_{k \neq i} x_k$. Then $x = x_i \vee t = y_j \vee t$ for some j by Theorem 4.9. By distributivity, we have $x_i = x_i \wedge x = x_i \wedge (y_j \vee t) = (x_i \wedge y_j) \vee (x_i \wedge t)$ and so, since x_i is join-irreducible, either $x_i = x_i \wedge y_j$ or $x_i = x_i \wedge t$, i.e. either $x_i \leqslant y_j$ or $x_i \leqslant t$. The latter is impossible since it contradicts the fact that $\bigvee\limits_{i=1}^{m} x_i$ is irredundant. Thus $x_i \leqslant y_j$. Likewise, if

$a = \bigvee_{k \neq j} y_k$ with $x = y_j \vee a = x_p \vee a$ then a similar argument gives $y_j \leqslant x_p$.
Thus $x_i \leqslant y_j \leqslant x_p$ and so, by irredundancy, we must have $p = i$ whence
$x_i = y_j$. Thus the elements appearing in both expressions are the same.   □

EXERCISES

5.12. Let L be a distributive lattice. If $a \in L$ is a join-irreducible element such
that $a \leqslant \bigvee_{i=1}^{n} x_i$ prove that $a \leqslant x_i$ for some i.

5.13. Show that in the 18-element distributive lattice

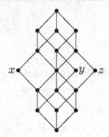

every element has a unique expression in terms of x, y, z.

Theorem 5.2 prompts us to take a closer look at *finite* distributive lattices.
Here, of course, the descending chain condition holds.

Theorem 5.3 *The length of a finite distributive lattice L is $|\mathcal{J}(L)|$.*

Proof Let C be a maximal chain in L. Observe that by Theorem 5.2 we can
define a surjective mapping $\vartheta : \mathcal{J}(L) \to C \backslash \{0\}$ by setting, for each $a \in \mathcal{J}(L)$,

$$\vartheta(a) = \min\{x \in C \mid a \leqslant x\}.$$

That ϑ is also injective can be seen as follows. Suppose that $\vartheta(a) = \vartheta(b)$ and
let $x \in C$ be such that $x \prec \vartheta(a)$. Since C is maximal and $a, b \not\leqslant x$ we must
have $x \vee a = x \vee b = \vartheta(a)$ whence

$$(a \wedge x) \vee (a \wedge b) = a \wedge (x \vee b) = a \wedge (x \vee a) = a.$$

Since a is join-irreducible we deduce that $a = a \wedge x$ or $a = a \wedge b$. But the
former gives the contradiction $a \leqslant x$, so we conclude that $a \leqslant b$. Similarly,
$b \leqslant a$ and the resulting equality shows that ϑ is also injective.   □

There is a remarkable duality that connects a finite distributive lattice L
with the down-set lattice of the set of join irreducible elements in L. This we
now describe.

Theorem 5.4 (1) *If E is a finite ordered set there is an order isomorphism*

$$E \simeq \mathcal{J}(\mathcal{O}(E)).$$

(2) *If L is a finite distributive lattice there is a lattice isomorphism*

$$L \simeq \mathcal{O}(\mathcal{J}(L)).$$

Proof (1) Observe that the join-irreducible elements of the lattice $\mathcal{O}(E)$ are the principal down-sets. The mapping $f : E \to \mathcal{J}(\mathcal{O}(E))$ given by $f(x) = x^{\downarrow}$ is then an order isomorphism.

(2) Consider the mapping $g : L \to \mathcal{O}(\mathcal{J}(L))$ given by

$$g(x) = \{a \in \mathcal{J}(L) \mid a \leqslant x\}.$$

We know that $x = \sup g(x)$ and it follows immediately from this that g is injective. To see that g is also surjective, let $A \in \mathcal{O}(\mathcal{J}(L))$ and let $x = \sup A$, so that $A \subseteq g(x)$. For every $y \in g(x)$ we have, by distributivity,

$$y = y \wedge x = y \wedge \sup A = \sup\{y \wedge z \mid z \in A\}.$$

Since by definition y is join-irreducible it follows that $y = y \wedge z$ for some $z \in A$, whence $y \in A$. Thus $g(x) \subseteq A$ and we have $g(x) = A$.

To complete the proof, it suffices to show that $g(x \vee y) \subseteq g(x) \cup g(y)$, the converse inequality being trivial, and invoke Theorem 2.9.

So suppose that $t \in g(x \vee y)$. We have $t = t \wedge (x \vee y) = (t \wedge x) \vee (t \wedge y)$ and therefore, since t is join-irreducible, either $t = t \wedge x$ or $t = t \wedge y$. This gives either $t \in g(x)$ or $t \in g(y)$, whence $t \in g(x) \cup g(y)$ and the result follows. $\square$

The above result shows that the structure of a finite distributive lattice is completely determined by its set of join-irreducible elements.

Example 5.11

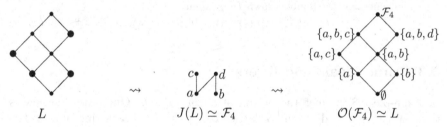

$$L \qquad\qquad \mathcal{J}(L) \simeq \mathcal{F}_4 \qquad\qquad \mathcal{O}(\mathcal{F}_4) \simeq L$$

EXERCISES

5.14. Determine the distributive lattice L for which $\mathcal{J}(L)$ is the ordered set

5.15. Determine the finite distributive lattices L for which the set $\mathcal{J}(L)$ is an antichain.

5.16. Let L be a finite distributive lattice and let $\mathcal{M}(L)$ be the set of meet-irreducible elements of L. Prove that if $x \in \mathcal{J}(L)$ then $L \backslash x^{\uparrow} = y^{\downarrow}$ where $y \in \mathcal{M}(L)$. Hence show that the ordered sets $\mathcal{J}(L)$ and $\mathcal{M}(L)$ are isomorphic.

5.17. Given a bounded ordered set E, consider the set $\text{Isomap}(E, \mathbf{2})$ of isotone mappings from E to the 2-element chain. By Example 5.6, this set forms a distributive lattice. Show that the meet-irreducible elements in this lattice are the residuated mappings $\beta_x : E \to \mathbf{2}$ defined for each $x \in E$ by

$$\beta_x(y) = \begin{cases} 0 & \text{if } y \leqslant x; \\ 1 & \text{, otherwise.} \end{cases}$$

5.18. Consider the sequence $(j_n)_{n \geqslant 0}$ given by

$$j_0 = j_1 = 1, \quad (\forall n \geqslant 2) \ j_n = 2j_{n-1} + j_{n-2}.$$

Prove that $\sum_{i=0}^{n} j_i = \frac{1}{2}(j_n + j_{n-1})$.

5.19. By a **double fence** we mean an ordered set DF_{2n} of the form

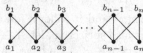

By revisiting the Berman–Köhler theorem and using the previous exercise prove that if $L(DF_{2n})$ is the distributive lattice whose set of join-irreducible elements is the double fence DF_{2n} then $|L(DF_{2n})| = j_{n+1}$.

5.20. For the **extended double fence** Z_{2n} with Hasse diagram

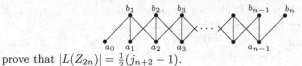

prove that $|L(Z_{2n})| = \frac{1}{2}(j_{n+2} - 1)$.

5.21. By a **double crown** we mean an ordered set DC_{2n} of the form

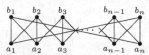

Prove that if $L(DC_{2n})$ is the distributive lattice whose set of join-irreducible elements is the double crown DC_{2n} then

$$|L(DC_{2n})| = |L(DF_{2n})| - 2|L(Z_{2(n-3)})| = 2j_n + 1.$$

5.3 Prime ideals and filters

In Example 5.8 we met the notion of a *ring of sets*. Our objective now is to show that every distributive lattice can be realised as a ring of sets. For this purpose, we shall require the following equivalent form of the Axiom of Choice :

Zorn's Axiom If $\mathcal{C}$ is a collection of sets that is inductively ordered, in the sense that $\sup_{\mathcal{C}} T$ exists for every chain T in $\mathcal{C}$, then $\mathcal{C}$ has a maximal element.

Definition We say that an ideal I of a lattice L is **prime** if $I \neq L$ and

$$a \wedge b \in I \implies a \in I \text{ or } b \in I.$$

Dually, we say that a filter F is **prime** if $F \neq L$ and

$$a \vee b \in F \implies a \in F \text{ or } b \in F.$$

Example 5.12 If L is distributive and $a \in L$ is $\vee$-irreducible then $L \setminus a^{\uparrow}$ is a prime ideal. In fact, $L \setminus a^{\uparrow} = \{x \in L \mid a \not\leqslant x\}$ so if $x \in L \setminus a^{\uparrow}$ and $y \leqslant x$ we have $y \in L \setminus a^{\uparrow}$. Now $a \leqslant x \vee y$ implies that $a = (a \wedge x) \vee (a \wedge y)$, whence $a = a \wedge x$ or $a = a \wedge y$, and so $a \leqslant x$ or $a \leqslant y$. It follows that $L \setminus a^{\uparrow}$ is an ideal of L. That it is prime follows from the fact that $x \wedge y \in a^{\uparrow}$ if and only if $a \leqslant x$ and $a \leqslant y$.

Now, as we have seen in Example 4.12, in an infinite distributive lattice there may be no elements that are $\vee$-irreducible. In the infinite case, the rôle of the $\vee$-irreducible elements is assumed by the prime ideals, their existence being assured by the following result.

Theorem 5.5 (Stone [104]) *Let L be a distributive lattice. Let I be an ideal and let F be a filter such that $I \cap F = \emptyset$. Then there is a prime ideal P that separates I and F, in the sense that $I \subseteq P$ and $P \cap F = \emptyset$.*

Proof Let $\mathcal{C}$ be the collection of ideals of L that contain I and are disjoint from F. Then $\mathcal{C}$ is inductively ordered and so, by Zorn's axiom, $\mathcal{C}$ has a maximal element, P say. We claim that the ideal P is prime. Suppose, by way of obtaining a contradiction, that P is not prime. Then there exist $a, b \in L$ such that $a, b \notin P$ and $a \wedge b \in P$. By the maximality of P, we have

$$(P \vee a^{\downarrow}) \cap F \neq \emptyset, \quad (P \vee b^{\downarrow}) \cap F \neq \emptyset$$

so we can choose $p, q \in P$ such that $p \vee a \in F$ and $q \vee b \in F$. Now let $x = (p \vee a) \wedge (q \vee b)$. Since F is a filter, we have $x \in F$. But since L is distributive we also have $x = (p \wedge q) \vee (p \wedge b) \vee (a \wedge q) \vee (a \wedge b) \in P$. Hence we have the contradiction $P \cap F \neq \emptyset$. Consequently, P is a prime ideal. It is clear by its construction that P separates I and F. $\square$

Definition We shall say that an ideal I of a lattice L **separates** elements a and b of L if I contains one of a, b but not the other.

Theorem 5.6 *A lattice L is distributive if and only if, for any two distinct elements x, y of L, there is a prime ideal that separates x and y.*

Proof Necessity follows from Theorem 5.5 and the observation that when $x \neq y$ we have either $x^{\downarrow} \cap y^{\uparrow} = \emptyset$ or $x^{\uparrow} \cap y^{\downarrow} = \emptyset$.

As for sufficiency, suppose that L is not distributive. Then L contains a sublattice of one of the forms

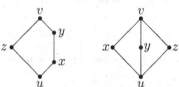

In either situation, consider a prime ideal P that contains the element x. Since $y \wedge z = u < x$ we have $y \wedge z \in P$ and so, since P is prime, either $y \in P$ or $z \in P$. Now if $z \in P$ then $v = x \vee z \in P$, and this gives $y \in P$. Thus every prime ideal that contains x also contains y. Consequently, if the condition holds then L must be distributive. $\square$

Theorem 5.7 (Birkhoff [8]) *Every distributive lattice is isomorphic to a ring of sets.*

Proof For every $x \in L$ let $\mathcal{F}(x)$ denote the set of prime filters of L that contain x. Then we have

$$F \in \mathcal{F}(x \wedge y) \iff x \wedge y \in F \iff x, y \in F \iff F \in \mathcal{F}(x) \cap \mathcal{F}(y);$$

$$F \in \mathcal{F}(x \vee y) \iff x \vee y \in F \iff x \in F \text{ or } y \in F \iff F \in \mathcal{F}(x) \cup \mathcal{F}(y).$$

Thus $\mathcal{F}(x \wedge y) = \mathcal{F}(x) \cap \mathcal{F}(y)$ and $\mathcal{F}(x \vee y) = \mathcal{F}(x) \cup \mathcal{F}(y)$. Consequently $\mathcal{F} = \{\mathcal{F}(x) \; ; \; x \in L\}$ is a ring of sets and the mapping described by $x \mapsto \mathcal{F}(x)$ is a lattice morphism of L onto $\mathcal{F}$. Since, by the dual of Theorem 5.6, this mapping is injective, we conclude that $L \simeq \mathcal{F}$.   □

EXERCISES

5.22. If L is a lattice prove that I is a prime ideal of L if and only if there is a surjective lattice morphism $f : L \to \mathbf{2}$ such that $I = \{x \in L \mid f(x) = 0\}$.

5.23. If L is a distributive lattice prove that every maximal ideal of L is prime.

5.24. Prove that every ideal of a distributive lattice is the intersection of all the prime ideals that contain it.

5.25. Identify the prime ideals of the distributive lattice $\mathbf{n} \times \mathbf{n}$.

5.4 Baer semigroups and distributivity

In order to obtain a coordinatisation of distributive lattices by Baer semigroups we introduce the following mappings.

Let L be a bounded lattice and let $a, b \in L$ be such that $a \parallel b$. Consider the mapping $\xi_{a,b} : L \to L$ given by

$$\xi_{a,b}(x) = \begin{cases} a \wedge x & \text{if } x \leqslant a \text{ or } a \wedge b \leqslant x \leqslant a \vee b; \\ a \wedge (b \vee x) & \text{if } x \not\leqslant a \text{ and } a \wedge b \not\leqslant x \leqslant a \vee b; \\ a & \text{if } x \not\leqslant a \vee b. \end{cases}$$

Observe that if L is distributive then for $a \wedge b \leqslant x \leqslant a \vee b$ we have $a \wedge x = (a \wedge b) \vee (a \wedge x) = a \wedge (b \vee x)$ and consequently $\xi_{a,b}$ coincides with the mapping $\vartheta_{a,b}$ that we encountered in the coordinatisation of modular lattices. Since L is modular, it follows by the Corollary to Theorem 4.11 that $\vartheta_{a,b}$ is residuated. Hence so is $\xi_{a,b}$. This establishes half of the following result.

Theorem 5.8 *The following conditions on a bounded lattice L are equivalent*:

(1) *L is distributive*;

(2) *for all $a, b \in L$ with $a \parallel b$, $\xi_{a,b}$ is residuated*.

Proof It suffices to establish (2) $\Rightarrow$ (1), and for this we first show that L is modular. Suppose, by way of obtaining a contradiction, that L were not modular. Then there exist $a, b \in L$ with $a \parallel b$ such that $M(b, a)$ fails. Then, by Theorem 4.3, L contains a sublattice of the form in which $y \leqslant a$. In this sublattice we have

$$\xi_{a,b}(b) \vee \xi_{a,b}(x) = (a \wedge b) \vee x = x < y \leqslant a \wedge (b \vee x) = \xi_{a,b}(b \vee x),$$

from which we see that $\xi_{a,b}$ is not a $\vee$-morphism. By Theorem 2.8, this contradicts the hypothesis that $\xi_{a,b}$ is residuated. Thus L is modular.

To show that L is distributive, it suffices by Theorem 5.1 to show that L has no sublattice of the form

Indeed, if such a sublattice exists then we have

$$\xi_{a,b}(b) \vee \xi_{a,b}(c) = (a \wedge b) \vee (a \wedge c) = e < a = a \wedge (b \vee c) = \xi_{a,b}(b \vee c),$$

which again contradicts the fact that $\xi_{a,b}$ is residuated. Thus we conclude that L is distributive. $\qquad\square$

As observed above, when L is distributive $\xi_{a,b}$ coincides with $\vartheta_{a,b}$. In what follows we shall use the latter notation.

Theorem 5.9 *Let L be a bounded distributive lattice and let $a, b, c \in L$ be such that $a \parallel b$ and $a \parallel c$. Then if $a \wedge c \leqslant a \wedge b$ we have $a \parallel b \vee c$ and*

$$\vartheta_{a,b\vee c} \circ \vartheta_{b,a} = \vartheta_{a\wedge b} = \vartheta_{b,a} \circ \vartheta_{a,b\vee c}.$$

Proof Clearly, we cannot have $a \geqslant b \vee c$. Moreover, if $a \leqslant b \vee c$ then the hypotheses give $a = a \wedge (b \vee c) = (a \wedge b) \vee (a \wedge c) = a \wedge b$ and we have the contradiction $a \leqslant b$. Consequently $a \parallel b \vee c$. A routine case-by-case analysis establishes the stated identities. $\qquad\square$

Theorem 5.10 (Blyth [20]) *For a bounded lattice L the following statements are equivalent:*

(1) *L is distributive;*

(2) *L can be coordinatised by a Baer semigroup S such that if $eS, fS, gS \in \mathcal{R}(S)$ with $eS \cap fS = eS \cap gS$ then there exist idempotents $\bar{e}, \bar{f}, \bar{g} \in S$ such that $\bar{e}S = eS$, $\bar{f}S = fS$, $\bar{g}S = gS$ and $\bar{e}$ commutes with $\bar{f}$ and $\bar{g}$;*

(3) *L can be coordinatised by a Baer semigroup S such that if $eS, fS, gS \in \mathcal{R}(S)$ with f, g idempotents and $eS \cap fS = eS \cap gS$ then there exists an idempotent $\bar{e} \in S$ such that $\bar{e}S = eS$ and $\bar{e}f = f\bar{e}f$, $\bar{e}g = g\bar{e}g$.*

Proof (1) $\Rightarrow$ (2): Suppose that L is distributive and let $S = \operatorname{Res} L$. Let $eS, fS, gS \in \mathcal{R}(S)$ and consider the relation

$$eS \cap fS = eS \cap gS. \tag{5.1}$$

There are several cases to examine.

(a) If $eS \subseteq fS$ then by $(\star)$ we have $eS \subseteq gS$. Now in the fundamental coordinatisation of Theorem 2.19 we have $\mathcal{R}(S) \simeq L$ under the mapping $eS \mapsto \bar{e}(1)$ where $\bar{e}$ is an idempotent generator of eS. Taking $\bar{e} = \vartheta_{e(1)}$,

$\overline{f} = \vartheta_{f(1)}, \overline{g} = \vartheta_{g(1)}$ we deduce that $e(1) = \overline{e}(1) \leqslant \overline{f}(1) = f(1)$ and likewise $e(1) \leqslant g(1)$. It follows by Theorem 4.12(1) that $\overline{e}$ commutes with $\overline{f}$ and $\overline{g}$.

(b) If $eS \supseteq fS$ then by (5.1) and the fundamental isomorphism we have $f(1) = e(1) \wedge g(1)$. Define $\overline{e} = \vartheta_{e(1),g(1)}, \overline{g} = \vartheta_{g(1),e(1)}, \overline{f} = \vartheta_{f(1)}$. Then by Theorem 4.12(2) we see that $\overline{e}\,\overline{g} = \overline{f}$. Since $\overline{e}$ and $\overline{g}$ commute, it follows that so also do $\overline{e}$ and $\overline{f}$.

(c) Suppose now that $eS \parallel fS$. Then, as observed prior to Theorem 4.13, we have $e(1) \parallel f(1)$ whence it follows by (5.1) that $eS \not\subseteq gS$. There are then two possibilities to consider.

If $eS \supset gS$ then again by (5.1) and the fundamental isomorphism we have $g(1) = e(1) \wedge f(1)$. The same argument as used in (b) with f, g interchanged establishes (2) in this case.

If $eS \parallel gS$ then $e(1) \parallel g(1)$ and in this case we obtain $e(1) \wedge f(1) = e(1) \wedge g(1)$. It now follows by Theorem 5.9 that $e(1) \parallel f(1) \vee g(1)$. Now let $\overline{e} = \vartheta_{e(1),f(1)\vee g(1)}, \overline{f} = \vartheta_{f(1),e(1)}, \overline{g} = \vartheta_{g(1),e(1)}$. Then two applications of Theorem 5.9 give $\overline{e}\,\overline{f} = \vartheta_{e(1)\wedge f(1)} = \overline{f}\,\overline{e}$ and $\overline{e}\,\overline{g} = \vartheta_{e(1)\wedge g(1)} = \overline{g}\,\overline{e}$. Thus (2) holds also in this case.

(2) $\Rightarrow$ (3): This follows as in the corresponding part of Theorem 4.13.

(3) $\Rightarrow$ (1): If (3) holds then, taking $f = g$, we see by Theorem 4.13 that L is modular. Suppose, by way of obtaining a contradiction, that L is not distributive. Then $\mathcal{R}(S) \simeq L$ contains a sublattice of the form

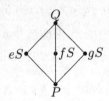

and $\mathcal{L}(S)$ has a corresponding sublattice

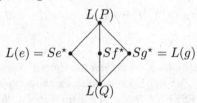

Now by (3) we have $\overline{e}f = f\overline{e}f$ and $\overline{e}g = g\overline{e}g$ and so $\overline{e}fS = eS \cap fS = P = eS \cap gS = \overline{e}gS$. Now let $x \in L(P)$. Then $x\overline{e}f = 0 = x\overline{e}g$, so

$$x\overline{e} \in L(f) \cap L(g) = Sf^* \cap Sg^* = L(Q) \subseteq Se^* = L(e)$$

whence $x\overline{e}e = 0$. But $\overline{e}S = eS$ and so $e = \overline{e}e$. Thus $xe = 0$ and $x \in L(e) = Se^*$. Consequently we have the contradiction $L(P) \subseteq Se^*$. We conclude that $\mathcal{R}(S)$, hence L, is distributive. □

6

Complementation; boolean algebras

6.1 Complemented elements

If L is a bounded lattice then we say that $y \in L$ is a **complement** of $x \in L$ if $x \wedge y = 0$ and $x \vee y = 1$. In this case we say that x is a **complemented element** of L. Clearly, every complement of a complemented element is itself complemented.

Example 6.1 In each of the lattices

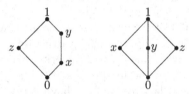

the first of which is non-modular and the second is modular but not distributive, the elements x and y are complements of z. Note that in general complements are not unique.

We say that a lattice L is **complemented** if every element of L is complemented.

Example 6.2 Let V be a vector space and consider the lattice $L(V)$ of subspaces of V. By Example 4.5, $L(V)$ is modular. It is also complemented. To establish this, we observe that if W is a subspace of V then any basis of W can be extended to a basis of V by means of a set $A = \{x_\alpha \mid \alpha \in I\}$ of elements of V. The subspace generated by A then serves as a complement of W in $L(V)$. The non-uniqueness of complements can easily be seen in this situation. For example, in the real vector space $\mathbb{R}^2$ consider the subspace $X = \{(x, 0) \mid x \in \mathbb{R}\}$. Then the subspaces $Y = \{(0, y) \mid y \in \mathbb{R}\}$ and $Z = \{(x, x) \mid x \in \mathbb{R}\}$ are each complements of X.

We say that a lattice L is **relatively complemented** if every interval $[x, y]$ of L is complemented. A complement in $[x, y]$ of $a \in [x, y]$ is called a **relative complement** of a.

Of the complemented lattices in Example 6.1, the non-modular lattice is not relatively complemented (the interval $[0, y]$ is not complemented), but the modular lattice is. This is true in general:

Theorem 6.1 *All complemented modular lattices are relatively complemented.*

Proof Let L be a complemented modular lattice. Given any $[a, b] \subseteq L$ and $x \in [a, b]$, let y be a complement of x in L. Consider the element

$$z = b \wedge (a \vee y) = (b \wedge y) \vee a.$$

Clearly, we have $z \in [a, b]$ and, by the modularity,

$$x \wedge z = x \wedge (y \vee a) = (x \wedge y) \vee a = 0 \vee a = a;$$
$$x \vee z = x \vee (b \wedge y) = (x \vee y) \wedge b = 1 \wedge b = b.$$

Thus z is a complement of x in $[a, b]$. □

Theorem 6.2 *In a distributive lattice all complements and relative complements that exist are unique.*

Proof This is an immediate consequence of the Birkhoff criterion for distributivity in Theorem 5.1. □

6.2 Uniquely complemented lattices

In view of Theorem 6.2, it is natural to consider complemented lattices in which complements are unique. In such a lattice we shall use the notation x' to denote the unique complement of x. Denoting likewise the complement of x' by x'' we then have, by the uniqueness, $x'' = x$. There is an interesting history concerning these lattices, it long having been suspected that every uniquely complemented lattice is distributive. However, this is not the case. In fact, R. P. Dilworth [44] established the remarkable result that *every lattice can be embedded in a uniquely complemented lattice*. The following results will illustrate the difficulty in seeking uniquely complemented lattices that are not distributive.

Definition If L is a lattice with bottom element 0 then by an **atom** of L we shall mean an element a such that $0 \prec a$. If for every $x \in L \setminus \{0\}$ there is an atom a such that $a \leqslant x$ then we say that L is **atomic**. Dually, we have the notions of a **coatom** and that of L being **coatomic**.

Theorem 6.3 (Birkhoff–Ward [14]) *Every uniquely complemented atomic lattice is distributive.*

Proof We establish the proof by means of the following sequence of non-trivial observations.

(1) *If $x > y$ then there is an atom p such that $p \leqslant x$ and $p \wedge y = 0$.*
In fact, $x > y$ gives $y' \vee x = 1$. We cannot therefore have $y' \wedge x = 0$ since this would imply, by the uniqueness of complements, that $x = y'' = y$. Thus

$y' \wedge x > 0$ and so there is an atom p such that $p \leqslant y' \wedge x$, whence $p \leqslant x$ and $p \leqslant y'$, the latter giving $p \wedge y = 0$.

(2) *If x and y contain the same atoms then $x = y$.*
In fact, if x and y contain the same atoms then so do x and $x \wedge y$. Suppose that $x \wedge y < x$. Then by (1) there would exist an atom p contained in x but not contained in $x \wedge y$. This contradiction shows that $x \wedge y = x$ whence $x \leqslant y$. Similarly, we have $y \leqslant x$ and so $x = y$.

(3) *The complement of an atom is a coatom.*
Let p be an atom. Then $p' \neq 1$ since otherwise we have the contradiction $p = p'' = 0$. Suppose that $p' < x < 1$. Then $p \vee x = 1$. But, p being an atom, either $p \wedge x = p$ or $p \wedge x = 0$. The former gives $p \leqslant x$ whence the contradiction $x = p \vee x = 1$. Thus $p \wedge x = 0$ and consequently $x = p'$. Thus we see that p' is a coatom.

(4) *If p and q are distinct atoms then $q \leqslant p'$.*
By (3), both p' and q' are coatoms. Suppose that $q \not\leqslant p'$. Then necessarily $q \vee p' = 1$ and $q \wedge p' = 0$, which gives the contradiction $q = p'' = p$.

(5) *If p is an atom then $p \wedge x = 0 \iff x \leqslant p'$.*
If $p \wedge x = 0$ then $p \not\leqslant x$ and so every atom q under x is distinct from p. Thus, by (4), every atom under x is an atom under $x \wedge p'$. Since the converse is also true, it follows by (2) that $x = x \wedge p'$, whence $x \leqslant p'$.

(6) *If p is an atom then $p \leqslant x \vee y$ if and only if $p \leqslant x$ or $p \leqslant y$.*
Suppose that $p \leqslant x \vee y$. Since p is an atom we have either $p \wedge x = p$ or $p \wedge x = 0$, i.e. either $p \leqslant x$ or, by (5), $x \leqslant p'$. Likewise, either $p \leqslant y$ or $y \leqslant p'$. Now we cannot have both $x \leqslant p'$ and $y \leqslant p'$ since then $x \vee y \leqslant p'$, which gives $p \leqslant x \vee y \leqslant p'$ and the contradiction $p = 0$. Thus we must have either $p \leqslant x$ or $p \leqslant y$.

With these technical details to hand, suppose now that $\mathcal{A}$ is the set of atoms of L. For every $x \in L$ let $\mathcal{A}_x = \{a \in \mathcal{A} \mid a \leqslant x\}$, and consider the mapping $f : L \rightarrow \mathbb{P}(\mathcal{A})$ given by the prescription $f(x) = \mathcal{A}_x$. It is clear from (2) that f is injective. Moreover, using (6) we see that $\mathcal{A}_{x \vee y} = \mathcal{A}_x \cup \mathcal{A}_y$ and so f is a $\vee$-morphism. Now clearly, for $p \in \mathcal{A}$, we have $p \leqslant x \wedge y$ if and only if $p \leqslant x$ and $p \leqslant y$. It follows that $\mathcal{A}_{x \wedge y} = \mathcal{A}_x \cap \mathcal{A}_y$ and so f is a lattice morphism. Thus $L \simeq \operatorname{Im} f$ where $\operatorname{Im} f$ is a sublattice of the distributive lattice $\mathbb{P}(\mathcal{A})$. Consequently, L is distributive. $\square$

Corollary *If L is complete then $L \simeq \mathbb{P}(\mathcal{A})$.*

Proof Suppose that L is complete and let $N = \{p_i \mid i \in I\}$ where each $p_i \in \mathcal{A}$. Let q be an atom with $q \leqslant \bigvee_{i \in I} p_i$. Then necessarily $q = p_i$ for some $i \in I$. In fact, suppose that $q \neq p_i$ for all $i \in I$. Then (4) gives $p_i \leqslant q'$ whence we have the contradiction $q \leqslant \bigvee_{i \in I} p_i \leqslant q'$. We conclude that $N = \mathcal{A}_x$ where $x = \bigvee_{i \in I} p_i$. Hence f is also surjective and $L \simeq \mathbb{P}(\mathcal{A})$. $\square$

Definition By the **width** of a lattice L we mean the supremum of the cardinalities of the antichains in L.

Theorem 6.4 *Every uniquely complemented lattice of finite width is distributive.*

Proof Let L be uniquely complemented and of finite width. We show as follows that L satisfies the descending chain condition, whence it is atomic and therefore, by Theorem 6.3, is distributive. By way of obtaining a contradiction, suppose that there is an infinite descending chain $x_1 > x_2 > x_3 > \cdots$ in L. Observe first that, for each i, $x_i \wedge x'_{i+1} \neq 0$. In fact, if $x_i \wedge x'_{i+1} = 0$ then x'_{i+1} would have two complements, namely x_{i+1} and x_i. For each i define $y_i = x_i \wedge x'_{i+1}$. Then for $i < j$ we have $y_j \leqslant x_j \leqslant x_{i+1}$ and consequently $y_i \wedge y_j \leqslant x_{i+1} \wedge x'_{i+1} = 0$. It follows that $\{y_i \mid i \geqslant 1\}$ forms an infinite antichain and provides the required contradiction. $\qquad\square$

Theorem 6.5 *In a uniquely complemented lattice L the following properties of complementation are equivalent:*

(1) $(\forall x, y \in L)\ x \leqslant y \Rightarrow y' \leqslant x'$;
(2) $(\forall x, y \in L)\ (x \wedge y)' = x' \vee y'$;
(3) $(\forall x, y \in L)\ (x \vee y)' = x' \wedge y'$.

Moreover, each implies that L is distributive.

Proof (1) $\Rightarrow$ (2): Suppose that (1) holds, i.e. that $x \mapsto x'$ is antitone. Then from $x \wedge y \leqslant x, y$ we obtain $x' \vee y' \leqslant (x \wedge y)'$ and consequently $x \wedge y = (x \wedge y)'' \leqslant (x' \vee y')'$. But likewise $x', y' \leqslant x' \vee y'$ gives $(x' \vee y')' \leqslant x'' \wedge y'' = x \wedge y$. Hence $x \wedge y = (x' \vee y')'$ and consequently $(x \wedge y)' = (x' \vee y')'' = x' \vee y'$.

(2) $\Rightarrow$ (1): This is clear.

A dual proof establishes the equivalence of (1) and (3).

As for the distributivity, suppose that any one of the above conditions holds. Then we have the property that

$$x \leqslant y \Rightarrow \begin{cases} y = x \vee (x' \wedge y); & (4) \\ x = (x \vee y') \wedge y. & (5) \end{cases}$$

In fact, if $x \leqslant y$ then since

$$[x \vee (x' \wedge y)]' \vee y \geqslant [x \vee (x' \wedge y)]' \vee x \vee (x' \wedge y) = 1$$

we have $[x \vee (x' \wedge y)]' \vee y = 1$; and, by (3),

$$[x \vee (x' \wedge y)]' \wedge y = x' \wedge (x' \wedge y)' \wedge y = 0.$$

Thus $y = [x \vee (x' \wedge y)]'' = x \vee (x' \wedge y)$ and so (4) holds. As for (5), using (4) we see that

$$x \leqslant y \Rightarrow y' \leqslant x' \Rightarrow x' = y' \vee (y \wedge x') \Rightarrow x = x'' = y \wedge (y' \vee x).$$

We now use (4) and (5) to show that L is distributive. For this purpose, suppose that $a, b, c \in L$ are such that

$$a \vee c = b \vee c = \alpha, \quad a \wedge c = b \wedge c = \beta.$$

Then on the one hand

$$\begin{aligned} a \vee \alpha' \vee (c \wedge \beta') &= a \vee \beta \vee \alpha' \vee (c \wedge \beta') && \text{since } \beta \leqslant a \\ &= a \vee c \vee \alpha' && \beta \leqslant c \text{ and (4)} \\ &= \alpha \vee \alpha' = 1, \end{aligned}$$

and similarly $b \vee \alpha' \vee (c \wedge \beta') = 1$. On the other hand,

$$
\begin{aligned}
a \wedge [\alpha' \vee (c \wedge \beta')] &= a \wedge \alpha \wedge [\alpha' \vee (c \wedge \beta')] &&\text{since } a \leqslant \alpha \\
&= a \wedge c \wedge \beta' &&c \wedge \beta' \leqslant c \leqslant \alpha \text{ and } (5) \\
&= \beta \wedge \beta' = 0,
\end{aligned}
$$

and similarly $b \wedge [\alpha' \vee (c \wedge \beta')] = 0$. The uniqueness of complements now gives $a = [\alpha' \vee (c \wedge \beta')]' = b$. It follows by Theorem 5.1 that L is distributive.   □

Definition The properties (2) and (3) in Theorem 6.5 are often referred to as the **de Morgan laws**.

Theorem 6.6 (von Neumann [88]) *Every uniquely complemented modular lattice is distributive.*

Proof We begin by noting that in such a lattice we have the property

$$
a \wedge b = 0 \Rightarrow a \leqslant b'.
$$

To see this observe that

$$
\begin{aligned}
[a \vee (a \vee b)'] \wedge b &= [a \vee (a \vee b)'] \wedge (a \vee b) \wedge b \\
&= \{a \vee [(a \vee b)' \wedge (a \vee b)]\} \wedge b &&\text{by modularity} \\
&= a \wedge b \\
&= 0.
\end{aligned}
$$

Since clearly $a \vee (a \vee b)' \vee b = 1$ it follows that $b' = a \vee (a \vee b)' \geqslant a$.

Suppose now, by way of obtaining a contradiction, that there exists in L a sublattice of the form

(a) Suppose first that $u = 0$. Then by the above observation we have $y \leqslant x'$ and $z \leqslant x'$ whence the contradiction $x \leqslant v = y \vee z \leqslant x'$.

(b) Suppose now that $u \neq 0$. Let $x^\star = u' \wedge x$, $y^\star = u' \wedge y$ and $z^\star = u' \wedge z$. Then $x^\star \wedge y^\star = y^\star \wedge z^\star = z^\star \wedge x^\star = 0$. Consider $u \vee v'$. We have, on the one hand,

$$
\begin{aligned}
(u \vee v') \wedge (x^\star \vee y^\star) &= (u \vee v') \wedge v \wedge (x^\star \vee y^\star) &&x^\star \vee y^\star \leqslant v \\
&= [u \vee (v \wedge v')] \wedge (x^\star \vee y^\star) &&\text{by modularity} \\
&= u \wedge (x^\star \vee y^\star) \\
&= 0 &&x^\star \vee y^\star \leqslant u',
\end{aligned}
$$

and, on the other,

$$
\begin{aligned}
u \vee v' \vee (x^\star \vee y^\star) &= u \vee v' \vee (u' \wedge x) \vee (u' \wedge y) \\
&= v' \vee x \vee y &&\text{by modularity} \\
&= 1 &&x \vee y = v.
\end{aligned}
$$

Hence $x^\star \vee y^\star$ is the complement of $u \vee v'$. Similarly, so also are $y^\star \vee z^\star$ and $z^\star \vee x^\star$. Hence we have $x^\star \vee y^\star = y^\star \vee z^\star = z^\star \vee x^\star$. This then takes us back to the situation in (a) and the subsequent contradiction completes the argument.   □

6.3 Boolean algebras and boolean rings

Although from the classic result of Dilworth we know that non-distributive uniquely complemented lattices exist, such algebraic structures have not so far played a significant role in mathematics. In complete contrast, however, the (uniquely) complemented distributive lattices lie at the very heart of lattice theory and have far-reaching applications. These we now consider.

Definition By a **boolean lattice** we mean a complemented distributive lattice. By a **boolean algebra** we mean a boolean lattice together with the unary operation of complementation.

Example 6.3 The archetypal example of a boolean algebra is, for every set E, the algebra $(\mathbb{P}(E); \cap, \cup, ')$. As a lattice, this algebra is complete.

The principal properties of complementation in a boolean algebra are the following.

Theorem 6.7 *If B is a boolean algebra then*

(1) [**de Morgan laws**] $(\forall x, y \in B)$ $\quad (x \wedge y)' = x' \vee y'$, $\quad (x \vee y)' = x' \wedge y'$;
(2) $(\forall x, y \in B)$ $\quad x \leqslant y \iff x' \geqslant y'$;
(3) $(\forall x, y, z \in B)$ $\quad x \wedge y \leqslant z \iff x \leqslant z \vee y'$;
(4) $(\forall x, y, z \in B)$ $\quad x \vee y \geqslant z \iff x \geqslant z \wedge y'$.

Proof (1) Observe that, by the distributivity,

$$(x \wedge y) \vee x' \vee y' = (x \vee x' \vee y') \wedge (y \vee x' \vee y') = 1 \wedge 1 = 1;$$
$$x \wedge y \wedge (x' \vee y') = (x \wedge y \wedge x') \vee (x \wedge y \wedge y') = 0 \vee 0 = 0,$$

so that the (unique) complement of $x \wedge y$ is $x' \vee y'$. The other law is established similarly.

(2) By (1) we have $x \leqslant y \iff x \wedge y = x \iff x' \vee y' = x' \iff y' \leqslant x'$.

(3) If $x \wedge y \leqslant z$ then $z \vee y' \geqslant (x \wedge y) \vee y' = (x \vee y') \wedge (y \vee y') = x \vee y' \geqslant x$; and if $x \leqslant z \vee y'$ then $x \wedge y \leqslant (z \vee y') \wedge y = (z \wedge y) \vee (y' \wedge y) = z \wedge y \leqslant z$.

(4) This is the dual of (3). □

An interesting consequence of Theorem 6.7 is the following.

Theorem 6.8 *In every complete boolean algebra both infinite distributive laws hold.*

Proof Clearly, we have $x \wedge \bigvee_{i \in I} y_i \geqslant \bigvee_{i \in I} (x \wedge y_i)$. To establish the reverse inequality, let $z = \bigvee_{i \in I} (x \wedge y_i)$. Then for every $i \in I$ we have $x \wedge y_i \leqslant z$ and so, by Theorem 6.7(3), $y_i \leqslant z \vee x'$ whence $\bigvee_{i \in I} y_i \leqslant z \vee x'$ and then, by Theorem 6.7(3) again, $x \wedge \bigvee_{i \in I} y_i \leqslant z = \bigvee_{i \in I} (x \wedge y_i)$.

The other infinite distributive law is established similarly using Theorem 6.7(4). □

Corollary *If B is a complete boolean algebra then for every family $(x_\alpha)_{\alpha \in A}$ of elements of B there hold the infinite de Morgan laws*

$$\Big(\bigwedge_{\alpha \in A} x_\alpha \Big)' = \bigvee_{\alpha \in A} x'_\alpha, \qquad \Big(\bigvee_{\alpha \in A} x_\alpha \Big)' = \bigwedge_{\alpha \in A} x'_\alpha.$$

Proof By the above we have $\bigwedge_{\alpha \in A} x_\alpha \wedge \bigvee_{\beta \in A} x'_\beta = \bigvee_{\beta \in A} \big(\bigwedge_{\alpha \in A} x_\alpha \wedge x'_\beta \big) = 0$ and $\bigwedge_{\alpha \in A} x_\alpha \vee \bigvee_{\beta \in A} x'_\beta = \bigwedge_{\alpha \in A} \big(x_\alpha \vee \bigvee_{\beta \in A} x'_\beta \big) = 1$. Hence we have the first equality. For the second, replace x_α by x'_α and take complements. $\qquad \square$

Example 6.4 A topological space $\mathcal{X} = (X, \tau)$ consists of a set X and a collection τ of subsets of X that is closed under arbitrary unions and finite intersections. From the lattice theoretic point of view, a topology on X is then a $\cup$-complete sublattice of $\mathbb{P}(X)$ that contains $\emptyset$ and X. We refer to the elements of τ as **open sets**. For example, $\emptyset$ together with the cofinite subsets of X form a topology on X. If A is a subset of X then by the **interior** A° of A we mean the union of all the open sets contained in A. The mapping $A \mapsto A^\circ$ is a dual closure on $\mathbb{P}(X)$ and satisfies the property that $(A \cap B)^\circ = A^\circ \cap B^\circ$. In fact, by isotonicity we clearly have $(A \cap B)^\circ \subseteq A^\circ \cap B^\circ$. On the other hand $A \cap B \supseteq A^\circ \cap B^\circ$ which is open and therefore $(A \cap B)^\circ \supseteq A^\circ \cap B^\circ$.

A topology on X is not necessarily closed under complementation. For example, if X is infinite then the complement of a cofinite subset of X is not cofinite. A subset A of X is said to be **closed** if its complement A' is an open subset. By the above Corollary the intersection of any family of closed subsets of X is closed. For every subset A of X we may therefore define the **closure** A^- of A to be the intersection of the closed subsets that contain A. Then dually $A \mapsto A^-$ is a closure on $\mathbb{P}(X)$ and $(A \cup B)^- = A^- \cup B^-$.

A subset A of X is said to be **regular open** if it coincides with the interior of its closure; i.e. if $A = A^{-\circ}$. Consider now the set R of regular open subsets of X. Since the mappings $A \mapsto A^\circ$ and $A \mapsto A^-$ are isotone it follows that for all $A, B \in R$ we have $(A \cap B)^{-\circ} \subseteq A^{-\circ} \cap B^{-\circ} = A \cap B$. Conversely, $A \cap B \subseteq (A \cap B)^-$ and $A \cap B$ open give $A \cap B = (A \cap B)^\circ \subseteq (A \cap B)^{-\circ}$. Hence $A \cap B = (A \cap B)^{-\circ} \in R$ and so R is a $\cap$-semilattice.

Observe now that for every subset A of X we have $A^\circ = A'^{-'}$. In fact, on the one hand $A \subseteq A^-$ gives $A^{-'} \subseteq A'$ whence, replacing A by A', we obtain $A'^{-'} \subseteq A'' = A$ and hence $A'^{-'} \subseteq A^\circ$. On the other hand, $A^\circ \subseteq A$ gives $A' \subseteq A^{\circ'}$ whence, replacing A by A', we obtain $A = A'' \subseteq A'^{\circ'}$ and hence $A^- \subseteq A'^{\circ'}$. It follows that $A'^\circ \subseteq A^{-'}$ whence, on replacing A by A', we obtain the reverse inclusion $A^\circ \subseteq A'^{-'}$. As a consequence, $A \mapsto A^{-\circ}$ is idempotent. In fact, on the one hand we have $A^{-\circ} = A^{-\circ\circ} \subseteq A^{-\circ-\circ}$; and on the other, $A^{-'-} = A''^{-'-} \subseteq A'^{-'-'-}$ whence $A'^{-'-'-'} \subseteq A^{-'-'}$, i.e. $A^{\circ-\circ} \subseteq A^{-\circ}$, and replacing A by A^- in this we obtain the reverse inclusion $A^{-\circ-\circ} \subseteq A^{-\circ}$.

We deduce from the above that R is a lattice in which $A \vee B = (A \cup B)^{-\circ}$. Then R is complemented, the complement of $A \in R$ being $A'^\circ \in R$.

We next observe that if $A, B \in R$ then $A \cap B^- \subseteq (A \cap B)^-$. In fact, let $x \in A \cap B^-$ and let F be a closed subset such that $A \cap B \subseteq F$. Then $B \subseteq F \cup A'$ and $F \cup A'$ is closed. Since $x \in B^-$ it follows that $x \in F \cup A'$ whence $x \in F$. Consequently, $x \in (A \cap B)^-$.

We use this fact as follows to show that R is distributive. If $A, B \in R$ then

$$
\begin{aligned}
A \cap (B \vee C) = A \cap (B \cup C)^{-\circ} &= A \cap (B^- \cup C^-)^{\circ} \\
&= A^{\circ} \cap (B^- \cup C^-)^{\circ} \\
&= [A \cap (B^- \cup C^-)]^{\circ} \\
&= [(A \cap B^-) \cup (A \cap C^-)]^{\circ} \\
&\subseteq [(A \cap B)^- \cup (A \cap C)^-]^{\circ} \\
&= [(A \cap B) \cup (A \cap C)]^{-\circ} \\
&= (A \cap B) \vee (A \cap C).
\end{aligned}
$$

Since the reverse inclusion always holds, it follows that R is distributive.

In summary therefore we see that the regular open subsets of a topological space X form a boolean algebra.

EXERCISES

6.1. Let L be a bounded distributive lattice. If $a \in L$ is complemented, prove that the mapping $f : L \to a^{\uparrow} \times a^{\downarrow}$ given by the prescription $f(x) = (a \vee x, a \wedge x)$ is an isomorphism.

6.2. Prove that the lattice of idempotents of a commutative ring (Exercise 5.1) is relatively complemented. When is it boolean?

6.3. Let n be a positive integer and let $D(n)$ be the set of positive divisors of n. Prove that the distributive lattice $(D(n); |)$ is boolean if and only if n is square-free, in the sense that it has no factor of the form p^2 where p is a prime.

6.4. Let L be a bounded lattice and for every $x \in L$ consider the translations $\lambda_x : y \mapsto x \wedge y$ and $\mu_x : y \mapsto x \vee y$. Prove that if L is boolean then λ_x is residuated with $\lambda_x^+ = \mu_{x'}$. Conversely, if λ_x is residuated for every $x \in L$ and there exists $y \in L$ such that $\lambda_x^+ = \mu_y$ prove that L is boolean.

6.5. Prove that every interval of a boolean algebra is a boolean algebra.

6.6. Let E be an infinite set and let X be the ideal of $\mathbb{P}(E)$ consisting of all finite subsets of E. Prove that X has no complement in the lattice of ideals of $\mathbb{P}(E)$. Deduce that if L is boolean then $I(L)$ need not be.

6.7. Prove that an ideal I of a boolean lattice B is prime if and only if I separates x and x' for every $x \in B$. Deduce that every prime ideal of a boolean lattice is maximal.

6.8. Prove that a bounded distributive lattice is boolean if and only if the set of prime ideals of L is an antichain.

[Hint. $\Leftarrow$: Suppose that $a \in L$ has no complement and consider the filter $D = \{x \in L \mid a \vee x = 1\}$. Observe that $0 \notin D \vee a^{\uparrow}$ and deduce that there is a prime ideal P such that $P \cap (D \vee a^{\uparrow}) = \emptyset$. Observe that $1 \notin a^{\downarrow} \vee P$ so that there is a prime ideal Q that contains $a^{\downarrow} \vee P$, whence the contradiction $P \subsetneq Q$.]

There is an interesting connection between boolean algebras and rings.

Definition By a **boolean ring** we mean a ring with an identity in which every element is idempotent.

Theorem 6.9 *Let* $(B; \wedge, \vee, ')$ *be a boolean algebra. Define a multiplication and an addition on* B *by setting*

$$(\forall x, y \in B) \quad xy = x \wedge y, \quad x + y = (x \wedge y') \vee (x' \wedge y).$$

Then $(B; \cdot, +)$ *is a boolean ring.*

Proof Clearly, $(B; \cdot)$ is a semigroup with an identity, namely the top element 1 of B. Moreover, for every $x \in B$ we have $x^2 = x \cdot x = x \wedge x = x$, so every element is idempotent. Now given $x, y, z \in B$ it is easy to verify that

$$(x + y) + z = (x \wedge y' \wedge z') \vee (x' \wedge y \wedge z') \vee (x' \wedge y' \wedge z) \vee (x \wedge y \wedge z)$$

which, being symmetric in x, y, z is also equal to $x + (y + z)$. Since $x + 0 = (x \wedge 0') \vee (x' \wedge 0) = (x \wedge 1) \vee 0 = x$ and $x + x = (x \wedge x') \vee (x' \wedge x) = 0 \vee 0 = 0$, we see that $(B; +)$ is an abelian group in which $-x = x$ for every $x \in B$. Finally, for all $x, y, z \in B$ we have

$$
\begin{aligned}
xy + xz &= [xy \wedge (xz)'] \vee [(xy)' \wedge xz] \\
&= [x \wedge y \wedge (x \wedge z)'] \vee [(x \wedge y)' \wedge x \wedge z] \\
&= [x \wedge y \wedge (x' \vee z')] \vee [(x' \vee y') \wedge x \wedge z] \\
&= (x \wedge y \wedge z') \vee (x \wedge y' \wedge z) \\
&= x \wedge [(y \wedge z') \vee (y' \wedge z)] \\
&= x(y + z).
\end{aligned}
$$

Thus $(B; \cdot, +)$ is a boolean ring. $\qquad\qquad\square$

We can also proceed in the opposite direction. Given a boolean ring $(B; \cdot, +)$ we can equip it with the structure of a boolean algebra. For this purpose we first note that in such a ring we have

$$x + y = (x + y)^2 = x^2 + xy + yx + y^2 = x + xy + yx + y$$

whence $xy + yx = 0$ and so $-xy = yx$. Taking $y = x$, we obtain $-x^2 = x^2$, i.e. $-x = x$, so that $x + x = 0$. Thus a boolean ring is of characteristic 2. Now since $x = -x$ for every x we have $xy = -xy = yx$, whence we see that a boolean ring is commutative.

Theorem 6.10 *Let* $(B; \cdot, +)$ *be a boolean ring. For all* $x, y, z \in B$ *define*

$$x \wedge y = xy, \quad x \vee y = x + y + xy, \quad x' = 1 + x.$$

Then $(B; \wedge, \vee, ')$ *is a boolean algebra.*

Proof It is clear from the above that $(B; \wedge)$ is an abelian semigroup. Also,

$$(x \vee y) \vee z = (x \vee y) + z + (x \vee y)z = x + y + xy + z + xz + yz + xyz,$$

the symmetry of which shows that $(B; \vee)$ is also a semigroup, again abelian. Since now $x \wedge (x \vee y) = x(x + y + xy) = x + xy + xy = x + 0 = x$ and $x \vee (x \wedge y) = x + xy + x^2 y = x + xy + xy = x + 0 = x$, it follows that $(B; \wedge, \vee)$ is a lattice. This lattice is distributive since

$$
\begin{aligned}
x \wedge (y \vee z) = x(y + z + yz) &= xy + xz + xyz \\
&= xy + xz + xyxz \\
&= xy \vee xz \\
&= (x \wedge y) \vee (x \wedge z).
\end{aligned}
$$

Now the order in this lattice is given by

$$x \leqslant y \iff x = x \wedge y = xy.$$

Hence the lattice is bounded with top element 1 and bottom element 0. Finally, $x \vee x' = x + x' + xx' = x + 1 + x + x(1 + x) = 1$ and $x \wedge x' = xx' = x(1 + x) = x + x = 0$, and so x' is a, hence the, complement of x. Thus $(B; \wedge, \vee,')$ is a boolean algebra. $\square$

EXERCISES

6.9. For a given boolean algebra B let $\mathcal{R}(B)$ be the associated boolean ring as described in Theorem 6.9; and for a given boolean ring R let $\mathcal{B}(R)$ be the associated boolean algebra as described in Theorem 6.10. Prove that $\mathcal{B}[\mathcal{R}(B)] = B$ and that $\mathcal{R}[\mathcal{B}(R)] = R$.

6.10. Prove that subalgebras of a boolean algebra correspond to subrings with a 1 of the associated boolean ring. Prove also that lattice ideals correspond to ring ideals.

6.4 Boolean algebras of subsets

We have seen in Theorem 5.7 that every distributive lattice is isomorphic to a ring of sets. The corresponding result for boolean lattices is as follows.

Definition By a **field of sets** we mean a collection $\mathcal{B}$ of subsets of a set E such that

$$(\forall X, Y \in \mathcal{B}) \quad X \cap Y \in \mathcal{B}, \quad X \cup Y \in \mathcal{B}, \quad X' \in \mathcal{B}.$$

Theorem 6.11 (Stone [103]) *Every boolean lattice is isomorphic to a field of sets.*

Proof Let B be a boolean lattice and let P be a prime filter of B. Then for every $x \in B$ we have $x \vee x' = 1 \in P$ and so either $x \in P$ or $x' \in P$. If, as in the proof of Theorem 5.7, $\mathcal{F}(x)$ denotes the set of prime filters that contain x then we have

$$F \in \mathcal{F}(x') \iff x' \in F \iff x \notin F \iff F \notin \mathcal{F}(x),$$

and so $\mathcal{F}(x') = [\mathcal{F}(x)]'$. Consequently, $\mathcal{F} = \{\mathcal{F}(x) \mid x \in B\}$ is a field of sets and, by Theorem 5.7, $x \mapsto \mathcal{F}(x)$ is an isomorphism of B onto $\mathcal{F}$. $\square$

The archetypal example of a boolean algebra being a power set algebra, we now proceed to obtain a characterisation of such algebras. Following this, we give examples of boolean algebras that are not power set algebras.

Theorem 6.12 (Lindenbaum–Tarski [109]) *For a boolean algebra B the following statements are equivalent*:

(1) $B \simeq \mathbb{P}(E)$ *for some set E*;
(2) B *is complete and atomic.*

Proof (1) $\Rightarrow$ (2): This is clear, the atoms of $\mathbb{P}(E)$ being the singleton subsets.

(2) $\Rightarrow$ (1): Suppose that B is complete and atomic. Let $\mathcal{A}$ be the set of atoms of B. Given $x \neq 0$ in B, let $X = \{a \in \mathcal{A} \mid a \leqslant x\}$ and let $y = \sup X$. Clearly, we have $y \leqslant x$. We first show that in fact $y = x$.

Suppose, by way of obtaining a contradiction, that $y < x$. Then there exists $z \neq 0$ such that $x = y \vee z$ and $y \wedge z = 0$; in fact, by Theorem 6.1, z is the relative complement of y in $[0, x]$. Now since B is atomic there is an atom b such that $b \leqslant z \leqslant x$. Then $b \in X$ and we have the contradiction that $b \leqslant y \wedge z = 0$.

Now define $\varphi : B \to \mathbb{P}(\mathcal{A})$ by

$$\varphi(x) = \begin{cases} \{a \in \mathcal{A} \mid a \leqslant x\} & \text{if } x \neq 0; \\ \emptyset & \text{otherwise,} \end{cases}$$

and define $\psi : \mathbb{P}(\mathcal{A}) \to B$ by

$$\psi(X) = \begin{cases} \sup X & \text{if } X \neq \emptyset; \\ 0 & \text{otherwise.} \end{cases}$$

By the above observation we see that

$$(\forall x \in B) \quad \psi[\varphi(x)] = x,$$

so that $\psi \circ \varphi = \text{id}$ and hence ψ is surjective.

Now suppose that $\psi(X) = \psi(Y) \neq 0$ and that $y \in Y$. Then by Theorem 6.8 we have

$$y = y \wedge \psi(Y) = y \wedge \psi(X) = y \wedge \sup X = \sup(y \wedge X)$$

where $y \wedge X = \{y \wedge x \mid x \in X\}$. Since y is an atom there exists (an atom) $x \in X$ such that $0 < y \wedge x \leqslant y$ whence $y \wedge x = y$, which gives $y \leqslant x$ and hence $y = x \in X$. Thus we see that $Y \subseteq X$. Similarly, $X \subseteq Y$ and hence ψ is injective, and therefore a bijection. From $\psi \circ \varphi = \text{id}$ it then follows on pre-composing with ψ^{-1} that $\varphi = \psi^{-1}$. Since φ and ψ are clearly isotone, we deduce that ψ is an isotone bijection whose inverse φ is also isotone. Consequently, $B \simeq \mathbb{P}(\mathcal{A})$. $\qquad\square$

Corollary *If B is a finite boolean algebra then B has 2^n elements where n is the number of atoms in B.*

Proof From the theorem we have $B \simeq \mathbb{P}(E)$ for some finite set E. Without loss of generality we may assume that $E = \{1, 2, \ldots, n\}$. Let $\mathbf{2}$ denote the two-element chain $0 < 1$ and consider the mapping $f : \mathbb{P}(E) \to \mathbf{2}^n$ given by $f(X) = (x_1, \ldots, x_n)$ where

$$x_i = \begin{cases} 1 & \text{if } i \in X; \\ 0 & \text{otherwise.} \end{cases}$$

Given $A, B \in \mathbb{P}(E)$, let $f(A) = (a_1, \ldots, a_n)$ and $f(B) = (b_1, \ldots, b_n)$. Then we have $A \subseteq B$ if and only if, for every i, $i \in A$ implies $i \in B$, which is equivalent to $a_i = 1$ implies $b_i = 1$, i.e. to $a_i \leqslant b_i$, which by the definition is equivalent to $f(A) \leqslant f(B)$. Moreover, given any $x = (x_1, \ldots, x_n) \in \mathbf{2}^n$ we have $f(C) = x$ where $C = \{i \mid x_i = 1\}$. It therefore follows by Theorem 1.10 that $\mathbb{P}(E) \simeq \mathbf{2}^n$.

Hence $B \simeq \mathbf{2}^n$ and so has 2^n elements. Moreover, since E has n elements, $B \simeq \mathbb{P}(E)$ has n atoms, these corresponding to the singleton subsets of E.   □

As the following two examples show, not all boolean algebras are power set algebras.

Example 6.5 *A boolean algebra that is not complete.* A subset X of a set E is said to be **cofinite** if its complement X' is finite. Consider the subset B of $\mathbb{P}(E)$ that consists of those subsets of E that are finite or cofinite. It is readily seen that B is a field of subsets of E and therefore is boolean. If E is finite then clearly $B = \mathbb{P}(E)$; but if E is infinite then we have a new example of a boolean lattice. For example, let B be the finite-cofinite algebra of $\mathbb{Z}$ and consider the singleton subsets $\{2n\}$ of even integers. The supremum of this family does not exist in B since the only candidate for this is $\bigcup_{n \in \mathbb{Z}} \{2n\} = 2\mathbb{Z}$ which is neither finite nor cofinite. Hence the finite-cofinite algebra of $\mathbb{Z}$ is a boolean algebra that is not complete.

Example 6.6 *A boolean algebra that is not atomic.* Let E be an infinite set and define a relation ϑ on $\mathbb{P}(E)$ by

$$(X, Y) \in \vartheta \iff \begin{cases} X \text{ and } Y \text{ differ in only a} \\ \text{finite number of elements} \end{cases}$$
$$\iff (X \cup Y) \setminus (X \cap Y) \text{ is finite.}$$

It is readily seen (draw a Venn diagram!) that ϑ is an equivalence relation on $\mathbb{P}(E)$. Now if $(X \cup Y) \setminus (X \cap Y)$ is finite then, as can also be seen from a Venn diagram, for all $Z \in \mathbb{P}(E)$ so is $[(X \cup Y) \cap Z] \setminus (X \cap Y \cap Z)$. Consequently we see that

$$(X, Y) \in \vartheta \Rightarrow (X \cap Z, Y \cap Z) \in \vartheta$$

and so ϑ is compatible with intersection. In a similar way we can see that $(X \cup Y \cup Z) \setminus [(X \cap Y) \cup Z]$ is finite and hence

$$(X, Y) \in \vartheta \Rightarrow (X \cup Z, Y \cup Z) \in \vartheta$$

so that ϑ is also compatible with union. Thus ϑ is a congruence on $\mathbb{P}(E)$. In the quotient lattice $L = \mathbb{P}(E)/\vartheta$ the lattice operations on the ϑ-classes are given by

$$[X]_\vartheta \wedge [Y]_\vartheta = [X \cap Y]_\vartheta, \quad [X]_\vartheta \vee [Y]_\vartheta = [X \cup Y]_\vartheta.$$

Since $\mathbb{P}(E)$ is distributive it is clear that so also is L.

Now the order in L is given by

$$[X]_\vartheta \leqslant [Y]_\vartheta \iff [X \cap Y]_\vartheta = [X]_\vartheta \wedge [Y]_\vartheta = [X]_\vartheta$$
$$\iff (X \cap Y, X) \in \vartheta$$
$$\iff X \setminus (X \cap Y) \text{ is finite}$$
$$\iff X \setminus Y \text{ is finite.}$$

Since for every $X \in \mathbb{P}(E)$ we have $X \setminus E = \emptyset$, we have that $[X]_\vartheta \leqslant [E]_\vartheta$; and for every finite subset F of E we have that $F \setminus X$ is finite, so $[\emptyset]_\vartheta = [F]_\vartheta \leqslant [X]_\vartheta$. Thus $[\emptyset]_\vartheta$ is the bottom element of L and $[E]_\vartheta$ is the top element of L. If now

X' is the complement of X then from $[X]_\vartheta \vee [X']_\vartheta = [X \cup X']_\vartheta = [E]_\vartheta$ and $[X]_\vartheta \wedge [X']_\vartheta = [X \cap X']_\vartheta = [\emptyset]_\vartheta$ we see that L is boolean, the complement of $[X]_\vartheta$ being $[X']_\vartheta$.

We now observe that L has no atoms. Indeed, if $A \in \mathbb{P}(E)$ is finite then $[A]_\vartheta = [\emptyset]_\vartheta$, so if $[X]_\vartheta$ is an atom in L then X must be infinite. But it is a property of an infinite set X that there exist infinite subsets Y and Z such that $X = Y \cup Z$ and $Y \cap Z = \emptyset$. Since $[Y]_\vartheta \neq [\emptyset]_\vartheta$ we must have $[\emptyset]_\vartheta < [Y]_\vartheta$. Also, $[Y]_\vartheta \leqslant [Y]_\vartheta \vee [Z]_\vartheta = [Y \cup Z]_\vartheta = [X]_\vartheta$, and indeed $[Y]_\vartheta < [X]_\vartheta$ since X and Y differ by Z which is infinite. Thus we have $[\emptyset]_\vartheta < [Y]_\vartheta < [X]_\vartheta$ which contradicts the assumption that $[X]_\vartheta$ is an atom of L.

EXERCISES

6.11. Let S be the set of infinite sequences whose elements are 0 or 1. Show that S forms a boolean algebra that is isomorphic to the power set algebra $\mathbb{P}(\mathbb{N})$.

6.12. Let T be the set of $(0,1)$-sequences that contain either a finite number of 0s or a finite number of 1s. Prove that T forms a boolean algebra that is isomorphic to the finite-cofinite algebra of a denumerably infinite set.

6.13. Let B be the set of $(0,1)$-sequences that have period 2^n for some positive integer n; i.e. are of the form

$$x_1, x_2, \ldots, x_{2^n}, x_1, x_2, \ldots, x_{2^n}, \ldots.$$

Prove that B forms a boolean algebra that is neither complete nor atomic.

6.14. Define a **boolean function** of n variables to be a mapping $f : \mathbf{2}^n \to \mathbf{2}$. Prove that the set of boolean functions of n variables is a boolean algebra having 2^{2^n} elements.

6.15. If $f : \mathbf{2}^n \to \mathbf{2}$ is a boolean function show that $f(x_1, \ldots, x_n)$ can be written in the form

$$[f(x_1, \ldots, 1, \ldots, x_n) \wedge x_i] \vee [f(x_1, \ldots, 0, \ldots, x_n) \wedge x_i'],$$

and also in the form

$$[f(x_1, \ldots, 0, \ldots, x_n) \vee x_i] \wedge [f(x_1, \ldots, 1, \ldots, x_n) \vee x_i'].$$

6.16. If $f : \mathbf{2}^n \to \mathbf{2}$ is a boolean function prove that

$$f(x_1, \ldots, x_n) = \bigvee_\sigma f(i_1, \ldots, i_n) \wedge x_1^{i_1} \wedge \cdots \wedge x_n^{i_n}$$

where the join is over all n-tuples $(i_1, \ldots, i_n) \in \mathbf{2}^n$ with

$$x_t^{i_t} = \begin{cases} x_t & \text{if } i_t = 0; \\ x_t' & \text{if } i_t = 1. \end{cases}$$

This is known as the **disjunctive normal form** of f.

6.17. In applications of boolean algebra it is common practice to employ what is often called *engineer's notation*. In this, $\wedge$ is written as multiplication and $\vee$ as addition (not to be confused with addition in the corresponding boolean ring). Using this notation, show that the disjunctive normal form of a boolean function of two variables is

$$f(x, y) = axy + bx'y + cxy' + dx'y',$$

and identify the constants a, b, c, d. Show also that the disjunctive normal form of the complement of f is

$$f'(x, y) = a'xy + b'x'y + c'xy' + d'x'y'.$$

6.18. Determine the disjunctive normal forms of the boolean functions given by
  (1) $f(x, y) = x + y'$;
  (2) $f(x, y, z) = 1$;
  (3) $f(x, y, z) = x(y + z') + (x' + y)z$.

6.19. Determine the boolean functions of two variables that satisfy the identity
  $f(x, f(y, x)) = 1$.

6.20. Determine the boolean functions of two variables that satisfy the identity
  $f(x, f'(y, x)) = 1$.

6.21. Develop Example 6.6 via boolean rings, using the fact that $(X, Y) \in \vartheta$ if and only if $X + Y$ is finite.

6.5 The Dedekind–MacNeille completion of a boolean algebra

A further interesting property of boolean algebras is the following.

Theorem 6.13 *The Dedekind–MacNeille completion of a boolean algebra is a boolean algebra.*

Proof Let B be a boolean algebra and let $L = \{A^{\uparrow\downarrow} \mid A \in \mathbb{P}(B)\}$ be its Dedekind–MacNeille completion as constructed in Theorem 2.16. For each $X \in \mathbb{P}(B)$ let $X^{\star} = \{x' \mid x \in X\}$. Then we make the following observations:

(1) $(X^{\star})^{\downarrow\uparrow} = (X^{\uparrow\downarrow})^{\star}$.

This follows from the fact that

$$x \in (X^{\star})^{\downarrow\uparrow} \iff x \geqslant y \text{ for all } y \in (X^{\star})^{\downarrow}$$
$$\iff x' \leqslant y' \text{ for all } y' \in X^{\uparrow}$$
$$\iff x' \in X^{\uparrow\downarrow}$$
$$\iff x \in (X^{\uparrow\downarrow})^{\star}.$$

(2) $X^{\uparrow\downarrow}$ *is a down-set of* L.

This is clear.

(3) *If* $X \in L$ *then* $X^{\uparrow} \cap X^{\star} = \{1\}$.

By (2) we have $0 \in X^{\uparrow\downarrow} = X$ and so $1 = 0' \in X^{\uparrow} \cap X^{\star}$. Suppose now that $x \in X^{\uparrow} \cap X^{\star}$. Then $x \geqslant z$ for all $z \in X$, and $x = y'$ for some $y \in X$. The latter gives $x' = y \in X$ and so, by the former, $x \geqslant x'$ whence $x = 1$.

(4) *If* $X \in L$ *then* $X \cap (X^{\star})^{\downarrow} = \{0\}$.

Clearly, $0 \in X \cap (X^{\star})^{\downarrow}$. If now $x \in X \cap (X^{\star})^{\downarrow}$ then $x \in X$ and $x \leqslant y'$ for all $y \in X$. Hence $x \leqslant x'$ and so $x = 0$.

(5) $(X^{\star})^{\downarrow}$ *is the unique complement of* X *in* L.

Since $(X^{\star})^{\downarrow} = (X^{\star})^{\downarrow\uparrow\downarrow} \in L$ it follows by (4) and Theorem 2.14 that

$$\inf\nolimits_L\{X, (X^{\star})^{\downarrow}\} = X \cap (X^{\star})^{\downarrow} = \{0\}.$$

Again by Theorem 2.14,

$$\sup_L\{X, (X^\star)^\downarrow\} = [X \cup (X^\star)^\downarrow]^{\uparrow\downarrow}$$
$$= [X^\uparrow \cap (X^\star)^{\downarrow\uparrow}]^\downarrow \quad \text{by Theorem 2.15}$$
$$= [X^\uparrow \cap (X^{\uparrow\downarrow})^\star]^\downarrow \quad \text{by (1)}$$
$$= (X^\uparrow \cap X^\star)^\downarrow \quad \text{since } X \in L$$
$$= \{1\}^\downarrow \quad \text{by (3)}$$
$$= B.$$

Thus we see that $(X^\star)^\downarrow$ is a complement of X in L.

Suppose now that $Y \in L$ is also a complement of X. Then $X \cap Y = \{0\}$ and $(X \cup Y)^{\uparrow\downarrow} = B$. It follows that

$$\{1\} = B^\uparrow = (X \cup Y)^{\uparrow\downarrow\uparrow} = (X \cup Y)^\uparrow = X^\uparrow \cap Y^\uparrow.$$

Now if $x \in X$ and $y \in Y$ then by (2) we have $x \wedge y \in X \cap Y = \{0\}$ and so $x \wedge y = 0$. Thus $y \leqslant x'$ and hence $Y \subseteq (X^\star)^\downarrow$. To obtain the reverse inclusion, let $z \in (X^\star)^\downarrow$. Then $z \leqslant x'$ for all $x \in X$ whence $z' \in X^\uparrow$. Now if $t \in Y^\uparrow$ then $z' \vee t \in X^\uparrow \cap Y^\uparrow = \{1\}$ whence $z' \vee t = 1$. It follows that $z = z \wedge (z' \vee t) = z \wedge t$, so that $z \leqslant t$ for all $t \in Y^\uparrow$. Thus $z \in Y^{\uparrow\downarrow} = Y$ and hence $(X^\star)^\downarrow \subseteq Y$. Thus we see that $(X^\star)^\downarrow$ is the unique complement of X in L.

To complete the proof that L is boolean it now suffices, by Theorem 6.5, to establish either of the de Morgan laws. For this, we have

$$\sup_L\{(X^\star)^\downarrow, (Y^\star)^\downarrow\} = [(X^\star)^\downarrow \cup (Y^\star)^\downarrow]^{\uparrow\downarrow}$$
$$= [(X^\star)^{\downarrow\uparrow} \cap (Y^\star)^{\downarrow\uparrow}]^\downarrow$$
$$= [(X^{\uparrow\downarrow})^\star \cap (Y^{\uparrow\downarrow})^\star]^\downarrow$$
$$= (X^\star \cap Y^\star)^\downarrow$$
$$= [(X \cap Y)^\star]^\downarrow$$
$$= [(\inf_L\{X, Y\})^\star]^\downarrow,$$

and since the complement of X in L is $(X^\star)^\downarrow$ this gives the required de Morgan law. □

In complete contrast to the above, the following ingenious example, due to Funayama [52], shows that the Dedekind–MacNeille completion of a distributive lattice need not even be modular.

Example 6.7 Consider the chains

$$\overline{A} \equiv \underbrace{a_1 < a_2 < a_3 \cdots \to \leftarrow \cdots < \overbrace{a_3^\star < a_2^\star < a_1^\star}^{A^\star}}_{A}$$

$$\overline{B} \equiv \underbrace{b_1 < b_2 < b_3 \cdots \to \leftarrow \cdots < \overbrace{b_3^\star < b_2^\star < b_1^\star}^{B^\star}}_{B}$$

$$C \equiv u < v$$

and the subset of the distributive cartesian product lattice $\overline{A} \times \overline{B} \times C$ that is defined by $M = (\overline{A}, B^\star, v) \cup (A, \overline{B}, u)$ where, for example, $(\overline{A}, B^\star, v) =$

$\{(x, y, v) \mid x \in \overline{A}, y \in B^\star\}$. It is readily seen that M is a sublattice of $\overline{A} \times \overline{B} \times C$ and so is distributive. Let L be the Dedekind–MacNeille completion of M and consider the following subsets:

$$D = (A, B, u);$$
$$S = (a_1, b_1^\star, v)^{\uparrow\downarrow} = (\overline{A}, b_1^\star, v)^\downarrow = (a_1, \overline{B}, C) \cap M = (a_1, B^\star, v) \cup (a_1, \overline{B}, u);$$
$$T = (a_1, b_1^\star, u)^{\uparrow\downarrow} = [(\overline{A}, b_1^\star, C) \cap M]^\downarrow = [(\overline{A}, b_1^\star, v) \cup (A, b_1^\star, u)]^\downarrow = (a_1, \overline{B}, u);$$
$$U = (a_1, B, u).$$

Clearly, both S and T belong to L. Since

$$D^{\uparrow\downarrow} = [(A^\star, B^\star, C) \cap M]^\downarrow = (A^\star, B^\star, v)^\downarrow = (A, B, C) \cap M = (A, B, u) = D$$

we see that $D \in L$. Moreover, $D \cap S = D \cap (a_1, \overline{B}, C) = U = D \cap T$, and

$$(D \cup S)^{\uparrow\downarrow} = (D^\uparrow \cap S^\uparrow)^\downarrow = [(A^\star, B^\star, C) \cap (\overline{A}, b_1^\star, v) \cap M]^\downarrow = (A^\star, b_1^\star, v)^\downarrow;$$
$$(D \cup T)^{\uparrow\downarrow} = (D^\uparrow \cap T^\uparrow)^\downarrow = [(A^\star, B^\star, C) \cap (\overline{A}, b_1^\star, C) \cap M]^\downarrow$$
$$= [(A^\star, b_1^\star, C) \cap M]^\downarrow = (A^\star, b_1^\star, v)^\downarrow.$$

Since $T \subset S$ it follows that L has a sublattice of the form

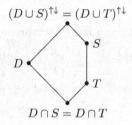

$$(D \cup S)^{\uparrow\downarrow} = (D \cup T)^{\uparrow\downarrow}$$

S

D

T

$$D \cap S = D \cap T$$

and consequently L is not modular.

6.6 Neutral and central elements

We now relate distributivity to given elements as follows.

Definition By a **neutral element** of a lattice L we shall mean an element z of L such that, for all $x, y \in L$,

(1) $z \wedge (x \vee y) = (z \wedge x) \vee (z \wedge y)$;
(2) $z \vee (x \wedge y) = (z \vee x) \wedge (z \vee y)$;
(3) if $z \wedge x = z \wedge y$ and $z \vee x = z \vee y$ then $x = y$.

Example 6.8 In a distributive lattice every element is neutral.

Example 6.9 Every bounded lattice contains neutral elements, for example 0 and 1.

Example 6.10 If A and B are bounded lattices then $(1_A, 0_B)$ and $(0_A, 1_B)$ are neutral elements of $A \times B$.

In a bounded lattice neutral elements can be characterised as follows.

Theorem 6.14 *If L is a bounded lattice and $z \in L$ then the following statements are equivalent*:

(1) *z is neutral*;

(2) *there are bounded lattices A and B and an embedding $\varphi : L \to A \times B$ such that $\varphi(z) = (1_A, 0_B)$*;

(3) *there are bounded lattices A and B and an embedding $\varphi : L \to A \times B$ such that $\varphi(z) = (0_A, 1_B)$*.

Proof (1) $\Rightarrow$ (2): If z is neutral then the mapping $\lambda_z^* : L \to z^{\downarrow}$ given by $\lambda_z^*(x) = z \wedge x$ is a surjective lattice morphism, as is the mapping $\mu_z^* : L \to z^{\uparrow}$ given by $\mu_z^*(x) = z \vee x$. Let $A = z^{\downarrow}$ and $B = z^{\uparrow}$, and consider the mapping $\varphi : L \to A \times B$ given by $\varphi(x) = (\lambda_z^*(x), \mu_z^*(x))$. Clearly, φ is a lattice morphism. Now since z is neutral we have that φ is injective and hence is an embedding. Moreover, $\varphi(z) = (z, z) = (1_A, 0_B)$.

(2) $\Rightarrow$ (1): Suppose that (2) holds and let $x, y \in L$. If $\varphi(x) = (a_1, b_1)$ and $\varphi(y) = (a_2, b_2)$ then a simple calculation using the morphism properties of φ shows that $\varphi[z \wedge (x \vee y)] = (a_1 \vee a_2, 0) = \varphi[(z \wedge x) \vee (z \wedge y)]$ and therefore, since φ is injective, we have $z \wedge (x \vee y) = (z \wedge x) \vee (z \wedge y)$. Similarly, $z \vee (x \wedge y) = (z \vee x) \wedge (z \vee y)$. Finally, if $z \wedge x = z \wedge y$ and $z \vee x = z \vee y$ then, applying φ to these equalities, we obtain $(1_A, b_1) = (1_A, b_2)$ and $(a_1, 0_B) = (a_2, 0_B)$ whence $\varphi(x) = \varphi(y)$ and therefore $x = y$. Thus we see that z is neutral.

(1) $\Leftrightarrow$ (3): This is established dually. $\square$

An important property of neutral elements is the following, in which $D(a, b, c)$ denotes the equality $a \wedge (b \vee c) = (a \wedge b) \vee (a \wedge c)$.

Theorem 6.15 *If at least one of $a, b, c \in L$ is neutral then $D(a, b, c)$ holds.*

Proof If a is neutral then it is clear that $D(a, b, c)$ holds. Suppose then that b or c is neutral. Choosing b for example, we let $\varphi : L \to L_1 \times L_2$ be an embedding such that $\varphi(b) = (1, 0)$, and let $\varphi(a) = (a_1, a_2)$ and $\varphi(c) = (c_1, c_2)$. Then a simple calculation that uses the morphism properties of φ shows that $\varphi[a \wedge (b \vee c)] = (a_1, a_2 \wedge c_2) = \varphi[(a \wedge b) \vee (a \wedge c)]$ whence $D(a, b, c)$ follows. $\square$

Definition By a **central element** of a bounded lattice L we mean an element z of L for which there are bounded lattices A and B and an isomorphism $\varphi : L \to A \times B$ such that $\varphi(z) = (1_A, 0_B)$ or $\varphi(z) = (0_A, 1_B)$.

Clearly, every central element is neutral. We can say more:

Theorem 6.16 *Every central element of a bounded lattice L has a unique complement which is also central.*

Proof Observe that the neutral elements $(1, 0)$ and $(0, 1)$ are complements of each other in $A \times B$. Consequently central elements in L are complemented and have a central complement. It follows by property (3) in the definition of neutral element that complements of central elements are unique. $\square$

Definition The set of central elements of a lattice L is called the **centre** of L and is denoted by $Z(L)$.

Theorem 6.17 *The centre of a bounded lattice L is a boolean sublattice of L.*

Proof Let $z_1, z_2 \in Z(L)$ with respective complements z_1', z_2' and consider the intervals $L_1 = [0, z_1 \wedge z_2]$ and $L_2 = [0, z_1' \vee z_2']$. Define $\varphi : L \to L_1 \times L_2$ by the prescription $\varphi(x) = \big(x \wedge z_1 \wedge z_2, \; x \wedge (z_1' \vee z_2')\big)$. It is clear that φ is isotone. For every $x \in L$ we have, using Theorem 6.15 and the fact that z_1, z_2, z_1', z_2' are central,

$$
\begin{aligned}
x &= x \wedge (z_1 \vee z_1') \\
&= (x \wedge z_1) \vee (x \wedge z_1') \\
&= [(x \wedge z_1) \vee (x \wedge z_1')] \wedge (z_2 \vee z_2') \\
&= (x \wedge z_1 \wedge z_2) \vee (x \wedge z_1 \wedge z_2') \vee (x \wedge z_1' \wedge z_2) \vee (x \wedge z_1' \wedge z_2') \\
&= (x \wedge z_1 \wedge z_2) \vee [(x \wedge z_1 \wedge z_2') \vee (x \wedge z_1' \wedge z_2')] \\
&\qquad \vee [(x \wedge z_1' \wedge z_2) \vee (x \wedge z_1' \wedge z_2')] \\
&= (x \wedge z_1 \wedge z_2) \vee (x \wedge z_2') \vee (x \wedge z_1') \\
&= (x \wedge z_1 \wedge z_2) \vee [x \wedge (z_1' \vee z_2')].
\end{aligned}
$$

It follows from this that if $\varphi(x) \leqslant \varphi(y)$ then $x \leqslant y$. Suppose now that $(a, b) \in L_1 \times L_2$. Then $a \leqslant z_1 \wedge z_2$ and $b \leqslant z_1' \vee z_2'$. Moreover,

$$(a \vee b) \wedge z_1 \wedge z_2 = (a \wedge z_1 \wedge z_2) \vee (b \wedge z_1 \wedge z_2).$$

Observing that $(z_1' \vee z_2') \wedge z_1 \wedge z_2 = (z_1' \wedge z_1 \wedge z_2) \vee (z_2' \wedge z_1 \wedge z_2) = 0$ we deduce that $b \wedge z_1 \wedge z_2 = 0$ and therefore $(a \vee b) \wedge z_1 \wedge z_2 = a \wedge z_1 \wedge z_2 = a$. In a similar way we can see that $(a \vee b) \wedge (z_1' \vee z_2') = b$ and consequently we have $\varphi(a \vee b) = (a, b)$, so that φ is surjective. It follows that φ is an isomorphism of L onto $L_1 \times L_2$. Finally, since $\varphi(z_1 \wedge z_2) = (z_1 \wedge z_2, 0)$ we see that $z_1 \wedge z_2$ is central. Thus $Z(L)$ is a $\wedge$-subsemilattice of L. A dual argument shows that it is also a $\vee$-subsemilattice. Thus $Z(L)$ is a sublattice and, by Theorems 6.15 and 6.16, is boolean. $\qquad\square$

EXERCISES

6.22. If L is a bounded modular lattice prove that $z \in L$ is neutral if and only if either λ_z or μ_z is a lattice morphism.

6.23. If L is a bounded distributive lattice prove that $z \in L$ is central if and only if it is complemented.

6.24. If L is a bounded lattice prove that for $z \in L$ the following statements are equivalent:

(1) z is central;

(2) z is neutral and complemented;

(3) there exists $z' \in L$ such that

$$(\forall x \in L) \quad x = (x \wedge z) \vee (x \wedge z') = (x \vee z) \wedge (x \vee z').$$

6.25. Let L be a bounded lattice such that, for every $x \in L$,

$$f(x) = \bigwedge \{z \in Z(L) \mid x \leqslant z\}$$

exists and is central. Prove that $x \mapsto f(x)$ is a **quantifier** in the sense that

(1) $f(0) = 0$;

(2) $(\forall x \in L) \; x \leqslant f(x)$;

(3) $(\forall x, y \in L) \; f[x \wedge f(y)] = f(x) \wedge f(y)$.

6.26. Let L be a complete lattice and let $z \in Z(L)$. Establish the distributive laws

$$z \wedge \bigvee_{i \in I} x_i = \bigvee_{i \in I}(z \wedge x_i) \quad \text{and} \quad z \vee \bigwedge_{i \in I} x_i = \bigwedge_{i \in I}(z \vee x_i).$$

6.7 Stone's representation theorem

Our objective now is to obtain a topological representation of boolean algebras. For this purpose, we require the following facts.

Theorem 6.18 *If L is a distributive lattice then*

(1) *every maximal ideal of L is prime;*

(2) *every proper ideal of L is an intersection of a family of prime ideals.*

Proof (1) Let I be a maximal ideal of L. To prove that I is prime, we show that if $x \wedge y \in I$ and $x \notin I$ then necessarily $y \in I$. For this purpose, consider the principal ideal $x^{\downarrow}$. Since I is maximal we have $x^{\downarrow} \vee I = L$. Then $y \leqslant x \vee i$ for some $i \in I$ whence, by the distributivity,

$$y = y \wedge (x \vee i) = (y \wedge x) \vee (y \wedge i) \in I.$$

(2) Let I be an ideal of L and let $x \notin I$. Consider the set $\mathcal{X}$ of ideals J of L such that $x \notin J$ and $I \subseteq J$. Clearly, $\mathcal{X} \neq \emptyset$. Consider now a chain $(J_\alpha)_{\alpha \in A}$ of elements of $\mathcal{X}$. Since the set-theoretic union of the family $(J_\alpha)_{\alpha \in A}$ also belongs to $\mathcal{X}$, it follows by Zorn's axiom that $\mathcal{X}$ has a maximal member M. We show as follows that M is a prime ideal of L. Suppose in fact that we had $v \notin M$, $w \notin M$ but $v \wedge w \in M$. By the maximality of M in $\mathcal{X}$ we have $v^{\downarrow} \vee M \notin \mathcal{X}$ whence $x \in v^{\downarrow} \vee M$ and $x \in w^{\downarrow} \vee M$, so there exist $m, n \in M$ such that $x \leqslant v \vee m$ and $x \leqslant w \vee n$ But then we have the contradiction

$$x \leqslant (v \vee m) \wedge (w \vee n) = (v \wedge w) \vee (v \wedge n) \vee (m \wedge w) \vee (m \wedge n) \in M.$$

The above shows that if I is a proper ideal of L then there is a prime ideal that contains I. Let $\{P_\alpha\}_{\alpha \in A}$ be the set of prime ideals that contain I. Then $I \subseteq \bigcap_{\alpha \in A} P_\alpha$, and if $x \notin I$ then $x \notin \bigcap_{\alpha \in A} P_\alpha$. It follows that $I = \bigcap_{\alpha \in A} P_\alpha$. $\square$

Theorem 6.19 *If L is a complemented lattice then every prime ideal of L is maximal.*

Proof Let P be a prime ideal of L. If $x \notin P$ then since $x \wedge x' = 0 \in P$ the fact that P is prime gives $x' \in P$. Consequently $1 = x \vee x' \in x^{\downarrow} \vee P$ and therefore $L = x^{\downarrow} \vee P$. It follows that P is maximal. $\square$

It follows immediately from Theorems 6.18 and 6.19 that in a boolean algebra the prime ideals and the maximal ideals coincide.

For the topological description that follows we require some further terminology. A subset A of a topological space (X, τ) is said to be **clopen** if it is both open and closed. The space is said to be **totally disconnected** if every open set is the union of a family of clopen sets. If $Y \subseteq X$ then $\{X_i \mid i \in I\} \subseteq \tau$

is an **open cover** of Y if $Y \subseteq \bigcup_{i \in I} X_i$. If I^* is a finite subset of I such that $Y \subseteq \bigcup_{i \in I^*} X_i$ then $\bigcup_{i \in I^*} X_i$ is a **finite subcover** of Y. A topological space (X, τ) is said to be **compact** if every open cover of Y has a finite subcover. Finally, (X, τ) is said to be **Hausdorff** if for distinct $x, y \in X$ there exist $X_1, X_2 \in \tau$ such that $x \in X_1, y \in X_2$ and $X_1 \cap X_2 = \emptyset$. These notions come together in the following concept.

Definition By a **boolean space** we shall mean a compact totally disconnected Hausdorff topological space.

Stone's celebrated representation theorem for boolean algebras is then the following.

Theorem 6.20 (Stone [105]) *For every boolean algebra B there is a boolean space $\mathcal{M}$ such that B is isomorphic to the algebra of clopen subsets of $\mathcal{M}$.*

Proof For each $x \in B$ let $\mathcal{M}_x$ be the set of maximal ideals of B that do *not* contain x. Then we observe first that

$$\mathcal{M}_x \cap \mathcal{M}_y = \mathcal{M}_{x \wedge y}, \quad \mathcal{M}_x \cup \mathcal{M}_y = \mathcal{M}_{x \vee y}.$$

In fact, if M is a maximal ideal and if $x \wedge y \notin M$ then clearly we cannot have $x \in M$ or $y \in M$. Thus if $M \in \mathcal{M}_{x \wedge y}$ then $M \in \mathcal{M}_x$ and $M \in \mathcal{M}_y$, so that $\mathcal{M}_{x \wedge y} \subseteq \mathcal{M}_x \cap \mathcal{M}_y$. On the other hand, since every maximal ideal is prime, if $x \notin M$ and $y \notin M$ then $x \wedge y \notin M$, so that $\mathcal{M}_x \cap \mathcal{M}_y \subseteq \mathcal{M}_{x \wedge y}$. The second equality is clear from the fact that for every ideal M we have $x \vee y \notin M$ if and only if $x \notin M$ or $y \notin M$.

If $\mathcal{M}$ is the set of maximal ideals of B we can now use $\{\mathcal{M}_x \mid x \in B\}$ as the open sets of a topology on $\mathcal{M}$. In this way a collection of maximal ideals of B is open if it can be expressed as the union of a family of sets of the form $\mathcal{M}_x$. Note from the above that $\mathcal{M}_x$ and $\mathcal{M}_{x'}$ are complementary in $\mathcal{M}$ and so each $\mathcal{M}_x$ is clopen. Thus the space $\mathcal{M}$ of maximal ideals is totally disconnected.

We now show that $\mathcal{M}$ is Hausdorff. To this end, let M and N be distinct maximal ideals of B. Then we may assume the existence of $x \in M$ such that $x \notin N$. Then $N \in \mathcal{M}_x$, and $M \notin \mathcal{M}_x$ whence $M \in \mathcal{M}_{x'}$. Thus $\mathcal{M}_x$ and $\mathcal{M}_{x'}$ are disjoint open sets such that $N \in \mathcal{M}_x$ and $M \in \mathcal{M}_{x'}$. Consequently, $\mathcal{M}$ is Hausdorff.

To show that $\mathcal{M}$ is compact, it suffices to prove that if $\bigcup_{i \in I} \mathcal{M}_{x_i} = \mathcal{M}$ then there is a finite subset J of I such that $\bigcup_{i \in J} \mathcal{M}_{x_i} = \mathcal{M}$. Let N be the ideal generated by $\{x_i \mid i \in I\}$. If $N \neq B$ then there is a maximal ideal M that contains N. But then $x_i \in M$ for all $i \in I$ which gives $M \notin \bigcup_{i \in I} \mathcal{M}_{x_i} = \mathcal{M}$, a contradiction. Thus we must have $N = B$ whence there exists a finite subset J of I such that $\bigvee_{i \in J} x_i = 1_B$. Consequently, from the first obsrvation above, we have $\bigcup_{i \in J} \mathcal{M}_{x_i} = \mathcal{M}_{1_B} = \mathcal{M}$.

Thus we see that $\mathcal{M}$ is a boolean space. To complete the proof, consider the mapping f from B to the algebra of clopen subsets of $\mathcal{M}$ given by $f(x) = \mathcal{M}_x$. By the first observation above, f is a lattice morphism. That it is injective follows from Theorem 5.6. To see that it is surjective it remains to show that if $\mathcal{K}$ is a clopen subset of $\mathcal{M}$ then there exists $x \in B$ such that $\mathcal{K} = \mathcal{M}_x$. Now since $\mathcal{K}$ is clopen we have $\mathcal{K} = \bigcup_{i \in I} \mathcal{M}_{x_i}$ and there exist $x_1, \ldots, x_n$ such that

$$\mathcal{K} = \bigcup_{j=1}^{n} \mathcal{M}_{x_j}. \text{ Hence } \mathcal{K} = \mathcal{M}_y \text{ where } y = \bigvee_{j=1}^{n} x_j. \qquad \square$$

6.8 Baer semigroups and complementation

We now turn our attention to a Baer semigroup coordinatisation of complemented modular lattices. For this purpose we recall that in Chapter 4 we introduced the notion of a modular lattice via that of a residuated mapping $f : E \to F$ being range-closed, in the sense that $\operatorname{Im} f$ is a down-set of F. We strengthen this notion as follows.

Definition If $f : E \to F$ is a residuated mapping then we shall say that f is **totally range-closed** if the image under f of every principal down-set of E is a principal down-set of F; equivalently, if $(\forall x \in E) \; f^{\to}(x^{\downarrow}) = [f(x)]^{\downarrow}$. Similarly, we say that f is **dually totally range-closed** if the image under f^+ of every principal filter of F is a principal filter of E.

Such residuated mappings on a lattice can be characterised as follows.

Theorem 6.21 *If L is a lattice then a residuated mapping $f : L \to L$ is totally range-closed if and only if*

$$(\forall x, y \in L) \quad f[f^+(x) \wedge y] = x \wedge f(y);$$

and is dually totally range-closed if and only if

$$(\forall x, y \in L) \quad f^+[f(x) \vee y] = x \vee f^+(y).$$

Proof Clearly, for all $x, y \in L$ we have $f[f^+(x) \wedge y] \leqslant x \wedge f(y)$. If now f is totally range-closed then for all $x, y \in L$ there exists $z \leqslant y$ such that $x \wedge f(y) = f(z)$. This gives $z \leqslant f^+(x) \wedge y$ and consequently $x \wedge f(y) = f(z) \leqslant f[f^+(x) \wedge y]$, whence we have the reverse inequality.

Suppose now that the first identity holds. Then we have $x \leqslant f(y)$ if and only if $x = f[f^+(x) \wedge y] \in f^{\to}(y^{\downarrow})$, and therefore f is totally range-closed.

The second identity holds by duality. $\qquad \square$

Definition A residuated mapping $f : E \to F$ will be called **strongly range-closed** if it is both totally and dually totally range-closed.

Theorem 6.22 *If L is a bounded lattice then the strongly range-closed residuated mappings on L form a semigroup.*

Proof Suppose that $f, g \in \operatorname{Res} L$ are strongly range-closed. Then for all $x, y \in L$ we have

$$
\begin{aligned}
fg[(fg)^+(x) \wedge y] &= fg[g^+ f^+(x) \wedge y] \\
&= f[f^+(x) \wedge g(y)] \qquad g \text{ totally range-closed} \\
&= x \wedge fg(y) \qquad\qquad\; f \text{ totally range-closed}
\end{aligned}
$$

whence we see that fg is totally range-closed. A dual argument completes the proof. □

In order to obtain a Baer semigroup coordinatisation of complemented modular lattices, we require the following technical results.

Theorem 6.23 *If L is a bounded lattice and if $f \in \operatorname{Res} L$ is a range-closed idempotent then $f^+(0) \wedge f(1) = 0$.*

Proof Since f is a range-closed idempotent, if $x \leqslant f(1)$ then $f(x) = x$ and therefore, since f^+ is also idempotent, we have

$$
\begin{aligned}
x = f(x) = f f^+ f^+ f(x) &\geqslant f f^+[x \vee f^+(0)] \\
&= [x \vee f^+(0)] \wedge f(1) \qquad \text{by Theorem 4.1} \\
&\geqslant x \wedge f(1) \\
&= x.
\end{aligned}
$$

Hence $x = [x \vee f^+(0)] \wedge f(1)$, and taking $x = 0$ we obtain the result. □

Corollary *If L is a bounded lattice and $f \in \operatorname{Res} L$ is a weakly regular idempotent then $f^+(0)$ and $f(1)$ are complementary.*

Proof This is immediate from the above result and its dual. □

Theorem 6.24 *Let L be a bounded lattice and suppose that $f, g \in \operatorname{Res} L$ are such that f is range-closed, g is dually range-closed, and $g \circ f$ is range-closed. Then $M^\star(f(1), g^+(0))$.*

Proof Let $a \geqslant g^+(0)$. Then since g is dually range-closed we have $a = g^+(b)$ for some $b \in L$. Consequently

$$
\begin{aligned}
a \wedge [f(1) \vee g^+(0)] &= a \wedge g^+ g f(1) \qquad\quad g \text{ dually range-closed} \\
&= g^+(b) \wedge g^+ g f(1) \\
&= g^+[b \wedge g f(1)] \\
&= [g^+ \circ g f \circ (gf)^+](b) \quad gf \text{ range-closed} \\
&= g^+ g f f^+ g^+(b) \\
&= g^+ g f f^+(a) \\
&= g^+ g[a \wedge f(1)] \qquad\quad f \text{ range-closed} \\
&= [a \wedge f(1)] \vee g^+(0) \qquad g \text{ dually range-closed.}
\end{aligned}
$$

Hence we see that $M^\star(f(1), g^+(0))$ holds. □

Suppose now that S is a Baer semigroup. For every idempotent $e \in S$ we have $eS \in \mathcal{R}(S)$ if and only if $eS = RL(e)$. As seen in Theorem 1.13, for every $z \in S$ we can define a residuated mapping $\varphi_z : \mathcal{R}(S) \to \mathcal{R}(S)$ by the prescription $\varphi_z(eS) = RL(ze)$. Similarly, we can define a residuated mapping $\eta_z : \mathcal{L}(S) \to \mathcal{L}(S)$ by the prescription $\eta_z(Sf) = LR(fz)$. From the proof of Theorem 1.13, we know that φ_z^+ is given by $\varphi_z^+[R(x)] = R(xz)$. We now observe that $\varphi_z^+ = R\eta_z L$ and $\eta_z^+ = L\varphi_z R$. In fact, if $R(x) = eS$ and $L(e) = Sf$ then $R\eta_z L(eS) = R\eta_z L(e) = R\eta_z(Sf) = RLR(fz) = R(fz)$. But since

$$fzt = 0 \iff zt \in R(f) = RL(e) = eS = R(x) \iff xzt = 0$$

we have that $R(fz) = R(xz)$. Consequently, $R\eta_z L(eS) = R(xz) = \varphi_z^+[R(x)] = \varphi_z^+(eS)$ and therefore $\varphi_z^+ = R\eta_z L$. Similarly, we have $\eta_z^+ = L\varphi_z R$.

Definition If S is a Baer semigroup then we shall say that an element z of S is range-closed, strongly range-closed, etc., if the associated residuated mapping $\varphi_z \in \operatorname{Res}\mathcal{R}(S)$ has the corresponding property; and we shall say that S itself is range-closed, strongly range-closed, etc., if every element of S has the property in question.

Theorem 6.25 *Let S be a Baer semigroup. If $xS \in \mathcal{R}(S)$ then x is a range-closed element of S.*

Proof Suppose that $xS \in \mathcal{R}(S)$, so that $xS = RL(x) = \varphi_x(1S)$. Let $yS \in \mathcal{R}(S)$ be such that $yS \subseteq xS$. Then $y = xz$ for some $z \in S$. Let $wS = RL(z)$ and observe that

$$txz = 0 \iff tx \in L(z) = L(w) \iff txw = 0$$

whence $L(xz) = L(xw)$. Then $yS = RL(y) = RL(xz) = RL(xw) = \varphi_x(wS)$ and therefore φ_x is range-closed in $\operatorname{Res}\mathcal{R}(S)$. Hence x is range-closed. $\square$

Theorem 6.26 (Janowitz [68]) *For a bounded lattice L the following statements are equivalent:*

(1) *L is a complemented modular lattice;*
(2) *L can be coordinatised by a strongly range-closed Baer semigroup;*
(3) *L can be coordinatised by a weakly regular Baer semigroup;*
(4) *L can be coordinatised by a range-closed Baer semigroup.*

Proof (1) $\Rightarrow$ (2): Suppose that (1) holds and let $a, b \in L$ be complementary. Consider the mapping $f_{a,b} : L \to L$ given by

$$(\forall x \in L) \quad f_{a,b}(x) = (x \vee a) \wedge b.$$

Observe that $f_{a,b} = \lambda_b \mu_a$ where $\lambda_b : x \mapsto x \vee b$ and $\mu_a : x \mapsto x \vee a$. Now, by the modularity of L and the fact that a, b are complementary we see that $[\lambda_b \mu_a \circ \mu_a \lambda_b](x) = [(x \wedge b) \vee a] \wedge b = x \wedge b \leqslant x$ and so $\lambda_b \mu_a \circ \mu_a \lambda_b \leqslant \operatorname{id}_L$. Similarly, $\mu_a \lambda_b \circ \lambda_b \mu_a \geqslant \operatorname{id}_L$. Thus we see that $f_{a,b} = \lambda_b \mu_a$ is residuated with $f_{a,b}^+ = \mu_a \lambda_b$. Again by modularity and the fact that a, b are complementary,

$$f_{a,b}[f_{a,b}^+(x) \wedge y] = \{\{[(x \wedge b) \vee a] \wedge y\} \vee a\} \wedge b$$
$$= [(x \wedge b) \vee a] \wedge (y \vee a) \wedge b$$
$$= x \wedge b \wedge (y \vee a)$$
$$= x \wedge f_{a,b}(y).$$

By Theorem 6.21, $f_{a,b}$ is totally range-closed. Combining this with a dual argument, we deduce that for all complementary pairs a, b the mapping $f_{a,b}$ is strongly range-closed. Since the identity $\{[(x \vee a) \wedge b] \vee a\} \wedge b = (x \vee a) \wedge b$ holds (in general) we also have that each $f_{a,b}$ is idempotent.

Observe now that if T is the semigroup formed by the strongly range-closed residuated mappings on L and if $g, h \in T$ then we have

$$gh = 0 \iff (\forall x \in L) \quad h(x) \leqslant g^+(0)$$
$$\iff (\forall x \in L) \quad h(x) = \big(h(x) \vee [g^+(0)]'\big) \wedge g^+(0) \quad \text{by modularity}$$
$$\iff h = f_{[g^+(0)]',g^+(0)}h$$
$$\iff h \in f_{[g^+(0)]',g^+(0)} \circ T$$

and so $R(g) = f_{[g^+(0)]',g^+(0)} \circ T$. Similarly, we see that $L(g) = T \circ f_{g(1),[g(1)]'}$. Consequently T is a Baer semigroup that coordinatises L, and we have (2).

(2) $\Rightarrow$ (3) and (3) $\Rightarrow$ (4) are clear.

(4) $\Rightarrow$ (1): Suppose that L can be coordinatised by a range-closed Baer semigroup S. If $eS, fS \in \mathcal{R}(S)$ and if e^* is an idempotent in S such that $Se^* = L(e)$ then, by the dual of Theorem 6.25, we have that e^* is dually range-closed. Since, by the hypothesis, e^*f is range-closed it follows by Theorem 6.24 that $M^*\big(\varphi_f(1S), \varphi_{e^*}^+(0S)\big)$ holds. But $\varphi_f(1S) = fS$ and $\varphi_{e^*}^+(0S) = R(e^*) = eS$ whence we conclude that $\mathcal{R}(S)$ is modular. Now since e^* is weakly regular it follows by the Corollary to Theorem 6.23 that $eS = \varphi_{e^*}^+(0S)$ has a complement in $\mathcal{R}(S)$. Consequently we have (1). □

For boolean lattices there exists a particularly nice coordinatisation.

Theorem 6.27 (Foulis [49]) *A bounded lattice is boolean if and only if it can be coordinatised by a commutative Baer semigroup.*

Proof $\Rightarrow$: Suppose that L is boolean, and consider the mappings $f_{a,b}$ as defined in the proof of Theorem 6.26. For every $a \in L$ we have $f_{a',a}(x) = (x \vee a') \wedge a = x \wedge a = \lambda_a(x)$. Consequently $\{\lambda_a \mid a \in L\}$ is a commutative Baer semigroup that coordinatises L.

$\Leftarrow$: Suppose that S is a commutative Baer semigroup that coordinatises L. Clearly, the conditions of Theorem 5.10(2) are satisfied. Consequently L is distributive. To see that it is also complemented, we use the fact that if $Se^* = L(e)$ and $Sf^* = L(f)$ then, since $\mathcal{R}(S)$ and $\mathcal{L}(S)$ are dually isomorphic, we have $eS \vee fS = R(Se^* \cap Sf^*)$. For this purpose, suppose that $Se^* = L(e)$ so that $eS = RL(e) = R(e^*)$ and therefore $eS \cap e^*S = 0S$. By the commutativity we then have $eS \vee e^*S = R(Se^* \cap Se) = R(0S) = 1S$. Hence e^*S is the complement of eS in $\mathcal{R}(S)$. Consequently $\mathcal{R}(S)$, hence L, is boolean. □

EXERCISE

6.27. Let (X, τ) be a topological space. Define a relation R on X to be **continuous** if it satisfies the property

$$A \text{ open } \Rightarrow R^d(A) \text{ open},$$

and let $CR(X)$ be the set of continuous relations on X. Show that, with respect to the law of composition defined in Example 2.25, $CR(X)$ is a semigroup whose zero element is $\emptyset$. With notation as in Example 2.25, prove that if A is open then I_A is continuous. Prove also that left annihilators in $CR(X)$ are given by

$$L(T) = CR(X) \cdot I_A \text{ where } A = (\operatorname{Im} T)^{-\prime}.$$

For B open define J_B by

$$(x, y) \in J_B \iff \begin{cases} \text{either} & x \notin B \text{ and } x = y; \\ \text{or} & x \in B \text{ and } y \notin B. \end{cases}$$

Prove that $J_B \in CR(X)$ and that right annihilators in $CR(X)$ are given by

$$R(T) = J_B \cdot CR(X) \text{ where } B = \operatorname{Dom} T.$$

Deduce that $CR(X)$ is a Baer semigroup that coordinatises the lattice of closed subsets of X.

6.9 Generalisations of boolean algebras

There are several lattice structures that arise through generalisations of boolean algebras. Since a boolean algebra is a complemented distributive lattice, significant generalisations can be achieved either by relaxing the distributivity and retaining complementation (which is then no longer unique), or by retaining the distributivity and relaxing the complementation.

As far as the first of these is concerned, an important class consists of lattices that are **orthomodular**. These can be described as follows. By an **involution lattice** we mean a bounded lattice L together with an antitone mapping $' : L \to L$ such that $x'' = x$ for every $x \in L$. An **ortholattice** is an involution lattice $(L; ')$ in which the involution is an **orthocomplementation** in the sense that $x \wedge x' = 0$ for every $x \in L$. If in addition the **orthomodular identity**

$$x \leqslant y \Rightarrow y = x \vee (y \wedge x')$$

holds then L is called an **orthomodular lattice**. Such lattices arise, for example, in connection with Hilbert spaces. If H is a Hilbert space then we say that $x, y \in H$ are **orthogonal**, and write $x \perp y$, if $\langle x|y \rangle = 0$. For every subspace X of H we let $X^\perp = \{y \in H \mid (\forall x \in X) \, y \perp x\}$, and say that X is **closed** if $X = X^{\perp\perp}$. Then the complete lattice of closed subspaces of a Hilbert space is orthomodular. Such lattices also arise in the underlying 'logic' of quantum physics. An excellent reference book that deals with such matters is that by Kalmbach [73]. In Blyth and Janowitz [24] there is developed a coordinatisation, due to Foulis [49], of orthomodular lattices in terms of the following type of semigroup.

An **involution semigroup** $(S; {}^\star)$ consists of a semigroup S together with a unary operation $x \mapsto x^\star$ such that $x^{\star\star} = x$ and $(xy)^\star = y^\star x^\star$. An idempotent e of an involution semigroup is called a **projection** if $e = e^\star$. The set of projections in S is denoted by $P(S)$.

A **Foulis semigroup** consists of a semigroup with 0 together with an involution $x \mapsto x^\star$ such that

$$(\forall x \in S)(\exists e_x \in P(S)) \ R(x) = e_x S.$$

It follows from this that $L(x) = S e_{x^\star}$, so we have a Baer semigroup (hence the original terminology of **Baer *-semigroup**). If $(L; {}')$ is an orthomodular lattice then Res L is a Foulis semigroup that coordinatises L. Here the involution on Res L is given by $f \mapsto f^\star$ where

$$(\forall x \in L) \quad f^\star(x) = [f^+(x')]'$$

and the idempotent generators of the right annihilators are the **Sasaki projections** φ_a given by

$$(\forall x \in L) \quad \varphi_a(x) = (x \vee a') \wedge a.$$

As to the second generalisation, a natural procedure is to consider a bounded distributive lattice with the complementation replaced by a dual endomorphism f, i.e. $f(0) = 1$, $f(1) = 0$ and the de Morgan type equalities $f(x \wedge y) = f(x) \vee f(y)$ and $f(x \vee y) = f(x) \wedge f(y)$ hold. The structure so obtained is called an **Ockham algebra**, a detailed account of which can be found in Blyth and Varlet [34] along with many references. An important aspect of this theory is that the Birkhoff duality for finite distributive lattices (Theorem 5.4) can be extended via **Priestley duality** to general bounded distributive lattices. An excellent account of this can be found in Davey and Priestley [42]. Basically, an ordered topological space $(X; \tau, \leqslant)$ is said to be **totally order-disconnected** if for $x, y \in X$ such that $x \not\leqslant y$ there exists a clopen down-set U such that $y \in U$ and $x \notin U$. A **Priestley space** is a compact totally order-disconnected space. If L is a bounded distributive lattice and $I_p(L)$ is the set of prime ideals of L then the **dual space** of L is $(I_p(L); \tau, \subseteq)$ where a sub-basis for the topology τ consist of the sets $\{X \in I_p(L) \mid a \in X\}$ and $\{X \in I_p(L) \mid a \notin X\}$ for every $a \in L$. Then $(I_p(L); \tau, \subseteq)$ is a Priestley space and $L \simeq \mathcal{O}(I_p(L))$ via $a \mapsto \{X \in I_p(L) \mid a \notin X\}$. Conversely, if P is a Priestley space then $\mathcal{O}(P)$ is a distributive lattice and $P \simeq (I_p(\mathcal{O}(P)); \tau, \subseteq)$.

This duality has far-reaching consequences. In particular, for Ockham algebras we define an **Ockham space** to be a Priestley space together with a continuous antitone mapping g. A dual equivalence between Ockham algebras and Ockham spaces was established by Urquhart [110] and is as follows. If $(X; g)$ is an Ockham space then $(\mathcal{O}(X); f)$ is an Ockham algebra where

$$(\forall A \in \mathcal{O}(X)) \quad f(A) = X \backslash g^{\leftarrow}(A).$$

Conversely if $(L; f)$ is an Ockham algebra then $(I_p(L); g)$ is an Ockham space where

$$(\forall X \in I_p(L)) \quad g(X) = \{a \in L \mid f(a) \notin X\}.$$

Pseudocomplementation; Stone and Heyting algebras

7.1 Pseudocomplements

We now consider a generalisation of the notion of a complement. If L is a lattice with bottom element 0 then we say that $x \in L$ is **pseudocomplemented** if there exists a biggest $x^* \in L$ that is disjoint from x, in the sense that $x \wedge x^* = 0$. Thus $x^* = \max\{y \in L \mid x \wedge y = 0\}$, and when it exists it is called the **pseudocomplement** of x. We say that L is **pseudocomplemented** if every element of L is pseudocomplemented.

Note that every pseudocomplemented lattice is necessarily bounded. In fact, since $0 \wedge y = 0$ for every $y \in L$ we have $y \leqslant 0^*$ for every $y \in L$ whence 0^* is the top element of L.

Example 7.1 Every finite distributive lattice is pseudocomplemented.

Example 7.2 Every complete lattice L in which the infinite distributive law

$$x \wedge \bigvee_{i \in I} y_i = \bigvee_{i \in I} (x \wedge y_i)$$

holds is pseudocomplemented. Clearly, for every $x \in L$ we have

$$x^* = \sup\{y \in L \mid x \wedge y = 0\}.$$

Example 7.3 Let B be a boolean algebra and define

$$B^{[2]} = \{(a, b) \in B \times B \mid a \leqslant b\}.$$

Then if $a \leqslant b$ and $x \leqslant y$ we have

$$(a, b) \wedge (x, y) = (0, 0) \iff a \wedge x = 0 = b \wedge y \iff x \leqslant y \leqslant b'.$$

It follows that $(a, b)^*$ exists and is (b', b'). Thus $B^{[2]}$ is pseudocomplemented.

Example 7.4 If L is a bounded distributive lattice then the lattice $I(L)$ of ideals of L is pseudocomplemented. In fact, for every ideal I of L let

$$I^* = \{x \in L \mid (\forall i \in I)\, x \wedge i = 0\}.$$

It is clear that I^* is an ideal of L and that $I \cap I^* = \{0\}$. Suppose now $J \in I(L)$ is such that $I \cap J = \{0\}$. If $j \in J$ then we have $i \wedge j = 0$ for every $i \in I$ and so $j \in I^*$. Thus $J \subseteq I^*$ and consequently I^* is the pseudocomplement of I.

Definition By a *p*-**algebra** we shall mean a pseudocomplemented lattice together with the unary operation of pseudocomplementation.

Our first task is to investigate the basic properties of the unary operation $x \mapsto x^*$.

Theorem 7.1 *If L is a p-algebra then*

(1) $x \mapsto x^*$ *is antitone;*
(2) $(\forall x \in L)\ x^* = x^{***}$;
(3) $x \mapsto x^{**}$ *is a closure mapping;*
(4) $(\forall x, y \in L)\ x \wedge (x \wedge y)^* = x \wedge y^*$;
(5) $(\forall x, y \in L)\ (x \wedge y)^{**} = x^{**} \wedge y^{**}$;
(6) $(\forall x, y \in L)\ (x \vee y)^* = x^* \wedge y^*$.

Proof (1) If $x \leqslant y$ then $y \wedge y^* = 0$ gives $x \wedge y^* = 0$ whence, by the definition of x^*, we have $y^* \leqslant x^*$.

(2) Since $x \wedge x^* = 0$ we have $x \leqslant x^{**}$ so it follows by (1) that $x^{***} \leqslant x^*$. But $x^* \wedge x^{**} = 0$ gives $x^* \leqslant x^{***}$. Hence we see that $x^* = x^{***}$.

(3) It is immediate from (1) that $x \mapsto x^{**}$ is isotone. As in (2), it is greater than or equal to the identity map. It is also idempotent since, by (2), we have $x^{**} = x^{****}$. Hence $x \mapsto x^{**}$ is a closure mapping on L.

(4) The equality $x \wedge y \wedge (x \wedge y)^* = 0$ gives $x \wedge (x \wedge y)^* \leqslant y^*$ and therefore $x \wedge (x \wedge y)^* \leqslant x \wedge y^*$. On the other hand, $x \wedge y \leqslant y$ together with (1) gives $(x \wedge y)^* \geqslant y^*$ whence the reverse inequality $x \wedge (x \wedge y)^* \geqslant x \wedge y^*$.

(5) From $x \wedge y \leqslant x, y$ we obtain $(x \wedge y)^{**} \leqslant x^{**} \wedge y^{**}$. As for the reverse inequality, from $x \wedge y \wedge (x \wedge y)^* = 0$ we have $x \wedge (x \wedge y)^* \leqslant y^* = y^{***}$ which gives $x \wedge y^{**} \wedge (x \wedge y)^* = 0$ and then $y^{**} \wedge (x \wedge y)^* \leqslant x^* = x^{***}$ which in turn gives $x^{**} \wedge y^{**} \wedge (x \wedge y)^* = 0$ and consequently $x^{**} \wedge y^{**} \leqslant (x \wedge y)^{**}$.

(6) By (1), (2), and (5) we have $(x \vee y)^* \leqslant x^* \wedge y^* = (x^* \wedge y^*)^{**}$. On the other hand, $(x^* \wedge y^*)^* \geqslant x^{**} \geqslant x$ and similarly $(x^* \wedge y^*)^* \geqslant y$ so that $(x^* \wedge y^*)^* \geqslant x \vee y$ and consequently $(x^* \wedge y^*)^{**} \leqslant (x \vee y)^*$. We thus see that $(x \vee y)^* = (x^* \wedge y^*)^{**} = x^{***} \wedge y^{***} = x^* \wedge y^*$. $\square$

Corollary *A bounded lattice L is a p-algebra if and only if there is a unary operation $x \mapsto x^*$ on L such that the following identities hold*:

(1) $x \wedge (x \wedge y)^* = x \wedge y^*$;
(2) $x \wedge 0^* = x$;
(3) $0^{**} = 0$.

Proof $\Rightarrow$: If $(L; ^*)$ is a p-algebra then (1) holds by Theorem 7.1(4). Also the definition of pseudocomplement gives $0^* = 1$ and $1^* = 0$ whence (2) and (3) are immediate.

$\Leftarrow$: Suppose conversely that (1), (2), (3) hold. Then on the one hand we have

$$x \wedge x^* = x \wedge (x \wedge 0^*)^* = x \wedge 0^{**} = x \wedge 0 = 0;$$

and on the other, if $x \wedge y = 0$ then

$$y \wedge x^* = y \wedge (y \wedge x)^* = y \wedge 0^* = y$$

and so $y \leqslant x^*$. Thus $x \mapsto x^*$ is the pseudocomplementation. $\square$

If L is a p-algebra then the set $S(L) = \{x^{\star\star} \mid x \in L\}$, which by Theorem 7.1(2) is the same as the set $\{x^\star \mid x \in L\}$, is the image of the closure map $x \mapsto x^{\star\star}$. It follows by Theorem 2.7 that $S(L)$ forms a lattice which is a $\wedge$-subsemilattice of L and in which, taking Theorem 7.1(6) into account, suprema are given by $x \sqcup y = (x \vee y)^{\star\star} = (x^\star \wedge y^\star)^\star$. The lattice $S(L)$ is called the **skeleton** of L. It has the following important property.

Theorem 7.2 (Glivenko [57]) $\big(S(L); \wedge, \sqcup, {}^\star, 0, 1\big)$ *is a boolean algebra.*

Proof To show that $S(L)$ is distributive, it suffices to prove that
$$(\forall x, y, z \in S(L)) \qquad z \wedge (x \sqcup y) \leqslant (z \wedge x) \sqcup (z \wedge y).$$
For this purpose, let $t = (z \wedge x) \sqcup (z \wedge y)$. Then $z \wedge x \leqslant t = t^{\star\star}$ gives $z \wedge x \wedge t^\star = 0$ and so $z \wedge t^\star \leqslant x^\star$. Similarly, $z \wedge t^\star \leqslant y^\star$ and therefore $z \wedge t^\star \leqslant x^\star \wedge y^\star = (x^\star \wedge y^\star)^{\star\star}$. It follows from this that $z \wedge t^\star \wedge (x^\star \wedge y^\star)^\star = 0$ and hence that $z \wedge (x \sqcup y) = z \wedge (x^\star \wedge y^\star)^\star \leqslant t^{\star\star} = t = (z \wedge x) \sqcup (z \wedge y)$.

To see that $S(L)$ is also complemented, observe that $1 = 0^\star \in S(L)$ and $0 = 1^\star \in S(L)$. Since for every $x \in S(L)$ we have $x \wedge x^\star = 0$ and $x \sqcup x^\star = (x^\star \wedge x^{\star\star})^\star = 0^\star = 1$, we see that a, hence the, complement of $x \in S(L)$ is $x^\star$. Hence $S(L)$ is boolean. $\qquad\square$

Definition If L is a p-algebra then $d \in L$ is said to be **dense** if $d^\star = 0$ or, equivalently, $d^{\star\star} = 1$.

Example 7.5 It follows from Theorem 7.1(6) that in a p-algebra L every element of the form $x \vee x^\star$ is dense.

The set of dense elements of L will be denoted by $D(L)$.

Theorem 7.3 $D(L)$ *is a filter of* L.

Proof Suppose that $d \in D(L)$ and that $x \geqslant d$. Then $x^\star \leqslant d^\star = 0$ gives $x^\star = 0$ and so $x \in D(L)$. Thus $D(L)$ is an up-set of L. Now if $x, y \in D(L)$ then by Theorem 7.1(5) we have $(x \wedge y)^{\star\star} = x^{\star\star} \wedge y^{\star\star} = 1 \wedge 1 = 1$ and so $x \wedge y \in D(L)$. Hence $D(L)$ is a filter of L. $\qquad\square$

EXERCISES

7.1. Let L be a lattice with a top element 1. Extend L to a lattice L^0 by adjoining a (new) bottom element. Prove that L^0 is pseudocomplemented and describe its skeleton and its dense filter.

7.2. Given a boolean algebra B construct a (non-boolean) pseudocomplemented lattice whose skeleton is isomorphic to B.

7.3. Let p, q be distinct primes. Consider the lattice consisting of the positive divisors of $p^m q^n$ under divisibility. Identify pseudocomplements in this lattice and hence describe its skeleton and its dense filter.

7.4. If L is a distributive p-algebra prove that $S(L)$ is the set of complemented elements in L.

7.5. Prove that in a distributive p-algebra L every $x \in L$ can be expressed as $x = a \wedge b$ where $a \in S(L)$ and $b \in D(L)$.

7.6. Let L be a distributive p-algebra. Prove that L is boolean if and only if $D(L) = \{1\}$.

7.7. For a boolean lattice B and $n \geqslant 2$, let
$$B^{[n]} = \{(x, y_1, \ldots, y_{n-1}) \mid (\forall i)\ x \leqslant y_i\}.$$
Show that $B^{[n]}$ is pseudocomplemented. Identify its skeleton and dense filter.

7.8. If L is a bounded lattice then $f : L \to L$ is said to be a **quantifier** if
   (a) $f(0) = 0$;
   (b) $\mathrm{id}_L \leqslant f$;
   (c) $(\forall x, y \in L)\ f[x \wedge f(y)] = f(x) \wedge f(y)$.
   If f is a quantifier on L prove that
   (1) $f(1) = 1$;
   (2) f is isotone;
   (3) f is idempotent;
   (4) $(\forall x, y \in L)\ f[x \vee f(y)] = f(x \vee y)$.
   If, moreover, L is pseudocomplemented prove that
   (5) $(\forall x \in L)\ [f(x)]^* = f([f(x)]^*)$.
   Give an example to show that $\mathrm{Im}\, f$ is not in general a sublattice of L.

7.9. Let L be a pseudocomplemented distributive lattice and let P be a prime ideal of L. Prove that the following conditions are equivalent:

   (1) P is minimal;
   (2) $x \in P \Rightarrow x^* \notin P$;
   (3) $x \in P \Rightarrow x^{**} \in P$;
   (4) $P \cap D(L) = \emptyset$.

7.10. Let L be the distributive lattice obtained by adjoining a new top element to the lattice $\mathbf{4} \times \mathbf{4}$. Show that L is pseudocomplemented and that the identity
$$(x \wedge y)^* \vee (x^* \wedge y)^* \vee (x \wedge y^*)^* = 1$$
holds in L.

7.2 Stone algebras

In general, in a p-algebra L the boolean algebra $S(L)$ is not a sublattice of L. For example, the distributive lattice

is pseudocomplemented with $b^* = c$, $c^* = b$, $a^* = 1^* = 0$, and $0^* = 1$. Its skeleton is then $\{0, b, c, 1\}$ which is not a sublattice.

Precisely when $S(L)$ is a sublattice is the substance of the following result.

Theorem 7.4 *Let L be a pseudocomplemented distributive lattice. Then the following statements are equivalent*:

(1) $S(L)$ *is a sublattice of L*;
(2) $(\forall x, y \in L)$ $(x \vee y)^{\star\star} = x^{\star\star} \vee y^{\star\star}$;
(3) $(\forall x, y \in L)$ $(x \wedge y)^{\star} = x^{\star} \vee y^{\star}$;
(4) $(\forall x \in L)$ $x^{\star} \vee x^{\star\star} = 1$.

Proof (1) $\Rightarrow$ (4): If (1) holds then, for every $x \in L$, $x^{\star} \vee x^{\star\star} = x^{\star} \sqcup x^{\star\star} = (x^{\star\star} \wedge x^{\star})^{\star} = 0^{\star} = 1$.

(4) $\Rightarrow$ (3): By the distributivity we have $x \wedge y \wedge (x^{\star} \vee y^{\star}) = 0$. If now $x \wedge y \wedge a = 0$ then $y \wedge a \leqslant x^{\star}$ and so $x^{\star\star} \wedge y \wedge a = 0$ and hence $x^{\star\star} \wedge a \leqslant y^{\star}$. It follows by (4) that $a = a \wedge 1 = a \wedge (x^{\star} \vee x^{\star\star}) = (a \wedge x^{\star}) \vee (a \wedge x^{\star\star}) \leqslant x^{\star} \vee y^{\star}$. Consequently the pseudocomplement of $x \wedge y$ is $x^{\star} \vee y^{\star}$.

(3) $\Rightarrow$ (2): By Theorem 7.1(6) and (3) we have $(x \vee y)^{\star\star} = (x^{\star} \wedge y^{\star})^{\star} = x^{\star\star} \vee y^{\star\star}$.

(2) $\Rightarrow$ (1): If (2) holds then for all $x, y \in S(L)$ we have $x \vee y = x^{\star\star} \vee y^{\star\star} = (x \vee y)^{\star\star} \in S(L)$ and consequently $S(L)$ is a sublattice. $\square$

Definition A pseudocomplemented distributive lattice that satisfies any of the equivalent properties of Theorem 7.4 is said to be a **Stone lattice**. A **Stone algebra** is a Stone lattice together with the unary operation of pseudo-complementation.

Example 7.6 Every bounded chain is a Stone lattice. In fact, it is clearly distributive and for every $x \neq 0$ we have $x^{\star\star} = 1$. Thus the skeleton is the sublattice $\{0, 1\}$.

Example 7.7 If $L_1, \ldots, L_n$ are Stone lattices then so is the cartesian product lattice $\underset{i=1}{\overset{n}{\times}} L_i$. This follows from the observation that in a cartesian product pseudocomplements are obtained componentwise, so that $(x_1, \ldots, x_n)^{\star} = (x_1^{\star}, \ldots, x_n^{\star})$. Note that in this case we have $S\left(\underset{i=1}{\overset{n}{\times}} L_i\right) = \underset{i=1}{\overset{n}{\times}} S(L_i)$.

Example 7.8 The lattice of positive divisors of a positive integer n is a Stone lattice. In fact, if n is prime then the lattice is a 2-element chain; if $n = p^{\alpha}$ where p is prime then the lattice is an $(\alpha + 1)$-chain; and if $n = \prod_{i=1}^{k} p_i^{\alpha_i}$ where $p_1, \ldots, p_k$ are distinct primes then the lattice is a cartesian product of k chains. The result then follows from the two preceding Examples.

Observe that if L is a Stone lattice then the equivalence relation associated with the closure $x \mapsto x^{\star\star}$, namely that given by

$$(x, y) \in G \iff x^{\star\star} = y^{\star\star} \iff x^{\star} = y^{\star},$$

is a congruence. This follows immediately from Theorem 7.1(6) and Theorem 7.4(3). The congruence G so defined is known as the **Glivenko congruence** on L.

EXERCISES

7.11. For the Stone lattice $\mathbf{2} \times \mathbf{3} \times \mathbf{3}$ determine the classes modulo the Glivenko congruence.

7.12. Let L be a Stone lattice and for every $a \in S(L)$ let $[a]_G = \{x \in L \mid x^{**} = a\}$ be the class of a relative to the Glivenko congruence. Prove that if $a, b \in S(L)$ are such that $a \leqslant b$ then the assignment $x \mapsto (x \vee a^{*}) \wedge b$ is an injective lattice morphism from $[a]_G$ to $[b]_G$.

7.13. A **double Stone algebra** is a Stone algebra $(L; ^{*})$ whose dual is also a Stone algebra. We denote the dual pseudocomplement of $x \in L$ by x^{+}. Prove that if L is a double Stone algebra then
   (1) $S(L) = \{x^{++} \mid x \in L\} = \{x^{+} \mid x \in L\}$;
   (2) $(\forall x \in L)\ x^{*} \leqslant x^{+}$;
   (3) $(\forall x \in L)\ x^{+*} = x^{++} \leqslant x \leqslant x^{**} = x^{*+}$.

7.14. If B is a boolean algebra prove that $B^{[2]} = \{(x, y) \in B^2 \mid x \leqslant y\}$ is a double Stone algebra in which $(x, y)^{*} = (y', y')$ and $(x, y)^{+} = (x', x')$.

7.15. Prove that a Stone algebra $(L; ^{*})$ can be made into a double Stone algebra if and only if the closure $x \mapsto x^{**}$ is residuated and its residual preserves suprema.

We shall now give a complete description of all finite Stone lattices. For this purpose we require the following result.

Theorem 7.5 *Let L be a bounded distributive lattice. If $z \in L$ is complemented then the mapping $f_z : L \to z^{\uparrow} \times z^{\downarrow}$ given by $f_z(x) = (z \vee x, z \wedge x)$ is a lattice isomorphism.*

Proof It is clear from the distributivity that f_z is a lattice morphism. If $f_z(x) = f_z(y)$ then x and y are relative complements of z in the same interval of L so, by Theorem 6.2, $x = y$ and hence f_z is injective. If now z' is the complement of z in L then given any $(x, y) \in z^{\uparrow} \times z^{\downarrow}$ it is readily seen that $f_z[(y \vee z') \wedge x] = (x, y)$, so that f_z is also surjective. Thus f_z is a lattice isomorphism.   □

Theorem 7.6 *Let L be a finite distributive lattice. Then L is a Stone lattice if and only if L is a (finite) cartesian product of (finite) distributive lattices each of which has a single atom.*

Proof $\Rightarrow$: If $S(L)$ reduces to $\{0, 1\}$ then since $x^{**} = 0$ implies that $x = 0$ we must have $x^{**} = 1$ for every $x \neq 0$, and therefore $D(L) = L \backslash \{0\}$. Since $D(L)$ is a filter it follows that L has a single atom which is the bottom element of $D(L)$.

If now $S(L) \neq \{0, 1\}$ then by Theorem 7.4 there exists $x \in L$ such that $x \notin \{0, 1\}$ and x^{*} is complemented. By Theorem 7.5 we then have $L \simeq x^{*\uparrow} \times x^{*\downarrow}$. Now since L is finite each of $x^{*\uparrow}$ and $x^{*\downarrow}$ is a pseudocomplemented distributive lattice and we may repeat the above observations; either the skeleton reduces to the two-element boolean algebra or there is a proper complemented element in the skeleton, in which case there is a further cartesian product decomposition. Continuing in this way we eventually arrive at a decomposition $L \simeq \bigtimes_{i=1}^{n} L_i$ in which each L_i is such that $S(L_i)$ is the two-element boolean algebra. Since pseudocomplements in a cartesian product arise componentwise, it follows that every L_i is a Stone lattice and has a single atom.

$\Leftarrow$: Clearly, if $L \simeq \bigtimes_{i=1}^{n} L_i$ where each L_i is a finite distributive lattice having a single atom then each L_i is a Stone lattice and so then is L. $\square$

The structure of an arbitrary Stone lattice is more complicated. In order to describe this, we require the following fact concerning the lattice $F(L)$ of filters of a distributive lattice L.

Theorem 7.7 *Let L be a distributive lattice. If F_1 and F_2 are filters of L such that the filters $F_1 \cap F_2$ and $F_1 \vee F_2$ are principal then F_1 and F_2 are principal.*

Proof Let $F_1 \cap F_2 = x^{\uparrow}$ and $F_1 \vee F_2 = y^{\uparrow}$. There exists $i \in F_1$ and $j \in F_2$ such that $y = i \wedge j$. Let $c = x \wedge i \in F_1$ and $b = x \wedge j \in F_2$. Then $c^{\uparrow} \subseteq F_1$ and $b^{\uparrow} \subseteq F_2$. Suppose, by way of obtaining a contradiction, that $c^{\uparrow} \neq F_1$. Then there exists $a \in F_1$ with $a < c$. Now $c \wedge b = x \wedge i \wedge j = x \wedge y = y \leqslant a \wedge b$ whence $a \wedge b = c \wedge b = y$. Moreover, $c \vee b = x \wedge (i \vee j) = x$; and $a \vee b \in F_1 \cap F_2 = x^{\uparrow}$ gives $a \vee b = x$ since $a < c < x$ and $b < x$. This gives $\{y, a, b, c, x\}$ as a forbidden five-element non-modular lattice. Thus $F_1 = c^{\uparrow}$, and similarly $F_2 = b^{\uparrow}$. $\square$

A general Stone lattice was first characterised by Chen and Grätzer [38] in terms of the boolean algebra $S(L)$, the dense filter $D(L)$, and a $(0,1)$-lattice morphism $\varphi_L : S(L) \to F(D(L))$. The following structure theorem is a variant. Here, and in later results, we shall use the following alternative terminology for a $\wedge$-morphism.

Definition If L is a lattice then a mapping $f : L \to L$ is said to be **multiplicative** if $f(x \wedge y) = f(x) \wedge f(y)$ for all $x, y \in L$.

Example 7.9 If L is a pseudocomplemented lattice then by Theorem 7.1(5) the closure $x \mapsto x^{\star\star}$ on L is multiplicative.

Theorem 7.8 (Blyth–Varlet [33]) *Let B be boolean algebra and let D be a distributive lattice with a top element 1_D. Let $x \mapsto \overline{x}$ be a multiplicative closure on D and let $\varphi : B \to F(D)$ be a $(0,1)$-lattice morphism. Then the subset of the cartesian product lattice $B \times F(D)^d$ given by*

$$B \times_{\varphi} F(D)^d = \{(a, \varphi(a') \vee x^{\uparrow}) \mid \overline{x} \in \varphi(a')\}$$

is a Stone algebra in which $(a, \varphi(a') \vee x^{\uparrow})^{\star} = (a', \varphi(a))$.

Moreover, every Stone algebra can be obtained in this manner. More precisely, if L is a Stone algebra then the mapping $\varphi_L : S(L) \to F(D(L))$ given by $\varphi_L(x) = x^{\star\uparrow} \cap D(L)$ is a $(0,1)$-lattice morphism, $x \mapsto x^{\star\star}$ is a multiplicative closure, and $L \simeq S(L) \times_{\varphi_L} F(D(L))^d$.

Proof We observe first that

$$(a, \varphi(a') \vee x^\uparrow) \wedge (b, \varphi(b') \vee y^\uparrow) = (a \wedge b, \varphi(a') \vee x^\uparrow \vee \varphi(b') \vee y^\uparrow)$$
$$= (a \wedge b, \varphi[(a \wedge b)'] \vee (x \wedge y)^\uparrow),$$

and that since $\overline{x} \in \varphi(a')$ and $\overline{y} \in \varphi(b')$ we have $\overline{x \wedge y} = \overline{x} \wedge \overline{y} \in \varphi(a') \vee \varphi(b') = \varphi[(a \wedge b)']$. Consequently $B \times_\varphi F(D)^d$ is a $\wedge$-subsemilattice of $B \times F(D)^d$.

To show that it is also a $\vee$-subsemilattice, we next observe that each of the filters $\varphi(a) \cap x^\uparrow$ of D is principal. In fact, let $P = \varphi(a) \cap x^\uparrow$ and consider the filter $Q = \varphi(a') \cap x^\uparrow$. We have $P \cap Q = \{1_D\} \cap x^\uparrow = \{1_D\}$ and, since $F(D)$ is distributive, $P \vee Q = D \cap x^\uparrow = x^\uparrow$. It then follows by Theorem 7.7 that P is principal. Now

$$(a, \varphi(a') \vee x^\uparrow) \vee (b, \varphi(b') \vee y^\uparrow) = (a \vee b, [\varphi(a') \vee x^\uparrow] \cap [\varphi(b') \vee y^\uparrow])$$
$$= (a \vee b, \varphi[(a \vee b)'] \vee t^\uparrow)$$

where $t^\uparrow = (\varphi(a') \cap y^\uparrow) \vee (\varphi(b') \cap x^\uparrow) \vee (x \vee y)^\uparrow$. By the above observation, we can write $t^\uparrow = p^\uparrow \vee q^\uparrow \vee (x \vee y)^\uparrow$, so that $t = p \wedge q \wedge (x \vee y)$ and consequently, since $\overline{p} \geqslant \overline{y}$ and $\overline{q} \geqslant \overline{x}$, we have $\overline{t} = \overline{p} \wedge \overline{q} \wedge \overline{x \vee y} \geqslant (\overline{p} \wedge \overline{x}) \vee (\overline{q} \wedge \overline{y})$. Since $\overline{p} \geqslant p \in \varphi(a')$ and by definition $\overline{x} \in \varphi(a')$ we have $\overline{p} \wedge \overline{x} \in \varphi(a')$, and similarly $\overline{q} \wedge \overline{y} \in \varphi(b')$. Consequently $\overline{t} \in \varphi(a') \cap \varphi(b') = \varphi[(a \vee b)']$ as required.

The bottom element of $B \times_\varphi F(D)^d$ being $(0_B, D)$, we have

$$(a, \varphi(a') \vee x^\uparrow) \wedge (b, \varphi(b') \vee y^\uparrow) = (0_B, D) \iff a \wedge b = 0_B,$$

whence $B \times_\varphi F(D)^d$ is pseudocomplemented with $(a, \varphi(a') \vee x^\uparrow)^* = (a', \varphi(a))$. Since then $(a, \varphi(a') \vee x^\uparrow)^{**} = (a, \varphi(a'))$ it follows by Theorem 7.4(4) that $B \times_\varphi F(D)^d$ is a Stone lattice.

To show that every Stone algebra arises in this manner, let L be a Stone algebra. Observe by Example 7.9 that the closure $x \mapsto x^{**}$ is multiplicative on L and reduces to the constant mapping $x \mapsto 1$ on $D(L)$. Consider the mapping $\varphi_L : S(L) \to F(D(L))$ given by

$$(\forall x \in S(L)) \quad \varphi_L(x) = x^{*\uparrow} \cap D(L).$$

By Theorem 7.1(6) we have

$$\varphi_L(a) \vee \varphi_L(b) = (a^{*\uparrow} \vee b^{*\uparrow}) \cap D(L) = (a^* \wedge b^*)^\uparrow \cap D(L)$$
$$= (a \vee b)^{*\uparrow} \cap D(L)$$
$$= \varphi_L(a \vee b);$$

and, by Theorem 7.4(3),

$$\varphi_L(a) \cap \varphi_L(b) = a^{*\uparrow} \cap b^{*\uparrow} \cap D(L) = (a^* \vee b^*)^\uparrow \cap D(L)$$
$$= (a \wedge b)^{*\uparrow} \cap D(L)$$
$$= \varphi_L(a \wedge b).$$

Since also $\varphi_L(0) = \{1\}$ and $\varphi_L(1) = D(L)$ it follows that φ_L is a $(0, 1)$-lattice morphism.

Now, by the distributivity, we have $x = x^{**} \wedge (x \vee x^*)$. It follows that

$$(\forall x \in L) \quad x^\uparrow = x^{**\uparrow} \vee (x^\uparrow \cap D(L)).$$

In fact, if $t \in x^\uparrow$ then $t = t \vee x = (t \vee x^{**}) \wedge (t \vee x \vee x^*)$ where $x \vee x^* \in D(L)$; and conversely if $t \in x^{**\uparrow} \vee (x^\uparrow \cap D(L))$ then $t \geqslant y \wedge z$ where $y \geqslant x^{**} \geqslant x$ and $z \geqslant x$ and thus $t \geqslant x$.

Consider now the mapping $\vartheta : L \to S(L) \times_{\varphi_L} F(D(L))^d$ given by

$$(\forall x \in L) \quad \vartheta(x) = (x^{**}, x^{\uparrow} \cap D(L)).$$

Clearly, if $x \leqslant y$ then $x^{**} \leqslant y^{**}$ and $y^{\uparrow} \subseteq x^{\uparrow}$ whence it follows that $\vartheta(x) \leqslant \vartheta(y)$. Conversely, if $\vartheta(x) \leqslant \vartheta(y)$ then by the above identity we have $x^{\uparrow} = x^{**\uparrow} \vee (x^{\uparrow} \cap D(L)) \supseteq y^{**\uparrow} \vee (y^{\uparrow} \cap D(L)) = y^{\uparrow}$ and consequently $x \leqslant y$. To see that ϑ is a lattice isomorphism it suffices to show that it is surjective. For this purpose, let $z \in S(L) \times_{\varphi_L} F(D(L))^d$, so that

$$z = (x^{**}, \varphi_L(x^*) \vee y^{\uparrow}) = (x^{**}, (x^{**} \wedge y)^{\uparrow} \cap D(L)).$$

Let $h = x^{**} \wedge y$. Then, since $x \mapsto x^{**}$ is multiplicative and by definition $y^{**} = 1$, we have $h^{**} = x^{**}$. Consequently $z = (h^{**}, h^{\uparrow} \cap D(L)) = \vartheta(h)$ and so ϑ is surjective. Finally, we have

$$\begin{aligned}
x^{\uparrow} \cap D(L) &= (x^{**\uparrow} \vee (x \vee x^*)^{\uparrow}) \cap D(L) \\
&= (x^{**\uparrow} \cap D(L)) \vee (x \vee x^*)^{\uparrow} \\
&= \varphi_L(x^*) \vee (x \vee x^*)^{\uparrow},
\end{aligned}$$

whence it follows that

$$\begin{aligned}
[\vartheta(x)]^* &= \left(x^{**}, x^{\uparrow} \cap D(L)\right)^* \\
&= \left(x^{**}, \varphi_L(x^*) \vee (x \vee x^*)^{\uparrow}\right)^* \\
&= (x^*, \varphi_L(x^{**})) \\
&= (x^*, x^{*\uparrow} \cap D(L)) \\
&= \vartheta(x^*),
\end{aligned}$$

and so ϑ is an isomorphism of Stone algebras. □

EXERCISES

7.16. Let L be a complete Stone algebra and for every ideal I of L let I^* be the pseudocomplement of I (Example 7.4). Prove that $I^* = (\sup_L I)^{*\downarrow}$ and hence show that the lattice $I(L)$ of ideals of L is also a complete Stone algebra.

7.17. Prove that a Stone algebra is complete if and only if $S(L)$ and $D(L)$ are complete.

7.18. Prove that Stone algebras L and M are isomorphic if and only if there are isomorphisms $\alpha : S(L) \to S(M)$ and $\beta : F(D(L)) \to F(D(M))$ such that the diagram

$$\begin{array}{ccc}
S(L) & \xrightarrow{\varphi_L} & F(D(L)) \\
\downarrow{\alpha} & & \downarrow{\beta} \\
S(M) & \xrightarrow{\varphi_M} & F(D(M))
\end{array}$$

is commutative.

7.3 Heyting algebras

In a pseudocomplemented distributive lattice the interval $[0, x]$ is pseudocomplemented, the pseudocomplement of $y \in [0, x]$ being $y^* \wedge x$. This prompts consideration of distributive lattices in which every interval is pseudocomplemented. Such a structure is called a **Heyting lattice**.

Example 7.10 Every boolean lattice is a Heyting lattice. In fact, if B is boolean then for every $x \in [a, b]$ the relative complement $x^\dagger = b \wedge (a \vee x') = a \vee (b \wedge x')$ is the pseudocomplement of x in $[a, b]$.

Heyting lattices can be characterised as follows.

Theorem 7.9 If L is a bounded lattice then the following conditions are equivalent:

(1) L is a Heyting lattice;
(2) for every $x \in L$ the translation $\lambda_x : y \mapsto x \wedge y$ is residuated.

Proof (1) $\Rightarrow$ (2): Suppose that (1) holds and let $x_a^\star$ denote the pseudocomplement of x in $a^\uparrow = [a, 1]$. Then for every $x \in a^\uparrow$ we have

$$x_a^\star = \max\{y \in a^\uparrow \mid x \wedge y = a\}.$$

In particular,

$$a_{a \wedge b}^\star = \max\{y \in (a \wedge b)^\uparrow \mid a \wedge y = a \wedge b\}.$$

Thus clearly $a \wedge a_{a \wedge b}^\star = a \wedge b \leqslant b$. Now suppose that $a \wedge x \leqslant b$. Then $x \vee (a \wedge b) \in (a \wedge b)^\uparrow$ and $a \wedge [x \vee (a \wedge b)] = (a \wedge x) \vee (a \wedge b) \leqslant b \vee (a \wedge b) = b$, so that $x \leqslant x \vee (a \wedge b) \leqslant a_{a \wedge b}^\star$. Thus λ_a is residuated with $\lambda_a^+(b) = a_{a \wedge b}^\star$.

(2) $\Rightarrow$ (1): Since, by Theorem 2.8, residuated mappings preserve suprema it is immediate from (2) that L is distributive. Suppose now that $c \in [a, b]$. Since $c \wedge a = a$ we have $a \leqslant \lambda_c^+(a)$ so $a \leqslant \lambda_c^+(a) \wedge b$ with $c \wedge \lambda_c^+(a) = a$. Consequently $\lambda_c^+(a) \wedge b \in [a, b]$ is such that $c \wedge \lambda_c^+(a) \wedge b = a \wedge b = a$. If now $x \in [a, b]$ is such that $c \wedge x = a$ then $x \leqslant \lambda_c^+(a)$ and $x \leqslant b$, so that $x \leqslant \lambda_c^+(a) \wedge b$. This shows that the pseudocomplement of c exists in $[a, b]$ and is $\lambda_c^+(a) \wedge b$. $\square$

Because of Theorem 7.9, a Heyting lattice is also known as a **residuated lattice** or a **relatively pseudocomplemented lattice**. Another common terminology is an **implicative lattice**.

Theorem 7.10 A complete lattice is a Heyting lattice if and only if it satisfies the infinite distributive law $x \wedge \bigvee_{i \in I} y_i = \bigvee_{i \in I} (x \wedge y_i)$.

Proof $\Rightarrow$: It suffices to prove that $x \wedge \bigvee_{i \in I} y_i \leqslant \bigvee_{i \in I} (x \wedge y_i)$. For this purpose, let $z = \bigvee_{I \in I} (x \wedge y_i)$. Then for every $i \in I$ we have $x \wedge y_i \leqslant z$ whence $y_i \leqslant \lambda_x^+(z)$ for every $i \in I$, and therefore $\bigvee_{I \in I} y_i \leqslant \lambda_x^+(z)$ which gives $x \wedge \bigvee_{i \in I} y_i \leqslant z$ as required.

$\Leftarrow$: Given $x, z \in L$ let $X = \{y_i \mid i \in I\}$ be the set of elements $y_i \in L$ such that $x \wedge y_i \leqslant z$. If $y = \sup_L X$ then $x \wedge y = x \wedge \bigvee_{i \in I} y_i = \bigvee_{i \in I} (x \wedge y_i) \leqslant z$ and consequently $\lambda_x^+(z)$ exists and is y. $\square$

Corollary *Every finite distributive lattice is a Heyting lattice.*   □

As we have seen above, in a Heyting lattice every translation $\lambda_x : y \mapsto x \wedge y$ is residuated. For notational convenience in what follows we shall write $\lambda_x^+(y)$ as $y:x$, and call this the **residual of** y **by** x. Thus we have

$$y:x = \max\{z \in L \mid x \wedge z \leqslant y\},$$

and in this notation the pseudocomplement $x^\star$ is $0:x$.

Definition A **Heyting algebra** is a Heyting lattice together with the binary operation $(x, y) \mapsto x:y$ of residuation.

The principal properties of residuation in a Heyting algebra are the following.

Theorem 7.11 *Let L be a Heyting algebra and let $x, y, z \in L$. Then*

(1) *if $x \leqslant y$ then $x:z \leqslant y:z$ and $z:y \leqslant z:x$;*
(2) $y \leqslant y:x$;
(3) $x \wedge (y:x) = x \wedge y$;
(4) $(z:y):x = z:(y \wedge x)$;
(5) $(y \wedge z):x = (y:x) \wedge (z:x)$;
(6) $z:(x \vee y) = (z:x) \wedge (z:y)$;
(7) $x \wedge (y:z) = x \wedge [(x \wedge y):(x \wedge z)]$.

Proof (1) If $x \leqslant y$ then from $z \wedge (x:z) \leqslant x \leqslant y$ we deduce that $x:z \leqslant y:z$; and from $x \wedge (z:y) \leqslant y \wedge (z:y) \leqslant z$ we deduce that $z:y \leqslant z:x$.

(2) This follows immediately from $x \wedge y \leqslant y$.

(3) Clearly, $x \wedge (y:x) \leqslant x \wedge y$, and the reverse inequality follows by (2).

(4) Clearly, we have $x \wedge y \wedge [(z:y):x] \leqslant y \wedge (z:y) \leqslant z$ and therefore $(z:y):x \leqslant z:(x \wedge y)$. On the other hand, since $x \wedge y \wedge [z:(x \wedge y)] \leqslant z$ we have $x \wedge [z:(x \wedge y)] \leqslant z:y$ and therefore $z:(x \wedge y) \leqslant (z:y):x$.

(5) By (1) we have $(y \wedge z):x \leqslant (y:x) \wedge (z:x)$. On the other hand, by (3), we have $x \wedge (y:x) \wedge (z:x) = x \wedge y \wedge z \leqslant y \wedge z$ and therefore $(y:x) \wedge (z:x) \leqslant (y \wedge z):x$.

(6) By (1) we have $z:(x \vee y) \leqslant (z:x) \wedge (z:y)$. On the other hand, by (3), $x \wedge (z:x) \wedge (z:y) \leqslant z \wedge (z:y) = z$ and likewise $y \wedge (z:x) \wedge (z:y) \leqslant z$. Since L is distributive it follows that $(x \vee y) \wedge (z:x) \wedge (z:y) \leqslant z$ and consequently $(z:x) \wedge (z:y) \leqslant z:(x \vee y)$.

(7) Since $x \wedge z \wedge x \wedge (y:z) \leqslant x \wedge y$ we have $x \wedge (y:z) \leqslant (x \wedge y):(x \wedge z)$ whence $x \wedge (y:z) \leqslant x \wedge [(x \wedge y):(x \wedge z)]$. Conversely, $x \wedge z \wedge [(x \wedge y):(x \wedge z)] \leqslant x \wedge y \leqslant y$ gives $x \wedge [(x \wedge y):(x \wedge z)] \leqslant y:z$ whence we obtain the reverse inequality.   □

Corollary *A bounded lattice L is a Heyting algebra if and only if there is a binary operation $(x, y) \mapsto x:y$ on L such that the following identities hold:*

(1) $x \wedge (y:x) = x \wedge y$;
(2) $x \wedge (y:z) = x \wedge [(x \wedge y):(x \wedge z)]$;
(3) $z \wedge (x:(x \wedge y)) = z$.

Proof ⇒: If L is a Heyting algebra then (1) and (2) follow from the above whereas (3) is immediate from the fact that $x:(x \wedge y) = 1$.

⇐: Conversely, if the identites hold then $x \wedge (y:x) = x \wedge y \leqslant y$; and if $x \wedge z \leqslant y$ then

$$z \wedge (y:x) = z \wedge [(z \wedge y):(z \wedge x)] = z \wedge [(y \wedge z):(x \wedge y \wedge z)] = z$$

and so $z \leqslant y:x$. Thus $(L;:)$ is a Heyting algebra. □

Example 7.11 If L is a distributive lattice then the ideal lattice $I(L)$ is a complete Heyting algebra in which residuals are given by

$$I:J = \{x \in L \mid x^{\downarrow} \cap J \subseteq I\}.$$

In fact, it is clear that $I:J$ so defined is an ideal of L and is such that $J \cap (I:J) \subseteq I$. Suppose now that $K \in I(L)$ is such that $J \cap K \subseteq I$. Then for every $x \in K$ we have $x^{\downarrow} \cap J \subseteq I$. Consequently, $x \in I:J$ and therefore $K \subseteq I:J$.

A basic property of Heyting algebras is the following.

Theorem 7.12 *If L is a Heyting algebra then so is every interval of L.*

Proof Given $a, b \in L$ let $c, d \in [a, b]$. Consider the element

$$\alpha = (d:c) \wedge b.$$

Since $a \leqslant d \leqslant d:c$ and $a \leqslant b$ we have $\alpha \in [a, b]$. Observe by Theorem 7.11(3) that $c \wedge \alpha = c \wedge (d:c) \wedge b = c \wedge d \wedge b \leqslant d$, and if $x \in [a, b]$ is such that $c \wedge x \leqslant d$ then $x \leqslant d:c$ and $x \leqslant b$, and consequently $x \leqslant \alpha$. Hence the residual $[d:c]_a^b$ of d by c in $[a, b]$ exists and is $\alpha = (d:c) \wedge b$. Thus every interval $[a, b]$ of L is a Heyting algebra. □

By definition, in a Heyting algebra every interval is pseudocomplemented. We now investigate bounded lattices in which every interval is a Stone lattice.

Theorem 7.13 *If L is a bounded lattice then the following statements are equivalent:*

(1) *every interval of L is a Stone lattice;*
(2) *L is a Heyting lattice in which*

$$(\forall x, y \in L) \quad (x:y) \vee (y:x) = 1.$$

Proof (1) ⇒ (2): If every interval of L is a Stone lattice then every interval is pseudocomplemented and so L is a Heyting algebra. Recalling from the proof of Theorem 7.9 that $x:y = y^*_{x \wedge y}$, we have

$$(x:y) \vee (y:x) = y^*_{x \wedge y} \vee x^*_{x \wedge y} = (y \wedge x)^*_{x \wedge y} = 1.$$

(2) ⇒ (1): Suppose now that L is a Heyting lattice that satisfies the identity in (2). Then L is a Stone lattice. To see this, take $y = 0:x$ in the identity and observe, using Theorem 7.11(4), that

$$1 = [x:(0:x)] \vee [(0:x):x] = [x:(0:x)] \vee (0:x).$$

But $x:(0:x) = 0:(0:x)$; for $(0:x) \wedge [x:(0:x)] = (0:x) \wedge x = 0$ and so $x:(0:x) \leqslant 0:(0:x)$ and the reverse inequality is trivial. Thus we have

$$1 = [0 : (0 : x)] \vee (0 : x) = x^{**} \vee x^*$$

and so L is a Stone lattice.

Since every principal filter of L also satisfies the identity in (2), and since by Theorem 7.12 every such filter is also a Heyting lattice, it follows that every principal filter of L is a Stone lattice.

Now let $a \leqslant b$ in L and let $x \in [a, b]$. If $x_a^\star$ is the pseudocomplement of x in $a^\uparrow$, consider the element

$$\beta = (x_a^\star \wedge b) \vee a.$$

Clearly, $\beta \in [a, b]$ and

$$x \wedge \beta = (x \wedge x_a^\star \wedge b) \vee (x \wedge a) = (a \wedge b) \vee (x \wedge a) = a.$$

If now $y \in [a, b]$ is such that $x \wedge y = a$ then $y \leqslant x_a^\star$ and therefore we have $y \leqslant (x_a^\star \wedge b) \vee a = \beta$. Thus β is the pseudocomplement of x in $[a, b]$.

Now let γ be the pseudocomplement of β in $[a, b]$. Then

$$\begin{aligned}
\gamma = (\beta_a^\star \wedge b) \vee a &= \big([(x_a^\star)_a^\star \vee b_a^\star] \wedge a_a^\star \wedge b\big) \vee a \\
&= [(x_a^\star)_a^\star \wedge b] \vee (b_a^\star \wedge b) \vee a \\
&= (x_a^\star)_a^\star \wedge b,
\end{aligned}$$

from which it follows, since $a^\uparrow$ is a Stone lattice, that

$$\begin{aligned}
\beta \vee \gamma = (x_a^\star \wedge b) \vee a \vee [(x_a^\star)_a^\star \wedge b] &= \big([x_a^\star \vee (x_a^\star)_a^\star] \wedge b\big) \vee a \\
&= b \vee a \\
&= b.
\end{aligned}$$

Consequently, $[a, b]$ is a Stone lattice.   □

A complete Heyting algebra (also known as a **frame**) is the underlying structure of the **intuitionist logic** constructed by Brouwer as an alternative to boolean logic, as formulated in the calculus of propositions. In intuitionist logic it is assumed that the propositions form a complete Heyting algebra $L(0, 1, \wedge, \vee, \neg, \rightarrow, \bigvee, \bigwedge)$ in which 0 is interpreted as an absurdity, 1 as a tautology, $\wedge$ as 'and', $\vee$ as 'or', $\neg$ as 'not', $\rightarrow$ as 'implies', $\bigvee$ as 'for some', and $\bigwedge$ as 'for all'. In order to marry the differences in notational practice it is necessary to read $\neg x$ as x^*, and $x \rightarrow y$ as $y : x$.

EXERCISES

7.19. In a Heyting algebra establish the identities
  (1) $x \wedge (y : z) = x \wedge [(x \wedge y) : (x \wedge z)]$;
  (2) $(x : y)^* = x^* \wedge y^{**}$;
  (3) $(x : y) \wedge (y : x) = (x \wedge y) : (x \vee y)$.

7.20. Prove that in a Heyting algebra $x : [x : (x : y)] = x : y$. Interpret this when $x = 0$.

7.21. Let L be a Heyting algebra. Prove that the following are equivalent:
  (1) L is boolean;
  (2) $(\forall x, y \in L)\quad x \vee y : x = 1$;
  (3) $(\forall x, y \in L)\quad x : (x : y) = x \vee y$;
  (4) $(\forall x, y \in L)\quad x : (y : x) = x$.

7.22. Describe residuals in the Heyting lattice consisting of the positive divisors of 12.

7.23. Let F be the set of real functions $f : [0,1] \to [0,1]$. Prove that F is a Heyting algebra and describe residuals therein.

7.24. Prove that the open sets of a topological space T form a Heyting algebra H in which residuals are given by

$$A:B = (A \cup B')^\circ$$

where X° denotes the interior of $X \in H$. Show also that the boolean algebra H^{**} consists of the regular open sets, and that the dense filter consists of those open sets A that are dense in the topological sense, namely $A^- = H$.

7.25. A **closure algebra** consists of a boolean algebra B together with a closure $x \mapsto x^c$ on B that is **additive** in the sense that $(x \vee y)^c = x^c \vee y^c$ for all $x, y \in B$, and is such that $0^c = 0$. Call $x \in B$ **open** if $x' \in B^c$, and let B° be the subset of open elements of B. Prove that B° is a sublattice of B and is a Heyting algebra in which residuals are given by

$$(\forall x, y \in B) \quad x:y = [(x' \wedge y)^c]'.$$

7.26. Let p, q be distinct primes. Prove that in the lattice of positive divisors of $p^m q^n$ every interval is a Stone lattice.

7.27. If L is a Heyting algebra prove that $f : L \to L$ is a multiplicative closure (also known as a **modal operator**) if and only if

$$(\forall x, y \in L) \quad f(x):y = f(x):f(y).$$

7.28. If L is a complete Heyting algebra prove that so also is the set $\mathrm{MC}(L)$ of multiplicative closures on L. Prove further that $\mathrm{MC}(L)$ is boolean if and only if every principal filter of L contains a smallest dense element.

7.4 Baer semigroups and residuation

We now seek to obtain a Baer semigroup coordinatisation of a Heyting lattice. For this purpose we first characterise the range-closed idempotent residuated mappings that are multiplicative.

Theorem 7.14 *Let L be a bounded lattice and let $f \in \mathrm{Res}\, L$ be a range-closed idempotent. Then f is multiplicative if and only if $f = \lambda_{f(1)}$.*

Proof $\Rightarrow$: By the hypotheses on f and Theorem 4.1 we have, for every $x \in L$,

$$f[f^+(x)] = (f \circ f \circ f^+)(x) = f[x \wedge f(1)] = f(x) \wedge f(1) = f(x),$$

and so $f = f \circ f^+$. Thus $f(x) = f[f^+(x)] = x \wedge f(1)$ and hence $f = \lambda_{f(1)}$.
$\Leftarrow$: This is clear. $\qquad \square$

We recall that if S is a Baer semigroup then for every $z \in S$ there is an induced residuated mapping $\varphi_z : \mathcal{R}(S) \to \mathcal{R}(S)$. If $eS \in \mathcal{R}(S)$ then φ_z is given by $\varphi_z(eS) = RL(ze)$. We shall say that an idempotent $e \in S$ is **multiplicative** if the induced idempotent residuated mapping $\varphi_e : \mathcal{R}(S) \to \mathcal{R}(S)$ is multiplicative. Furthermore, we shall say that S is **weakly multiplicative** if every element of $\mathcal{R}(S)$ is generated by a unique multiplicative idempotent.

For the purpose in hand, we also require the following result.

Theorem 7.15 *Let E be a bounded ordered set and let $f \in \operatorname{Res} E$. If $R(f) = g \circ \operatorname{Res} E$ with $g^2 = g$ then $g(1) = f^+(0)$.*

Proof Since $f \circ g = 0$ we have that $g(1) \leqslant f^+(0)$. Consider now, for every $e \in E$, the mapping $\alpha_e \in \operatorname{Res} E$ given (as in Theorem 1.12) by

$$\alpha_e(x) = \begin{cases} 0 & \text{if } x = 0; \\ e & \text{otherwise,} \end{cases} \qquad \alpha_e^+(x) = \begin{cases} 1 & \text{if } e \leqslant x; \\ 0 & \text{otherwise,} \end{cases}$$

Observe that $f \circ \alpha_{f^+(0)} = 0$ and therefore $\alpha_{f^+(0)} = g \circ \alpha_{f^+(0)}$ whence $f^+(0) = \alpha_{f^+(0)}(1) = g[\alpha_{f^+(0)}(1)] \leqslant g(1)$. □

Theorem 7.16 (Blyth [19]) *A bounded lattice is a Heyting lattice if and only if it can be coordinatised by a weakly multiplicative Baer semigroup.*

Proof ⇒: If $f \in \operatorname{Res} L$ we recall from Theorem 2.19 that $R(f) = \vartheta_{f^+(0)} \circ \operatorname{Res} L$. Now since every translation λ_x on L is residuated and, as is readily seen, we have $\lambda_x \circ \vartheta_x = \vartheta_x$ and $\vartheta_x \circ \lambda_x = \lambda_x$, it follows that $\vartheta_x \circ \operatorname{Res} L = \lambda_x \circ \operatorname{Res} L$. Consequently, for every $f \in \operatorname{Res} L$ we have

$$R(f) = \vartheta_{f^+(0)} \circ \operatorname{Res} L = \lambda_{f^+(0)} \circ \operatorname{Res} L.$$

Suppose now that $R(f) = g \circ \operatorname{Res} L$ where g is idempotent. Then g is range-closed by Theorem 6.25. Moreover, by Theorem 7.15, $g(1) = f^+(0)$. If now g is multiplicative then by Theorem 7.14 we have $g = \lambda_{g(1)} = \lambda_{f^+(0)}$. Consequently $\lambda_{f^+(0)}$ is the unique multiplicative idempotent generator of $R(f)$. Hence S is weakly multiplicative.

⇐: Suppose that S is weakly multiplicative and let $eS \in \mathcal{R}(S)$ with $e \in S$ a multiplicative idempotent. Then the range-closed idempotent φ_e is multiplicative and so, by Theorem 7.14, we have $\varphi_e = \lambda_{\varphi_e(1S)}$. Since $\varphi_e(1S) = RL(e) = eS$ we thus have $\varphi_e = \lambda_{eS}$ whence λ_{eS} is residuated. Consequently $L \simeq \mathcal{R}(S)$ is a Heyting lattice. □

Definition By a **double Heyting lattice** we shall mean a Heyting lattice L whose dual L^d is also a Heyting lattice.

In a double Heyting lattice L every translation $\lambda_x : y \mapsto x \wedge y$ is residuated and every translation $\mu_x : y \mapsto x \vee y$ is dually residuated. The latter means that there is a unique isotone mapping μ_x^- such that

$$\mu_x \mu_x^- \geqslant \operatorname{id}_L \geqslant \mu_x^- \mu_x.$$

It follows that $\mu_x^- \in \operatorname{Res} L$ with $(\mu_x^-)^+ = \mu_x$.

Example 7.12 By Theorem 7.10 and its dual, a complete lattice in which both infinite distributive laws are valid is a double Heyting lattice.

Example 7.13 By Theorem 6.8, every complete boolean algebra is a double Heyting algebra.

Example 7.14 Every finite distributive lattice is a double Heyting lattice.

EXERCISES

7.29. Let L be a double Heyting lattice and let $Z(L)$ be the subset consisting of the complemented elements of L. Prove that for, every $x \in L$, both $\vartheta(x) = \min\{x^\uparrow \cap Z(L)\}$ and $\varphi(x) = \max\{x^\downarrow \cap Z(L)\}$ exist. Prove that $\vartheta : x \mapsto \vartheta(x)$ is a residuated dual closure on L with $\vartheta^+ = \varphi : x \mapsto \varphi(x)$. Prove also that ϑ is a **quantifier**, in the sense that

$$(\forall x, y \in L) \qquad \vartheta[x \wedge \vartheta(y)] = \vartheta(x) \wedge \vartheta(y).$$

7.30. If L is the lattice $\mathbf{2} \times \mathbf{3} \times \mathbf{4}$, illustrate the various points of the previous exercise using the Hasse diagram of L.

In order to obtain a Baer semigroup coordinatisation of a double Heyting lattice we require the following notion.

Definition By a **decreasing mapping** on an ordered set E we mean a mapping $f : E \to E$ such that $f \leqslant \mathrm{id}_E$. By a **decreasing Baer semigroup** we mean a Baer semigroup S in which each induced residuated mapping φ_z on $\mathcal{R}(S)$ is decreasing.

Theorem 7.17 (Blyth–Janowitz [23]) *A lattice is a double Heyting lattice if and only if it can be coordinatised by a decreasing Baer semigroup.*

Proof $\Rightarrow$: If L is a double Heyting lattice then each translation $\lambda_x : y \mapsto x \wedge y$ is residuated and each translation $\mu_x : y \mapsto x \vee y$ is dually residuated. Let ν_x be the unique residuated mapping such that $\nu_x^+ = \mu_x$; i.e. $\nu_x = \mu_x^-$. Since $\mu_x \geqslant \mathrm{id}_L$ we have $\mathrm{id}_L \geqslant \nu_x \mu_x \geqslant \nu_x$ and so ν_x is decreasing.

Consider now the subsemigroup S of $\mathrm{Res}\, L$ that consists of the decreasing residuated mappings on L. Then $\lambda_x, \nu_x \in S$ for every $x \in L$. If now $f, g \in S$ then clearly $g \in R(f)$ if and only if $g(1) \leqslant f^+(0)$. But if $g(1) \leqslant f^+(0)$ then $g = \lambda_{f^+(0)} \circ g$; and if $g = \lambda_{f^+(0)} \circ g$ then $g(1) = \lambda_{f^+(0)}[g(1)] = f^+(0) \wedge g(1) \leqslant f^+(0)$. Consequently $R(f) = \lambda_{f^+(0)} \circ S$. Likewise we have $L(f) = S \circ \nu_{f(1)}$. Hence S is a decreasing Baer semigroup that coordinatises L.

$\Leftarrow$: Let S be a decreasing Baer semigroup that coordinatises L. Let $eS \in \mathcal{R}(S)$ with $e^2 = e$. Then for every $z \in S$ we have $ze \in RL(ze) = \varphi_z(eS) \subseteq eS$, which gives $zeS \subseteq eS$. Hence the right ideal eS is also a left ideal of S. Consequently, from $e \in eS$ we obtain $xe \in eS$ for every $x \in S$ and hence $xe = exe$.

Now let $fS \in \mathcal{R}(S)$. Then $fe = efe$ and so $fe \in eS \cap fS$ whence $RL(fe) \subseteq RL(eS) \cap RL(fS) = eS \cap fS$. On the other hand, if $x \in eS \cap fS$ then $x = ex = fx = fex$. Then for every $y \in L(fe)$ we have $yx = yfex = 0$ which gives $x \in R(y)$. Hence $x \in RL(fe)$. Thus we see that $eS \cap fS = RL(fe) = \varphi_f(eS)$. This gives $\lambda_{fS} = \varphi_f$ and therefore $\lambda_{fS} \in \mathrm{Res}\, \mathcal{R}(S)$.

If now we let $Se^\star = L(e)$ and $Sf^\star = L(f)$ then a symmetric argument to the above shows that the translation $\lambda_{Sf^\star} \in \mathrm{Res}\, \mathcal{L}(S)$. Taking right annihilators, we see that the translation μ_{fS} is dually residuated on $\mathcal{R}(S)$. Hence $\mathcal{R}(S)$, and therefore L, is a double Heyting lattice. $\square$

Congruences; subdirectly irreducible algebras

8.1 More on lattice congruences

In the light of the various types of lattice that we have considered, we take up again the notion of a congruence as described in Chapter 3. We recall from Theorem 3.9 that for every lattice L the set $\text{Con}\,L$ of congruences on L forms a complete lattice which is a sublattice of the complete lattice of equivalence relations on L. The lattice operations in $\text{Con}\,L$ are described in Example 2.19.

Theorem 8.1 (Funayama–Nakayama [53]) *If L is a lattice then $\text{Con}\,L$ is a complete Heyting lattice.*

Proof By virtue of Theorem 7.10, it suffices to show that for every congruence S and every family $(R_\alpha)_{\alpha \in A}$ of congruences on L there holds the inequality

$$\bigvee_{\alpha \in A} (S \wedge R_\alpha) \geqslant S \wedge \bigvee_{\alpha \in A} R_\alpha.$$

Now if $(x, y) \in S \wedge \bigvee_{\alpha \in A} R_\alpha$ then we have

$$x = z_0 \; R_{\alpha_1} \; z_1 \; R_{\alpha_2} \; z_2 \; \cdots \; z_{n-1} \; R_{\alpha_n} \; z_n = y.$$

For each i let $v_i = [(x \wedge y) \vee z_i] \wedge (x \vee y)$. Then clearly $v_0 = x$, and $v_n = y$. Moreover, for $i = 1, \ldots, n$ we have

$$v_{i-1} = [(x \wedge y) \vee z_{i-1}] \wedge (x \vee y) \; R_{\alpha_i} \; [(x \wedge y) \vee z_i] \wedge (x \vee y) = v_i.$$

Now since $x \wedge y \leqslant v_i \leqslant x \vee y$ for all i, and $(x \wedge y, x \vee y) \in S$, we deduce that $(v_{i-1}, v_i) \in S$ for all i, whence $(v_{i-1}, v_i) \in S \wedge R_{\alpha_i}$ for all i, and therefore $(x, y) \in \bigvee_{\alpha \in A} (S \wedge R_\alpha)$ as required. $\qquad\square$

In the case of groups, rings, modules, etc., every congruence is uniquely determined by any one of its classes. This is not the case for lattices (nor, more generally, for semigroups). For example, consider the chain $0 < x < 1$. The partition $\{\{0\}, \{x, 1\}\}$ defines a congruence which is distinct from the trivial congruence but has a class in common with it. There is, however, a class of lattices in which it is true.

Theorem 8.2 (Hashimoto [61]) *If L is a relatively complemented lattice then every congruence on L is determined by any one of its classes.*

Proof Let $\equiv$ and $\sim$ be congruences on L. Suppose first that E is a $\equiv$-class, that F is a $\sim$-class, and that $E \subseteq F$. Then we make the following observations:

(1) *If $x, y \geqslant a \in E$ then $x \equiv y \Rightarrow x \sim y$.*
In fact, let z be a complement of $x \wedge y$ in $[a, x \vee y]$. Then we have $z = z \wedge (x \vee y) \equiv z \wedge x \equiv z \wedge x \wedge y = a$, so that $z \in E$. Consequently $z \in F$ with $z \sim a$ and $x \vee y = z \vee (x \wedge y) \sim a \vee (x \wedge y) = x \wedge y$. It follows by the convexity of the $\sim$-classes that $x \sim y$.

(2) *If $x, y \leqslant a$ for some $a \in E$ then $x \equiv y \Rightarrow x \sim y$.*
This is dual to (1).

(3) *If $b \leqslant a \in E$ and $G = [b]_\equiv$ then $p \sim q$ for all $p, q \in G$.*
In fact, if $p \in G$ then $a \vee p \equiv a \vee b = a$ and $a \wedge p \equiv a \wedge b = b$. By (1) and (2) we deduce that $a \vee p \sim a$ and $a \wedge p \sim b$. Hence $p = (a \vee p) \wedge p \sim a \wedge p \sim b$. Thus, for all $p, q \in G$ we have $p \sim b \sim q$.

Suppose now that $x, y \in L$ are such that $x \equiv y$. Choose an element $a \in E$ and let $G = [a \wedge x \wedge y]_\equiv$. By (3) we have $p \sim q$ for all $p, q \in G$. Since $x, y \geqslant a \wedge x \wedge y$ it follows by (1), with G playing the role of E, that $x \sim y$.

In a similar way we see that if $F \subseteq E$ then $x \sim y$ implies $x \equiv y$. Thus, if $E = F$ then $\equiv$ and $\sim$ coincide. $\qquad\square$

There is another property of congruences that holds for groups but not in general for lattices. To describe this, let ϑ and φ be congruences on an algebraic structure A. Then the **product** $\vartheta\varphi$ is the relation defined by

$$(x, y) \in \vartheta\varphi \iff (\exists t \in A) \quad x \overset{\vartheta}{\equiv} t \overset{\varphi}{\equiv} y.$$

In general, the relation $\vartheta\varphi$ is not transitive. However, if ϑ and φ commute in the sense that $\vartheta\varphi = \varphi\vartheta$ then it is transitive, for in this case $x \overset{\vartheta}{\equiv} t \overset{\varphi}{\equiv} y \overset{\vartheta}{\equiv} s \overset{\varphi}{\equiv} z$ implies the existence of $u \in A$ such that $x \overset{\vartheta}{\equiv} t \overset{\vartheta}{\equiv} u \overset{\varphi}{\equiv} s \overset{\varphi}{\equiv} z$, so that $x \overset{\vartheta}{\equiv} u \overset{\varphi}{\equiv} z$. In the case where ϑ, φ commute it is clear that $\vartheta\varphi$ is also a congruence on A and coincides with the transitive product $\vartheta \vee \varphi$.

Suppose now that ϑ_H and ϑ_K are congruences on a group G where H and K are the associated normal subgroups (the respective cosets that contain 1_G). Suppose that $(x, z) \in \vartheta_H\vartheta_K$, so that there exists $y \in G$ with $x \, \vartheta_H \, y \, \vartheta_K \, z$. Then $xy^{-1} \in H$ and $yz^{-1} \in K$ and so, since H and K are normal subgroups, $xz^{-1} = xy^{-1} \cdot yz^{-1} \in HK = KH$. It follows that $xz^{-1} = kh$ for some $k \in K$ and $h \in H$. Now let $t \in G$ be such that $x = kt$. Then $xt^{-1} = k$ and $tz^{-1} = h$. Consequently $x \, \vartheta_K \, t \, \vartheta_H \, z$ and so $(x, z) \in \vartheta_K\vartheta_H$. Thus we see that every pair of congruences on a group commutes.

This does not carry over to lattices in general. To see this, consider again the chain $0 < x < 1$. Let $\equiv$ and $\sim$ be the congruences with the respective partitions $\{\{0, x\}, \{1\}\}$ and $\{\{0\}, \{x, 1\}\}$. We have $0 \equiv x \sim 1$ but there is no y such that $0 \sim y \equiv 1$. Thus $\equiv$ and $\sim$ do not commute.

There is, however, a class of lattices in which the property does hold.

Theorem 8.3 (Dilworth [45]) *On a relatively complemented lattice any two congruences commute.*

Proof Let L be relatively complemented and let $\equiv$ and $\sim$ be congruences on L. Suppose that $a \equiv x \sim b$. Then $a = a \vee a \equiv a \vee x \sim a \vee b$ and so $a \equiv a \vee x \sim a \vee x \vee b$. Let y be a complement of $a \vee x$ in $[a, a \vee b \vee x]$. Then we have

$$a = (a \vee x) \wedge y \sim (a \vee b \vee x) \wedge y = y = a \vee y \equiv a \vee x \vee y = a \vee b \vee x$$

so that $a \sim y \equiv a \vee b \vee x$. Interchanging a, b and $\equiv, \sim$ we deduce similarly that $b \equiv z \sim a \vee b \vee x$ where z a complement of $b \vee x$ in $[b, a \vee b \vee x]$. Consequently,

$$y = y \wedge (a \vee b \vee x) \sim y \wedge z \equiv (a \vee b \vee x) \wedge z = z.$$

Since $a \sim y$ and $z \equiv b$ we deduce that $a \sim y \wedge z \equiv b$. □

8.2 Congruence kernels

Suppose now that ϑ is a congruence on a lattice L. If the quotient lattice L/ϑ has a bottom element then we shall denote it by $[0]_\vartheta$ (which is what we would write if L itself had a bottom element). We shall call $[0]_\vartheta$ the **kernel** of the congruence ϑ and denote it by $\operatorname{Ker}\vartheta$. Clearly, $\operatorname{Ker}\vartheta$ is an ideal of L.

Definition By a **kernel ideal** of a lattice L we shall mean an ideal of L that is the kernel of some congruence on L.

Example 8.1 Consider the lattice

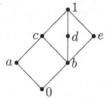

The subset $I = \{0, a, b, c\}$ is an ideal, but is not a kernel ideal. In fact, if I were the kernel of a congruence $\equiv$ then we would have $d = d \vee 0 \equiv d \vee c = 1$ and $e = e \vee 0 \equiv e \vee c = 1$ whence the contradiction $0 \equiv b = d \wedge e \equiv 1$.

The existence in the above example of a non-distributive sublattice prompts a consideration of the distributive case. In so doing, we obtain the following characterisation of distributive lattices.

Theorem 8.4 *A lattice L is distributive if and only if every ideal of L is a kernel ideal.*

Proof $\Rightarrow$: Suppose that L is distributive and let I be an ideal of L. Define the relation ϑ_I on L by

$$(a, b) \in \vartheta_I \iff (\exists i \in I) \quad a \vee i = b \vee i.$$

Clearly, ϑ_I is an equivalence relation on L that is compatible with $\vee$. To see that it is also compatible with $\wedge$, suppose that $(a, b) \in \vartheta_I$. Then $a \vee i = b \vee i$ for some $i \in I$ and so, for every $x \in L$,

$$(a \wedge x) \vee (i \wedge x) = (a \vee i) \wedge x = (b \vee i) \wedge x = (b \wedge x) \vee (i \wedge x).$$

Since $i \wedge x \in I$, it follows that $(a \wedge x, b \wedge x) \in \vartheta_I$. Thus ϑ_I is a congruence on L. Now if $i \in I$ then $(a, i) \in \vartheta_I$ implies that, for some $j \in I$, $a \vee j = i \vee j \in I$ and hence $a \in I$. It follows that I is a ϑ_I-class, and is indeed the bottom element of L/ϑ_I. Thus I is the kernel of ϑ_I.

$\Leftarrow$: Conversely, suppose that every ideal of L is a kernel ideal. If L were not distributive then by Theorem 5.1 it would contain as a sublattice one of the lattices

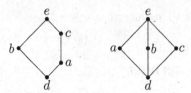

In either case, let $I = a^{\downarrow}$ and let $\equiv$ be a congruence of which I is the kernel. Then $e = a \vee b \equiv d \vee b = b$ from which we obtain $c = c \wedge e \equiv c \wedge b = d \equiv a$, which shows that $c \in I = a^{\downarrow}$, a contradiction in each case. Thus L has no such sublattices and is therefore distributive.   □

The congruence ϑ_I considered in the proof of Theorem 8.4 has a more significant property.

Theorem 8.5 *If I is an ideal of a distributive lattice then ϑ_I is the smallest congruence with kernel I.*

Proof Suppose that $(x, y) \in \vartheta_I$, so that $x \vee i = y \vee i$ for some $i \in I$. Then for any congruence φ with kernel I we have

$$[x]_\varphi = [x]_\varphi \vee [i]_\varphi = [x \vee i]_\varphi = [y \vee i]_\varphi = [y]_\varphi \vee [i]_\varphi = [y]_\varphi$$

so that $(x, y) \in \varphi$ and consequently $\vartheta_I \subseteq \varphi$.   □

The notion of a **cokernel filter** in a lattice is dual to that of a kernel ideal; i.e. is a filter that is the top class $[1]_\vartheta$ modulo some congruence. If F is a filter of a distributive lattice L then, dual to the above, the relation φ_F defined on L by

$$(a, b) \in \varphi_F \iff (\exists j \in F) \quad a \wedge j = b \wedge j$$

is a congruence and is the smallest congruence on L with cokernel F.

Theorem 8.6 *If I is an ideal and F is a filter of a lattice L then, when non-empty, $I \cap F$ is a convex sublattice of L. Conversely, every convex sublattice is such an intersection. Moreover, if I is a kernel ideal and F is a cokernel filter then, when non-empty, $I \cap F$ is a congruence class.*

Proof It is clear that, when non-empty, $I \cap F$ is closed under $\wedge$ and $\vee$ and so is a sublattice. That it is convex follows from the observation that if $a \leqslant x \leqslant b$ with $a, b \in I \cap F$ then $x \leqslant b \in I$ and I an ideal gives $x \in I$; and $x \geqslant a \in F$ and F a filter gives $x \in F$.

Conversely, let C be a convex sublattice of L. Then $F = \bigcup_{c \in C} c^{\uparrow}$ is a filter of L. For, if $a \in F$ then $a \geqslant c$ for some $c \in C$, so $x \geqslant a$ implies $x \geqslant c$ so that $x \in F$; and if $a, b \in F$ then $a \geqslant c_1$ and $b \geqslant c_2$ for some $c_1, c_2 \in C$ whence $a \wedge b \geqslant c_1 \wedge c_2 \in C$ and so $a \wedge b \in F$. Dually, we see that $I = \bigcup_{c \in C} c^{\downarrow}$ is an ideal of L. Now clearly we have $C \subseteq I \cap F$. But if $x \in I \cap F$ then $d_1 \leqslant x \leqslant d_2$ for some $d_1, d_2 \in C$. Since C is convex, it follows that $x \in C$ and consequently $C = I \cap F$.

Finally, suppose that I is the kernel ideal of a congruence ϑ and that F is the cokernel filter of a congruence φ on L. Let $H = L/\vartheta \times L/\varphi$ and let $\natural :$ $L \to H$ be the lattice morphism given by the prescription $\natural(x) = ([x]_\vartheta, [x]_\varphi)$. Consider now the congruence $\ker \natural$, namely that given by

$$(x, y) \in \ker \natural \iff \natural(x) = \natural(y).$$

Since $x \in I \cap F$ if and only if $\natural(x) = ([0]_\vartheta, [1]_\varphi)$ it follows that $I \cap F$ is a $\ker \natural$-class. $\square$

Theorem 8.7 *A lattice is distributive if and only if every convex sublattice is a congruence class.*

Proof $\Rightarrow$: If L is distributive then by Theorem 8.4 and its dual every ideal of L is a kernel ideal and every filter is a cokernel filter. It now follows by Theorem 8.6 that every convex sublattice of L is a congruence class.

$\Leftarrow$: If every convex sublattice of L is a congruence class then every ideal I, being a convex sublattice, is a ϑ-class for some congruence ϑ on L. Since $x \wedge i \in I$ for all $x \in L$ and all $i \in I$, it follows that in L/ϑ we have $I \leqslant [x]_\vartheta$. Thus I is the bottom element of L/ϑ and so is the kernel of ϑ. It now follows by Theorem 8.4 that L is distributive. $\square$

Combining the above characterisations of distributive lattices with Theorems 8.1 and 8.2, we can establish the following result.

Theorem 8.8 *The following statements concerning a lattice L are equivalent:*

(1) *L is distributive and relatively complemented;*
(2) *every ideal of L is the kernel of a unique congruence;*
(3) *every filter of L is the cokernel of a unique congruence;*
(4) *every convex sublattice of L is a class modulo a unique congruence;*
(5) *L is distributive and any two congruences on L commute.*

Proof We prove that $(1) \Leftrightarrow (5)$ and that $(1) \Rightarrow (4) \Rightarrow (2) \Rightarrow (1)$. By duality, $(4) \Rightarrow (3) \Rightarrow (1)$.

$(1) \Rightarrow (5)$: This is immediate from Theorem 8.3.

$(5) \Rightarrow (1)$: Suppose that L is distributive and that any two congruences on L commute. Given any interval $[a, b]$ consider $c \in [a, b]$. If $I = c^{\downarrow}$ then, by Theorem 8.4, I is the kernel of the congruence ϑ_I given by

$$(x, y) \in \vartheta_I \iff (\exists i \leqslant c) \quad x \vee i = y \vee i;$$

and if $F = c^\uparrow$ then, by the dual of Theorem 8.4, F is the kernel filter of the congruence φ_F given by

$$(x, y) \in \varphi_F \iff (\exists j \geqslant c) \quad x \wedge j = y \wedge j.$$

Now it is clear that $a \, \vartheta_I \, c \, \varphi_F \, b$ and so, by the hypotheses, there exists $d \in L$ such that $a \, \varphi_F \, d \, \vartheta_I \, b$. Thus there exist $p \in F = c^\uparrow$ and $q \in I = c^\downarrow$ such that $a \wedge p = d \wedge p$ and $b \vee q = d \vee q$. But $a \leqslant c \leqslant p$ and $q \leqslant c \leqslant b$. Consequently we have $a = d \wedge c = d \wedge p \leqslant d \leqslant d \vee q = d \vee c = b$ and therefore $d \in [a, b]$ and is a complement in $[a, b]$ of c. Hence L is relatively complemented.

(1) $\Rightarrow$ (4): This is immediate from Theorems 8.2 and 8.7.

(4) $\Rightarrow$ (2): Ideals are convex sublattices.

(2) $\Rightarrow$ (1): Suppose that every ideal is the kernel of a unique congruence. Then L is distributive by Theorem 8.4. To show that it is also relatively complemented, let $[a, b]$ be any interval of L. Choose $c \in [a, b]$ and consider $I = \{i \in L \mid c \wedge i \leqslant a\}$. Since L is distributive, I is an ideal and so is the kernel of the congruence ϑ_I. Now the quotient lattice $L^\star = L/\vartheta_I$ is also distributive and so, for $[c] \in L^\star$, the filter $[c]^\uparrow$ is the cokernel of the congruence $\varphi = \varphi_{[c]^\uparrow}$ on $L^\star$ given by

$$([x], [y]) \in \varphi \iff (\exists [a] \geqslant [c]) \quad [x] \wedge [a] = [y] \wedge [a].$$

Let $\natural_\varphi : L^\star \to L^\star/\varphi$ be the natural morphism and let $h = \natural_\varphi \circ \natural_{\vartheta_I}$. Consider $J = h^\leftarrow\{[0_{L^\star}]_\varphi\}$. Since the bottom element of $L^\star$ is I, we have $I \subseteq J$.

Suppose now that $t \in J$. Then $[t] \, \varphi \, [0]$ and so, for some $[z] \geqslant [c]$ we have $[t] \wedge [z] = [0] \wedge [z]$ whence $[t] \wedge [c] = [0]$. Thus $t \wedge c \in I$ and so, by the definition of I, we have $t \wedge c \leqslant a$. Again by the definition of I, it follows that $t \in I$. This then shows that $J \subseteq I$.

Now since $I = J$ it follows by the hypotheses that $\vartheta_I = \ker h$. From $c \leqslant b$ we then have $[c] \leqslant [b]$ and so $[c] \, \varphi \, [b]$ whence $h(c) = h(b)$ and therefore $c \, \vartheta_I \, b$. Thus $c \vee i = b \vee i$ for some $i \in I$ and consequently $b = b \wedge (b \vee i) = b \wedge (c \vee i) = (b \wedge c) \vee (b \wedge i) = c \vee (b \wedge i)$, i.e., $b = c \vee j$ where $j = b \wedge i \in I$.

We complete the proof by showing that $a \vee j$ is a complement of c in $[a, b]$. Now on the one hand we have $c \vee a \vee j = c \vee j = b$; and on the other, by the definition of I, we have $a = a \wedge (a \vee j) \leqslant c \wedge (a \vee j) \leqslant a$, and so $c \wedge (a \vee j) = a$. Thus L is relatively complemented. $\quad\square$

Corollary *For a lattice L the following statements are equivalent:*

(1) *every ideal of L is the kernel of a unique congruence and every congruence on L has a kernel;*

(2) *L is distributive, relatively complemented, and has a bottom element.*

Proof It suffices to observe that if every congruence has a kernel then in particular this is so for the trivial congruence, whence L has a bottom element. Conversely, the existence of a bottom element implies that every quotient lattice of L has a bottom element. $\quad\square$

A relatively complemented distributive lattice with 0 is often called a **generalised boolean algebra**.

For a bounded lattice we obtain from the above the following characterisation of boolean lattices.

Theorem 8.9 *For a bounded lattice L the following statements are equivalent*:

(1) *every ideal of L is the kernel of a unique congruence*;
(2) *L is boolean*.

Proof It suffices to observe that if (1) holds then L is relatively complemented. But if (1) holds then in particular $[0, 1] = L$ is complemented and hence L is boolean. $\qquad\qquad\square$

EXERCISES

8.1. If L is a distributive lattice and F is a filter of L prove that F is a cokernel filter and that the smallest congruence with cokernel F is given by

$$(x, y) \in \varphi_F \iff (\exists f \in F) \quad x \wedge f = y \wedge f.$$

Prove also that if F and G are filters of L then φ_F and φ_G commute.

8.2. If L is a bounded distributive lattice prove that the biggest congruence with kernel I is given by ϑ^I where

$$(a, b) \in \vartheta^I \iff \{x \in L \mid x \wedge a \in I\} = \{x \in L \mid x \wedge b \in I\}.$$

8.3. Let ϑ and φ be congruences on a lattice L and suppose that $\vartheta \leqslant \varphi$. Prove that there is a unique surjective lattice morphism $f : L/\vartheta \to L/\varphi$ such that $f \circ \natural_\vartheta = \natural_\varphi$.

8.4. If I is an ideal of a distributive lattice L prove that

$$(x, y) \in \vartheta_I \iff (\exists i \in I) \quad x \vee y = (x \wedge y) \vee i.$$

8.5. If B is a boolean lattice and I is an ideal of B prove that ϑ_I can be described by

$$(x, y) \in \vartheta_I \iff (x' \wedge y) \vee (x \wedge y') \in I.$$

Show also that if $(x, y) \in \vartheta_I$ then $(x', y') \in \vartheta_I$.

8.6. Prove that a principal ideal $z^\downarrow$ of a bounded relatively complemented lattice is the kernel of a congruence if and only if z is central.

8.7. If ϑ and φ are congruences on a lattice L prove that

$$\text{Ker}(\vartheta \wedge \varphi) = \text{Ker}\,\vartheta \cap \text{Ker}\,\varphi.$$

Deduce that the set of congruences on L that possess kernels forms a filter of $\text{Con}\,L$.

8.8. Let L be a lattice with bottom element 0. For every $\vartheta \in \text{Con}\,L$ let

$$k(\vartheta) = \bigwedge\{\varphi \in \text{Con}\,L \mid \text{Ker}\,\varphi = \text{Ker}\,\vartheta\}.$$

Prove that k is a residuated dual closure on $\text{Con}\,L$ with k^+ given by

$$k^+(\vartheta) = \bigvee\{\varphi \in \text{Con}\,L \mid \text{Ker}\,\varphi = \text{Ker}\,\vartheta\}.$$

8.3 Principal congruences

If L is a distributive lattice and $a, b \in L$ then by Theorem 8.6 the interval
$[a \wedge b, a \vee b]$ is a ϑ-class for some congruence ϑ on L. Such a congruence is said
to **identify a and b**, in the sense that $[a]_\vartheta = [b]_\vartheta$. Conversely, by virtue of the
compatibility, every congruence ϑ that identifies a and b contains the interval
$[a \wedge b, a \vee b]$ in one of its classes. Now we know by Theorem 8.1 that $\mathrm{Con}\, L$
is a complete lattice, so by considering the infimum of all the congruences
that identify a and b we obtain the smallest congruence to do so. We shall
denote this congruence by $\vartheta(a, b)$ and call it a **principal congruence**. It can
be described as follows.

Theorem 8.10 *If L is a distributive lattice and $a, b \in L$ then the smallest
congruence on L that identifies a and b is given by*

$$(x, y) \in \vartheta(a, b) \iff \begin{cases} x \vee a \vee b = y \vee a \vee b, \\ x \wedge a \wedge b = y \wedge a \wedge b. \end{cases}$$

Proof Let ψ be the relation defined on L by

$$(x, y) \in \psi \iff x \vee a \vee b = y \vee a \vee b \quad \text{and} \quad x \wedge a \wedge b = y \wedge a \wedge b.$$

Then we observe first that ψ is a congruence on L. To see this, let $I = (a \vee b)^\downarrow$
and $F = (a \wedge b)^\uparrow$. Then from what has gone before we have

$$\psi = \vartheta_I \wedge \varphi_F \in \mathrm{Con}\, L.$$

Now clearly, we have $(a, b) \in \psi$ and so $\vartheta(a, b) \subseteq \psi$. It remains to show that in
fact $\vartheta(a, b) = \psi$. For this purpose, it suffices to prove that every congruence
that identifies a and b contains ψ. Suppose then that α is such a congruence.
If $(x, y) \in \psi$ then from

$$x \vee a \vee b = y \vee a \vee b, \quad x \wedge a \wedge b = y \wedge a \wedge b, \quad (a, b) \in \alpha$$

we deduce that $(x \vee a, y \vee a) \in \alpha$ and $(x \wedge a, y \wedge a) \in \alpha$. In the quotient lattice
L/α this implies that

$$[x]_\alpha \vee [a]_\alpha = [y]_\alpha \vee [a]_\alpha, \quad [x]_\alpha \wedge [a]_\alpha = [y]_\alpha \wedge [a]_\alpha$$

whence, since L/α is distributive, we have $[x]_\alpha = [y]_\alpha$ and so $(x, y) \in \alpha$. Hence
$\psi \leqslant \alpha$ and consequently $\psi = \vartheta(a, b)$. $\square$

Corollary *If L is distributive and $a, b \in L$ with $a \leqslant b$ then the principal
congruence $\vartheta(a, b)$ is given by*

$$(x, y) \in \vartheta(a, b) \iff x \wedge a = y \wedge a \quad \text{and} \quad x \vee b = y \vee b.$$

In particular, if L is bounded then $\vartheta(0, b) = \ker \mu_b$ and $\vartheta(a, 1) = \ker \lambda_a$. $\square$

Principal congruences on a distributive lattice have the following interesting properties.

Theorem 8.11 *If L is distributive then the intersection of two pincipal congruences on L is a principal congruence. More precisely, if $a \leqslant b$ and $c \leqslant d$
then*

$$\vartheta(a, b) \wedge \vartheta(c, d) = \vartheta\big(b \wedge d \wedge (a \vee c), b \wedge d\big).$$

Proof By Theorem 8.10 it is readily seen that
$$\big(b \wedge d \wedge (a \vee c),\, b \wedge d\big) \in \vartheta(a,b) \wedge \vartheta(c,d)$$
and so $\vartheta\big(b \wedge d \wedge (a \vee c),\, b \wedge d\big) \leqslant \vartheta(a,b) \wedge \vartheta(c,d)$. To establish the reverse inequality, suppose that $(x,y) \in \vartheta(a,b) \wedge \vartheta(c,d)$, Then, by Theorem 8.10 again, we have
$$x \wedge a = y \wedge a, \quad x \vee b = y \vee b, \quad x \wedge c = y \wedge c, \quad x \vee d = y \vee d.$$
A simple computation now gives $x \wedge b \wedge d \wedge (a \vee c) = y \wedge b \wedge d \wedge (a \vee c)$ and $x \vee (b \wedge d) = y \vee (b \wedge d)$, whence we have $(x,y) \in \vartheta\big(b \wedge d \wedge (a \vee c),\, b \wedge d\big)$. $\square$

Theorem 8.12 *If L is a bounded distributive lattice then every principal congruence on L is complemented; for $a \leqslant b$ in L, the complement of $\vartheta(a,b)$ in $\operatorname{Con} L$ is $\varphi(a,b) = \vartheta(0,a) \vee \vartheta(b,1)$.*

Proof Denote the smallest element of $\operatorname{Con} L$ (the relation of equality) by ω, and the biggest element (the universal congruence on L) by ι. Since $\operatorname{Con} L$ is distributive we have, using Theorem 8.11,
$$\begin{aligned}
\vartheta(a,b) \wedge \varphi(a,b) &= [\vartheta(a,b) \wedge \vartheta(0,a)] \vee [\vartheta(a,b) \wedge \vartheta(b,1)] \\
&= \vartheta(b \wedge a, b \wedge a) \vee \vartheta(b,b) \\
&= \omega \vee \omega = \omega.
\end{aligned}$$
On the other hand, consider the subchain $0 \leqslant a \leqslant b \leqslant 1$ of L. Here we have
$$(0,a) \in \vartheta(0,a), \quad (a,b) \in \vartheta(a,b), \quad (b,1) \in \vartheta(b,1).$$
It follows that $(0,1) \in \vartheta(0,a) \vee \vartheta(a,b) \vee \vartheta(b,1)$ and so $\vartheta(a,b) \vee \psi(a,b) = \iota$. Hence the congruence $\varphi(a,b)$ is the complement of $\vartheta(a,b)$ in $\operatorname{Con} L$. $\square$

We can now identify the complemented elements of $\operatorname{Con} L$.

Theorem 8.13 *If L is a bounded distributive lattice then the complemented elements of $\operatorname{Con} L$ are those congruences that can be expressed as finite joins of principal congruences.*

Proof By Theorem 8.12, every principal congruence is complemented. Since $\operatorname{Con} L$ is distributive, it follows that if ϑ is the join of finitely many principal congruences then ϑ is complemented.

Conversely, suppose that ψ is a complemented congruence on L and let ψ' be its complement. Then $(0,1) \in \iota = \psi \vee \psi'$ implies the existence of $a_0, a_1, \ldots, a_n \in L$ such that $a_0 = 0$, $a_n = 1$, and a_{i-1} congruent to a_i modulo ψ or ψ' for $i = 1, \ldots, n$. Clearly, we have $\iota = \bigvee_{i=1}^{n} \vartheta(a_{i-1}, a_i)$. Since $\operatorname{Con} L$ is distributive, it follows that
$$\psi = \bigvee_{i=1}^{n} \big(\psi \wedge \vartheta(a_{i-1}, a_i)\big).$$
But for each i either $a_{i-1} \,\psi\, a_i$, in which case $\psi \wedge \vartheta(a_{i-1}, a_i) = \vartheta(a_{i-1}, a_i)$, or $a_{i-1} \,\psi'\, a_i$, in which case $\psi \wedge \vartheta(a_{i-1}, a_i)$ reduces to ω. It follows therefore that
$$\psi = \bigvee \{\vartheta(a_{i-1}, a_i) \mid (a_{i-1}, a_i) \in \psi\},$$
a finite join of principal congruences. $\square$

EXERCISES

8.9. Let z be a central element of a bounded lattice L and let z' be its unique complement. Prove that $\vartheta(0, z')$ is the complement of $\vartheta(0, z)$ in $\operatorname{Con} L$.

8.10. Consider the infinite chain C described by

$$a_0 < a_1 < \cdots < a_n < a_{n+1} < \cdots \cdots < b_{m+1} < b_m < \cdots < b_1 < b_0.$$

If $x, y \in C$, what are the $\vartheta(x, y)$-classes? If

$$\varphi = \bigvee_{i>0} \vartheta(a_0, a_i) \vee \bigvee_{i>0} \vartheta(b_i, b_0),$$

what are the φ-classes? Prove that

$$\varphi \vee \bigwedge_{i \geqslant 0} \vartheta(a_i, b_i) \neq \bigwedge_{i \geqslant 0} \big(\varphi \vee \vartheta(a_i, b_i) \big).$$

8.11. Let B be a boolean lattice and let $a, b \in B$. Prove that the $\vartheta(a, b)$-class of $x \in B$ is the interval $[x \wedge (a \vee b'), x \vee (a' \wedge b)]$.

8.12. If L is a complete lattice then $a \in L$ is said to be **compact** if $a \leqslant \sup_L X$ for some $X \subseteq L$ implies that $a \leqslant \sup_L X^*$ for some finite $X^* \subseteq X$. A complete lattice is said to be **algebraic** if every element is a join of compact elements. Prove that every principal congruence on a lattice L is compact in $\operatorname{Con} L$, and deduce that $\operatorname{Con} L$ is algebraic.

We know from Theorem 8.1 that if L is a lattice then $\operatorname{Con} L$ is a complete Heyting lattice. Our objective now is to determine precisely when $\operatorname{Con} L$ is boolean. For this purpose, we require the following results.

Theorem 8.14 *Let B be a complete boolean lattice. Then B is finite if and only if every ideal of B is principal.*

Proof $\Rightarrow$: This is clear.

$\Leftarrow$: Suppose that every ideal of B is principal. Then B has no infinite ascending chains. Indeed, if $x_1 < x_2 < x_3 < \cdots$ were such a chain in B then $\bigcup_{i \geqslant 1} x_i^{\downarrow}$ is an ideal of B that is not principal, a contradiction. Since complementation is a dual isomorphism on B, it follows that B also has no infinite descending chains. Consequently, every non-zero element of B contains an atom. Now B contains only finitely many atoms. To see this, suppose that $a_1, a_2, a_3, \cdots$ are distinct atoms of B. Then for all $n \geqslant 1$ we have $\bigvee_{i=1}^{n} a_i < \bigvee_{i=1}^{n+1} a_i$; for otherwise we have equality for some n whence $a_{n+1} \leqslant \bigvee_{i=1}^{n} a_i$ and there follows the contradiction $a_{n+1} = a_{n+1} \wedge \bigvee_{i=1}^{n} a_i = \bigvee_{i=1}^{n} (a_{n+1} \wedge a_i) = 0$. From this observation we obtain the infinite ascending chain

$$a_1 < a_1 \vee a_2 < a_1 \vee a_2 \vee a_3 < \cdots$$

in contradiction to the above. Now since B is complete and atomic we have by Theorem 6.12 that $B \simeq \mathbb{P}(\mathcal{A})$ where $\mathcal{A}$ is the set of atoms of B. Since $\mathcal{A}$ is finite it then follows that so is B. $\qquad \square$

Theorem 8.15 *If L is a bounded distributive lattice then there is a boolean lattice B such that $\operatorname{Con} L \simeq I(B) \simeq \operatorname{Con} B$.*

Proof Let B be the set of complemented elements in Con L. By Theorem 8.13, these are precisely the congruences that can be expressed as finite joins of principal congruences. Clearly, B is a boolean lattice. For every $\vartheta \in \text{Con}\,L$ let $J(\vartheta) = \{\varphi \in B \mid \varphi \leqslant \vartheta\}$. Then $J(\vartheta)$ is a principal ideal of B, and

$$J(\vartheta_1) \subseteq J(\vartheta_2) \iff \vartheta_1 \leqslant \vartheta_2.$$

In fact, if $J(\vartheta_1) \subseteq J(\vartheta_2)$ then

$$(a,b) \in \vartheta_1 \implies \vartheta(a,b) \leqslant \vartheta_1 \implies \vartheta(a,b) \in J(\vartheta_1) \implies (a,b) \in \vartheta_2;$$

and, clearly, if $\vartheta_1 \leqslant \vartheta_2$ then $J(\vartheta_1) \subseteq J(\vartheta_2)$.

If $I(B)$ is the lattice of ideals of B, let $J : \text{Con}\,L \to I(B)$ be the mapping given by $\vartheta \mapsto J(\vartheta)$. Given any $I \in I(B)$, let ϑ be the join in Con L of the elements of I. Then clearly $I \subseteq J(\vartheta)$. On the other hand, if $\vartheta(a,b) \in J(\vartheta)$ then $(a,b) \in \vartheta$ implies the existence of finitely many elements $\vartheta(a_i, b_i)$ of I such that $(a,b) \in \bigvee_{i=1}^{n} \vartheta(a_i, b_i)$. This shows that $\vartheta(a,b) \in I$ and consequently $I = J(\vartheta)$, whence J is surjective.

It follows from the above that J is an isomorphism from Con L to $I(B)$. The conclusion now follows from Theorem 8.9, since the assignment $I \mapsto \vartheta_I$ is an isomorphism from $I(B)$ to Con B. $\qquad\qquad\square$

We can now establish the following remarkable result.

Theorem 8.16 (Hashimoto [60]) *Let L be a bounded distributive lattice. Then Con L is boolean if and only if L is finite.*

Proof As in the proof of Theorem 8.15, let B be the boolean lattice consisting of the complemented elements in Con L.

$\Rightarrow$: Suppose that Con L is boolean. Then we have $B = \text{Con}\,L \simeq I(B)$. Given $I \in I(B)$, let

$$I^\star = \{x \in B \mid (\forall i \in I) \quad x \wedge i = 0\}.$$

Then $I^\star$ is the complement of I in the boolean lattice $I(B)$. In fact, it is clear that $I \cap I^\star = \{0\}$; and if X is the complement of I in $I(B)$ then $I \cap X = \{0\}$ gives $X \subseteq I^\star$, whence $I \vee X = B$ gives $I \vee I^\star = B$. It follows from this that $1 = i \vee x$ for some $i \in I$ and $x \in I^\star$. Since $i \wedge x = 0$ it follows that $i' = x \in I^\star$ and so $i' \wedge j = 0$ for every $j \in I$. Then $j \leqslant i'' = i$ for every $j \in I$ and therefore I is a principal ideal. It now follows by Theorem 8.14 that B is finite.

$\Leftarrow$: If L is finite then so is B. Every ideal of B is then principal and $B \simeq I(B)$ under the assignment $x \mapsto x^{\downarrow}$. That Con L is boolean now follows by the proof of Theorem 8.15. $\qquad\qquad\square$

Remark In fact, Hashimoto proved a stronger result, namely that a distributive lattice has a boolean lattice of congruences if and only if it is **locally finite** in the sense that all of its intervals are finite.

EXERCISES

8.13. Describe the lattice of congruences of each of the lattices

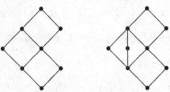

8.14. Let L be a complete relatively complemented lattice. Given $\vartheta \in \operatorname{Con} L$ and $I = \operatorname{Ker}\vartheta$, let $z = \sup_L I$ and $z^* = \sup_L \{x \in L \mid x^\downarrow \cap I = \{0\}\}$. Prove that z is central with z^* as its complement. Prove also that $\vartheta(0, z^*)$ is the pseudocomplement of ϑ in $\operatorname{Con} L$. Deduce that $\operatorname{Con} L$ is a Stone lattice.

8.4 Congruences on p-algebras

In order to deal with congruences on lattices with additional operations (for example, pseudocomplementation) we require some of the language of **universal algebra**. Fundamental to this is the notion of an **n-ary operation** on a set A which is defined to be a mapping $f : A^n \to A$. We shall be concerned only with the cases where $n = 0, 1, 2$ which give respectively a **nullary operation** (this simply fixes an element of A), a **unary operation** (such as $x \mapsto x'$), and a **binary** operation (such as $(x, y) \mapsto x \wedge y$).

An **algebra of type** $(n_1, \ldots, n_k)$ is a pair $(A; F)$ where A is a non-empty set and F is a k-tuple $(f_1, \ldots, f_k)$ such that each f_i is an n_i-ary operation on A. Thus, for example, a lattice $(L; \wedge, \vee)$ is an algebra of type $(2, 2)$; a bounded lattice $(L; \wedge, \vee, 0, 1)$ is an algebra of type $(2, 2, 0, 0)$; a boolean lattice $(B; \wedge, \vee, 0, 1)$ is an algebra of type $(2, 2, 0, 0)$; a boolean algebra $(B; \wedge, \vee,', 0, 1)$ is an algebra of type $(2, 2, 1, 0, 0)$.

A **subalgebra** of an algebra A is a non-empty subset B that is closed under all the operations of A. For example, a subalgebra of a Stone lattice is just a sublattice, whereas a subalgebra of a Stone algebra is a sublattice that is closed under $a \mapsto a^*$. It is clear that an algebra and its subalgebras are of the same type.

If A and B are algebras of the same type then a **morphism** from A to B is a mapping $g : A \to B$ which is such that, for each n-ary operation f on A,

$$g[f(a_1, \ldots, a_n)] = f\big(g(a_1), \ldots, g(a_n)\big).$$

For example, if A and B are Stone algebras then a (Stone) morphism is a lattice morphism $g : A \to B$ such that $g(x^*) = [g(x)]^*$ for every $x \in A$.

A **congruence** on an algebra A is an equivalence relation ϑ on A that satisfies the **compatibility property** for each operation in A; more precisely, for each n-ary operation f of A,

$$(a_i, b_i) \in \vartheta \ (i = 1, \ldots, n) \ \Rightarrow \ \big(f(a_1, \ldots, a_n), f(b_1, \ldots, b_n)\big) \in \vartheta.$$

If ϑ is a congruence on A then the set A/ϑ of ϑ-classes can be made into an algebra of the same type by defining for each n-ary operation f on A an n-ary operation $\overline{f}$ on A/ϑ by

$$\overline{f}([a_1]_\vartheta, \ldots, [a_n]_\vartheta) = [f(a_1, \ldots, a_n)]_\vartheta.$$

The mapping $\natural_\vartheta : A \to A/\vartheta$ given by $\natural_\vartheta(x) = [x]_\vartheta$ is then a morphism, called the **natural morphism**, and its kernel is ϑ.

It follows from the above that, for example, in a p-algebra an equivalence relation ϑ is a congruence if and only if

$(a, b) \in \vartheta$ and $(c, d) \in \vartheta$ imply $(a \wedge c, b \wedge d) \in \vartheta$;
$(a, b) \in \vartheta$ and $(c, d) \in \vartheta$ imply $(a \vee c, b \vee d) \in \vartheta$;
$(a, b) \in \vartheta$ implies $(a^\star, b^\star) \in \vartheta$.

i.e., if ϑ is a lattice congruence that satisfies the additional requirement

$$(a, b) \in \vartheta \;\Rightarrow\; (a^\star, b^\star) \in \vartheta.$$

In this situation, and in others to follow, every congruence is in particular a lattice congruence and it is essential to distinguish these two types. In order to do so, we shall use the subscript 'lat' to denote a lattice congruence.

Example 8.2 Let $\mathcal{L} = (L; \wedge, \vee, ^\star, 0, 1)$ be the p-algebra whose lattice reduct $(L; \wedge, \vee, 0, 1)$ is

The equivalence relation ϑ whose partition is

$$\{\{0, a\}, \{b, c\}, \{1\}\}$$

is such that $\vartheta \in \mathrm{Con}_{\mathrm{lat}}\, \mathcal{L}$. Clearly, $(0, a) \in \vartheta$ but $(0^\star, a^\star) = (1, b) \notin \vartheta$, and so $\vartheta \notin \mathrm{Con}\, \mathcal{L}$.

In a p-algebra there is a simple criterion for a lattice congruence to be a congruence.

Theorem 8.17 *If $\mathcal{L}$ is a p-algebra then a lattice congruence ϑ on $\mathcal{L}$ is a congruence if and only if*

$$(0, x) \in \vartheta \;\Rightarrow\; (x^\star, 1) \in \vartheta.$$

Proof The condition is clearly necessary. Conversely, suppose that the condition holds and that ϑ is a lattice congruence with $(x, y) \in \vartheta$. Then $0 = x^\star \wedge x \, \vartheta \, x^\star \wedge y$ and consequently $1 \, \vartheta \, (x^\star \wedge y)^\star$. Using Theorem 7.1(4), we deduce that $x^\star \, \vartheta \, x^\star \wedge (x^\star \wedge y)^\star = x^\star \wedge y^\star$. Similarly, we have $y^\star \, \vartheta \, x^\star \wedge y^\star$ and consequently $(x^\star, y^\star) \in \vartheta$. $\square$

The notion of the kernel of a lattice congruence extends to congruences in general. In a distributive p-algebra the kernel ideals are characterised as follows.

Theorem 8.18 *If $\mathcal{L} = (L; \wedge, \vee, ^\star, 0, 1)$ is a distributive p-algebra then an ideal I of L is a kernel ideal of $\mathcal{L}$ if and only if*

$$i \in I \;\Rightarrow\; i^{\star\star} \in I.$$

Proof $\Rightarrow$: If I is the kernel of $\vartheta \in \operatorname{Con} \mathcal{L}$ and $i \in I$ then from $i \vartheta 0$ we have $i^* \vartheta 1$ and then $i^{**} \vartheta 0$, whence $i^{**} \in I$.

$\Leftarrow$: Let I be an ideal of L such that $i \in I$ implies $i^{**} \in I$ and consider the relation ϑ_I given by

$$(x, y) \in \vartheta_I \iff (\exists i \in I) \quad x \wedge i^* = y \wedge i^*.$$

By the distributivity, we have $\vartheta_I \in \operatorname{Con}_{\operatorname{lat}} L$. Now we have

$$\begin{aligned}
x \, \vartheta_I \, 0 &\iff (\exists i \in I) \quad x \wedge i^* = 0 \\
&\iff (\exists i \in I) \quad x \leqslant i^{**} \in I; \\
x^* \, \vartheta_I \, 1 &\iff (\exists i \in I) \quad x^* \wedge i^* = i^* \\
&\iff (\exists i \in I) \quad x^* \geqslant i^* \\
&\iff (\exists i \in I) \quad x \leqslant x^{**} \leqslant i^{**} \in I.
\end{aligned}$$

These observations show on the one hand, by Theorem 8.17, that $\vartheta_I \in \operatorname{Con} \mathcal{L}$, and on the other that I is the kernel of ϑ_I. $\qquad\square$

EXERCISES

8.15. Show that if $\mathcal{L}$ is a distributive p-algebra then an alternative description of the congruence ϑ_I in the proof of Theorem 8.18 is

$$(x, y) \in \vartheta_I \iff (\exists i \in I) \quad x \vee i^{**} = y \vee i^{**}.$$

Show further that ϑ_I is the smallest congruence on $\mathcal{L}$ with kernel I.

8.16. If $\mathcal{L}$ is a distributive p-algebra prove that a principal ideal $x^{\downarrow}$ is a kernel ideal if and only if $x = x^{**}$.

8.17. Prove that in a distributive p-algebra $\mathcal{L}$ the following statements are equivalent:

(1) every ideal is a kernel ideal;
(2) every principal ideal is a kernel ideal;
(3) $\mathcal{L}$ is boolean.

As for the principal congruences on a distributive p-algebra, these can be described as follows. Here again we must distinguish them from the principal lattice congruences. We do so in denoting the latter by $\vartheta_{\operatorname{lat}}(a, b)$.

Theorem 8.19 (Lakser [77]) *Let $\mathcal{L}$ be a distributive p-algebra. If $a, b \in \mathcal{L}$ with $a \leqslant b$ then $\vartheta(a, b) = \vartheta_{\operatorname{lat}}(a, b) \vee \vartheta_{\operatorname{lat}}((a^* \wedge b)^*, 1)$.*

Proof Let $\vartheta = \vartheta_{\operatorname{lat}}(a, b) \vee \vartheta_{\operatorname{lat}}((a^* \wedge b)^*, 1)$. Since $\vartheta_{\operatorname{lat}}(a, b)$ identifies a and b it follows that so also does the equivalence relation ϑ.

Now if $\varphi \in \operatorname{Con} \mathcal{L}$ identifies a and b then from $b \varphi a$ we obtain $a^* \wedge b \varphi 0$ and then $(a^* \wedge b)^* \varphi 1$, so that φ also identifies $(a^* \wedge b)^*$ and 1. It follows from this that $\vartheta \leqslant \varphi$. It suffices therefore to show that $\vartheta \in \operatorname{Con} \mathcal{L}$; it will then be the smallest congruence that identifies a and b.

For this purpose, observe that for every $y \in \mathcal{L}$ we have $\vartheta_{\operatorname{lat}}(y, 1) \in \operatorname{Con} \mathcal{L}$. In fact, if $(x, 0) \in \vartheta_{\operatorname{lat}}(y, 1)$ then $x \wedge y = 0$ whence $y \leqslant x^*$ and then $(x^*, 1) \in \vartheta_{\operatorname{lat}}(y, 1)$ and the conclusion follows by Theorem 8.17. Thus in particular we have $\vartheta_{\operatorname{lat}}((a^* \wedge b)^*, 1) \in \operatorname{Con} \mathcal{L}$.

If now $(x, y) \in \vartheta$ then there exist $x = x_0, x_1, \ldots, x_n = y$ such that

$$(x_i, x_{i+1}) \in \vartheta_{\text{lat}}(a, b) \quad \text{or} \quad (x_i, x_{i+1}) \in \vartheta_{\text{lat}}((a^\star \wedge b)^\star, 1).$$

In the latter case, as we have just seen, we have $(x_i^\star, x_{i+1}^\star) \in \vartheta_{\text{lat}}((a^\star \wedge b)^\star, 1)$. In the former, $x_i \wedge a = x_{i+1} \wedge a$ and $x_i \vee b = x_{i+1} \vee b$ from which we obtain

$$x_i^{\star\star} \wedge a^{\star\star} = x_{i+1}^{\star\star} \wedge a^{\star\star}, \qquad x_i^\star \wedge b^\star = x_{i+1}^\star \wedge b^\star. \tag{8.1}$$

Consider now in a boolean algebra the conditions $x \wedge p = y \wedge p$ and $x' \wedge q = y' \wedge q$. Taking the join with p' in the first of these, we obtain $x \vee p' = y \vee p'$ whence $x' \wedge p = y' \wedge p$. This, together with the second, gives $x' \wedge (p' \wedge q')' = x' \wedge (p \vee q) = (x' \wedge p) \vee (x' \wedge q) = (y' \wedge p) \vee (y' \wedge q) = y' \wedge (p' \wedge q')'$. Applying this observation to (8.1), we deduce that $x_i^\star \wedge (a^\star \wedge b^{\star\star})^\star = x_{i+1}^\star \wedge (a^\star \wedge b^{\star\star})^\star$. But $(a^\star \wedge b^{\star\star})^\star = (a^\star \wedge b)^{\star\star\star} = (a^\star \wedge b)^\star$, so the above becomes $x_i^\star \wedge (a^\star \wedge b)^\star = x_{i+1}^\star \wedge (a^\star \wedge b)^\star$, i.e., $(x_i^\star, x_{i+1}^\star) \in \vartheta_{\text{lat}}((a^\star \wedge b)^\star, 1)$.

Thus we see that $(x, y) \in \vartheta$ implies $(x^\star, y^\star) \in \vartheta$ and so $\vartheta \in \text{Con}\,\mathcal{L}$ as required. $\qquad\square$

In a Stone algebra the principal congruences have a simpler description.

Theorem 8.20 *If $\mathcal{S}$ is a Stone algebra and $a \leqslant b$ then*

$$\vartheta(a, b) = \vartheta_{\text{lat}}(a \vee b^\star, b \vee a^\star).$$

Proof Since $(a, b) \in \vartheta(a, b)$ and $(a^\star, b^\star) \in \vartheta(a, b)$ we have $(a \vee b^\star, b \vee a^\star) \in \vartheta(a, b)$ and so $\vartheta_{\text{lat}}(a \vee b^\star, b \vee a^\star) \leqslant \vartheta(a, b)$. To obtain the reverse inequality, observe that $a \wedge (a \vee b^\star) = a$ and $b \wedge (a \vee b^\star) = b \wedge a = a$, and therefore $(a, b) \in \vartheta_{\text{lat}}(a \vee b^\star, b \vee a^\star)$. But since $\mathcal{S}$ is a Stone algebra we also have $(a^\star \wedge b)^\star = a^{\star\star} \vee b^\star \geqslant a \vee b^\star$ and $a^{\star\star} \vee b^\star \vee b \vee a^\star = 1$ whence we see that $((a^\star \wedge b)^\star, 1) \in \vartheta_{\text{lat}}(a \vee b^\star, b \vee a^\star)$. These observations together with Theorem 8.19 give the reverse inequality. $\qquad\square$

As for boolean algebras, the situation is even simpler.

Theorem 8.21 *In a boolean algebra every lattice congruence is a congruence.*

Proof Let $\vartheta \in \text{Con}_{\text{lat}}\,\mathcal{B}$ and let $(x, y) \in \vartheta$. Then $0 = x \wedge x' \;\vartheta\; y \wedge x'$ whence $y' \;\vartheta\; y' \vee (y \wedge x') = y' \vee x'$. Similarly, we have $x' \;\vartheta\; x' \vee y'$. Thus $(x', y') \in \vartheta$ and consequently $\vartheta \in \text{Con}\,\mathcal{B}$. $\qquad\square$

EXERCISES

8.18. Prove that every lattice congruence φ is such that

$$(x, y) \in \vartheta_{\text{lat}}(a, b) \vee \varphi \iff ([x]_\varphi, [y]_\varphi) \in \vartheta_{\text{lat}}([a]_\varphi, [b]_\varphi).$$

Deduce that in a distributive p-algebra

$$(x, y) \in \vartheta(a, b) \iff \begin{cases} x \wedge a = y \wedge a; \\ (x \vee b) \wedge (a^\star \wedge b)^\star = (y \vee b) \wedge (a^\star \wedge b)^\star. \end{cases}$$

8.19. Let L be the distributive lattice with Hasse diagram

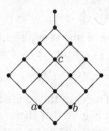

Let $\mathcal{L}$ be the p-algebra whose lattice reduct is L. Describe diagrammatically the $\vartheta(a,c)$-classes and the $\vartheta(b,c)$-classes. Prove that $\vartheta(a,c) \wedge \vartheta(b,c)$ is not a principal congruence on $\mathcal{L}$.

8.20. If S is a Stone algebra prove that $\vartheta(a,b)$ is complemented in $\operatorname{Con} S$ if and only if a is complemented in $[0,b]$. When it exists, show that the complement of $\vartheta(a,b)$ is $\vartheta_{\mathrm{lat}}(a^* \wedge b^{**}, 1)$.

8.21. Let $S = (L; \wedge, \vee, {}^*, 0, 1)$ be a Stone algebra. Given $\varphi \in \operatorname{Con} S(L)$ define $\overline{\varphi} = \bigvee\{\vartheta_{\mathrm{lat}}(x^{**}, y^{**}) \mid (x^{**}, y^{**}) \in \varphi\}$. Prove that $\overline{\varphi}$ is the smallest lattice congruence on L that extends φ. Prove also that $\overline{\varphi} \in \operatorname{Con} S$ and that $\varphi \mapsto \overline{\varphi}$ is a lattice morphism.

8.22. Let S be a Stone algebra. Prove that if $\vartheta \in \operatorname{Con} S$ then

$$(x^{**}, y^{**}) \in \vartheta \iff (x,y) \in \vartheta \vee G$$

where G is the Glivenko congruence. Prove that the relation ξ defined on $\operatorname{Con} S$ by

$$(\vartheta_1, \vartheta_2) \in \xi \iff \vartheta_1 \vee G = \vartheta_2 \vee G$$

is a lattice congruence and that $(\operatorname{Con} S)/\xi \simeq [G, \iota]$. Show that every ξ-class is of the form $[\varphi|_{S(L)}, \varphi]$ for a unique congruence $\varphi \in [G, \iota]$. Deduce that $\varphi \mapsto \overline{\varphi|_{S(L)}}$ is a residuated dual closure on $\operatorname{Con} S$.

8.23. In a double Stone algebra (Exercise 7.13) prove that

$$\vartheta(a,b) = \vartheta_{\mathrm{lat}}\big((a \vee b^*) \wedge b^+, (b \vee a^*) \wedge a^+\big).$$

Prove also that $\vartheta(a,b)$ is complemented if and only if $b \wedge b^+ \leqslant a \vee a^*$ and that, when it exists, the complement of $\vartheta(a,b)$ is

$$\vartheta_{\mathrm{lat}}\big((a^* \wedge b^{**}) \vee (a^+ \wedge b^{++}), 1\big).$$

8.5 Congruences on Heyting algebras

We now turn our attention to congruences on Heyting algebras. For this purpose we first highlight the morphisms. If $\mathcal{H}$ and $\mathcal{K}$ are Heyting algebras then a Heyting morphism $f : \mathcal{H} \to \mathcal{K}$ is a lattice morphism with the property

$$(\forall x, y \in \mathcal{H}) \quad f(x\!:\!y) = f(x)\!:\!f(y).$$

Observe that if $f : \mathcal{H} \to \mathcal{K}$ is a lattice morphism then

$$f(y) \wedge f(x\!:\!y) = f[y \wedge (x\!:\!y)] = f(y \wedge x) \leqslant f(x)$$

so that we always have the inequality $f(x\!:\!y) \leqslant f(x)\!:\!f(y)$.

Example 8.3 If $\mathcal{H}$ is a Heyting algebra with a Stone lattice reduct then $x \mapsto x^{**}$ is a Heyting morphism. To see this, observe by Theorem 7.11(4) that

$$x^* : y = (0 : x) : y = 0 : (x \wedge y) = (x \wedge y)^*$$

and consequently $x^{**} : y^{**} = (x^* \wedge y^{**})^*$. But $x \leqslant x : y$ gives $(x : y)^* \leqslant x^*$ and $y^* = 0 : y \leqslant x : y$ gives $(x : y)^* \leqslant y^{**}$, so that $(x : y)^* \leqslant x^* \wedge y^{**}$. Combining these observations, we obtain $x^{**} : y^{**} = (x^* \wedge y^{**})^* \leqslant (x : y)^{**}$. In view of the above remark, we thus see that the lattice morphism $x \mapsto x^{**}$ is such that $(x : y)^{**} = x^{**} : y^{**}$ and so is a Heyting morphism.

Heyting congruences can be described as follows.

Theorem 8.22 *Let $\mathcal{H}$ be a Heyting algebra. If F is a filter of $\mathcal{H}$ and φ_F is the associated smallest lattice congruence with cokernel F then $\varphi_F \in \mathrm{Con}\,\mathcal{H}$.*

Proof In the quotient lattice $\mathcal{H}/\varphi_F$ we have

$$
\begin{aligned}
[x] \leqslant [y] &\iff [x] = [x \wedge y] \\
&\iff (\exists z \in F) \quad z \wedge x = z \wedge x \wedge y \\
&\iff (\exists z \in F) \quad z \wedge x \leqslant y.
\end{aligned}
$$

It follows that

$$
\begin{aligned}
[a] \wedge [x] \leqslant [b] &\iff (\exists z \in F) \quad z \wedge a \wedge x \leqslant b \\
&\iff (\exists z \in F) \quad z \wedge x \leqslant b : a \\
&\iff [x] \leqslant [b : a].
\end{aligned}
$$

Thus we see that residuals exist in $\mathcal{H}/\varphi_F$ and are given by $[b] : [a] = [b : a]$. It follows that $\natural : \mathcal{H} \to \mathcal{H}/\varphi_F$ is a Heyting morphism with kernel φ_F and consequently $\varphi_F \in \mathrm{Con}\,\mathcal{H}$. $\square$

That every Heyting congruence arises in this way is the substance of the following result.

Theorem 8.23 *Let $\mathcal{H}$ be a Heyting algebra. Then an equivalence relation ϑ on $\mathcal{H}$ is a Heyting congruence if and only if there is a filter F of $\mathcal{H}$ such that $\vartheta = \varphi_F$.*

Proof $\Rightarrow$: If ϑ is a Heyting congruence, consider its cokernel, namely the set

$$F = \{x \in \mathcal{H} \mid (x, 1) \in \vartheta\}.$$

Clearly, F is a filter of $\mathcal{H}$. Now on the one hand we have

$$
\begin{aligned}
(x, y) \in \varphi_F &\Rightarrow (\exists z \in F = [1]_\vartheta) \quad x \wedge z = y \wedge z \\
&\Rightarrow [x]_\vartheta = [x]_\vartheta \wedge [1]_\vartheta = [y]_\vartheta \wedge [1]_\vartheta = [y]_\vartheta \\
&\Rightarrow (x, y) \in \vartheta,
\end{aligned}
$$

and so $\varphi_F \leqslant \vartheta$. On the other hand, if $(x, y) \in \vartheta$ then from $[x]_\vartheta = [y]_\vartheta$ we obtain $[1]_\vartheta = [y : x]_\vartheta = [x : y]_\vartheta$ whence $(x : y) \wedge (y : x) \in F$. Since, by Theorem 7.11(3),

$$x \wedge (x : y) \wedge (y : x) = x \wedge y = y \wedge (x : y) \wedge (y : x),$$

it therefore follows that $(x, y) \in \varphi_F$ and consequently we have the reverse inequality $\vartheta \leqslant \varphi_F$.

$\Leftarrow$: This is Theorem 8.22. $\square$

Corollary *If $\mathcal{H}$ is a Heyting algebra then the lattice of filters of $\mathcal{H}$ is isomorphic to* $\mathrm{Con}\,\mathcal{H}$.

Proof It follows from the above that $F \mapsto \varphi_F$ is a bijection. Clearly, $F \subseteq G$ if and only if $\varphi_F \leqslant \varphi_G$ and so we have a lattice isomorphism.   □

We can now determine the principal congruences on a Heyting algebra.

Theorem 8.24 *If $\mathcal{H}$ is a Heyting algebra and $a \leqslant b$ then*

$$\vartheta(a,b) = \vartheta_{\mathrm{lat}}(a\!:\!b,1) = \varphi_{(a:b)^\uparrow}.$$

Proof Observe that $a \wedge (a\!:\!b) = a$ and that, by Theorem 7.11(3), we have $b \wedge (a\!:\!b) = b \wedge a = a$. Consequently, $\vartheta_{\mathrm{lat}}(a\!:\!b,1)$ identifies a and b. Moreover, by definition and Theorem 8.22, we have

$$\vartheta_{\mathrm{lat}}(a\!:\!b,1) = \varphi_{(a:b)^\uparrow} \in \mathrm{Con}\,\mathcal{H}.$$

Suppose now that $\psi \in \mathrm{Con}\,\mathcal{H}$ identifies a and b. Then from $a\,\psi\,b$ we obtain $a\!:\!b\,\psi\,b\!:\!b = 1$, so that ψ identifies $a\!:\!b$ and 1. It follows that $\psi \geqslant \vartheta_{\mathrm{lat}}(a\!:\!b,1)$. Thus we see that $\vartheta(a,b) = \vartheta_{\mathrm{lat}}(a\!:\!b,1)$.   □

EXERCISES

8.24. Let $\mathcal{H}$ be a Heyting algebra. Prove that an equivalence relation ϑ on $\mathcal{H}$ is a congruence if and only if there is a filter F of $\mathcal{H}$ such that

$$(x,y) \in \vartheta \iff (x\!:\!y) \wedge (y\!:\!x) \in F.$$

8.25. Let $\mathcal{H}$ be the Heyting algebra whose lattice reduct is the lattice of Exercise 8.19. Describe the $\vartheta(a,c)$-classes and the $\vartheta(b,c)$-classes. Is $\vartheta(a,c) \wedge \vartheta(b,c)$ a principal congruence of $\mathcal{H}$?

8.26. If $\mathcal{H}$ is a Heyting algebra and F is a filter of $\mathcal{H}$ prove that $(^{**})^{\rightarrow}(F)$ is a filter of $\mathcal{H}^{**}$; and conversely that if K is a filter of $\mathcal{H}^{**}$ then $(^{**})^{\leftarrow}(K)$ is a filter of $\mathcal{H}$.

8.27. A maximal filter is often called an **ultrafilter**. Show that in the previous Exercise 'filter' may be replaced by 'ultrafilter'.

8.28. Let $\mathcal{H}$ be a Heyting algebra. If F is a filter of $\mathcal{H}$, prove that the following statements are equivalent:

(1) $\mathcal{H}/\varphi_F$ is a boolean algebra;
(2) $F = (^{**})^{\leftarrow}[(^{**})^{\rightarrow}(F)]$;
(3) F is an intersection of ultrafilters;
(4) F contains the dense filter D.

8.29. Let $\mathcal{H}$ be a Heyting algebra. If F is a filter of $\mathcal{H}$, prove that the following statements are equivalent:

(1) F is an ultrafilter;
(2) $\mathcal{H}/\varphi_F$ is the simple boolean algebra $\mathbf{2}$.

Deduce that the dense filter D of $\mathcal{H}$ is an ultrafilter if and only if $\mathcal{H}^{**}$ is simple.

8.6 Subdirectly irreducible algebras

For our purposes now we borrow some more terminology from universal algebra. If $(A_i)_{i \in I}$ is a family of non-empty sets then the cartesian product $\underset{i \in I}{\times} A_i$ is the set of all mappings $f = (x_i)_{i \in I} : I \to \underset{i \in I}{\cup} A_i$ such that $x_i = f(i) \in A_i$ for every $i \in I$. If each A_i is an algebra of type $(n_1, \ldots, n_k)$ then we can make $\underset{i \in I}{\times} A_i$ into an algebra of the same type; for every $(g_1, \ldots, g_{n_i}) \in \left(\underset{i \in I}{\times} A_i \right)^{n_i}$ we define $f_i(g_1, \ldots, g_{n_i})$ by setting, for every $j \in I$,

$$[f_i(g_1, \ldots, g_{n_i})](j) = f_i\big(g_1(j), \ldots, g_{n_i}(j)\big).$$

Suppose now that $(A_i)_{i \in I}$ is a family of algebras, all of the same type. If A is an algebra also of this type and if $(g_i)_{i \in I}$ is a family of morphisms $g_i : A \to A_i$ then we can define in a natural way a morphism $g : A \to \underset{i \in I}{\times} A_i$ by the prescription $g(x) = \big(g_i(x)\big)_{i \in I}$. We first determine when this morphism is injective, in which case we call it an **embedding**.

Definition A family $(g_i)_{i \in I}$ of morphisms with $g_i : A \to A_i$ for each $i \in I$ is said to be **separating** if, whenever $x, y \in A$ are such that $x \neq y$, there exists $i \in I$ such that $g_i(x) \neq g_i(y)$.

Theorem 8.25 *If $(g_i)_{i \in I}$ is a family of morphisms with $g_i : A \to A_i$ for each $i \in I$ then the following statements are equivalent:*

(1) *the induced morphism $g : A \to \underset{i \in I}{\times} A_i$ is an embedding;*

(2) $\underset{i \in I}{\bigcap} \ker g_i = \omega$;

(3) $(g_i)_{i \in I}$ *is separating.*

Proof Let $x, y \in A$ be such that $x \neq y$.

(1) $\Rightarrow$ (2): From $g(x) \neq g(y)$ we obtain, for some $i \in I$, $g_i(x) \neq g_i(y)$ and so $(x, y) \notin \ker g_i$. It follows from this that (2) holds.

(2) $\Rightarrow$ (3): Since $(x, y) \notin \omega = \underset{i \in I}{\bigcap} \ker g_i$ there exists $i \in I$ such that $(x, y) \notin \ker g_i$ whence $g_i(x) \neq g_i(y)$.

(3) $\Rightarrow$ (1): Since $g_i(x) \neq g_i(y)$ for some $i \in I$ we have $g(x) \neq g(y)$ and so g is injective. $\square$

Definition Let $(A_i)_{i \in I}$ be a family of algebras, all of the same type. For every $j \in I$ the mapping $\pi_j : \underset{i \in I}{\times} A_i \to A_j$ given by $\pi_j\big((x_i)_{i \in I}\big) = x_j$ is a surjective morphism, called the j-**th projection**. By a **subdirect product** of $(A_i)_{i \in I}$ we mean a subalgebra A of $\underset{i \in I}{\times} A_i$ such that, for every $j \in I$, the j-th projection maps A onto A_j.

Example 8.4 If L is a lattice then $M = \{(x, x) \mid x \in L\}$ is a sublattice of $L \times L$ which is a subdirect product.

Example 8.5 If L is a lattice then $L \times \{1\} = \{(x,1) \mid x \in L\}$ is a sublattice of $L \times \mathbf{2}$, but it is not a subdirect product since the projection onto the second component does not map $L \times \{1\}$ onto $\mathbf{2}$.

Definition An embedding $\alpha : A \to \underset{i \in I}{\mathsf{X}} A_i$ is said to be **subdirect** if $\operatorname{Im}\alpha$ is a subdirect product of $\underset{i \in I}{\mathsf{X}} A_i$.

Theorem 8.26 *If $(\vartheta_i)_{i \in I}$ is a family of congruences on an algebra A such that $\bigwedge_{i \in I} \vartheta_i = \omega$ then the natural morphism $\natural : A \to \underset{i \in I}{\mathsf{X}} A/\vartheta_i$, given by the prescription $\natural(x) = ([x]_{\vartheta_i})_{i \in I}$, is a subdirect embedding.*

Proof Let $\natural_{\vartheta_i} : A \to A/\vartheta_i$ be the natural morphism. In Theorem 8.25, take

$$A_i = A/\vartheta_i, \quad g_i = \natural_{\vartheta_i}, \quad g = \natural.$$

Then we deduce immediately that $\natural$ is an embedding. Now for every $i \in I$ and every $[x]_{\vartheta_j} \in A/\vartheta_j$ we have $[x]_{\vartheta_j} = \pi_j\big[([x]_{\vartheta_i})_{i \in I}\big] = \pi_j[\natural(x)]$, so that π_j maps $\operatorname{Im}\natural$ onto L/ϑ_j. Thus $\operatorname{Im}\natural$ is a subdirect product of $(A/\vartheta_i)_{i \in I}$ and hence $\natural$ is a subdirect embedding. $\qquad\square$

Definition An algebra A is **subdirectly irreducible** if for every subdirect embedding $\alpha : A \to \underset{i \in I}{\mathsf{X}} A_i$ there exists $i \in I$ such that $\pi_i \circ \alpha : A \to A_i$ is an isomorphism.

Example 8.6 If $A = \{a\}$ is a trivial algebra and $\alpha : A \to \underset{i \in I}{\mathsf{X}} A_i$ is a subdirect embedding then, since $\pi_i \circ \alpha$ is surjective, every A_i is trivial and consequently every $\pi_i \circ \alpha$ is an isomorphism. Thus every trivial algebra is subdirectly irreducible.

Example 8.7 Every two-element algebra is also subdirectly irreducible. If $A = \{a,b\}$ then since α is injective there exists $i \in I$ such that $(\alpha(a))_i \neq (\alpha(b))_i$ in A_i. Since $\pi_i \circ \alpha$ is surjective it follows that $|\operatorname{Im}\pi_i \circ \alpha| = 2$ whence $\pi_i \circ \alpha$ is an isomorphism.

Example 8.8 The three-element chain $\mathbf{3}$ is not a subdirectly irreducible lattice. In fact, we can embed $\mathbf{3}$ into the lattice $\mathbf{2} \times \mathbf{2}$ as follows:

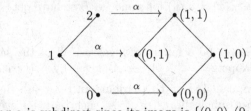

This embedding α is subdirect since its image is $\{(0,0),(0,1),(1,1)\}$ which is a subdirect product of $\{0,1\} \times \{0,1\}$, but neither $\pi_1 \circ \alpha$ nor $\pi_2 \circ \alpha$ is injective.

A useful characterisation of the non-trivial subdirectly irreducible algebras, which is often taken as a definition, is the following.

Theorem 8.27 *A non-trivial algebra A is subdirectly irreducible if and only if it has a smallest non-trivial congruence.*

Proof $\Rightarrow$: Suppose that $L = \operatorname{Con} A \backslash \{\omega\}$ does not have a smallest element. Then $\inf_{\operatorname{Con} L} L = \omega$ and it follows by Theorem 8.26 that the natural morphism $\natural : A \to \underset{\vartheta_i \in L}{\mathsf{X}} A/\vartheta_i$ is a subdirect embedding. Since $\natural_{\vartheta_i} : A \to A/\vartheta_i$ is not injective for any $\vartheta_i \in L$, and since $\pi_i \circ \natural = \natural_{\vartheta_i}$, it follows that A is not subdirectly irreducible.

$\Leftarrow$: Let ϑ be the bottom element of $\operatorname{Con} A \backslash \{\omega\}$ and let $a, b \in A$ be such that $a \neq b$ and $(a, b) \in \vartheta$. If $\alpha : A \to \underset{i \in I}{\mathsf{X}} A_i$ is a subdirect embedding then $\alpha(a) \neq \alpha(b)$ and so, for some $i \in I$, we have $(\alpha(a))_i \neq (\alpha(b))_i$. It follows that $(\pi_i \circ \alpha)(a) \neq (\pi_i \circ \alpha)(b)$ and therefore $(a, b) \notin \ker(\pi_i \circ \alpha)$. Thus we see that $\vartheta \not\leq \ker(\pi_i \circ \alpha)$. It follows from this that $\ker(\pi_i \circ \alpha) = \omega$ whence $\pi_i \circ \alpha : A \to A_i$ is injective. Since it is surjective by hypothesis, it is therefore an isomorphism. Consequently, A is subdirectly irreducible. $\square$

Definition If A is a subdirectly irreducible algebra then the smallest element of $\operatorname{Con} A \backslash \{\omega\}$ is called the **monolith** of $\operatorname{Con} A$.

We now consider subdirectly irreducible algebras in particular classes of algebras.

Definition A class of algebras is said to be **equational** if it can be defined by a set of identities.

It is a celebrated theorem of Birkhoff [10] that *a class of algebras is equational if and only if it is closed under the formation of subalgebras, quotient algebras, and cartesian products.*

Example 8.9 The classes of modular and distributive lattices are equational.

Example 8.10 The class of upper semi-modular lattices is not equational. For example, in the upper semi-modular lattice

the sublattice formed by the outer hexagon with one of the atoms removed is not upper semi-modular, so in particular this class is not closed under the formation of subalgebras.

Example 8.11 By the Corollary to Theorem 7.1, the class of p-algebras is equational.

Example 8.12 By the Corollary to Theorem 7.11, the class of Heyting algebras is equational.

Concerning subdirectly irreducible algebras in an equational class, we have the following fundamental result.

Theorem 8.28 (Birkhoff [12]) *If* **C** *is an equational class of algebras then every algebra in* **C** *can be embedded in a subdirect product of subdirectly irreducible algebras in* **C**.

Proof As we have seen above, every trivial algebra in **C** is subdirectly irreducible, so let A be a non-trivial algebra in **C**. Given $a, b \in A$ with $a \neq b$, let K be the set of all congruences ϑ on A such that $(a, b) \notin \vartheta$. Then K is inductively ordered and so by Zorn's axiom K has a maximal element, say $\vartheta_{a,b}$. We show first that $A/\vartheta_{a,b}$ is subdirectly irreducible.

Let φ be a non-trivial congruence on $A/\vartheta_{a,b}$ and define $\varPhi$ on A by

$$(x, y) \in \varPhi \iff ([x]_{\vartheta_{a,b}}, [y]_{\vartheta_{a,b}}) \in \varphi.$$

Then clearly $\varPhi$ is a congruence on A and $\vartheta_{a,b} \leqslant \varPhi$. We now observe that $([a]_{\vartheta_{a,b}}, [b]_{\vartheta_{a,b}}) \in \varphi$; for otherwise $(a, b) \notin \varPhi$ whence, by the maximality of $\vartheta_{a,b}$, we have $\vartheta_{a,b} = \varPhi$ and this contradicts the hypothesis that φ is non-trivial on $A/\vartheta_{a,b}$. Thus we see that $([a]_{\vartheta_{a,b}}, [b]_{\vartheta_{a,b}})$ belongs to every non-trivial congruence on $A/\vartheta_{a,b}$. Since $[a]_{\vartheta_{a,b}} \neq [b]_{\vartheta_{a,b}}$, it follows that the intersection of all the non-trivial congruences on $A/\vartheta_{a,b}$ is non-trivial, whence it is the monolith of $A/\vartheta_{a,b}$. It now follows by Theorem 8.27 that $A/\vartheta_{a,b}$ is subdirectly irreducible.

Now for all $a, b \in A$ with $a \neq b$ the natural morphism $\natural_{\vartheta_{a,b}} : A \to A/\vartheta_{a,b}$ is surjective, and by the definition of $\vartheta_{a,b}$ it separates a and b. We conclude by Theorem 8.25 that $\natural : A \to \underset{a \neq b}{\times} A/\vartheta_{a,b}$ is a subdirect embedding. $\square$

We now proceed to determine the subdirectly irreducible algebras in various classes of lattice-ordered algebras.

Theorem 8.29 *In the equational class of distributive lattices the only non-trivial subdirectly irreducible algebra is the two-element chain.*

Proof The distributive lattice **2** has only two congruences, so $\mathrm{Con}\, \mathbf{2} = \{\omega, \iota\}$. It follows by Theorem 8.27 that **2** is subdirectly irreducible.

Conversely, suppose that L is a non-trivial distributive lattice that is subdirectly irreducible. By way of obtaining a contradiction, suppose that $L \neq \mathbf{2}$. Then there exist $a, b, x \in L$ such that $a < x < b$. Now by the Corollary to Theorem 8.10 we see that $\vartheta_{\mathrm{lat}}(a, x) \wedge \vartheta_{\mathrm{lat}}(x, b) = \omega$ and so, by Theorem 8.27, either $\vartheta(a, x) = \omega$ or $\vartheta(x, b) = \omega$. Thus either $a = x$ or $x = b$, in each case a contradiction. $\square$

Corollary *For every non-trivial distributive lattice L there is a subdirect embedding of L into a cartesian product of copies of* **2**. $\square$

Theorem 8.30 (Lakser [76]) *In the equational class of distributive p-algebras the non-trivial subdirectly irreducible algebras are those whose lattice reduct consists of a boolean lattice with a new top element adjoined.*

Proof Suppose first that L is a distributive p-algebra such that $B = L\setminus\{1_L\}$ is boolean. If $(x, y) \in \vartheta_{\mathrm{lat}}(1_B, 1_L)$ then from $x \wedge 1_B = y \wedge 1_B$ we see that the classes modulo $\vartheta_{\mathrm{lat}}(1_B, 1_L)$ consist of $\{1_B, 1_L\}$ and singletons. It follows by Theorem 8.17 that $\vartheta_{\mathrm{lat}}(1_B, 1_L)$ is a congruence on L.

Suppose now that ϑ is any non-trivial congruence on L. Then there exist $x, y \in L$ such that $(x, y) \in \vartheta$ and $x < y$. There are two cases to consider.

(1) $y = 1_L$.

Clearly, $x \leqslant 1_B$ and in this case we have $(1_B, 1_L) \in \vartheta$.

(2) $y \leqslant 1_B$.

Here again we have $x \leqslant 1_B$. If $x = 0$ then $x^\star = 1_L$; otherwise $x^\star = x' \in B$. In each situation we have $0 < x^\star \wedge y \in B$ whence $(x^\star \wedge y)^\star = (x^\star \wedge y)'$. Then $(x, y) \in \vartheta$ gives $(0, x^\star \wedge y) \in \vartheta$ whence $\bigl(1_L, (x^\star \wedge y)'\bigr) \in \vartheta$ and since $(x^\star \wedge y)' \in B$ it follows that $(1_L, 1_B) \in \vartheta$.

Thus we see that in all cases $(1_B, 1_L) \in \vartheta$ and so $\vartheta_{\mathrm{lat}}(1_B, 1_L) \leqslant \vartheta$. Consequently $\vartheta_{\mathrm{lat}}(1_B, 1_L)$ is the monolith of $\mathrm{Con}\, L$ and therefore L is subdirectly irreducible.

Conversely, suppose that L is a distributive p-algebra that is subdirectly irreducible, and so has a monolith ϑ. Let $(p, q) \in \vartheta$ with $p < q$. Then necessarily we have $q = 1_L$; for otherwise the associated congruence $\varphi_{q\uparrow}$ is non-trivial, and $(p, q) \in \vartheta \leqslant \varphi_{q\uparrow}$ gives $p \wedge r = q \wedge r$ for some $r \in q^\uparrow$ whence the contradiction $p \wedge q = q$.

We now show that 1_L covers p. Again by way of obtaining a contradiction, suppose that there exists $x \in L$ with $p < x < 1_L$. Then $\varphi_{x\uparrow} \neq \omega$ and $(p, 1_L) = (p, q) \in \vartheta \leqslant \varphi_{x\uparrow}$ gives the contradiction $p \wedge x = x$.

Next, we observe that 1_L is $\vee$-irreducible. In fact, suppose that we had $1_L = x \vee y$ where $x < 1_L$ and $y < 1_L$. Then $\vartheta \leqslant \varphi_{x\uparrow}$ and $\vartheta \leqslant \varphi_{y\uparrow}$ give $p \wedge x = x$ and $p \wedge y = y$ whence the contradiction

$$1_L = x \vee y = (p \wedge x) \vee (p \wedge y) = p \wedge (x \vee y) = p \wedge 1_L = p.$$

From the above we thus have $L\setminus\{1_L\} = p^\downarrow$.

It remains to show that $p^\downarrow$ is boolean. This we achieve by showing that for $0 < x \leqslant p$ we have $x \vee x^\star = p$. Once more by way of obtaining a contradiction, suppose that for some $x \in p^\downarrow\setminus\{0\}$ we have $x \vee x^\star < p$. Consider the congruence $\psi = \vartheta_{\mathrm{lat}}(x \vee x^\star, p)$. We have

$$(y, 0) \in \psi \Rightarrow y \wedge (x \vee x^\star) = 0 \Rightarrow y \leqslant (x \vee x^\star)^\star = x^\star \wedge x^{\star\star} = 0,$$

and so it follows by Theorem 8.17 that ψ is a non-trivial congruence on L. Now clearly $(p, 1_L) \notin \psi$ and hence $\vartheta \nleqslant \psi$, a contradiction. Hence $x \vee x^\star = p$ for every $x \in p^\downarrow\setminus\{0\}$, as required. □

Theorem 8.31 *In the equational class of Stone algebras the only non-trivial subdirectly irreducible algebras are* **2** *and* **3**.

Proof We have seen above that **2** is subdirectly irreducible. Since $\mathrm{Con}\, \mathbf{3}$ is a three-element chain, it follows by Theorem 8.27 that **3** is also subdirectly irreducible.

Suppose now that L is a non-trivial subdirectly irreducible Stone algebra. Then, by Theorem 8.30, L can be described as a boolean algebra with a new top element adjoined. Denote the boolean algebra $L \setminus \{1_L\}$ by B. Then we observe that $|B| \leqslant 2$; for otherwise we must have $|B| \geqslant 4$ and, taking any $x \in B \setminus \{0_L, 1_B\}$, we obtain the contradiction $1_L = x^\star \vee x^{\star\star} = x^\star \vee x = 1_B$. We thus conclude that either $L \simeq \mathbf{2}$ or $L \simeq \mathbf{3}$. $\square$

Corollary *For every non-trivial Stone lattice L there is a subdirect embedding of L into a cartesian product of copies of $\mathbf{2}$ and $\mathbf{3}$.* $\square$

Theorem 8.32 *In the equational class of Heyting algebras the non-trivial subdirectly irreducible algebras are those whose lattice reduct consists of a Heyting lattice with a new top element adjoined.*

Proof By the Corollary to Theorem 8.23 we know that the lattice of filters of a Heyting algebra $\mathcal{H}$ is isomorphic to $\operatorname{Con} \mathcal{H}$. Thus $\mathcal{H}$ is subdirectly irreducible if and only if it has a smallest non-trivial filter.

Now if we take a Heyting algebra $\mathcal{H}$ and adjoin to it a new top element 1 then the resulting structure is also a Heyting algebra, and it has a smallest filter, namely $\{1_\mathcal{H}, 1\}$. It follows that this Heyting algebra is subdirectly irreducible.

Conversely, if $\mathcal{H}$ is a subdirectly irreducible Heyting algebra let its smallest non-trivial filter be F. Clearly, we must have $|F| = 2$ and so $F = a^\uparrow$ for some $a \neq 1$. Then for every $x \neq 1$ we have $a^\uparrow = F \subseteq x^\uparrow$ which gives $x \leqslant a$. Thus we see that a is the unique coatom of H, from which observation the result follows. $\square$

9

Ordered groups

9.1 Ordering groups

We now broaden our horizons to include some aspects of algebraic structures that are equipped with an order that relates to the algebraic operations.

Definition Let S be a set endowed with a law of composition that is written multiplicatively. By a **compatible order** on S we mean an order $\leqslant$ with respect to which all translations $y \mapsto xy$ and $y \mapsto yx$ are isotone.

We begin by considering the case where $(S; \cdot)$ is a group.

Definition By an **ordered group** we shall mean a group on which there is defined a compatible order.

Example 9.1 Under addition and the natural order each of $\mathbb{Z}$, $\mathbb{Q}$, $\mathbb{R}$ is an ordered abelian group.

Example 9.2 If G is an ordered group then, under the component-wise operations, so is the cartesian product group G^n.

Example 9.3 Let $C[0,1]$ be the set of all real continuous functions on the interval $[0,1]$. Defining an order on $C[0,1]$ as in Example 1.7, we see that $(C[0,1]; +)$ is an ordered abelian group.

In what follows, we shall normally write the group operation as multiplication. Thus the identity element will be denoted by 1_G, or simply 1 if there is no confusion.

Definition If G is an ordered group then $x \in G$ is said to be a **positive element** if $x \geqslant 1_G$; and a **negative element** if $x \leqslant 1_G$. The subset P_G of positive elements of G is called the **positive cone** of G, and the subset N_G of negative elements the **negative cone**.

An elementary but important consequence of translations being isotone is that the following statements concerning $x, y \in G$ are equivalent:

$$x \leqslant y; \quad xy^{-1} \in N_G; \quad y^{-1}x \in N_G; \quad x^{-1}y \in P_G; \quad yx^{-1} \in P_G; \quad y^{-1} \leqslant x^{-1}.$$

Thus an order on G is uniquely determined by either of the associated cones.

EXERCISES

9.1. Let G be an ordered group with $|G| > 1$. Prove that G cannot have a top element or a bottom element.

9.2. Prove that in an ordered group every translation is residuated and hence is an order isomorphism.

9.3. Let G be an ordered group such that $P_G \neq \{1\}$. Prove that every $x \in P_G$ with $x \neq 1$ is of infinite order.

If X is a subset of an ordered group then we shall use the notation

$$X^{-1} = \{x^{-1} \mid x \in X\}.$$

We then have the following result.

Theorem 9.1 *A subset P of a group G is the positive cone relative to some compatible order on G if and only if*

(1) $P \cap P^{-1} = \{1_G\}$;

(2) $P^2 = P$;

(3) $(\forall x \in G)\ xPx^{-1} = P$.

Moreover, this order is a total order if, in addition, $P \cup P^{-1} = G$.

Proof $\Rightarrow$: Let $\leqslant$ be a compatible order on G and let P_G be the associated positive cone. If $x \in P_G \cap P_G^{-1}$ then on the one hand $x \geqslant 1_G$, and on the other $x = y^{-1}$ for some $y \geqslant 1_G$. The latter gives $1_G \geqslant y^{-1} = x$. Hence $x = 1_G$ and so (1) holds. If now $x, y \in P_G$ then $x \geqslant 1_G$ and $y \geqslant 1_G$ whence $xy \geqslant 1_G$ and consequently $P_G^2 \subseteq P_G$. Then (2) follows from the fact that $P_G = P_G 1_G \subseteq P_G^2$. As for (3), if $y \in P_G$ then for every $x \in G$ we have $xyx^{-1} \geqslant x1_G x^{-1} = 1_G$ and so $xyx^{-1} \in P_G$. Thus $xP_G x^{-1} \subseteq P_G$. Since this holds for all $x \in G$ we can replace x by x^{-1} to obtain $x^{-1}P_G x \subseteq P_G$ which gives the reverse inclusion $P_G \subseteq xP_G x^{-1}$.

$\Leftarrow$: Suppose that P is a subset of G that satisfies properties $(1), (2), (3)$. Define the relation $\leqslant$ on G by

$$x \leqslant y \iff yx^{-1} \in P.$$

Clearly, $\leqslant$ is reflexive on G. If now $x \leqslant y$ and $y \leqslant x$ then $yx^{-1} \in P$ and $(yx^{-1})^{-1} = xy^{-1} \in P$. It follows by (1) that $yx^{-1} = 1_G$ whence $y = x$ and so $\leqslant$ is anti-symmetric. To prove that $\leqslant$ is transitive, let $x \leqslant y$ and $y \leqslant z$. Then $yx^{-1} \in P$ and $zy^{-1} \in P$ and it follows by (2) that $zx^{-1} = zy^{-1}yx^{-1} \in P$ whence $x \leqslant z$. Thus $\leqslant$ is an order on G. To see that it is compatible, let $x \leqslant y$. Then $yx^{-1} \in P$ and it follows by (3) that, for all $a, b \in G$,

$$ayb(axb)^{-1} = aybb^{-1}x^{-1}a^{-1} = a \cdot yx^{-1} \cdot a^{-1} \in P,$$

which shows that $axb \leqslant ayb$ whence it follows that $\leqslant$ is compatible. Finally, note that $1_G \leqslant y$ if and only if $y \in P$, so P is the associated positive cone.

Suppose now that $P \cup P^{-1} = G$. Then for all $x, y \in G$ we have either $xy^{-1} \in P$ or $xy^{-1} \in P^{-1}$, so that $xy^{-1} \geqslant 1_G$ or $xy^{-1} \leqslant 1_G$. In the former case we have $x \geqslant y$, and in the latter $x \leqslant y$. Consequently G is totally ordered. Conversely, if G is totally ordered then for every $x \in G$ we have either $x \geqslant 1_G$ or $x \leqslant 1_G$ so that $x \in P$ or $x \in P^{-1}$ whence $G = P \cup P^{-1}$. $\square$

From the above we see that the positive cone P_G of an ordered group G is a particular subsemigroup. Precisely which semigroups can be the positive cone of an ordered group is the substance of the following result.

Theorem 9.2 (Birkhoff [11]) *A semigroup P is the positive cone of some ordered group if and only if*

(1) *the cancellation laws hold in P;*
(2) *P has an identity element e;*
(3) *$(\forall x, y \in P)\ xy = e \Rightarrow x = y = e$;*
(4) *$(\forall x \in P)\ Px = xP$.*

Proof The necessity of the conditions is immediate from Theorem 9.1.

To establish sufficiency, we first show that a semigroup P with the stated properties can be embedded in a group. For this purpose, we observe that for any given $a, x \in P$ there exists, by (1) and (4), precisely one $x_a \in P$ such that $xa = ax_a$. Then we have

(A) $a_a = a$.
In fact, by definition, $aa = aa_a$ and so, by cancellation, $a = a_a$.

(B) $(\forall x, y \in P)\ (xy)_a = x_a y_a$.
This follows from $a(xy)_a = xya = xay_a = ax_a y_a$.

(C) $(\forall x \in P)\ (x_a)_b = x_{ab}$.
This follows from $abx_{ab} = xab = ax_a b = ab(x_a)_b$.

Consider now the set $P \times P$ and define a multiplication by

$$(a, b)(c, d) = (ac_b, db).$$

Then $P \times P$ becomes a semigroup. In fact, the equality

$$[(a, b)(c, d)](g, h) = (a, b)[(c, d)(g, h)]$$

is an immediate consequence of the equality

$$(ac_b g_{db}, hdb) = (a(cg_d)_b, hdb).$$

Consider now the relation $\equiv$ defined on $P \times P$ by

$$(a, b) \equiv (c, d) \iff ad_b = cb.$$

We observe first that this is an equivalence relation:

(α) Reflexivity is immediate from (A).

(β) If $(a, b) \equiv (c, d)$ then $ad_b = cb$ and so $ad_b d = cbd$. Now, by (C), we have $d_b d = d(d_b)_d = dd_{bd}$ and, by (B), we have $bd = (bd)_{bd} = b_{bd} d_{bd}$. Consequently, $add_{bd} = cb_{bd} d_{bd}$ and so, by cancellation, $ad = cb_{bd} = c(b_b)_d = cb_d$. Thus $(c, d) \equiv (a, b)$ and so $\equiv$ is symmetric.

(γ) As for transitivity, let $(c, d) \equiv (a, b)$ and $(a, b) \equiv (g, h)$. Then $ah_b = gb$ gives $ah_b d_b = gbd_b$. The right-hand side is gdb, whereas the left-hand side is

$$ad_b(h_b)_{d_b} = ad_b h_{bd_b} = ad_b h_{db} = cbh_{db} = ch_{db}.$$

It follows by cancellation that $ch_d = gd$, i.e. that $(c, d) \equiv (g, h)$.

We now observe that $\equiv$ is a congruence. In fact, if $(a, b) \equiv (c, d)$ then for all $(g, h) \in P \times P$ we require $(ag_b, hb) \equiv (cg_d, hd)$ and $(ga_h, bh) \equiv (gc_h, dh)$. For the first of these, we must establish

(δ) $ag_b(hd)_{hb} = cg_d hb$.

Observe that from $ad_b = cb$ we have $ad_b g_{db} h_b = cb g_{db} h_b$. The right-hand side of this is $cg_{db} bh_b = cg_d hb$ which is the right-hand side of (δ); and the left-hand side is $a(dg_d h)_b = a(gdh)_b = ag_b d_b h_b = ag_b h_b d_{hb} = ag_b(hd)_{hb}$ which is the left-hand side of (δ).

Thus we see that $\equiv$ is compatible on the right with the multiplication. In a similar way (the verification of which we leave to the reader), it is compatible on the left and is therefore a congruence. Consider now the quotient semigroup $(P \times P)/{\equiv}$, the elements of which we shall denote by $[a, b]$. It has an identity element, namely $[a, a]$ for every $a \in P$. In fact, we have $[a, a][c, d] = [ac_a, da] = [ca, da] = [c, d]$ the final equality being a consequence of $c(da)_d = cda_d = cad$; and similarly $[c, d][a, a] = [c, d]$. Moreover, every element of this quotient semigroup has an inverse, that of $[a, b]$ being $[b, a]$. In fact, $[a, b][b, a] = [ab_b, ab] = [ab, ab]$.

We thus see that the $\equiv$-classes form a group. Moreover, $a \mapsto [a, e]$ is an embedding of P into this group. To see this, observe that on taking $a = e$ in the equality $xa = ax_a$ we obtain $x = x_e$ for every $x \in P$, so that $[x, e][y, e] = [xy, e]$, and the morphism $x \mapsto [x, e]$ is injective.

If we identify P with $\{[x, e] \mid x \in P\}$, then property (3) implies that P satisfies property (1) of Theorem 9.1. Moreover, property (2) of Theorem 9.1 is immediate. As for property (3) of Theorem 9.1, note that

$$[a, b][x, e][a, b]^{-1} = [ax_b, b][b, a] = [ax_b b, ab].$$

But by property (4), given $a, b, x \in P$ there exists $y \in P$ such that $ax_b = ya$. Then $[ax_b b, ab] = [yab, ab] = [y, e]$. Consequently P also satisfies property (3) of Theorem 9.1 and the conclusion is reached. $\square$

EXERCISES

9.4. Prove that $2\mathbb{N}$ and $\mathbb{N} \setminus \{1\}$ are the positive cones of compatible orders on the additive abelian group $\mathbb{Z}$ and sketch the corresponding Hasse diagrams.

9.5. Prove that, under composition, the linear mappings $f_{a,b} : \mathbb{R} \to \mathbb{R}$ given by $f_{a,b}(x) = ax + b$ $(a, b \in \mathbb{R}, a > 0)$ form a non-abelian group G in which $f_{a,b}^{-1} = f_{a^{-1}, -ba^{-1}}$. If $P \subseteq G$ is defined by

$$f_{a,b} \in P \iff \begin{cases} a > 1; \\ \text{or } a = 1 \text{ and } b \geqslant 0, \end{cases}$$

prove that P is the positive cone of a compatible order on G and describe this order. Show further that it is a total order.

9.6. Consider the subset P of $\mathbb{C}$ given by

$$x + iy \in P \iff \begin{cases} x > 0; \\ \text{or } x = 0 \text{ and } y \geqslant 0. \end{cases}$$

Prove that P is the positive cone of a compatible order on $\mathbb{C}$ and describe this order. Show further that it is a total order.

9.7. Let G be the multiplicative group of upper triangular real matrices of the form

$$\begin{bmatrix} 1 & x & y \\ 0 & 1 & z \\ 0 & 0 & 1 \end{bmatrix}.$$

Let P be the subset that consists of those matrices in which *either* $x > 0$, *or* $x = 0$ and $y > 0$, *or* $x = y = 0$ and $z \geqslant 0$. Show that P is the positive cone of a compatible order on G.

9.8. Let G be the additive group $\mathbb{Z}/m\mathbb{Z} \times \mathbb{Z}$. Show that $P_G = \{([0], n) \mid n \geqslant 0\}$ is the positive cone of a compatible order on G.

9.9. Show that it is impossible to define a compatible total order on $\mathbb{C}$.

9.2 Convex subgroups

Definition If G is an ordered group then by a **convex subgroup** of G we shall mean a subgroup which, under the order of G, is a convex subset.

Theorem 9.3 *If H is a subgroup of an ordered group G then $P_H = H \cap P_G$. Moreover, the following statements are equivalent:*

(1) *H is convex;*
(2) *P_H is a down-set of P_G.*

Proof Since $1_H = 1_G$ it is clear that $P_H = H \cap P_G$.

(1) $\Rightarrow$ (2): Suppose that $1_H \leqslant y \leqslant x$ where $1_H, x \in P_H \subseteq H$. Then (1) gives $y \in H \cap P_G = P_H$ and so P_H is a down-set of P_G.

(2) $\Rightarrow$ (1): Suppose now that $x \leqslant y \leqslant z$ where $x, z \in H$. Then $1_H \leqslant x^{-1}y \leqslant x^{-1}z$ with $1_H, x^{-1}z \in P_H$. It follows by (2) that $x^{-1}y \in P_H \subseteq H$ whence $y \in xH = H$. Hence H is convex. $\square$

If G is an ordered group and H is a normal subgroup of G then a natural candidate for a positive cone of G/H is $\natural_H^{\rightarrow}(P_G)$. Precisely when this occurs is the substance of the following result.

Theorem 9.4 *Let G be an ordered group and let H be a normal subgroup of G. Then $\natural_H^{\rightarrow}(P_G) = \{pH \mid p \in P_G\}$ is the positive cone of a compatible order on the quotient group G/H if and only if H is convex.*

Proof Suppose that $Q = \{pH \mid p \in P_G\}$ is the positive cone of a compatible order on G/H. To show that H is convex, suppose that

$$1_G = 1_H \leqslant b \leqslant a \text{ with } a \in P_H.$$

Then $(bH)^{-1} = (bH)^{-1} \cdot aH = b^{-1}aH \in Q$ whence $bH \in Q \cap Q^{-1} = \{H\}$ and consequently $b \in H$. Thus we see that P_H is a down-set of P_G. It follows by Theorem 9.3 that H is convex.

Conversely, suppose that H is convex and let $Q = \{pH \mid p \in P_G\}$. It is clear that $Q^2 = Q$. Suppose now that $xH \in Q \cap Q^{-1}$. Then $xH = pH = q^{-1}H$ where $p, q \in P_G$. These equalities also give $pq \in H$. Consequently we have $1_H \leqslant p \leqslant pq \in H$ whence the convexity of H gives $p \in H$. It follows that $xH = H$ and hence $Q \cap Q^{-1} = \{H\}$. Finally, since P_G is a normal subsemigroup of G it is clear that $Q = \natural_H^{\rightarrow}(P_G)$ is a normal subsemigroup of G/H. It now follows by Theorem 9.1 that Q is the positive cone of a compatible order on G/H. $\square$

If H is a convex normal subgroup of an ordered group G then the order $\leqslant_H$ on G/H that corresponds to the positive cone $\{pH \mid p \in P_G\}$ can be described as in the proof of Theorem 9.1. We have

$$xH \leqslant_H yH \iff yx^{-1}H \in Q$$
$$\iff (\exists p \in P_G)\ yx^{-1} \in pH$$
$$\iff (\exists p \in P_G)(\exists h \in H)\ yx^{-1} = ph.$$

Now this last condition gives $yx^{-1} \geqslant h$ whence $y \geqslant hx$. Conversely, if $y \geqslant hx$ for some $h \in H$ then $y = y(hx)^{-1}hx$ gives $yx^{-1} = y(hx)^{-1}h$ where $y(hx)^{-1} \in P_G$. Thus we see that $\leqslant_H$ can be described by

$$xH \leqslant_H yH \iff (\exists h \in H)\ hx \leqslant y.$$

In referring to the **ordered quotient group** G/H we shall implicitly infer that the order is $\leqslant_H$ as described above.

EXERCISES

9.10. If G is an ordered group prove that the intersection of any family of convex subgroups of G is a convex subgroup of G.

9.11. If H is a subgroup of an ordered group G let $\langle H \rangle$ be the convex subgroup generated by H; i.e. the smallest convex subgroup that contains H. Prove that $\langle H \rangle = HP_G \cap HN_G$.

We now characterise group morphisms that are isotone.

Theorem 9.5 *Let G and H be ordered groups. If $f : G \to H$ is a group morphism then f is isotone if and only if $f^{\to}(P_G) \subseteq P_H$.*

Proof $\Rightarrow$: If $x \geqslant 1_G$ then $f(x) \geqslant f(1_G) = 1_H$.

$\Leftarrow$: If $x \leqslant y$ in G then $yx^{-1} \in P_G$ gives $f(y)[f(x)]^{-1} = f(yx^{-1}) \in P_H$ and hence $f(x) \leqslant f(y)$. $\square$

Corollary *If G is an ordered group and H is a convex normal subgroup of G then the natural morphism $\natural_H : G \to G/H$ is isotone.*

Proof If $yH \in \natural_H^{\to}(P_G)$ then for some $g \in P_G$ we have $yH = gH$ whence there exists $h \in H$ such that $y = gh \geqslant h$. From the above, this implies that $1H \leqslant_H yH$ and therefore $yH \in P_{G/H}$. $\square$

Definition If G and H are ordered groups then a mapping $f : G \to H$ is said to be **exact** if $f^{\to}(P_G) = P_H$.

Definition If G and H are ordered groups then we shall say that they are **isomorphic** if there is a group isomorphism $f : G \to H$ that is also an order isomorphism.

We shall indicate that ordered groups G and H are isomorphic by writing $G \stackrel{o}{\simeq} H$.

Theorem 9.6 *If G and H are ordered groups then the following statements are equivalent*:

(1) $G \overset{o}{\simeq} H$;

(2) *there is an exact group isomorphism $f : G \to H$.*

Proof (1) $\Rightarrow$ (2): If (1) holds then there is a group isomorphism $f : G \to H$ that is also an order isomorphism. Now Theorem 9.5 gives $f^{\to}(P_G) \subseteq P_H$. If $g = f^{-1}$ then since g is also isotone we have likewise $g^{\to}(P_H) \subseteq P_G$, which gives $P_H \subseteq g^{\leftarrow}(P_G) = f^{\to}(P_G)$, and consequently $f^{\to}(P_G) = P_H$.

(2) $\Rightarrow$ (1): If (2) holds then $P_G = f^{\leftarrow}(P_H)$ and by Theorem 9.5 we see that f is an isotone bijection whose inverse is also isotone. Hence the group isomorphism f is also an order isomorphism. $\qquad\qquad\square$

The first isomorphism theorem for ordered groups is the following.

Theorem 9.7 *Let G and H be ordered groups. If $f : G \to H$ is an exact group morphism then $\operatorname{Im} f \overset{o}{\simeq} G/\operatorname{Ker} f$.*

Proof Observe first that $\operatorname{Ker} f$ is clearly a convex normal subgroup of G, so we can form the ordered quotient group $G/\operatorname{Ker} f$. From the first isomorphism theorem of group theory, we have $G/\operatorname{Ker} f \simeq \operatorname{Im} f$. Writing $K = \operatorname{Ker} f$, we recall that an isomorphism $\vartheta : G/K \to \operatorname{Im} f$ is given by $\vartheta(xK) = f(x)$.

Now, using the above description of $\leqslant_K$, we have

$$xK \in P_{G/K} \iff 1_G K \leqslant_K xK \iff (\exists k \in K)\ k \leqslant x,$$

and this implies that $1_H = f(k) \leqslant f(x)$ whence $f(x) \in P_{\operatorname{Im} f}$. Conversely, if $f(x) \in P_{\operatorname{Im} f} \subseteq P_H$ then, since f is exact, there exists $g \in P_G$ such that $f(x) = f(g)$. Consequently $xK = gK$ and so there exists $k \in K$ such that $x = gk \geqslant k$. Thus we see that

$$xK \in P_{G/K} \iff \vartheta(xK) = f(x) \in P_{\operatorname{Im} f},$$

so ϑ is exact. It now follows by Theorem 9.6 that $G/\operatorname{Ker} f \overset{o}{\simeq} \operatorname{Im} f$. $\qquad\square$

EXERCISES

9.12. Let G and H be ordered groups, and let $f : G \to H$ be an isotone group monomorphism. Prove that $f^{\leftarrow}(P_H)$ is the positive cone of a compatible order on G.

9.13. Consider the additive abelian group $\mathbb{R} \times \mathbb{R}$. If the subset P is defined by

$$(x,y) \in P \iff \begin{cases} x = y = 0; \\ \text{or } x \geqslant 0 \text{ and } y > 0, \end{cases}$$

prove that P is the positive cone of a compatible order on $\mathbb{R} \times \mathbb{R}$. Consider now the convex subgroups $A = \{(x,0) \mid x \in \mathbb{R}\}$ and $B = \{(0,y) \mid y \in \mathbb{R}\}$.

Prove that the quotient group $A/(A \cap B) \overset{o}{\simeq} A$ is trivially ordered whereas the quotient group $(A \vee B)/B$ is a chain. Deduce that the second isomorphism theorem for groups does not hold for ordered groups.

9.14. Prove that $\mathbb{N}$ and $2\mathbb{N}$ are positive cones of compatible orders $\leqslant_{\mathbb{N}}$ and $\leqslant_{2\mathbb{N}}$ on the additive abelian group $\mathbb{Z}$. Show also that $\operatorname{id}_{\mathbb{Z}} : (\mathbb{Z}; \leqslant_{2\mathbb{N}}) \to (\mathbb{Z}; \leqslant_{\mathbb{N}})$ is not exact.

9.3 Lattice-ordered groups

If we impose the condition that an ordered group be a semilattice with respect to its order then the following remarkable situation develops.

Theorem 9.8 *If an ordered group $(G; \cdot, \leqslant)$ is such that $(G; \leqslant)$ is a semilattice then*

(1) $(G; \leqslant)$ *is a lattice;*
(2) *every translation in G is a complete lattice morphism.*

Proof We suppose in what follows that $(G; \leqslant)$ is a $\vee$-semilattice.

We observe first that every right translation $\lambda_x : y \mapsto yx$ is residuated (with residual λ_x^{-1}) and so, by Theorem 2.8, is a complete $\vee$-morphism. Thus

$$(\forall a, b, x \in G) \qquad (a \vee b)x = ax \vee bx. \tag{9.1}$$

Likewise, every left translation is also a complete $\vee$-morphism and so

$$(\forall a, b, x \in G) \qquad x(a \vee b) = xa \vee xb, \tag{9.2}$$

We now show that G is a lattice in which the meet operation is given by

$$a \wedge b = a(a \vee b)^{-1}b = b(a \vee b)^{-1}a. \tag{9.3}$$

For this purpose, observe that $a \leqslant a \vee b$ gives $(a \vee b)^{-1} \leqslant a^{-1}$ and therefore $a(a \vee b)^{-1}b \leqslant b$; and similarly $b \leqslant a \vee b$ gives $a(a \vee b)^{-1}b \leqslant a$. Thus $a(a \vee b)^{-1}b$ is a lower bound of $\{a, b\}$. Suppose now that $x \in G$ is any lower bound of $\{a, b\}$. Then $x \leqslant a, b$ gives $a^{-1} \vee b^{-1} \leqslant x^{-1}$ and therefore, by (9.1) and (9.2), $b^{-1}(a \vee b)a^{-1} = a^{-1} \vee b^{-1} \leqslant x^{-1}$ whence $x \leqslant [b^{-1}(a \vee b)a^{-1}]^{-1} = a(a \vee b)^{-1}b$. Consequently $\inf\{a, b\}$ exists and is $a \wedge b = a(a \vee b)^{-1}b$. Since $a \wedge b = b \wedge a$ this can also be written $a \wedge b = b(a \vee b)^{-1}a$. Thus we see that G is a lattice.

Next, we note that, when they exist,

$$\Big(\bigwedge_{\alpha \in I} x_\alpha \Big)^{-1} = \bigvee_{\alpha \in I} x_\alpha^{-1}, \quad \Big(\bigvee_{\alpha \in I} x_\alpha \Big)^{-1} = \bigwedge_{\alpha \in I} x_\alpha^{-1}. \tag{9.4}$$

In fact, if $y = \bigwedge_{\alpha \in I} x_\alpha$ then for every $\alpha \in I$ we have $y \leqslant x_\alpha$ and so $x_\alpha^{-1} \leqslant y^{-1}$ whence $\bigvee_{\alpha \in I} x_\alpha^{-1} \leqslant y^{-1} = \Big(\bigwedge_{\alpha \in I} x_\alpha \Big)^{-1}$. But if $z = \bigvee_{\alpha \in I} x_\alpha^{-1}$ then for all $\alpha \in I$ we have $z \geqslant x_\alpha^{-1}$ whence $z^{-1} \leqslant x_\alpha$ and so $z^{-1} \leqslant \bigwedge_{\alpha \in I} x_\alpha$ which gives $\Big(\bigwedge_{\alpha \in I} x_\alpha \Big)^{-1} \leqslant z = \bigvee_{\alpha \in I} x_\alpha^{-1}$. This then establishes the first equality; the second follows from this on replacing elements by their inverses.

Using (9.4) and the fact that every translation is a complete $\vee$-morphism, we now observe that

$$y\Big(\bigwedge_{\alpha \in I} x_\alpha \Big) = y\Big(\bigvee_{\alpha \in I} x_\alpha^{-1} \Big)^{-1} = \Big[\Big(\bigvee_{\alpha \in I} x_\alpha^{-1} \Big)y^{-1} \Big]^{-1}$$
$$= \Big(\bigvee_{\alpha \in I} x_\alpha^{-1}y^{-1} \Big)^{-1} = \bigwedge_{\alpha \in I} yx_\alpha.$$

Thus every left translation is also a complete $\wedge$-morphism; and similarly so is every right translation. It now follows that every translation is a complete lattice morphism. □

Definition By a **lattice-ordered group** we shall mean an ordered group $(G; \cdot, \leqslant)$ such that $(G; \leqslant)$ is a lattice.

A simple criterion for an ordered group to be lattice-ordered is provided by the following result.

Theorem 9.9 *An ordered group G is lattice-ordered if and only if* $\sup\{x, 1_G\}$ *exists for every $x \in G$.*

Proof The necessity of the condition is clear. To establish sufficiency, let $a, b \in G$ and consider the element $x = (ab^{-1} \vee 1_G)b$. Clearly, $x \geqslant ab^{-1}b = a$ and $x \geqslant 1_G b = b$, so x is an upper bound of $\{a, b\}$. Now let $y \in G$ be such that $y \geqslant a$ and $y \geqslant b$. Then $yb^{-1} \geqslant ab^{-1}$ and $yb^{-1} \geqslant 1_G$ whence $yb^{-1} \geqslant ab^{-1} \vee 1_G$ and consequently $y \geqslant (ab^{-1} \vee 1_G)b$. Hence $a \vee b$ exists in G and is $(ab^{-1} \vee 1_G)b$. Thus G is a $\vee$-semilattice and so, by Theorem 9.8, is a lattice. In this, the infimum of $\{a, b\}$ is given by $a \wedge b = (a^{-1} \vee b^{-1})^{-1}$. $\square$

Definition By a **Brouwer lattice** we shall mean a lattice L in which every translation $\lambda_x : y \mapsto x \wedge y$ is a complete $\vee$-morphism.

Example 9.4 By Theorems 2.8 and 7.9 we see that every Heyting lattice is a Brouwer lattice.

Example 9.5 By Theorem 7.10 we see that a complete lattice is a Brouwer lattice if and only if it is a Heyting lattice.

By definition, a Brouwer lattice is distributive. The relevance to lattice-ordered groups is the substance of the following result.

Theorem 9.10 *If $(G; \cdot, \leqslant)$ is a lattice-ordered group then the lattice $(G; \leqslant)$ is both brouwerian and dually brouwerian.*

Proof Suppose that $\{x_\alpha \mid \alpha \in I\} \subseteq G$ is such that $x = \bigvee\limits_{\alpha \in I} x_\alpha$ exists. Then for every $\alpha \in I$ we have $1_G \leqslant xx_\alpha^{-1}$. It follows that, for every $y \in G$, $y \wedge x = 1_G y \wedge x \leqslant xx_\alpha^{-1} y \wedge x = xx_\alpha^{-1}(y \wedge x_\alpha)$. Consequently, since $y \wedge x_\alpha \leqslant y \wedge x$ we have $1_G \leqslant (y \wedge x)(y \wedge x_\alpha)^{-1} \leqslant xx_\alpha^{-1}$. But, by Theorem 9.8(2), we have $1_G = x\left(\bigvee\limits_{\alpha \in I} x_\alpha\right)^{-1} = x\left(\bigwedge\limits_{\alpha \in I} x_\alpha^{-1}\right) = \bigwedge\limits_{\alpha \in I} xx_\alpha^{-1}$. It follows that

$$1_G = \bigwedge\limits_{\alpha \in I} (y \wedge x)(y \wedge x_\alpha)^{-1} = (y \wedge x)\left[\bigvee\limits_{\alpha \in I} (y \wedge x_\alpha)\right]^{-1}$$

which gives $\bigvee\limits_{\alpha \in I} (y \wedge x_\alpha) = y \wedge x = y \wedge \bigvee\limits_{\alpha \in I} x_\alpha$.

The dual is established similarly. $\square$

Theorem 9.11 *Let G be a lattice-ordered group and suppose that $x, y \in G$ commute. Then, for all $n > 1$,*

$$x^n \leqslant y^n \implies x \leqslant y.$$

Proof If x and y commute then so do x and y^{-1}. Thus, if $x^n \leqslant y^n$ then

$$(xy^{-1})^n = x^n(y^{-1})^n \leqslant y^n(y^{-1})^n = 1_G.$$

Now let $z = xy^{-1}$ and observe that, since z and 1_G commute, we have

$$\begin{aligned}
(z \vee 1_G)^n &= z^n \vee z^{n-1} \vee \cdots \vee z \vee 1_G \\
&= z^{n-1} \vee \cdots \vee z \vee 1_G \quad \text{since } z^n \leqslant 1_G \\
&= (z \vee 1_G)^{n-1}.
\end{aligned}$$

It follows by cancellation that $z \vee 1_G = 1_G$ whence we have $xy^{-1} = z \leqslant 1_G$ and consequently $x \leqslant y$. □

Corollary *If $x \in G$ is such that $x^n \in P_G$ with $n > 1$ then $x \in P_G$.* □

Theorem 9.12 *Lattice-ordered groups are torsion-free.*

Proof Suppose that G is a lattice-ordered group and let $x \in G$. Suppose that $x^n = 1_G$ for some positive integer n. Since $x, 1_G$ commute, we can apply Theorem 9.11 to $x^n \leqslant 1_G$ to obtain $x \leqslant 1_G$; and again to $1_G \leqslant x^n$ to obtain $1_G \leqslant x$. Thus $x = 1_G$ and the result follows. □

The positive cone of a lattice-ordered group has the following property.

Theorem 9.13 (Riesz [95]) *Let G be a lattice-ordered group. If $b_1, \ldots, b_n \in P_G$ and if $1_G \leqslant a \leqslant \prod_{i=1}^{n} b_i$ then there exist $a_1, \ldots, a_n \in P_G$ such that $a_i \leqslant b_i$ for each i and $a = \prod_{i=1}^{n} a_i$.*

Proof The result is established by induction. If $n = 1$ the result is trivial. For the inductive step, suppose that $b_1, \ldots, b_{n+1} \in P_G$ and that $1_G \leqslant a \leqslant \prod_{i=1}^{n+1} b_i$.

Then $1_G \vee ab_{n+1}^{-1} \leqslant \prod_{i=1}^{n} b_i$ and so, by the induction hypothesis, there exist $a_1, \ldots, a_n \in P_G$ with $a_i \leqslant b_i$ for each i such that

$$a(a \wedge b_{n+1})^{-1} = 1_G \vee ab_{n+1}^{-1} = \prod_{i=1}^{n} a_i.$$

Consequently, $a = \prod_{i=1}^{n+1} a_i$ where $a_{n+1} = a \wedge b_{n+1} \leqslant b_{n+1}$. □

Corollary *If $x, y, z \in P_G$ then $x \wedge yz \leqslant (x \wedge y)(x \wedge z)$.*

Proof From $1_G \leqslant x \wedge yz \leqslant yz$ and the above there exist $a, b \in P_G$ with $a \leqslant y$ and $b \leqslant z$ such that $x \wedge yz = ab$. Since $a \leqslant ab \leqslant x$ we have $a \leqslant x \wedge y$. Similarly $b \leqslant x \wedge z$ and the required inequality follows. □

EXERCISES

9.15. Consider the ordered cartesian product group $G = \mathbb{Z} \times \mathbb{Z} \times \mathbb{Z}$. Prove that the subset $H = \{(a,b,c) \mid c = a+b\}$ is a subgroup of G in which
$$\sup_H\{(a,b,c),(a',b',c')\} = \big(a \vee a',\, b \vee b',\, (a \vee a') + (b \vee b')\big).$$
Deduce that H is a lattice but is not a sublattice of G.

9.16. Let G be a lattice-ordered group and let H be a subgroup of G. Prove that the following statements are equivalent:
    (1) H is a sublattice of G;
    (2) $x \in H \Rightarrow x \vee 1_G \in H$.

9.17. Let G be a lattice-ordered group and let H be a convex normal subgroup of G. Prove that the following statements are equivalent:
    (1) G/H is lattice-ordered and $\natural_H : G \to G/H$ is a lattice morphism;
    (2) H is a sublattice of G.

9.18. If G is a lattice-ordered group and $x,y \in G$ are such that $x^n = y^n$ prove that $y = z^{-1}xz$ where $z = x^{n-1} \vee x^{n-2}y \vee \cdots \vee xy^{n-2} \vee y^{n-1}$.

9.19. An ordered group G is said to be **directed** if $x^\uparrow \cap y^\uparrow \neq \emptyset$ and $x^\downarrow \cap y^\downarrow \neq \emptyset$ for all $x,y \in G$. Prove that the following statements are equivalent:
    (1) G is directed;
    (2) for every $x \in G$ there exists an upper bound of $\{x, 1_G\}$;
    (3) for all $g,h \in G$ there exist $x,y \geqslant g$ such that $h = xy^{-1}$.
Deduce that an ordered group G is directed if and only if P_G generates G.

9.4 Absolute values and orthogonality

Definition If G is a lattice-ordered group then for every $x \in G$ we define the **positive part** of x by $x_+ = x \vee 1_G \in P_G$, and dually the **negative part** by $x_- = x \wedge 1_G \in N_G$.

Theorem 9.14 *If G is a lattice-ordered group then, for all $x,y \in G$,*
    (1) $(x_+)^{-1} = (x^{-1})_-$ *and* $(x_-)^{-1} = (x^{-1})_+$;
    (2) $x \vee y = (yx^{-1})_+x$ *and* $x \wedge y = x(x^{-1}y)_-$;
    (3) $x = x_+x_- = x_-x_+$;
    (4) $x \leqslant y$ *if and only if* $x_+ \leqslant y_+$ *and* $x_- \leqslant y_-$.

Proof (1) This is clear from the definitions.

(2) Observe that, as established in the proof of Theorem 9.9, $x \vee y = (yx^{-1} \vee 1_G)x = (yx^{-1})_+x$. The second expression follows from the equality $x \wedge y = (x^{-1} \vee y^{-1})^{-1}$ and (1).

(3) Observe that in general we have
$$xy \wedge yx \leqslant (x \vee y)(x \wedge y) \leqslant xy \vee yx.$$
In fact, by Theorem 9.8, $(x \vee y)(x \wedge y) = x(x \wedge y) \vee y(x \wedge y) \leqslant xy \vee yx$ and similarly $(x \vee y)(x \wedge y) = (x \vee y)x \wedge (x \vee y)y \geqslant yx \wedge xy$. Taking $y = 1_G$ in these inequalities, we obtain $x = (x \vee 1_G)(x \wedge 1_G) = x_+x_-$. A similar argument shows that $x = x_-x_+$.

(4) If $x \leqslant y$ then clearly $x_+ \leqslant y_+$ and $x_- \leqslant y_-$. The converse is immediate by (3). $\qquad\square$

Definition If G is a lattice-ordered group then for every $x \in G$ we define the **absolute value** of x by $|x| = x \vee x^{-1}$.

Example 9.6 The lattice-ordered group $(\mathbb{Z}; +, \leqslant)$, where $\leqslant$ is the natural order, is a chain. Here $x \vee y = \max\{x, y\}$, $x \wedge y = \min\{x, y\}$, $x^{-1} = -x$, so

$$|x| = x \vee (-x) = \begin{cases} x & \text{if } x \geqslant 0; \\ -x & \text{if } x \leqslant 0. \end{cases}$$

The basic properties of absolute values are listed in the following result.

Theorem 9.15 *If G is a lattice-ordered group then, for all $x, y, z \in G$,*

(1) $|x| = |x^{-1}|$;

(2) $|x| \in P_G$;

(3) $|x| = 1_G \iff x = 1_G$;

(4) $|x| \leqslant |y| \iff |y|^{-1} \leqslant x \leqslant |y|$;

(5) $|xy^{-1}| = (x \vee y)(x \wedge y)^{-1} = |yx^{-1}|$;

(6) $|xy| \leqslant |x|\,|y|\,|x|$;

(7) $|x| = x_+(x_-)^{-1}$;

(8) $|(x \vee z)(y \vee z)^{-1}| \cdot |(x \wedge z)(y \wedge z)^{-1}| = |xy^{-1}|$;

(9) $|x \vee y| \leqslant |x| \vee |y| \leqslant |x|\,|y|$.

Proof (1) This is clear from the definition.

(2) We have

$$1_G = (x \vee x^{-1})(x \vee x^{-1})^{-1} = (x \vee x^{-1})(x^{-1} \wedge x) \leqslant (x \vee x^{-1})^2 = |x|^2.$$

It therefore follows by Theorem 9.11 that $1_G \leqslant |x|$ and so $|x| \in P_G$.

(3) If $1_G = |x| = x \vee x^{-1}$ then $x \leqslant 1_G$ and $x^{-1} \leqslant 1_G$, the latter giving $1_G \leqslant x$. Hence $x = 1_G$.

(4) Observe that $|x|^{-1} = x \wedge x^{-1} \leqslant x \leqslant x \vee x^{-1} = |x|$. Thus, if $|x| \leqslant |y|$ then $|y|^{-1} \leqslant |x|^{-1}$ and we have $|y|^{-1} \leqslant x \leqslant |y|$. Conversely, if $|y|^{-1} \leqslant x \leqslant |y|$ then clearly also $|y|^{-1} \leqslant x^{-1} \leqslant |y|$, so that $x \vee x^{-1} \leqslant |y|$ and therefore $|x| \leqslant |y|$.

(5) We have

$$\begin{aligned}
|xy^{-1}| = xy^{-1} \vee (xy^{-1})^{-1} &= xy^{-1} \vee yx^{-1} \\
&= xy^{-1} \vee yx^{-1} \vee 1_G \qquad \text{by (2)} \\
&= (xy^{-1} \vee 1_G)(yx^{-1} \vee 1_G) \\
&= (x \vee y)y^{-1}y(x^{-1} \vee y^{-1}) \\
&= (x \vee y)(x \wedge y)^{-1},
\end{aligned}$$

whence we have the first equality. The second follows by (1).

(6) We have

$$|x|^{-1}|y|^{-1}|x|^{-1} \leqslant |x|^{-1}|y|^{-1} \leqslant xy \leqslant |x|\,|y| \leqslant |x|\,|y|\,|x|$$

from which the result follows by (4).

(7) We have $x_+(x_-)^{-1} = (x \vee 1_G)(x \wedge 1_G)^{-1} = (x \vee 1_G)(x^{-1} \vee 1_G) = 1_G \vee x \vee x^{-1} = 1_G \vee |x| = |x|$ by (2).

(8) Observe first that for all $x, y, z \in G$ we have

$$(x \wedge y) \vee (y \wedge z) \vee (z \wedge x) = (x \wedge y) \vee [(x \vee y) \wedge z]$$
$$= (x \wedge y)(x \wedge y \wedge z)^{-1}[(x \vee y) \wedge z] \quad \text{by (9.3)}$$
$$= (x \wedge y)[(x \wedge y)^{-1} \vee z^{-1}][(x \vee y) \wedge z]$$
$$= [1_G \vee (x \wedge y)z^{-1}][(x \vee y) \wedge z]$$
$$= [(x \wedge y) \vee z]z^{-1}[(x \vee y) \wedge z].$$

By the distributivity, $(x \wedge y) \vee (y \wedge z) \vee (z \wedge x)$ is self-dual. Hence

$$[(x \wedge y) \vee z]z^{-1}[(x \vee y) \wedge z] = [(x \vee y) \wedge z]z^{-1}[(x \wedge y) \vee z].$$

It therefore follows by (5) that

$$|(x \vee z)(y \vee z)^{-1}| \cdot |(x \wedge z)(y \wedge z)^{-1}|$$
$$= (x \vee y \vee z)[(x \wedge y) \vee z]^{-1} \cdot [(x \vee y) \wedge z](x \wedge y \wedge z)^{-1}$$
$$= \underline{(x \vee y)[(x \vee y) \wedge z]^{-1}z}[(x \wedge y) \vee z]^{-1}[(x \vee y) \wedge z]z^{-1}\underline{[(x \wedge y) \vee z](x \wedge y)^{-1}}$$
$$= (x \vee y)1_G(x \wedge y)^{-1} \quad \text{by the above observation}$$
$$= |xy^{-1}|.$$

(9) We have

$$|x \vee y| = x \vee y \vee (x \vee y)^{-1} \leqslant x \vee y \vee x^{-1} = |x| \vee y \leqslant |x| \vee |y|.$$

Moreover, $|x| \leqslant |x| \, |y|$ and $|y| \leqslant |x| \, |y|$ give $|x| \vee |y| \leqslant |x| \, |y|$.   □

Theorem 9.16 (Kalman [72]) *For a lattice-ordered group G the following statements are equivalent*:

(1) *G is commutative*;
(2) *P_G is commutative*;
(3) *$(\forall x, y \in G) \quad |xy| \leqslant |x| \, |y|$.*

Proof (1) $\Rightarrow$ (3): For all $x, y \in G$ we clearly have $xy \leqslant |x| \, |y|$, and $(xy)^{-1} = y^{-1}x^{-1} \leqslant |y^{-1}| \, |x^{-1}| = |y| \, |x|$. If now (1) holds then $|y| \, |x| = |x| \, |y|$ and so $xy \vee (xy)^{-1} \leqslant |x| \, |y|$, whence (3) follows.

(3) $\Rightarrow$ (2): Suppose that (3) holds and let $x, y \in P_G$. Then $xy \in P_G$ and the hypothesis gives $xy = |xy| = |(xy)^{-1}| = |y^{-1}x^{-1}| \leqslant |y^{-1}| \, |x^{-1}| = |y| \, |x| = yx$; and similarly $yx \leqslant xy$. Hence x and y commute and so the subsemigroup P_G is abelian.

(2) $\Rightarrow$ (1): If (2) holds then for all $x, y \in G$ we have that y_+ commutes with $(x^{-1})_+$ and therefore commutes with $[(x^{-1})_+]^{-1} = x_-$. Likewise, $(x^{-1})_+ = (x_-)^{-1}$ commutes with $(y^{-1})_+ = (y_-)^{-1}$ and so x_- commutes with y_-. Thus, by Theorem 9.14(3) we see that $xy = x_+x_-y_+y_- = y_+y_-x_+x_- = yx$, whence (1) holds.   □

EXERCISES

9.20. Prove that, in a lattice-ordered group G,
(1) $(xy)_+ \leqslant x_+y_+$ and $(xy)_- \leqslant x_-y_-$.
(2) $|x| = x_+(x^{-1})_+$.
(3) $x_+ \wedge (x^{-1})_+ = 1_G$.

9.21. If G is a lattice-ordered additive abelian group prove that, for all $x, y \in G$,
$$||x| - |y|| \leqslant |x - y|.$$

9.22. If G is a lattice-ordered group then $x \in G$ is called a **component** of $g \in G$ if $|x| \wedge |gx^{-1}| = 1_G$. Prove that if x is a component of g then $|x|$ is a component of $|g|$. The **absolute components** of g are defined to be the components of $|g|$. Prove that the absolute components of g form a boolean subalgebra of P_G.

Definition If G is a lattice-ordered group then we say that $x, y \in G$ are **orthogonal**, or **disjoint**, if $|x| \wedge |y| = 1_G$.

Example 9.7 If G is a lattice-ordered group then $x, y, z \in G$ are such that $z = x \wedge y$ if and only if xz^{-1} and yz^{-1} are orthogonal.

A simple, but important, property of orthogonal pairs is the following.

Theorem 9.17 *Let G be an ordered group. If $x, y, z \in P_G$ are such that x, y are orthogonal and x, z are orthogonal then so are x, yz.*

Proof Clearly, $yz \in P_G$ and so $x \wedge yz \in P_G$. On the other hand, we have $1_G = (x \wedge y)(x \wedge z) = x^2 \wedge xz \wedge yx \wedge yz \geqslant x \wedge yz$. Hence $x \wedge yz = 1_G$. □

Corollary *If $x, y \in P_G$ are orthogonal then so are x^m, y^n for all $m, n \geqslant 1$.*

Proof Take $z = y$ and use induction to obtain x, y^n orthogonal, and then similarly x^m, y^n orthogonal. □

Theorem 9.18 *If G is a lattice-ordered group then, for every $x \in G$ and every positive integer n, the elements $(x_+)^n$ and $(x_-)^{-n}$ are orthogonal.*

Proof Using Theorems 9.14 and 9.15 we have $x_+ \wedge (x_-)^{-1} = x_+ \wedge (x^{-1})_+ = (x \vee 1_G) \wedge (x^{-1} \vee 1_G) = (x \wedge x^{-1}) \vee 1_G = |x|^{-1} \vee 1_G = 1_G$. The result now follows by the Corollary to Theorem 9.17. □

Theorem 9.19 *If G is a lattice-ordered group and if $x, y \in P_G$ are orthogonal then $(xy^{-1})_+ = x$ and $(xy^{-1})_- = y^{-1}$.*

Proof We have $(xy^{-1})_+ = xy^{-1} \vee 1_G = x(y^{-1} \vee x^{-1}) = x(y \wedge x)^{-1} = x1_G = x$; and similarly $(xy^{-1})_- = xy^{-1} \wedge 1_G = (x \wedge y)y^{-1} = 1_G y^{-1} = y^{-1}$. □

Theorem 9.20 *If G is a lattice-ordered group then, for all $x \in G$ and all positive integers n, $(x_+)^n = (x^n)_+$ and $(x_-)^n = (x^n)_-$.*

Proof By Theorem 9.18, $(x_+)^n$ and $(x_-)^{-n}$ are orthogonal and so, by Theorem 9.19 we have $[(x_+)^n(x_-)^n]_+ = (x_+)^n$. But since x_+ and x_- commute by Theorem 9.14(3), we have $(x_+)^n(x_-)^n = (x_+x_-)^n = x^n$. Hence $(x^n)_+ = (x_+)^n$; and similarly $(x^n)_- = (x_-)^n$. □

These results lead to the following.

Theorem 9.21 *If G is a lattice-ordered group and $x, y \in G$ commute then, for all positive integers n,*

$$(x \vee y)^n = x^n \vee y^n \quad and \quad (x \wedge y)^n = x^n \wedge y^n.$$

Proof Since x, y commute, so do $xy^{-1} \vee 1_G$ and y. Then

$$
\begin{aligned}
(x \vee y)^n = (xy^{-1} \vee 1_G)^n y^n &= [(xy^{-1})_+]^n y^n \\
&= [(xy^{-1})^n]_+ y^n \quad \text{by Theorem 9.20} \\
&= [(xy^{-1})^n \vee 1_G] y^n \\
&= [x^n y^{-n} \vee 1_G] y^n \\
&= x^n \vee y^n.
\end{aligned}
$$

Similarly, it can be shown that $(x \wedge y)^n = x^n \wedge y^n$.   □

Theorem 9.22 *If G is a lattice-ordered group and $x_1, \ldots, x_n \in P_G$ are pairwise orthogonal then $x_1 \vee \cdots \vee x_n = x_1 \cdots x_n$. Consequently, mutually orthogonal elements of P_G commute.*

Proof If $x_1, x_2 \in P_G$ are orthogonal then since, by (9.3),

$$a \vee b = a(a \wedge b)^{-1} b,$$

we see that $x_1 \vee x_2 = x_1 1_G x_2 = x_1 x_2$. Now by Theorem 9.17 and induction x_n and $x_1 \cdots x_{n-1}$ are orthogonal. We deduce, using the same formula, that $(x_1 \cdots x_{n-1}) \vee x_n = x_1 \cdots x_{n-1} \cdot 1_G \cdot x_n = x_1 \cdots x_n$. An inductive argument with the hypothesis that, for $n \geqslant 2$, $x_1 \vee \cdots \vee x_{n-1} = x_1 \cdots x_{n-1}$ now establishes the result.   □

In a lattice-ordered group G the positive cone P_G is a sublattice with bottom element 1_G. The atoms of P_G, when they exist, are characterised as follows.

Theorem 9.23 *If G is a lattice-ordered group then $p \in P_G \setminus \{1_G\}$ is an atom of P_G if and only if*

$$(\forall x, y \in P_G) \qquad p \leqslant xy \Rightarrow p \leqslant x \ or \ p \leqslant y.$$

Proof $\Rightarrow$: Suppose that p is an atom of P_G and that $p \leqslant xy$ but neither $p \leqslant x$ nor $p \leqslant y$ holds. Then p and x are orthogonal, as are p and y. It follows by Theorem 9.17 that p and xy are orthogonal, a contradiction.

$\Leftarrow$: If $p > 1_G$ has the stated property and if $1_G \leqslant x < p$ then from $p = x \cdot x^{-1} p$ we obtain $p \leqslant x^{-1} p$ whence $x \leqslant pp^{-1} = 1_G$ and therefore $x = 1_G$. Thus p is an atom of P_G.   □

The subgroup generated by a subset of atoms of P_G has the following description.

Theorem 9.24 *If G is a lattice-ordered group then the subgroup generated by a set S of atoms in P_G is the free abelian group on S.*

Proof Observe first that distinct atoms of P_G are orthogonal and therefore, by Theorem 9.22, commute. Thus the subgroup generated by S is abelian. From group theory this subgroup is the free abelian group on S if and only if, for $x_1, \ldots, x_n \in S$, whenever $x_1^{k_1} \cdots x_n^{k_n} = 1_G$ we have each $k_i = 0$. In order to establish the latter, let $p_1, \ldots, p_n$ be atoms of P_G and suppose that $x = p_1^{k_1} \cdots p_n^{k_n} \geqslant 1_G$. Without loss of generality this inequality can be expressed in the form

$$p_1^{k_1} \cdots p_m^{k_m} = a \geqslant b = p_{m+1}^{-k_{m+1}} \cdots p_n^{-k_n}$$

in which all exponents are greater than or equal to 0. Now by Theorem 9.17 we see that p_1 and b are orthogonal, whence likewise a and b are orthogonal. It follows from this that $b = 1_G$ and consequently $x = a$ in which all the exponents are greater than or equal to 0. We deduce from this observation that if $x = 1_G$ then each $k_i = 0$ and the conclusion follows. $\square$

EXERCISES

9.23. If G is a lattice-ordered group and $x, y \in P_G$ are orthogonal, prove that there exist $a, b \in P_G$ such that $x = (a \wedge b)x$ and $y = (a \wedge b)y$.

9.24. Let G and H be lattice-ordered groups and let $f : G \to H$ be a group morphism. Prove that the following statements are equivalent:
   (1) f is a lattice morphism;
   (2) $(\forall g \in G)\ f(|g|) = |f(g)|$;
   (3) if $g, h \in P_G$ are orthogonal then so also are $f(g), f(h) \in P_H$;
   (4) $(\forall g \in G)\ f(g_+) = [f(g)]_+$.

9.25. If G is a lattice-ordered group such that the sublattice P_G has the descending chain condition, prove that every $x \in P_G$ is a product of finitely many atoms.

9.5 Convex ℓ-subgroups

In general, a subgroup of a lattice-ordered group need not be a sublattice. For example, in the lattice-ordered additive abelian group $G = \mathbb{Z} \times \mathbb{Z}$ the subset $H = \{(n, -n) \mid n \in \mathbb{Z}\}$ is a subgroup but is not a sublattice since $(0,0) \vee (1,-1) = (1,0) \notin H$.

Definition By an **ℓ-subgroup** of a lattice-ordered group G we shall mean a subgroup H of G that is also a sublattice of G.

Theorem 9.25 *A subgroup H of a lattice-ordered group G is an ℓ-subgroup of G if and only if $x \vee 1_G \in H$ for every $x \in H$.*

Proof The condition is clearly necessary. Conversely, if it holds then for all $x, y \in H$ we have $x \vee y = (xy^{-1} \vee 1_G)y \in H$ whence also $x \wedge y \in H$ by (9.3) and so H is a sublattice of G. $\square$

In the theory of lattice-ordered groups an important part is played by those ℓ-subgroups that are also convex subsets. We shall denote by $\mathcal{C}(G)$ the set of convex ℓ-subgroups of G. These subgroups can be characterised as follows.

Theorem 9.26 *Let G be a lattice-ordered group and let A be a convex sub-semigroup of P_G that contains 1_G. Then the subgroup $\langle A \rangle$ generated by A is given by $\langle A \rangle = \{xy^{-1} \mid x, y \in A\}$ and is a convex ℓ-subgroup of G. Moreover, every convex ℓ-subgroup of G arises in this manner.*

Proof Consider the set $A^\star = \{xy^{-1} \mid x, y \in A\}$. Clearly, $A \subseteq A^\star \subseteq \langle A \rangle$, and if $x \in A^\star$ then $x^{-1} \in A^\star$. To prove that $A^\star = \langle A \rangle$ it therefore suffices to show that $A^\star$ is a subsemigroup. For this purpose, let $xy^{-1}, gh^{-1} \in A^\star$ and consider the product $xy^{-1}gh^{-1}$. Let $\alpha = (y \wedge g)^{-1}y$ and $\beta = (y \wedge g)^{-1}g$. Then we have $\alpha \wedge \beta = 1_G$ and so α and β commute, by Theorem 9.22. Consequently α^{-1} and β commute and we have $xy^{-1}gh^{-1} = x\alpha^{-1}(y \wedge g)^{-1}(y \wedge g)\beta h^{-1} = x\alpha^{-1}\beta h^{-1} = x\beta\alpha^{-1}h^{-1} \in A^\star$ as required.

To show that $\langle A \rangle$ is convex it suffices by Theorem 9.3 to show that its positive cone is a down-set of P_G. Suppose then that $1_G \leqslant g \leqslant xy^{-1} \in \langle A \rangle$. Then since $y^{-1} \leqslant 1_G$ we have $1_G \leqslant g \leqslant x \in A$ whence, since A is convex and contains 1_G, we obtain $g \in A \subseteq \langle A \rangle$. Thus $P_{\langle A \rangle}$ is a down-set of P_G.

Since for all $x, y \in A$ we have $1_G \leqslant xy^{-1} \vee 1_G \leqslant x \vee 1_G = x \in A$ it follows also that $xy^{-1} \vee 1_G \in A \subseteq \langle A \rangle$ and so, by Theorem 9.25, $\langle A \rangle$ is a sublattice of G and is therefore an ℓ-subgroup.

To see that every convex ℓ-subgroup H of G arises in this way, we note from Theorem 9.14 the identity $x = x_+[(x^{-1})_+]^{-1}$ which by the above gives $H = \langle P_H \rangle$. $\qquad\square$

Corollary $\mathcal{C}(G)$ *is order isomorphic to the set of convex subsemigroups of P_G that contain 1_G.*

Proof Let $\mathcal{C}(P_G)$ be the set of convex subsemigroups of P_G that contain 1_G, and consider the isotone mappings $\eta : \mathcal{C}(P_G) \to \mathcal{C}(G)$ and $\zeta : \mathcal{C}(G) \to \mathcal{C}(P_G)$ given by the prescriptions

$$\eta(A) = \langle A \rangle, \qquad \zeta(H) = P_H.$$

From the above we have $\eta\zeta(H) = \langle P_H \rangle = H$. On the other hand, if $x \in P_{\langle A \rangle}$ then $1_G \leqslant x = ab^{-1}$ where $a, b \in A$, whence $1_G \leqslant x \leqslant a \in A$ which gives $x \in A$. Conversely, if $x \in A$ then clearly $x \in P_{\langle A \rangle}$. It follows that $\zeta\eta(A) = P_{\langle A \rangle} = A$. Thus we see that η, ζ are mutually inverse order isomorphisms. $\qquad\square$

Convex ℓ-subgroups that are also normal subgroups are naturally related to the following notion.

Definition If G and H are lattice-ordered groups then by an ℓ-**morphism** from G to H we shall mean a mapping $f : G \to H$ that is both a group morphism and a lattice morphism.

Clearly, the kernel of an ℓ-morphism $f : G \to H$ is a normal convex ℓ-subgroup of G and contains each of $x_-, x_+, |x|$ whenever it contains $x \in G$.

Theorem 9.27 *If K is a normal convex subgroup of a lattice-ordered group G then the following statements are equivalent:*

(1) *K is an ℓ-subgroup;*

(2) *the quotient group G/K is a lattice-ordered group and the natural mapping $\natural_K : G \to G/K$ is an ℓ-morphism.*

Proof If K is a normal convex subgroup of G then we can form the ordered quotient group G/K and, by the Corollary to Theorem 9.5, the group morphism $\natural_K$ is isotone.

(1) $\Rightarrow$ (2): Suppose that K is an ℓ-subgroup. If $[x]_K = [y]_K$ then $xy^{-1} \in K$ and for all $z \in K$ we have, by Theorem 9.15(8),

$$|(x \vee z)(y \vee z)^{-1}| \leqslant |xy^{-1}| \in K$$

whence, by the convexity, $|(x \vee z)(y \vee z)^{-1}| \in K$. It follows by Theorem 9.15(4) that we have $(x \vee z)(y \vee z)^{-1} \geqslant |xy^{-1}|^{-1} \in K$. So there exists $k = |xy^{-1}|^{-1} \in K$ such that $k(y \vee z) \leqslant x \vee z$ and consequently $[y \vee z]_K \leqslant [x \vee z]_K$. Similarly we can show that $[x \vee z]_K \leqslant [y \vee z]_K$ whence we have equality. Likewise we can see that $[x \wedge z]_K = [y \wedge z]_K$. Thus G/K is a lattice and $\natural_K$ is an ℓ-morphism.

(2) $\Rightarrow$ (1): If (2) holds then $[x \vee 1_G]_K = [x]_K \vee [1_G]_K$ for all $x \in G$. It follows that if $x \in K$ then $x \vee 1_G \in K$ whence K is an ℓ-subgroup. $\quad\square$

Corollary *A subgroup K of a lattice-ordered group G is a normal convex ℓ-subgroup of G if and only if it is the kernel of an ℓ-morphism $f : G \to H$.* $\square$

The first isomorphism theorem for lattice-ordered groups is the following.

Theorem 9.28 *Let G and H be lattice-ordered groups and let $f : G \to H$ be an ℓ-morphism. Then $\operatorname{Im} f$ is an ℓ-subgroup of H and*

$$G/\operatorname{Ker} f \overset{o}{\simeq} \operatorname{Im} f.$$

Proof It is clear that $\operatorname{Im} f$ is an ℓ-subgroup of H. Moreover, if $K = \operatorname{Ker} f$ then the group isomorphism $\vartheta : G/K \to \operatorname{Im} f$ given by $\vartheta([x]_K) = f(x)$ is also a lattice morphism. $\quad\square$

EXERCISES

9.26. In the lattice-ordered additive abelian group $G = \mathbb{R} \times \mathbb{R}$ consider the subgroup $H = \{(x, -x) \mid x \in \mathbb{R}\}$. Show that the quotient group G/H is totally ordered but $\natural_H$ is not a lattice morphism.

9.27. Prove that the second isomorphism theorem holds for lattice-ordered groups, namely that if H and K are normal convex ℓ-subgroups of G then

$$HK/K \overset{o}{\simeq} H/(H \cap K).$$

9.28. Prove that the third isomorphism theorem holds for lattice-ordered groups, namely that if H and K are normal convex ℓ-subgroups of G such that $H \subseteq K$ then K/H is a normal convex ℓ-subgroup of G/H and

$$(G/H)/(K/H) \overset{o}{\simeq} G/K.$$

The set $\mathcal{C}(G)$ of convex ℓ-subgroups of a lattice-ordered group has the following important property.

Theorem 9.29 $\mathcal{C}(G)$ *is a complete Heyting sublattice of the lattice of subgroups of G.*

Proof Let $(H_i)_{i \in I}$ be a family of convex ℓ-subgroups of G. Clearly, $\bigcap_{i \in I} H_i$ is an ℓ-subgroup of G. Since each H_i is convex, it follows by Theorem 9.3 that P_{H_i} is a down-set of P_G for each i, whence so is $\bigcap_{i \in I} P_{H_i} = P_{\bigcap_{i \in I} H_i}$ and hence $\bigcap_{i \in I} H_i$ is also convex. Hence $\mathcal{C}(G)$ is a complete $\cap$-semilattice.

Now let H be the subgroup of G that is generated by $\bigcup_{i \in I} H_i$. Suppose that

$$1_G \leqslant g \leqslant \prod_{i=1}^n a_i \in H \text{ where each } a_i \in \bigcup_{i \in I} H_i. \text{ Then } 1_G \leqslant g \leqslant \prod_{i=1}^n (a_i \vee 1_G).$$

Now by Theorem 9.13 there exist $b_1, \ldots, b_n \in G$ such that $1_G \leqslant b_i \leqslant a_i \vee 1_G$ for each i and $g = \prod_{i=1}^n b_i$. By the convexity of the H_i we have that $b_i \in \bigcup_{i \in I} H_i$ and therefore $g \in H$. Consequently H is convex.

If now $h \in H$ then $h = \prod_{i=1}^m c_i$ where each $c_i \in \bigcup_{i \in I} H_i$, and $1_G \leqslant h \vee 1_G \leqslant \prod_{i=1}^m (c_i \vee 1_G) \in H$. Since H is convex, it follows by Theorem 9.25 that H is an ℓ-subgroup of G.   •

It follows from the above that $\mathcal{C}(G)$ is a complete sublattice of the lattice of subgroups of G. To see that it is a Heyting lattice, let $(Y_i)_{i \in I}$ be a family of convex ℓ-subgroups of G. Then it suffices to show that, for every convex ℓ-subgroup X of G,

$$A = X \cap \bigvee_{i \in I} Y_i \subseteq \bigvee_{i \in I} (X \cap Y_i) = B.$$

Suppose then that $a \in P_A$. Then $1_G \leqslant a \in P_X$ and $a = \prod_{i=1}^t y_i$ where each $y_i \in P_{Y_i}$. Thus $1_G \leqslant y_i \leqslant a \in X$ for each i and so each $y_i \in X$, whence it follows that $a = \prod_{i=1}^t y_i \in P_B$. Thus we see that $P_A \subseteq P_B$. It now follows by the Corollary of Theorem 9.26 that $A \subseteq B$ as required.   □

Theorem 9.30 *If G is a lattice-ordered group then residuals in the Heyting lattice $\mathcal{C}(G)$ are given by*

$$A : B = \{g \in G \mid (\forall b \in B) \ |g| \wedge |b| \in A\}.$$

Proof Consider the set $Z = \{g \in G \mid (\forall b \in B) \ |g| \wedge |b| \in A\}$. Clearly, Z contains 1_G. If now $y, z \in Z$ then by Theorem 9.15(1),(6) we have $|yz^{-1}| \leqslant |y| |z^{-1}| |y| = |y| |z| |y|$ and so, by the Corollary of Theorem 9.13, for all $b \in B$,

$$1_G \leqslant |yz^{-1}| \wedge |b| \leqslant |y| |z| |y| \wedge |b| \leqslant (|y| \wedge |b|)(|z| \wedge |b|)(|y| \wedge |b|) \in A.$$

Since A is convex it follows that $yz^{-1} \in Z$ and so Z is a subgroup of G.

That Z is convex follows from the fact that if $1_G \leqslant g \leqslant z \in Z$ then, for all $b \in B$, $1_G \leqslant |g| \wedge |b| \leqslant |z| \wedge |b| \in A$ whence $g \in Z$. To see that Z is a sublattice of G we observe, using Theorem 9.15(5), that

$$|yz^{-1}| = (yz^{-1} \vee 1_G)(yz^{-1} \wedge 1_G)^{-1} = |yz^{-1} \vee 1_G| |yz^{-1} \wedge 1_G|$$
$$= |(y \vee z)z^{-1}| |(y \wedge z)z^{-1}|.$$

Suppose now that $y, z \in Z$. Then from $yz^{-1} \in Z$ and the convexity of Z we deduce that both $(y \vee z)z^{-1}$ and $(y \wedge z)z^{-1}$ belong to Z whence so also do $y \vee z$ and $y \wedge z$. Thus we see that $Z \in \mathcal{C}(G)$.

We now observe that $Z \cap B \subseteq A$. In fact, if $g \in Z \cap B$ then $g \in B$ and $|g| \wedge |b| \in A$ for all $b \in B$, so that $|g| \in A$, which gives $g \vee g^{-1} = |g| \in A$ and consequently $g \wedge g^{-1} = |g|^{-1} \in A$. It now follows by the convexity of A that $g \in A$.

Suppose now that $X \in \mathcal{C}(G)$ is such that $X \cap B \subseteq A$. If $x \in X$ then for every $b \in B$ we have $1_G \leqslant |x| \wedge |b| \in X \cap B \subseteq A$, and the definition of Z gives $x \in Z$.

This then shows that Z is the residual $A:B$ in $\mathcal{C}(G)$. $\square$

Corollary *Pseudocomplements in $\mathcal{C}(G)$ are given by*

$$H^\star = \{g \in G \mid (\forall h \in H) \ |g| \wedge |h| = 1_G\}.$$

Proof This is immediate from $H^\star = \{1_G\}:H$. $\square$

9.6 Polars

The above description of pseudocomplements in $\mathcal{C}(G)$ can be extended to subsets of G.

Definition If X is a non-empty subset of a lattice-ordered group G then by the **polar** $X^\perp$ of X we mean the set of elements of G that are orthogonal to every element of X; symbolically,

$$X^\perp = \{g \in G \mid (\forall x \in X) \ |g| \wedge |x| = 1_G\}.$$

When $X = \emptyset$ we define $\emptyset^\perp = G$.

We shall denote the set of polars of G by $\mathcal{P}(G)$.

Theorem 9.31 *The mapping $f : X \mapsto X^\perp$ is antitone with $\mathrm{id} \leqslant f^2$ and $f = f^3$. Consequently, $X \mapsto X^{\perp\perp}$ is a closure on $\mathbb{P}(G)$.*

Proof It is clear that f is antitone and that $\mathrm{id} \leqslant f^2$. These properties give $f = f^3$, whence it follows that f^2 is a closure. $\square$

Corollary *$\mathcal{P}(G)$ is a complete lattice.*

Proof This is immediate from Theorem 2.14. $\square$

On replacing Z by $X^\perp$ in the first part of the proof of Theorem 9.30, the reader will quickly verify that *for every non-empty subset X of G the polar $X^\perp$ is a convex ℓ-subgroup of G.*

We can therefore assert the following.

Theorem 9.32 *If G is a lattice-ordered group then $(\mathcal{P}(G); \cap, \vee, ^\perp, \{1_G\}, G)$ is a complete boolean algebra in which $H_1 \vee H_2 = \langle H_1 \cup H_2 \rangle^{\perp\perp}$.*

Proof This is an immediate consequence of the above and Theorem 7.2.   □

In the case of a singleton subset $\{x\}$ we shall write $\{x\}^{\perp}$ as simply $x^{\perp}$. Then for every subset X we have $X^{\perp} = \bigcap_{x \in X} x^{\perp}$.

It is clear from the above that for every subset X of G there exists a smallest convex ℓ-subgroup of G that contains X. We shall denote this by $G(X)$. In particular we shall write $G(\{g\})$ as simply $G(g)$ and call this a **principal convex ℓ-subgroup** of G.

Theorem 9.33 *If G is a lattice-ordered group and $X \subseteq G$ then $G(X) = G(\langle X \rangle)$. In particular, for every $g \in G$,*

$$G(g) = G(|g|) = \{x \in G \mid (\exists n \in \mathbb{Z})\ |x| \leqslant |g|^n\}.$$

Proof Since $X \subseteq \langle X \rangle$ we have $G(X) \subseteq G(\langle X \rangle)$. But since $G(X)$ is an ℓ-subgroup that contains X we also have the reverse inclusion.

Since $|g| \in G(g)$ we have $G(|g|) \subseteq G(g)$. Likewise, $g \in G(|g|)$ gives $G(g) \subseteq G(|g|)$, whence we have equality. Finally, note that $\langle |g| \rangle = \{|g|^n \mid n \in \mathbb{Z}\}$ is a totally ordered ℓ-subgroup of G and consequently we have $G(|g|) = G(\langle |g| \rangle) = \{x \in G \mid (\exists m, n \in \mathbb{Z})\ |g|^m \leqslant x \leqslant |g|^n\} = \{x \in G \mid (\exists n \in \mathbb{Z})\ |x| \leqslant |g|^n\}$.   □

Corollary *If $g, h \in P_G$ then*

$$G(g \wedge h) = G(g) \cap G(h) \quad and \quad G(g \vee h) = G(g) \vee G(h).$$

Proof If $x \in G(g \wedge h)$ then $|x| \leqslant (g \wedge h)^n \leqslant g^n, h^n$ and so $x \in G(g) \cap G(h)$. Conversely, if $x \in G(g) \cap G(h)$ then $|x| \leqslant g^n \wedge h^m$ and so, by the Corollary of Theorem 9.13, $|x| \leqslant (g^n \wedge h)^m \leqslant (g \wedge h)^{nm}$ whence $x \in G(g \wedge h)$.

If now $x \in G(g \vee h)$ then $|x| \leqslant (g \vee h)^n = [g(g \wedge h)^{-1}h]^n \leqslant (gh)^n$ and so $x \in G(gh) \subseteq G(g) \vee G(h)$ since $gh \in G(g) \vee G(h)$. Conversely, if $x \in G(g) \vee G(h)$ then $x = \prod_{i=1}^{k} x_i$ where each $x_i \in G(g) \cup G(h)$. Since $|x_i| \leqslant g^n$ or $|x_i| \leqslant h^m$ for each i, it follows that there exists a positive integer t such that $|x| \leqslant (g \vee h)^t$. Consequently, $x \in G(g \vee h)$.   □

EXERCISES

9.29. Prove that the centre of a lattice-ordered group is an ℓ-subgroup.

9.30. Consider the lattice-ordered additive abelian group $G = \overset{\infty}{\underset{-\infty}{\mathsf{X}}} \mathbb{Z}$. By considering the subgroup $H = \overset{\infty}{\underset{-\infty}{\bigoplus}} \mathbb{Z} = \{g \in G \mid g_i = 0 \text{ for all but finitely many } i\}$, and the subgroups $K_i = \{g \in G \mid g_i = 0\}$, show that $\mathcal{C}(G)$ is not dually brouwerian.

9.31. Determine $\mathcal{P}(G)$ where G is the additive abelian group $\mathbb{Z} \times \mathbb{Z}$.

9.32. If $X \subseteq P_G$ prove that

$$G(X) = \{g \in G \mid (\exists x_1, \ldots, x_n \in X)\ |g| \leqslant \prod_{i=1}^{n} x_i\}.$$

9.33. If G is a lattice-ordered group prove that, for all $x \in G$, $G(x)^{\perp} = x^{\perp}$ and $G(x) \subseteq x^{\perp\perp}$.

9.34. If G is a lattice-ordered group prove that, for all $x, y \in G$,
$$x^{\perp\perp} \cap y^{\perp\perp} = (|x| \wedge |y|)^{\perp\perp}, \quad x^{\perp\perp} \vee y^{\perp\perp} = (|x| \vee |y|)^{\perp\perp}.$$

9.35. If G is a lattice-ordered group prove that the following are equivalent:
    (1) $(\forall x \in G)\ G(x) = x^{\perp\perp}$;
    (2) $(\forall x, y \in G)\ G(x) = G(y) \iff x^{\perp\perp} = y^{\perp\perp}$;
    (3) $(\forall A \in \mathcal{C}(G))\ a \in A \Rightarrow a^{\perp\perp} \subseteq A$.

9.36. Prove that a finitely generated convex ℓ-subgroup of a lattice-ordered group is principal.

9.37. Prove that the principal convex ℓ-subgroups of a lattice-ordered group G are the compact elements of $\mathcal{C}(G)$ (cf. Exercise 8.12).

9.7 Coset ordering; prime subgroups

If G is a lattice-ordered group then in order to deal with a convex ℓ-subgroup H that is not necessarily a normal subgroup we may focus on the set $\mathcal{R}(H)$ of right cosets of H in G. We can in fact define an order on $\mathcal{R}(H)$ in a way that is similar to that for quotient groups. Consider the relation $\leqslant$ defined on $\mathcal{R}(H)$ by

$$Hx \leqslant Hy \iff (\exists h \in H)\ hx \leqslant y.$$

This relation is clearly reflexive and transitive. That it is also anti-symmetric follows from the observation that if $Hx \leqslant Hy$ and $Hy \leqslant Hx$ then there exist $h, k \in H$ such that $hx \leqslant y$ and $ky \leqslant x$. Consequently $h \leqslant yx^{-1} \leqslant k^{-1}$ whence the convexity of H gives $yx^{-1} \in H$ and therefore $Hy = Hx$. This order is often called the (right) **coset order**.

Theorem 9.34 *If H is a subgroup of a lattice-ordered group G then the following statements are equivalent*:

    (1) *H is a convex ℓ-subgroup*;
    (2) *$(\mathcal{R}(H); \leqslant)$ is a lattice in which*
$$Hx \vee Hy = H(x \vee y) \ \text{ and } \ Hx \wedge Hy = H(x \wedge y).$$

Proof (1) $\Rightarrow$ (2): Suppose that (1) holds and that $x, y \in G$. Then $1_G x = x \leqslant x \vee y$ gives $Hx \leqslant H(x \vee y)$, and similarly $Hy \leqslant H(x \vee y)$. Suppose now that $z \in G$ is such that $Hx \leqslant Hz$ and $Hz \leqslant Hy$. Then there exist $a, b \in H$ such that $ax \leqslant z$ and $by \leqslant z$. Consequently

$$(a \wedge b)(x \vee y) = (a \wedge b)x \vee (a \wedge b)y \leqslant ax \vee by \leqslant z$$

and so $H(x \vee y) \leqslant Hz$. Thus we see that $\mathcal{R}(H)$ is a join-semilattice in which $Hx \vee Hy = H(x \vee y)$. Dually, $\mathcal{R}(H)$ is a meet-semilattice in which $Hx \wedge Hy = H(x \wedge y)$.

    (2) $\Rightarrow$ (1): Suppose that (2) holds and let $1_G \leqslant g \leqslant h \in H$. Then on the one hand $Hg = H(1_G \vee g) = H \vee Hg$ whence $H \leqslant Hg$; and on the other $hg \leqslant h^2$ so that $Hg \leqslant Hh^2 = H$. Thus $Hg = H$ and so $g \in H$ and (1) follows.　□

Corollary *If H is a convex ℓ-subgroup then the lattice $\mathcal{R}(H)$ is distributive.*

Proof This follows from the above and the fact that G is distributive. ☐

The theory of ordered groups is vast, and in this a significant part is played by those that are totally ordered. These may be characterised as follows.

Theorem 9.35 *Let G be a lattice-ordered group. Then the following statements are equivalent:*

(1) *G is totally ordered;*
(2) *every subgroup of G is an ℓ-subgroup;*
(3) $(\forall x, y \in G)\ x, y > 1_G \Rightarrow x \wedge y > 1_G.$

Proof (1) ⇒ (2): This is clear.

(2) ⇒ (3): Suppose that every subgroup is an ℓ-subgroup. By way of obtaining a contradiction, suppose that there exist $x, y > 1_G$ with $x \wedge y = 1_G$. Then by Theorem 9.22 we have $x \vee y = xy = yx$. Let $g = xy^{-1}$ and consider the subgroup $\langle g \rangle$. By hypothesis, $\langle g \rangle$ is an ℓ-subgroup and so $g^{-1} \vee 1_G \in \langle g \rangle$. But $g^{-1} \vee 1_G = yx^{-1} \vee 1_G = (y \vee x)x^{-1} = yxx^{-1} = y$. Hence $y = g^n$ for some $n \geqslant 1$, and $x = gy = g^{n+1}$. Now since x and y are orthogonal it follows by the Corollary to Theorem 9.17 that so also are x^n and y^{n+1}. But $x^n = g^{n^2+n} = y^{n+1}$ and so $g^{n^2+n} = 1_G$. Thus the ℓ-subgroup $\langle g \rangle$ is finite. But this contradicts the fact that a non-trivial lattice-ordered group cannot have either a top or a bottom element.

(3) ⇒ (1): Observe from Theorem 9.18 that for every $x \in G$ we have $x_+ \wedge (x_-)^{-1} = 1_G$. Thus, if (3) holds then $x_+ = 1_G$ or $x_- = 1_G$ whence $x \leqslant 1_G$ or $x \geqslant 1_G$. Consequently, G is totally ordered. ☐

Definition A convex ℓ-subgroup H of a lattice-ordered group G is said to be **prime** if $\mathcal{R}(H)$ is totally ordered.

The prime subgroups of a lattice-ordered group can be identified as follows.

Theorem 9.36 *If G is a lattice-ordered group and H is a convex ℓ-subgroup of G then the following statements are equivalent:*

(1) *H is prime;*
(2) *if $K_1, K_2 \in \mathcal{C}(G)$ are such that $H \subseteq K_1 \cap K_2$ then either $K_1 \subseteq K_2$ or $K_2 \subseteq K_1$;*
(3) *H is meet-irreducible in the lattice $\mathcal{C}(G)$;*
(4) *if $x, y \in G$ are orthogonal then either $x \in H$ or $y \in H$.*

Proof (1) ⇒ (2): Suppose that H is prime and that $K_1, K_2 \in \mathcal{C}(G)$ are such that $H \subseteq K_1 \cap K_2$. By way of obtaining a contradiction, suppose that $K_1 \parallel K_2$ in $\mathcal{C}(G)$. Let $x, y \in G$ be such that $1_G < x \in K_1 \setminus K_2$ and $1_G < y \in K_2 \setminus K_1$. By (1) we can assume that $Hx \leqslant Hy$, so that $hx \leqslant y$ for some $h \in H$. Then $1_G < x \leqslant h^{-1}y \in K_2$ whence by the convexity we have the contradiction $x \in K_2$.

(2) ⇒ (3): If (2) holds and $H = K_1 \cap K_2$ then necessarily $K_1 = H$ or $K_2 = H$.

(3) $\Rightarrow$ (4): Suppose that (3) holds and that $x, y \in G$ are orthogonal. Then by Theorem 9.33, its Corollary, and Theorem 9.34 we have

$$
\begin{aligned}
[H \vee G(x)] \cap [H \vee G(y)] &= H \vee [G(x) \cap G(y)] \\
&= H \vee [G(|x|) \cap G(|y|)] \\
&= H \vee G(|x| \wedge |y|) \\
&= H \vee G(1_G) \\
&= H.
\end{aligned}
$$

Then (3) gives $H \vee G(x) = H$ or $H \vee G(y) = H$ whence $x \in H$ or $y \in H$.

(4) $\Rightarrow$ (1): Observe that for all $x, y \in G$ we have

$$
x(x \wedge y)^{-1} \wedge y(x \wedge y)^{-1} = (x \wedge y)(x \wedge y)^{-1} = 1_G
$$

and so by (4) either $x(x \wedge y)^{-1} \in H$ or $y(x \wedge y)^{-1} \in H$. The former gives $Hx = H(x \wedge y) \leqslant Hy$ and the latter gives $Hy = H(x \wedge y) \leqslant Hx$. Consequently $\mathcal{R}(H)$ is totally ordered and so H is prime. $\quad\square$

Corollary 1 *In a lattice-ordered group G the intersection of a chain of prime subgroups of G is a prime subgroup of G.*

Proof Let $(H_i)_{i \in I}$ be a chain of prime subgroups of G and let $g, h \in G$ be orthogonal. By the above we have to show that g or h belongs to $\bigcap_{i \in I} H_i$.
Suppose, by way of obtaining a contradiction, that neither does so. Then there exist H_i and H_j with $H_i \subseteq H_j$ such that $g \notin H_i$ and $h \notin H_j$. Since the latter implies that $h \notin H_i$ we see that neither belongs to H_i and this contradicts the hypothesis that H_i is prime. $\quad\square$

Corollary 2 *In a lattice-ordered group every prime subgroup contains a minimal prime subgroup.* $\quad\square$

If G is a lattice-ordered group and $g \in G$ is such that $g \neq 1_G$ then by Zorn's axiom the set of convex ℓ-subgroups that do not contain g has maximal members. A maximal convex ℓ-subgroup that does not contain a particular element of G is called a **regular subgroup** of G. When that element is g such a subgroup is said to be a **value** of g. We shall now identify the regular subgroups of G. For this purpose we require the following notion.

Definition If L is a complete lattice then $x \in L$ is said to be **completely meet-irreducible** if $x \neq 1$ and, for every subset S of G, whenever $x = \inf_L S$ we have $x \in S$.

Clearly, every completely meet-irreducible element is meet-irreducible. The regular subgroups are identified as follows,

Theorem 9.37 *Let G be a lattice-ordered group and let H be a convex ℓ-subgroup of G. Then H is a regular subgroup of G if and only if H is completely meet-irreducible in the lattice $\mathcal{C}(G)$.*

Proof $\Rightarrow$: Suppose that H is regular, so that H is a value of some $g \in G$. Suppose that $(H_i)_{i \in I}$ is a family of convex ℓ-subgroups of G such that $H = \bigcap_{i \in I} H_i$. If $H \neq H_i$ for each $i \in I$ then by the maximality of H with respect to not containing g we have $g \in H_i$ for every $i \in I$ whence $g \in \bigcap_{i \in I} H_i = H$, a contradiction. Hence we must have $H = H_i$ for some $i \in I$ and so H is completely meet-irreducible in $\mathcal{C}(G)$.

$\Leftarrow$: Conversely, suppose that H is completely meet-irreducible in $\mathcal{C}(G)$. Then H is covered in $\mathcal{C}(G)$ by

$$\widehat{H} = \bigcap\{K \in \mathcal{C}(G) \mid H \subset K\}$$

and so there exists $g \in \widehat{H} \backslash H$. Suppose that H is not a value of g. Then there is a value K of g such that $H \subset K$. But $\widehat{H} \subseteq K$ gives the contradiction $g \in K$. Hence H is a value of g and so H is regular. $\square$

Corollary *Every regular subgroup is prime.* $\square$

It follows from the above that if $g \in G \backslash \{1_G\}$ then every value of g contains a minimal prime subgroup P_g. Since $g \notin P_g$ it follows that $\bigcap_{g \neq 1_G} P_g = \{1_G\}$. Consequently the intersection of all minimal prime subgroups of G is $\{1_G\}$.

Minimal prime subgroups can be described in terms of polars.

Theorem 9.38 *Let G be a lattice-ordered group and let P be a convex ℓ-subgroup of G. Then the following statements are equivalent:*
(1) P is a minimal prime subgroup of G;
(2) $P = \bigcup \{g^\perp \mid g \in P_G \backslash P\}$.

Proof (1) $\Rightarrow$ (2): Suppose that P is a minimal prime subgroup. If $x, y \in P_G \backslash P$ then by Theorem 9.36 we have $x \wedge y \neq 1_G$. It follows from this observation that there is a maximal filter F of P_G that contains $P_G \backslash P$. Consider now $F^\star = \bigcup \{g^\perp \mid g \in F\}$. Since $g \in F$ gives $g^\perp \subseteq P_G \backslash F \subseteq P$, we see that $F^\star \subseteq P$. We establish the result by showing that $F^\star = P$ and $F = P_G \backslash P$.

For this purpose, we observe that $F^\star \in \mathcal{C}(G)$. In fact, if $a, b \in F^\star$ then $a \in F^\star$ and $b \in g^\perp$ for some $f, g \in F$. Using Theorem 9.15(6) and the Corollary to Theorem 9.13 we have

$$1_G \leqslant |ab^{-1}| \wedge f \wedge g \leqslant |a| \, |b| \, |a| \wedge f \wedge g \leqslant (|a| \wedge f)(|b| \wedge g)(|a| \wedge f) = 1_G$$

and likewise, by Theorem 9.15(9),

$$1_G \leqslant |a \vee b| \wedge f \wedge g \leqslant |a| \, |b| \wedge f \wedge g \leqslant (|a| \wedge f)(|b| \wedge g) = 1_G$$

whence both ab^{-1} and $a \vee b$ belong to $F^\star$. Thus $F^\star$ is an ℓ-subgroup of G. To see that $F^\star$ is convex, suppose that $x < y < z$ with $x, z \in F^\star$. Then $x \in g^\perp$ and $y \in h^\perp$ for some $g, h \in F$. Now

$$|y| = y \vee y^{-1} \leqslant z \vee x^{-1} \leqslant |z \vee x^{-1}| \leqslant |z| \, |x|$$

and therefore

$$1_G \leqslant |y| \wedge g \wedge h \leqslant |z| \, |x| \wedge g \wedge h \leqslant (|z| \wedge h)(|x| \wedge g) = 1_G$$

whence $y \in (g \wedge h)^\perp$ and so $y \in F^\star$.

We now use Theorem 9.36 to show that F^* is a prime subgroup. Suppose that $x \wedge y = 1_G$ and that $x \notin F^*$. Then $x \wedge g > 1_G$ for all $g \in F$. Since F is a maximal filter of P_G it follows that $x \in F$. Hence $y \in x^\perp \subseteq F^*$ and so F^* is prime.

Finally, since $F^* \subseteq P$ and P is a minimal prime it follows that $F^* = P$. Moreover, if $x \in P = F^*$ then $x \in g^\perp$ for some $g \in F$ whence it follows that $x \notin F$. Thus we see that $F \subseteq P_G \setminus P$, whence we have equality.

$(2) \Rightarrow (1)$: Suppose that $P \in \mathcal{C}(G)$ is such that

$$P = \bigcup \{g^\perp \mid g \in P_G \setminus P\}.$$

Let Q be a minimal prime subgroup contained in P. If there exists $g \in P_G \setminus Q$ then it follows from $(1) \Rightarrow (2)$ applied to Q that we have the contradiction $g^\perp \subseteq Q \subset P$ with $g \in P$. $\qquad\square$

EXERCISES

9.38. Let G be a lattice-ordered group and let $A, B \in \mathcal{C}(G)$ be such that $A \subseteq B$. Prove that A is prime in B if and only if $A:B$ is prime in G.

9.39. Let G be a lattice-ordered group and let $A \in \mathcal{C}(G)$. Prove that the assignment $P \mapsto P \cap A$ describes an isotone bijection between the set of prime subgroups of G that do not contain A and the set of prime subgroups of A.

9.40. If G is a lattice-ordered group and P is a non-trivial convex ℓ-subgroup of G prove that P is totally ordered if and only if $P^\perp$ is a minimal prime subgroup.

9.41. If G is an ordered group then a filter F of G is said to be **prime** if $xy \in F$ implies that $x \in F$ or $y \in F$. Prove that every maximal filter of P_G is prime.

9.42. Prove that there is an antitone bijection between the set of prime subgroups of a lattice-ordered group G and the set of prime filters of G.

9.43. Prove that every convex ℓ-subgroup of a lattice-ordered group G is the intersection of the regular subgroups that contain it.

9.44. Let G be a lattice-ordered group. Prove that G is totally ordered if and only if the lattice $\mathcal{C}(G)$ is a chain.

9.45. An ℓ-endomorphism f of a lattice-ordered group is said to be **polar-preserving** if $x \wedge y = 1_G$ implies that $f(x) \wedge y = 1_G$. Prove that the following are equivalent:
   (1) f is polar-preserving;
   (2) $f^\rightarrow(P_G) \subseteq P_G$ and $f^\rightarrow(P) \subseteq P$ for all polars P;
   (3) $f^\rightarrow(P_G) \subseteq P_G$ and $f^\rightarrow(M) \subseteq M$ for all minimal prime subgroups M.

9.8 Representable groups

We now consider a class of ordered groups that has several interesting properties.

Definition A lattice-ordered group is said to be **representable** if it is isomorphic to a subdirect product of totally ordered groups.

Theorem 9.39 (Lorenzen [78]) *If G is a lattice-ordered group then the following statements are equivalent:*

(1) *G is representable;*
(2) *there is a family $(N_\alpha)_{\alpha \in A}$ of normal prime subgroups with*

$$\bigcap_{\alpha \in A} N_\alpha = \{1_G\};$$

(3) $(\forall a, b \in G)\ (a \wedge b)^2 = a^2 \wedge b^2$;
(4) $(\forall a, b \in G)\ a^+ \wedge b^{-1}(a^{-1})^+ b = 1_G$;
(5) $(\forall a, b, c \in G)\ a \wedge b = 1_G \Rightarrow a \wedge c^{-1}bc = 1_G$;
(6) *all polars of G are normal subgroups;*
(7) *all minimal prime subgroups of G are normal.*

Proof (1) $\Rightarrow$ (2): Let $f : G \to \underset{\alpha \in A}{\Large\times}\, N_\alpha$ be an ℓ-monomorphism of G into a cartesian ordered cartesian product of totally ordered groups. For each $\alpha \in A$ let $\mathrm{pr}_\alpha : \underset{\alpha \in A}{\Large\times}\, N_\alpha \to N_\alpha$ be the canonical projection and let $P_\alpha = \mathrm{Ker}(\mathrm{pr}_\alpha \circ f)$. Then P_α is normal. Moreover, $G/P_\alpha \simeq \mathrm{Im}(\mathrm{pr}_\alpha \circ f)$ is totally ordered, so P_α is prime. If now $g \in \underset{\alpha \in A}{\bigcap}\, P_\alpha$ then for all $\alpha \in A$ we have $\mathrm{pr}_\alpha f(g) = 1_G$. By the universal property of products, it follows from this that $f(g) = 1_G$ and hence $g = 1_G$.

(2) $\Rightarrow$ (1): Let $(N_\alpha)_{\alpha \in A}$ be a family of normal prime subgroups of G such that $\underset{\alpha \in A}{\bigcap}\, N_\alpha = \{1_G\}$. Each quotient G/N_α being totally ordered, let

$$f : G \to \underset{\alpha \in A}{\Large\times}\, G/N_\alpha$$

be given by the prescription

$$f(g) = (N_\alpha g)_{\alpha \in A}.$$

Clearly, f is an ℓ-morphism. Moreover, $f(g) = f(h)$ gives $gh^{-1} \in N_\alpha$ for every $\alpha \in A$, whence $g = h$. Hence f is an embedding.

(1) $\Rightarrow$ (3): In every totally ordered group we clearly have the identity $(a \wedge b)^2 = a^2 \wedge b^2$. If (1) holds then this identity also holds in G.

(3) $\Rightarrow$ (4): If (3) holds then for all $a, b \in G$ we have $(ba)^2 \wedge b^2 = (ba \wedge b)^2 \leqslant ba \cdot b$ whence $a \wedge b^{-1}a^{-1}b \leqslant 1_G$ and consequently

$$1_G = (a \wedge b^{-1}a^{-1}b) \vee 1_G = a^+ \wedge b^{-1}(a^{-1})^+ b.$$

(4) $\Rightarrow$ (5): If $a \wedge b = 1_G$ then by Theorem 9.22 we have $a = abb^{-1} = (a \vee b)b^{-1} = ab^{-1} \vee 1_G = (ab^{-1})^+$, and similarly $b = (ba^{-1})^+$. Now if (4) holds then for every $c \in G$ we have

$$1_G = (ab^{-1})^+ \wedge c^{-1}(ba^{-1})^+ c = a \wedge c^{-1}bc.$$

(5) $\Rightarrow$ (6): This is clear.

(6) $\Rightarrow$ (7): Let P be a minimal prime subgroup of G. Then by Theorem 9.38 we have $P = \bigcup\{g^\perp \mid g \in P_G \backslash P\}$. If now (6) holds then, for every $h \in G$,

$$h^{-1}Ph = \bigcup\{h^{-1}g^\perp h \mid g \in P_G \backslash P\} = \bigcup\{g^\perp \mid g \in P_G \backslash P\} = P$$

and so P is normal.

(7) $\Rightarrow$ (2): This follows from the earlier observation that the intersection of all minimal prime subgroups of G is $\{1_G\}$. $\square$

EXERCISES

9.46. Prove that a lattice-ordered group G is representable if and only if
$$(\forall x, y \in G)(\forall n \geqslant 0) \ (x \wedge y)^n = x^n \wedge y^n.$$

9.47. Prove that a lattice-ordered group G is representable if and only if
$$(\forall x, y \in G)(\forall n \geqslant 0) \ (x \vee y)^n = x^n \vee y^n.$$

9.48. Prove that a lattice-ordered group G fails to be representable if and only if there exist $x, y > 1_G$ such that $x \wedge y^{-1}xy = 1_G$.

There is an extensive theory of ordered groups, and here we have merely scratched the surface. Significant references are the books by Bigard, Keimel and Wolfenstein [7], Conrad [39], Darnell [41], Fuchs [51], Glass [55], Glass and Holland [56], and Ribenboim [93]. In particular, [41] and [56] contain excellent comprehensive bibliographies.

In the next chapter we give consideration to ordered rings and fields. In particular we include a characterisation of the field $\mathbb{R}$ of real numbers as, to within isomorphism, the only totally ordered field that is Dedekind complete; and, using concepts from the present chapter, we derive the fact that every archimedean lattice-ordered group is commutative.

Archimedean ordered structures

10.1 Totally ordered rings and fields

We begin this chapter by recalling some basic algebra.

By an **integral domain** we shall mean a non-trivial ring with a 1 in which there are no zero divisors. By a **division ring** we shall mean an integral domain D in which the multiplicative subsemigroup $D\setminus\{0\}$ is a group. A commutative division ring is called a **field**.

If F is a field and $x \in F$ then for $n \in \mathbb{Z}$ we define nx as follows: if $n > 0$ then $nx = x+x+\cdots+x$ in which there are n summands; if $n = 0$ then $nx = 0$; and if $n < 0$ then $nx = -|n|x$. For every subset X of F the intersection of all the subfields of F that contain X is also a subfield of F. It is then the smallest subfield of F that contains X and is called the subfield **generated** by X. In particular, the subfield generated by $\{1_F\}$ is

$$\{\frac{m1_F}{n1_F} \mid m, n \in \mathbb{Z},\ n1_F \neq 0\}.$$

If D is a commutative integral domain then by a **field of quotients** of D we shall mean a field F together with a ring monomorphism $f : D \to F$ such that, for every division ring X and every ring monomorphism $g : D \to X$, there is a unique ring monomorphism $h : F \to X$ such that the diagram

is commutative, in the sense that $h \circ f = g$.

Theorem 10.1 *If D is a commutative integral domain then there exists, to within isomorphism, a unique field of quotients (F, f) of D. Moreover, F is generated by $\operatorname{Im} f$.*

Proof We sketch the proof, leaving the finer details to the reader.

Let $D^\star = D\setminus\{0\}$ and define addition and multiplication on $D \times D^\star$ by

$$(x_1, y_1) + (x_2, y_2) = (x_1y_2 + x_2y_1, y_1y_2); \quad (x_1, y_1)(x_2, y_2) = (x_1x_2, y_1y_2),$$

each being commutative and associative. The relation ϑ defined on $D \times D^\star$ by

$$(x_1, y_1) \overset{\vartheta}{\equiv} (x_2, y_2) \iff x_1 y_2 = x_2 y_1$$

is a congruence. Let $F = (D \times D^\star)/\vartheta$ and denote the ϑ-class $[(x, y)]_\vartheta$ by $\dfrac{x}{y}$. Then F has the induced laws

$$\frac{x_1}{y_1} + \frac{x_2}{y_2} = \frac{x_1 y_2 + x_2 y_1}{y_1 y_2}; \qquad \frac{x_1}{y_1} \cdot \frac{x_2}{y_2} = \frac{x_1 x_2}{y_1 y_2}.$$

It is readily seen that F is a commutative ring with identity element $\dfrac{x}{x}$ for every $x \in D^\star$, and zero element $\dfrac{0}{y}$ for every $y \in D^\star$. Finally, if $\dfrac{x}{y}$ is a non-zero element of F then it has an inverse, namely $\dfrac{y}{x}$. Thus F is a field.

Consider now $f : D \to F$ given by the prescription $f(x) = \dfrac{x}{1}$. This is a ring monomorphism. To see that (F, f) is a field of quotients of D, suppose that X is a division ring and that $g : D \to X$ is a ring monomorphism. If $(a, b) \overset{\vartheta}{\equiv} (c, d)$ then $ad = bc$ whence $g(a)g(d) = g(b)g(c)$ and consequently we can define a mapping $h : F \to X$ by the prescription

$$h\left(\frac{a}{b}\right) = g(a)[g(b)]^{-1}.$$

It is readily seen that h is a ring monomorphism with $h \circ f = g$. To establish the uniqueness of h, suppose that $k : F \to X$ is also a ring monomorphism such that $k \circ f = g$. Then we have

$$\begin{aligned}
k\left(\frac{x}{y}\right) = k\left(\frac{x}{1} \cdot \frac{1}{y}\right) = k\big(f(x)[f(y)]^{-1}\big) &= (k \circ f)(x)[(k \circ f)(y)]^{-1} \\
&= g(x)[g(y)]^{-1} \\
&= h\left(\frac{x}{y}\right)
\end{aligned}$$

and therefore $k = h$.

This then establishes the existence of a field of quotients. As for uniqueness up to isomorphism, suppose that $(F^\star, f^\star)$ is also a field of quotients of D. Then there are unique monomorphisms j, k such that the diagrams

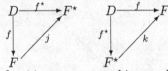

are commutative. Then $k \circ j$ is a monomorphism such that the diagram

is commutative. Now since (F, f) is a field of quotients there is a unique monomorphism that makes this last diagram commutative; and id_F does precisely that. Hence $k \circ j = \mathrm{id}_F$, and likewise $j \circ k = \mathrm{id}_{F^\star}$. Thus j is an isomorphism whose inverse is k.

Finally, to show that $\operatorname{Im} f$ generates F let the subfield of F that is generated by $\operatorname{Im} f$ be H. Consider the diagram

$$D \xrightarrow{\;f^+\;} \operatorname{Im} f \xrightarrow{\;\iota_1\;} H \xrightarrow{\;\iota_2\;} F$$

with $f : D \to F$ below.

in which $f^+ : D \to \operatorname{Im} f$ is the epimorphism given by $x \mapsto f(x)$ and ι_1, ι_2 are the canonical injections. There exists a unique monomorphism $\alpha : F \to H$ such that $\alpha \circ f = \iota_1 \circ f^+$ whence the mapping $\beta = \iota_2 \circ \alpha : F \to F$ is a monomorphism such that $\beta \circ f = \iota_2 \circ \iota_1 \circ f^+$. But only one such monomorphism β can exist, and clearly this must be id_F. Consequently $\iota_2 \circ \alpha = \beta = \mathrm{id}_F$ and so the canonical injection ι_2 is also surjective. Hence $H = F$ as claimed. $\square$

The above result shows that all fields of quotients of D are isomorphic. We can therefore agree to identify all such fields and thus talk of *the* field of quotients of D, taking as a model for this the field constructed in the above proof. If (F, f) is the field of quotients of D then it is common practice to identify D with the subdomain $\operatorname{Im} f$ of F. In the particular case where D is the integral domain $\mathbb{Z}$ then the field of quotients of $\mathbb{Z}$ is denoted by $\mathbb{Q}$ and is called the field of **rational numbers**. The identification of $\mathbb{Z}$ as a subdomain of $\mathbb{Q}$ is obtained by identifying $n \in \mathbb{Z}$ with $\frac{n}{1} \in \mathbb{Q}$.

If R is a ring with a 1 then the subring $\mathrm{Rg}\{1\}$ of R that is generated by $\{1\}$ is given by $\{n1 \mid n \in \mathbb{Z}\}$. The mapping $\zeta : \mathbb{Z} \to \mathrm{Rg}\{1\}$ described by $n \mapsto n1$ is a ring epimorphism. Since $\mathbb{Z}$ is a principal ideal domain and $\operatorname{Ker} \zeta$ is an ideal, we have $\operatorname{Ker} \zeta = p\mathbb{Z}$ for some $p \in \mathbb{Z}_+$. This integer p is called the **characteristic** of R. A commutative integral domain D is of characteristic 0 if and only if every non-zero element of D is of infinite order in the additive group of D.

If X is a division ring then the intersection of all the subfields of X is a subfield of X. It is therefore the smallest subfield of X and is called the **prime subfield** of X. It is readily seen that this is precisely the subfield generated by $\{1\}$, namely $P = \{\dfrac{m1}{n1} \mid m, n \in \mathbb{Z}, \ n1 \neq 0\}$.

Theorem 10.2 *If X is a division ring of characteristic 0 then its prime subfield is isomorphic to $\mathbb{Q}$.*

Proof Since X is of characteristic 0 we have $n1 = 0$ if and only if $n = 0$, so the mapping $\zeta : n \mapsto n1$ is a monomorphism. Since $\mathbb{Q}$ is the field of quotients of $\mathbb{Z}$ there is a unique monomorphism $h : \mathbb{Q} \to X$ such that the diagram

is commutative, in which ι denotes the canonical embedding of $\mathbb{Z}$ into $\mathbb{Q}$.

Now from the proof of Theorem 10.1 we know that the monomorphism h is given by $h(\frac{m}{n}) = g(m)[g(n)]^{-1} = \frac{m1}{n1}$. Thus the prime subfield of X, which is $\operatorname{Im} h$, is isomorphic to $\mathbb{Q}$.   □

Corollary *Every field of characteristic 0 contains a copy of $\mathbb{Q}$.*   □

We now consider such algebraic structures when equipped with a compatible order.

Definition By an **ordered ring** we shall mean a ring $(R; +, \cdot)$ which is equipped with an order $\leqslant$ such that $(R; +, \leqslant)$ is an ordered (abelian) group whose positive cone R_+ is a subsemigroup of $(R; \cdot)$.

Note that R_+ a subsemigroup of $(R; \cdot)$ is equivalent to the property

$$(\forall x, y \in R) \qquad x, y \geqslant 0 \Rightarrow xy \geqslant 0.$$

The following is the ring analogue of Theorem 9.1.

Theorem 10.3 *A non-empty subset P of an ordered ring R is the positive cone relative to some compatible order on R if and only if*

(1) $P \cap (-P) = \{0\}$;
(2) $P + P \subseteq P$;
(3) $PP \subseteq P$.

Moreover, this order is a total order if and only if, in addition, $P \cup (-P) = R$.

Proof This is immediate from Theorem 9.1 (with additive notation). Property (3) of Theorem 9.1 holds automatically here since $(R; +)$ is abelian. As for the above (3), this is clearly equivalent to the property

$$(\forall x, y \in R) \qquad x, y \geqslant 0 \Rightarrow xy \geqslant 0.$$

which in turn is equivalent to R_+ being a subsemigroup of $(R; \cdot)$.   □

Example 10.1 In the integral domain $\mathbb{Z}$ the subset $\mathbb{N}$ is the positive cone of a total order on $\mathbb{Z}$. This, the 'usual' order, is given by $m \leqslant n \iff n - m \in \mathbb{N}$. It is in fact the only compatible total order possible on the ring $\mathbb{Z}$. To see this, suppose that some compatible total order is defined by the positive cone P. Then since $\mathbb{Z} = P \cup (-P)$ we have either $1 \in P$ or $1 \in -P$, the latter being equivalent to $-1 \in P$. Now since $P \cap (-P) = \{0\}$ we cannot have both; and since $PP \subseteq P$ we cannot have $-1 \in P$ since otherwise we have the contradiction $1 = (-1)(-1) \in P$. Hence we must have $1 \in P$. It follows that $\mathbb{Z}_+ \subseteq P$ whence $-\mathbb{Z}_+ \subseteq -P$. Since

$$\{-\mathbb{Z}_+ \backslash \{0\}, \{0\}, \mathbb{Z}_+ \backslash \{0\}\}, \qquad \{-P \backslash \{0\}, \{0\}, P \backslash \{0\}\}$$

are each partitions of $\mathbb{Z}$ we conclude that $P = \mathbb{Z}_+$ and so only one compatible total order is possible.

We now show that there is a unique compatible total order $\preceq$ on $\mathbb{Q}$ that extends that on $\mathbb{Z}$, in the sense that if $m, n \in \mathbb{Z}$ then $m \preceq n$ in $\mathbb{Q}$ if and only if $m \leqslant n$ in $\mathbb{Z}$.

Theorem 10.4 *Let $(D; \leqslant)$ be a totally ordered commutative integral domain and let F be its field of quotients. Then there is one and only one compatible total order $\preceq$ on F that extends $\leqslant$.*

Proof Consider the subset F_+ of F given by

$$F_+ = \{xy^{-1} \mid x \in D_+, \, y \in D_+ \setminus \{0\}\}.$$

We show that F_+ is the positive cone of a compatible total order on F. The laws of composition in F show that F_+ so defined is closed under addition and multiplication, and so F satisfies (2) and (3) of Theorem 10.3. As for (1), let $a \in F_+ \cap -F_+$. Then there exist $x, p \in D$ and $y, q \in D \setminus \{0\}$ such that $a = \dfrac{x}{y} = -\dfrac{p}{q}$ whence $0 \leqslant xq = -yp \leqslant 0$, giving $xq = 0$. Since D is an integral domain we deduce that $x = 0$ whence $a = 0$. We now note that if $a = \dfrac{x}{y} \in F$ then there is no loss of generality in supposing that $y \in D_+ \setminus \{0\}$. This being the case, if $x \in D_+$ then we have $a \in F_+$; and if $x \notin D_+$ then, D being totally ordered, $-x \in D_+$ whence $-a = \dfrac{-x}{y} \in F_+$ and consequently $a = -(-a) \in -F_+$. We conclude that $F_+ \cup (-F_+) = F$.

Since, as is readily seen, $D \cap F_+ = D_+$ the compatible total order thus defined by F_+ is an extension of that on D_+.

As for the uniqueness, let $\preceq$ be an order on F with respect to which F is a totally ordered field in which $\preceq$ extends the order $\leqslant$ on D. Let the positive cone of F relative to $\preceq$ be $F_\oplus = \{x \in F \mid 0 \preceq x\}$. We show that $F_\oplus = F_+$ whence the uniqueness follows. Now both

$$\{-F_\oplus \setminus \{0\}, \, \{0\}, \, F_\oplus \setminus \{0\}\}, \qquad \{-F_+ \setminus \{0\}, \, \{0\}, \, F_+ \setminus \{0\}\}$$

are partitions of F. But if $a \in F_\oplus$, say $a = \dfrac{x}{y}$ where $x \in D$ and (without loss of generality) $y \in D_+ \setminus \{0\}$, then we have $x = ay$ where $a \succeq 0$ and, since $D_+ = D \cap F_\oplus$, also $y \succeq 0$. Since $(F, \preceq)$ is an ordered ring we have $x = ay \succeq 0$ and so $x \geqslant 0$. Consequently we have $a = \dfrac{x}{y} \in F_+$ whence $F_\oplus \subseteq F_+$. We conclude from the above partitions that $F_\oplus = F_+$ as required. □

Corollary *There is a unique compatible total order on $\mathbb{Q}$ which extends that on $\mathbb{Z}$.* □

The properties listed in Theorem 10.3 give rise to the following facts that will be used throughout what follows.

Theorem 10.5 *Let $(R; \leqslant)$ be an ordered ring. Then, for all $x, y \in R$,*

(1) $x \leqslant y \Rightarrow (\forall z \geqslant 0)\ xz \leqslant yz$ *and* $zx \leqslant zy$;
(2) $x \leqslant y \Rightarrow (\forall z \leqslant 0)\ xz \geqslant yz$ *and* $zx \geqslant zy$.

Moreover, if R is totally ordered,

(3) $(\forall x \in R)\ x^2 \geqslant 0$;
(4) *if $x \in R$ is invertible then $x > 0 \iff x^{-1} > 0$.*

Proof (1) If $x \leqslant y$ then $y - x \in R_+$ so for all $z \in R_+$ we have $zy - zx = z(y - x) \in R_+ R_+ \subseteq R_+$ and so $zx \leqslant zy$; and likewise $xz \leqslant yz$.

(2) If $x \leqslant y$ then $y - x \in R_+$ so for all $z \leqslant 0$ we have $-z \in R_+$ and consequently $-zy + zx = -z(y - x) \in R_+ R_+ \subseteq R_+$ whence $zy \leqslant zx$; and likewise $yz \leqslant xz$.

(3) If $x \geqslant 0$ then $x^2 \in R_+ R_+ \subseteq R_+$; and if $x \leqslant 0$ then again $x^2 = (-x)(-x) \in R_+ R_+ \subseteq R_+$.

(4) If x is invertible then we have $x = x^2 x^{-1}$. Thus if $x^{-1} \in R_+$ we have, by (3), $x \in R_+$. For the converse implication, replace x by x^{-1}.   □

Theorem 10.6 *Every totally ordered commutative integral domain D is of characteristic* 0.

Proof By Theorem 10.5(3), for every $x \in D$ we have $x^2 \geqslant 0$. Since D has no zero divisors, it follows that $x^2 > 0$ for all $x \in D \setminus \{0\}$. In particular, $1 = 1^2 > 0$. Since $D_+ + D_+ \subseteq D_+$ it follows by induction that $n1 > 0$ for all positive integers n. Thus the identity of D has infinite order, whence so does every non-zero element of D. Consequently D is of characteristic 0.   □

Theorem 10.7 *If D is a totally ordered commutative integral domain then there is an isotone monomorphism $f : \mathbb{Z} \to D$. If F is a totally ordered field then there is an isotone monomorphism $f : \mathbb{Q} \to F$.*

Proof By Theorem 10.6, D is of characteristic 0 and so the mapping $\zeta : \mathbb{Z} \to D$ given by $\zeta(n) = n1$ is a ring monomorphism. Since

$$m \leqslant n \iff n - m \geqslant 0 \iff n1 - m1 = (n - m)1 \geqslant 0 \iff m1 \leqslant n1$$

it follows that ζ is also isotone.

To prove the corresponding result for fields, we observe first that in any totally ordered field X we have $\dfrac{m}{n} \leqslant \dfrac{p}{q} \iff mq \leqslant np$. In fact, without loss of generality we may assume that $n > 0$ and $q > 0$ so that $nq \in X_+ X_+ \subseteq X_+$ whence, since X has no zero divisors, $nq > 0$. Thus, if $\dfrac{m}{n} \leqslant \dfrac{p}{q}$ then by Theorem 10.5(1) we have $mq = nq(\dfrac{m}{n}) \leqslant nq(\dfrac{p}{q}) = np$. Conversely, suppose that $mq \leqslant np$. Since $nq > 0$ it follows by Theorem 10.5(4) that $\dfrac{1}{nq} > 0$ so that $\dfrac{m}{n} = \dfrac{1}{nq}(mq) \leqslant \dfrac{1}{nq}(np) = \dfrac{p}{q}$.

Suppose now that F is a totally ordered field. Since $\mathbb{Q}$ is the field of quotients of $\mathbb{Z}$ the monomorphism $\zeta : \mathbb{Z} \to F$ given by $\zeta(n) = n1$ extends to a unique monomorphism $h : \mathbb{Q} \to F$. As seen in Theorem 10.1, this monomorphism is given by the prescription $h(\dfrac{m}{n}) = \dfrac{m1}{n1}$. To complete the proof, we must show that h is isotone. Now by applying the previous observation to both $\mathbb{Q}$ and F we obtain, for $n, q > 0$,

$$\dfrac{m1}{n1} \leqslant \dfrac{p1}{q1} \iff (mq)1 = (m1)(q1) \leqslant (n1)(p1) = (np)1$$

$$\iff mq \leqslant np \iff \dfrac{m}{n} \leqslant \dfrac{p}{q}$$

from which the result follows.   □

10.2 Archimedean ordered fields

Having established these preliminaries, we now consider the following class of ordered algebraic structures.

Definition If G is an ordered group and $x, y \in G$ then we say that x is **infinitely smaller** than y, and write $x \ll y$, if $x^n \leq y$ for all $n \in \mathbb{Z}$. We say that G is **archimedean** if $1_G \leq x \ll y$ implies $x = 1_G$. If $(R; +, \cdot, \leq)$ is an ordered ring (field) then we say that R is **archimedean** if the underlying ordered group $(R; +, \leq)$ is archimedean.

Note that a *totally* ordered group G is archimedean if and only if, whenever $x, y \in G_+$ with $1_G < x \leq y$, there exists $n \in \mathbb{Z}$ such that $x^n \leq y < x^{n+1}$.

Example 10.2 The field $\mathbb{Q}$ of rationals, together with the unique total order which extends that on $\mathbb{Z}$, is archimedean. For example, if $0 < \dfrac{m}{n} < \dfrac{p}{q}$ in $\mathbb{Q}$ with $n, q > 0$ then $(n^2 p + 1)\dfrac{m}{n} > n^2 p \dfrac{m}{n} = npm \geq np \geq \dfrac{np}{nq} = \dfrac{p}{q}$. Here of course we are using additive notation, so that x^n translates into nx.

Since, by Theorems 10.7 and 10.2, every totally ordered field contains a copy of $\mathbb{Q}$ as its prime subfield, it follows that $\mathbb{Q}$ is, to within isomorphism, the smallest archimedean totally ordered field. Our objective now is to establish the existence and uniqueness up to isomorphism of an archimedean totally ordered field that contains a copy of every archimedean totally ordered field. This will provide formal characterisations of the field $\mathbb{R}$ of real numbers.

If F is a totally ordered field and $x \in F$ then clearly $|x| = \max\{x, -x\}$. The principal properties of absolute values are summarised as follows.

Theorem 10.8 *If F is a totally ordered field then, for all $x, y \in F$,*

(1) $|x| = 0 \iff x = 0$;
(2) $x \leq |x|$;
(3) $|x| = |-x|$;
(4) $|xy| = |x| \, |y|$;
(5) $|x + y| \leq |x| + |y|$;
(6) $\big||x| - |y|\big| \leq |x - y|$;
(7) $|x| < y \iff -y < x < y$;
(8) *if $x \neq 0$ then $|x^{-1}| = |x|^{-1}$.*

Proof $(1), (2), (3), (4)$ are clear, whereas (5) follows from the additive version of Theorem 9.16. As for (6), by (5) we have $|x| = |x - y + y| \leq |x - y| + |y|$ and so $|x| - |y| \leq |x - y|$ whence (6) follows. Property (7) follows immediately from (2) and (3); and (8) follows from (4) on taking $y = x^{-1}$. $\square$

Definition Let F be a totally ordered field. By a **sequence** of elements of F we mean a mapping $a : \mathbb{N} \to F$. We shall say that a sequence a of elements of F is

(1) a **bounded sequence** if there exists $b \in F$ such that $|a(n)| \leq b$ for every $n \in \mathbb{N}$;

(2) a **Cauchy sequence** if, for every $\epsilon > 0$ in F, there is a positive integer a_ϵ such that $|a(p) - a(q)| < \epsilon$ for all $p, q \geqslant a_\epsilon$;

(3) a **null sequence** if, for every $\epsilon > 0$ in F, there is a positive integer a_ϵ such that $|a(p)| < \epsilon$ for all $p \geqslant a_\epsilon$.

We shall denote by $B(F)$, $C(F)$, $N(F)$ respectively the sets of all bounded, Cauchy, null sequences.

Theorem 10.9 $N(F) \subset C(F) \subset B(F)$.

Proof If $a \in N(F)$ then for any $\epsilon > 0$ in F we have, for $p, q \geqslant a_{\epsilon/2}$,

$$|a(p) - a(q)| \leqslant |a(p)| + |-a(q)| = |a(p)| + |a(q)| < \frac{\epsilon}{2} + \frac{\epsilon}{2} + \epsilon$$

whence $a \in C(F)$ and so $N(F) \subseteq C(F)$. That the containment is strict follows from the observation that the sequence a with $a(n) = 1$ for every n belongs to $C(F)$ but not to $N(F)$.

If now $a \in C(F)$ then taking $\epsilon = 1 > 0$ we have $|a(p) - a(q)| < 1$ for all $p, q > a_1$. Let $b = 1 + \max\{|a(1)|, \ldots, |a(a_1)|\}$. Then $|a(p)| \leqslant b$ for $p \leqslant a_1$, and for $p > a_1$ we have

$$|a(p)| = |a(p) - a(a_1) + a(a_1)| \leqslant |a(p) - a(a_1)| + |a(a_1)| < 1 + |a(a_1)| \leqslant b.$$

Hence $a \in B(F)$ and so $C(F) \subseteq B(F)$. That the containment is strict follows from the observation that the sequence a with $a(n) = (-1)^n$ for every n belongs to $B(F)$ but not to $C(F)$. □

Theorem 10.10 *The set $C(F)$ of Cauchy sequences in a totally ordered field F forms a commutative ring with an identity in which $a + b$ and ab are given by the prescriptions $(a+b)(n) = a(n) + b(n)$, $(ab)(n) = a(n)b(n)$. In this ring the set $N(F)$ of null sequences is an ideal and $C(F)/N(F)$ is a field.*

Proof If $a, b \in C(F)$ then for any given $\epsilon > 0$ in F there exist positive integers a_ϵ b_ϵ such that $|a(p) - a(q)| < \epsilon$ and $|b(p) - b(q)| < \epsilon$ for all $p, q \geqslant \max\{a_\epsilon, b_\epsilon\}$. Thus for $p, q \geqslant \max\{a_{\epsilon/2}, b_{\epsilon/2}\}$ we have

$$|a(p) + b(p) - [a(q) + b(q)]| \leqslant |a(p) - a(q)| + |b(p) - b(q)| < \frac{\epsilon}{2} + \frac{\epsilon}{2} = \epsilon.$$

Hence $a + b \in C(F)$. Since $C(F) \subset B(F)$ by Theorem 10.9, there exists $d > 0$ in F such that $|a(n)|, |b(n)| < d$ for all n. If $p, q \geqslant \max\{a_{\epsilon/2d}, b_{\epsilon/2d}\}$ then

$$|a(p)b(p) - a(q)b(q)| = |a(p)||b(p) - b(q)| + [a(p) - a(q)]|b(q)|$$
$$\leqslant |a(p)|\,|b(p) - b(q)| + |a(p) - a(q)|\,|b(q)|$$
$$< d\frac{\epsilon}{2d} + \frac{\epsilon}{2d}d = \epsilon.$$

Hence $ab \in C(F)$. It is now easy to see that $C(F)$, equipped with the above laws of composition, forms a commutative ring. This ring has an identity, namely the constant sequence $\mathbf{1}$ such that $\mathbf{1}(n) = 1$ for all n.

To show that $N(F)$ is an ideal of $C(F)$, we note first that if $a, b \in N(F)$ then $a - b \in N(F)$. Suppose now that $a \in N(F)$ and $b \in C(F)$. Since b is bounded there exists $d > 0$ in F such that $|b(n)| \leqslant d$ for every n. Given any $\epsilon > 0$ in F, for $p \geqslant a_{\epsilon/d}$ we have

$$|a(p)b(p)| = |a(p)|\,|b(p)| \leqslant |a(p)|d < \frac{\epsilon}{d}d = \epsilon$$

and consequently $ab \in N(F)$. Thus $N(F)$ is an ideal of $C(F)$. That it is a proper ideal is clear from Theorem 10.9.

To show that $C(F)/N(F)$ is a field, we must show that for every $a \in C(F)$ with $a \notin N(F)$ there exists $x \in C(F)$ such that $ax - 1 \in N(F)$. Now to say that a is a null sequence we mean that

$$(\forall \epsilon > 0)(\exists a_\epsilon > 0)(\forall n \geqslant a_\epsilon) \quad |a(n)| < \epsilon.$$

Since, by hypothesis, a is not a null sequence we therefore have the existence of $\epsilon > 0$ such that, for every positive integer t, there is some integer $m \geqslant t$ for which $|a(m)| \geqslant \epsilon$. Choosing $t = a_{\epsilon/2}$ we then have, for all $p \geqslant a_{\epsilon/2}$,

$$\epsilon \leqslant |a(m)| = |a(m) - a(p) + a(p)| \leqslant |a(m) - a(p)| + |a(p)| < \frac{\epsilon}{2} + |a(p)|$$

and hence $|a(p)| > \dfrac{\epsilon}{2}$. This shows that $a(p) \neq 0$. Now define the sequence x by

$$x(n) = \begin{cases} \dfrac{1}{a(n)} & \text{if } n \geqslant a_{\epsilon/2}; \\ 1 & \text{otherwise.} \end{cases}$$

To see that $x \in C(F)$ we observe that, for any given $\delta > 0$ and all $p, q \geqslant \max\{a_{\epsilon/2}, a_{\epsilon^2\delta/4}\}$,

$$|x(p) - x(q)| = \left| \frac{1}{a(p)} - \frac{1}{a(q)} \right| = \left| \frac{a(q) - a(p)}{a(p)a(q)} \right| < \frac{\epsilon^2\delta/4}{(\epsilon/2)^2} = \delta,$$

and so $x \in C(F)$. Finally, we have

$$a(n)x(n) - 1 = \begin{cases} 0 & \text{if } n \geqslant a_{\epsilon/2}; \\ a(n) - 1 & \text{otherwise,} \end{cases}$$

which shows that $ax - 1 \in N(F)$ as required. $\qquad\square$

Corollary $N(F)$ is a maximal ideal of $C(F)$. $\qquad\square$

For the purpose of defining a compatible total order on $C(F)/N(F)$ we require the following notion.

Definition A sequence a of elements in a totally ordered field F is said to be a **positive sequence** if

$$(\exists \epsilon_a \in F_+\setminus\{0\})(\exists n_a \in \mathbb{N})(\forall n \geqslant n_a) \quad a(n) > \epsilon_a.$$

Theorem 10.11 *If F is a totally ordered field then $C(F)/N(F)$ is a totally ordered field that contains a copy of F.*

Proof Suppose that $a, b \in C(F)$ are such that $b - a$ is a positive sequence; i.e.,

$$(\exists \epsilon_{a,b} \in F_+\setminus\{0\})(\exists n_{a,b} \in \mathbb{N})(\forall n \geqslant n_{a,b}) \quad b(n) - a(n) > \epsilon_{a,b}. \tag{10.1}$$

If $a^\star \in C(F)$ is such that $a^\star - a \in N(F)$ then

$$(\exists n_{a,a^\star} \in \mathbb{N})(\forall n \geqslant n_{a,a^\star}) \quad |a^\star(n) - a(n)| \leqslant \epsilon_{a,b}/4.$$

Likewise, if $b^\star \in C(F)$ is such that $b - b^\star \in N(F)$ then

$$(\exists n_{b,b^\star} \in \mathbb{N})(\forall n \geqslant n_{b,b^\star}) \quad |b(n) - b^\star(n)| \leqslant \epsilon_{a,b}/4.$$

Let $n^\star = \max\{n_{a,b}, n_{a,a^\star}, n_{b,b^\star}\}$. Then for all $n \geqslant n^\star$ we have

$$\begin{aligned}
b^\star(n) - a^\star(n) &= b^\star(n) - b(n) + b(n) - a(n) + a(n) - a^\star(n) \\
&\geqslant -|b^\star(n) - b(n)| + b(n) - a(n) - |a(n) - a^\star(n)| \\
&> -\epsilon_{a,b}/4 + \epsilon_{a,b} - \epsilon_{a,b}/4 \\
&= \epsilon_{a,b}/2.
\end{aligned}$$

Thus we see that $b^\star - a^\star$ is a positive sequence. It follows from this that we can define a binary relation $\prec$ on the quotient field $C(F)/N(F)$ by writing $[x] \prec [y]$ if and only if there exist $a \in [x]$ and $b \in [y]$ such that $b - a$ is a positive sequence.

The relation $\prec$ so defined is transitive. To see this, suppose that $[x] \prec [y]$ and $[y] \prec [z]$. Then there exist $a \in [x], b \in [y], c \in [z]$ such that a, b and b, c satisfy (10.1) above. Let $\epsilon = \epsilon_{b,c} + \epsilon_{a,b}$ and let $n^\star = \max\{n_{a,b}, n_{b,c}\}$. Then for all $n \geqslant n^\star$ we have

$$c(n) - a(n) = c(n) - b(n) + b(n) - a(n) > \epsilon_{b,c} + \epsilon_{a,b} = \epsilon$$

which shows that a, c also satisfy (10.1) and hence that $[x] \prec [z]$.

To show that $C(F)/N(F)$ is totally ordered under $\preceq$ it suffices to show that for any $[x], [y]$ precisely one of the statements $[x] \prec [y], [x] = [y], [y] \prec [x]$ holds. Clearly, no two of these can hold simultaneously so it will suffice to show that if $[x] \prec [y]$ and $[y] \prec [x]$ are each false then we must have $[x] = [y]$. Now given $a \in [x]$ and $b \in [y]$ we have, since each of these is a Cauchy sequence,

$$(\forall \epsilon > 0)(\exists n^\star \in \mathbb{N})(\forall m, n \geqslant n^\star) \quad |a(m) - a(n)| < \frac{\epsilon}{3}, \; |b(m) - b(n)| < \frac{\epsilon}{3}.$$

Since $[y] \prec [x]$ is false we obtain from the negation of (10.1) the existence of $p \geqslant n^\star$ such that $a(p) - b(p) \leqslant \frac{\epsilon}{3}$. Then for all $n \geqslant n^\star$ we have

$$\begin{aligned}
a(n) - b(n) &= a(n) - a(p) + a(p) - b(p) + b(p) - b(n) \\
&\leqslant |a(n) - a(p)| + a(p) - b(p) + |b(p) - b(n)| \\
&< \frac{\epsilon}{3} + \frac{\epsilon}{3} + \frac{\epsilon}{3} = \epsilon.
\end{aligned}$$

Since $[x] \prec [y]$ is also false we deduce similarly that $b(n) - a(n) < \epsilon$ for all $n \geqslant n^\star$. Hence $|a(n) - b(n)| < \epsilon$ for all $n \geqslant n^\star$ and so $a - b \in N(F)$. Thus $[x] = [y]$ as required.

As to compatibility, it follows readily from (10.1) that if $[a] \preceq [b]$ then for any $[c]$ we have $[a] + [c] \preceq [b] + [c]$; and for any $[d]$ with $[0] \preceq [d]$ we have $[a][d] \preceq [b][d]$. Thus $(C(F)/N(F); \preceq)$ is a totally ordered field. Finally, we observe that if, to every $x \in F$, we associate the constant sequence $\mathbf{x}$ (namely, the sequence every element of which is x) then the assignment $x \mapsto [\mathbf{x}]$ is a monomorphism and

$$x < y \iff x - y > 0 \iff [\mathbf{x}] \prec [\mathbf{y}].$$

Thus $C(F)/N(F)$ contains a copy of F. $\qquad\square$

Definition If F is a totally ordered field then we say that $t \in F$ is a **limit** of a sequence a of elements of F if

$$(\forall \epsilon \in F_+ \setminus \{0\})(\exists m \in \mathbb{N})(\forall n \geqslant m) \quad |a(n) - t| < \epsilon.$$

We say that a sequence of elements of F is **convergent** if it has a limit in F.

Theorem 10.12 *If a is a convergent sequence in a totally ordered field F then a has a unique limit and is a Cauchy sequence.*

Proof Suppose that t_1 and t_2 are each limits of a. We show that $t_1 < t_2$ is impossible. Suppose in fact that $t_1 < t_2$. Then taking $\epsilon = (t_2 - t_1)/2$ there exist $m, p \in \mathbb{N}$ such that $(\forall n \geqslant m)$ $|a(n) - t_1| < (t_2 - t_1)/2$ and $(\forall n \geqslant p)$ $|a(n) - t_2| < (t_2 - t_1)/2$. If $q = \max\{m, p\}$ we have the contradiction

$$t_2 - t_1 = t_2 - a(n) + a(n) - t_1 < |a(n) - t_2| + |a(n) - t_1| < t_2 - t_1.$$

It follows from this that $t_1 = t_2$ and so the sequence a has a unique limit which we shall denote by t.

Now there exists $m \in \mathbb{N}$ such that, for all $n \geqslant m$, $|a(n) - t| < \epsilon/2$. If $n, p \geqslant m$ we then have

$$|a(n) - a(p)| = |a(n) - t + t - a(p)| \leqslant |a(n) - t| + |a(p) - t| < \frac{\epsilon}{2} + \frac{\epsilon}{2} = \epsilon$$

and so a is a Cauchy sequence. □

The unique limit of a convergent sequence a is often written $\lim_{n \to \infty} a(n)$. In what follows we shall write simply $\lim a$.

EXERCISES

10.1. Prove that if F is a totally ordered field then the sequence b given by $b(n) = 2^{-n}$ is a null sequence in F. Deduce that any sequence a which is such that $0 \leqslant a(n) - a(n+1) \leqslant 2^{-(n+1)}$ is a Cauchy sequence in F.

10.2. For every positive integer n let $a(n)$ be the unique integer such that $[a(n)]^2 \leqslant 2.10^{2n} < [a(n) + 1]^2$. Show that, for each n,

$$2 - \left(\frac{a(n)}{10^n}\right)^2 < \frac{4}{10^n} + \frac{1}{10^{2n}}.$$

Deduce that the sequence b given by $b(n) = \left(a(n)/10^n\right)^2$ is convergent with $\lim b = 2$. Hence show that the sequence c given by $c(n) = a(n)/10^n$ is a Cauchy sequence in $\mathbb{Q}$ but that it does not converge in $\mathbb{Q}$.

10.3. If a is a Cauchy sequence in a totally ordered field F prove that precisely one of the following holds:

(1) $\lim a = 0$;

(2) a is a positive sequence;

(3) $-a$ is a positive sequence.

In Theorem 10.12 we have seen that every convergent sequence is a Cauchy sequence. The converse of this is not true in general; see for example Exercise 10.2. However, ordered fields in which the converse does hold are of especial importance as we shall see.

Definition A totally ordered field F is said to be **Cauchy complete** if every Cauchy sequence in F is convergent.

In order to provide an important example of a totally ordered field that is both archimedean and Cauchy complete we require the following characterisation of an archimedean totally ordered field in terms of its prime subfield.

Definition A non-empty subset S of a totally ordered field F is said to be **dense** if, whenever $a < b$ in F, there exists $x \in S$ such that $a < x < b$.

Theorem 10.13 *A totally ordered field F is archimedean if and only if its prime subfield is dense.*

Proof Since the prime subfield of F is isomorphic to $\mathbb{Q}$ we shall identify it with $\mathbb{Q}$ for convenience. Suppose then that $\mathbb{Q}$ is dense in F and let $a, b \in F$ be such that $0 < a < b$. By the density there exist integers $m, n > 0$ such that $0 < \dfrac{m}{n} < \dfrac{a}{b}$. Consequently $na = bn\dfrac{a}{b} > bn\dfrac{m}{n} = bm \geqslant b$ and so F is archimedean.

Conversely, suppose that F is archimedean and let $a, b \in F$ be such that $a < b$. There are several cases to consider:

(1) $a < 0 < b$: in this case $0 \in \mathbb{Q}$ satisfies the requirements.

(2) $0 < a < b$: in this case $b - a > 0$ and so $\dfrac{1}{b-a} > 0$ whence there exists $n > 0$ such that $n1 > \dfrac{1}{b-a}$ and hence $\dfrac{1}{n} < b - a$. Since F is archimedean and since $\mathbb{N}$ is well-ordered, there is a smallest $m \in \mathbb{N}$ such that $b \leqslant m\dfrac{1}{n} = \dfrac{m}{n}$. We then have $b > \dfrac{m-1}{n} = \dfrac{m}{n} - \dfrac{1}{n} > b - (b-a) = a$ and so in this case $\dfrac{m-1}{n}$ serves as the required element of $\mathbb{Q}$.

(3) $a < b < 0$: in this case we have $0 < -b < -a$ and so by case (2) there exists $x \in \mathbb{Q}$ with $-b < x < -a$ whence $a < -x < b$.   □

Theorem 10.14 *The totally ordered field $C(\mathbb{Q})/N(\mathbb{Q})$ is archimedean and Cauchy complete.*

Proof For convenience, we write $C(\mathbb{Q})/N(\mathbb{Q})$ as R and the elements of R as $[x]$ where $x \in C(\mathbb{Q})$. Moreover, $\mathbf{q}$ will denote the constant sequence each term of which is the rational q.

To show that R is archimedean we shall prove that the prime subfield of R is dense in R and appeal to Theorem 10.13. Now the prime subfield of R consists of the elements $[\mathbf{q}]$ where $q \in \mathbb{Q}$. Our objective therefore is to show that for any $[x], [y] \in R$ with $[x] \prec [y]$ there exists $q \in \mathbb{Q}$ such that $[x] \prec [\mathbf{q}] \prec [y]$. Now given $a \in [x]$ and $b \in [y]$ we have, since $b - a$ is a positive sequence,

$$(\exists \epsilon_{a,b} \in F_+ \setminus \{0\})(\exists n_{a,b} \in \mathbb{N})(\forall n \geqslant n_{a,b}) \quad b(n) - a(n) > \epsilon_{a,b}.$$

Since a and b are Cauchy sequences we also have

$$(\exists n_{a,b}^\star \in \mathbb{N})(\forall m, n \geqslant n_{a,b}^\star) \quad |a(m) - a(n)|, |b(m) - b(n)| < \epsilon_{a,b}/3.$$

Let $n^\star = \max\{n_{a,b}, n^\star_{a,b}\}$ and let $p = a(n^\star + 1) + b(n^\star + 1) \in \mathbb{Q}$. Then for all $n \geqslant n^\star$ we have

$$\frac{p}{2} - a(n) = \frac{1}{2}[b(n^\star + 1) - a(n^\star + 1)] + a(n^\star + 1) - a(n) > \frac{\epsilon_{a,b}}{2} - \frac{\epsilon_{a,b}}{3} = \frac{\epsilon_{a,b}}{6};$$

$$b(n) - \frac{p}{2} = \frac{1}{2}[b(n^\star + 1) - a(n^\star + 1)] + b(n) - b(n^\star + 1) > \frac{\epsilon_{a,b}}{2} - \frac{\epsilon_{a,b}}{3} = \frac{\epsilon_{a,b}}{6},$$

and consequently, taking $q = p/2$, we obtain $[x] \prec [\mathbf{q}] \prec [y]$.

We now show that R is Cauchy complete. For this purpose we show first that

$$a \in C(\mathbb{Q}) \Rightarrow \lim[\mathbf{a}(\mathbf{n})] = [a].$$

From the above, given any $[\delta] \succ [0]$ there exists $\epsilon \in \mathbb{Q}$ such that $[0] \prec [\epsilon] \prec [\delta]$. Since a is a Cauchy sequence we have

$$(\forall m, n \geqslant a_{\epsilon/2}) \quad |a(m) - a(n)| < \epsilon/2.$$

Let $n \geqslant a_{\epsilon/2}$ be fixed. Then for $p \geqslant a_{\epsilon/2}$ we have $a(p) - a(n) < \epsilon/2$ and $a(n) - a(p) < \epsilon/2$ so that $\epsilon - \big(a(p) - a(n)\big) > \epsilon/2$ and $\epsilon - \big(a(n) - a(p)\big) > \epsilon/2$. It follows that, in R,

$$[a] - [\mathbf{a}(\mathbf{n})] \prec [\epsilon] \prec [\delta], \quad [\mathbf{a}(\mathbf{n})] - [a] \prec [\epsilon] \prec [\delta].$$

Consequently

$$(\forall n \geqslant a_{\epsilon/2}) \quad |[a] - [\mathbf{a}(\mathbf{n})]| \prec [\delta]$$

and so $\lim[\mathbf{a}(\mathbf{n})] = [a]$.

Suppose now that $A^\star$ is a Cauchy sequence in R, so that the elements of $A^\star$ are of the form $A^\star(m) = [A(m)]$ where $A(m) \in C(\mathbb{Q})$ for each m. As there is nothing to prove if $A^\star$ is a constant sequence, we can assume without loss of generality that all the terms of $A^\star$ are distinct. By the above, for every p we can choose $a(p) \in \mathbb{Q}$ such that

$$|[\mathbf{a}(\mathbf{p})] - [A(p)]| < |[A(p+1)] - [A(p)]|.$$

The inequality

$$|[\mathbf{a}(\mathbf{p})] - [\mathbf{a}(\mathbf{q})]| \preceq |[\mathbf{a}(\mathbf{p})] - [A(p)]| + |[A(p)] - [A(q)]| + |[A(q)] - [\mathbf{a}(\mathbf{q})]|$$

and the isotone isomorphism $[\mathbf{x}] \mapsto x$ then show that the sequence given by $a : p \mapsto a(p)$ belongs to $C(\mathbb{Q})$. Finally, since $[a] = \lim[\mathbf{a}(\mathbf{n})]$ and

$$|[A(n)] - [a]| \preceq |[A(n)] - [\mathbf{a}(\mathbf{n})]| + |[\mathbf{a}(\mathbf{n})] - [a]|$$

we see that $[a] = \lim[A(n)]$. $\qquad\square$

Definition An ordered field is said to be **Dedekind complete** if every non-empty subset which is bounded above has a supremum.

Example 10.3 As an example of a totally ordered field that is not Dedekind complete we highlight $\mathbb{Q}$. Here, for example, $\sup\{x \in \mathbb{Q} \mid x^2 < 2\}$ does not exist in $\mathbb{Q}$. To see this, suppose that $t \in \mathbb{Q}$ is the supremum. Then since 2 is not the square of any rational we have $(t + 1)^2 > 2$. It follows that $2 - t^2 < 2t + 1$. Let $r = \dfrac{2 - t^2}{2t + 1}$. Then $0 < r < 1$ and so $r^2 < r$. Consequently $2tr + r^2 < 2tr + r = 2 - t^2$ and we have $(t + r)^2 < 2$ which contradicts the assumption that t is the supremum.

We shall in fact see that there is, to within isomorphism, only one Dedekind complete totally ordered field, and only one archimedean and Cauchy complete totally ordered field; moreover, they are isomorphic.

Definition By a **universal archimedean field** we shall mean a totally ordered field R that is archimedean and such that, if X is any archimedean totally ordered field, there is a unique monomorphism $h : X \to R$ such that the diagram

$$
\begin{array}{ccc}
Q & \xrightarrow{\iota_2} & X \\
{\scriptstyle \iota_1} \downarrow & \swarrow {\scriptstyle h} & \\
R & &
\end{array}
$$

is commutative, where ι_1 and ι_2 are the canonical embeddings.

We now establish the main result of this section.

Theorem 10.15 *To within isomorphism, there exists a unique universal archimedean field. Moreover, the following conditions on a totally ordered field F are equivalent*:

(1) *F is a universal archimedean field*;
(2) *F is Dedekind complete*;
(3) *F is archimedean and Cauchy complete*.

Proof We shall show that $(3) \Rightarrow (2) \Rightarrow (1) \Rightarrow (3)$, the first assertion being established in $(1) \Rightarrow (3)$.

$(3) \Rightarrow (2)$: Suppose that F is archimedean and Cauchy complete. Let A be a non-empty subset of F. For every positive integer p define

$$E_p = \{k \in \mathbb{Z} \mid 2^{-p}k \in A^\uparrow\}.$$

That $E_p \neq \emptyset$ follows from the fact that if $x \in A^\uparrow \cap \mathbb{N}$ then $2^p x \in E_p$. Now given any $x \in A$ we see that if $k \leqslant 2^p(x-1)$ then $2^{-p}k \leqslant x-1 < x$ whence $k \notin E_p$. Thus E_p is bounded below and hence has a smallest element, $k(p)$ say. Define a sequence a of elements of F by the prescription $a(p) = 2^{-p}k(p)$. Then $a(p) \in A^\uparrow$ and $a(p) - 2^{-p} = 2^{-p}[k(p) - 1] \notin A^\uparrow$. Equivalently, we have $2^{-(p+1)}[2k(p)] \in A^\uparrow$ and $2^{-(p+1)}[2k(p) - 2] \notin A^\uparrow$. It follows from this that either (i) $k(p+1) = 2k(p)$, or (ii) $k(p+1) = 2k(p) - 1$. Correspondingly,

$$
a(p+1) = 2^{-(p+1)}k(p+1) =
\begin{cases}
2^{-p}k(p) = a(p) & \text{(i)}; \\
2^{-(p+1)}[2k(p) - 1] = a(p) - 2^{-(p+1)} & \text{(ii)}.
\end{cases}
$$

Thus a is such that $0 \leqslant a(p) - a(p+1) \leqslant 2^{-(p+1)}$. We leave to the reader the task of showing that a is a Cauchy sequence (see Exercise 10.1). By the standing hypothesis that F is Cauchy complete, it then follows that $\alpha = \lim a$ exists in F. We now prove that $\alpha = \sup A$, whence (2) follows.

We note first that $\alpha \leqslant a(p)$ for every p. [In fact, if there existed p such that $\alpha > a(p)$ then $\alpha - a(p+1) > a(p) - a(p+1) \geqslant 0$ whence, by induction, $\alpha > a(t) \geqslant a(t+1)$ for all $t \geqslant p$; and this contradicts the fact that $\alpha = \lim a$.] We also note that $\alpha \in A^\uparrow$. [In fact, if $\alpha \notin A^\uparrow$ then there exists $x \in A$ such that $x > \alpha$ and so, for some p we have $a(p) - \alpha = |a(p) - \alpha| < x - \alpha$

whence $a(p) < x$; and this contradicts the fact that $a(p) \in A^{\uparrow}$.] Suppose now that $\beta \in A^{\uparrow}$ with $\beta < \alpha$. Since F is archimedean we can choose p such that $2^{-p} < \alpha - \beta$ whence $a(p) - 2^{-p} \geqslant \alpha - 2^{-p} > \beta$ and consequently we have the contradiction $a(p) - 2^{-p} \in A^{\uparrow}$. We conclude therefore that every $\beta \in A^{\uparrow}$ is such that $\beta \geqslant \alpha$ and hence that $\alpha = \sup A$.

(2) $\Rightarrow$ (1): Suppose now that F is Dedekind complete. If $0 < a < b$ in F let $E = \{na \mid n \in \mathbb{N}, \ n \neq 0\}$. If now we had $na \leqslant b$ for all $n \in \mathbb{N}$ then b would be an upper bound for E and so, by the hypothesis, E would have a supremum α. Since $0 < a$ we have $\alpha - a < \alpha$ whence there exists $n > 0$ such that $\alpha - a < na$ [for if no such n existed then $\alpha - a$ would be an upper bound which is strictly less than the least upper bound α, which is not possible]. We then have $\alpha = \alpha - a + a < na + a = (n+1)a < \alpha$. This contradiction therefore shows that for some n we must have $na > b$. Hence F is archimedean.

Suppose now that X is any archimedean totally ordered field. For every $x \in X$ define the subset L_x of F by

$$L_x = \{q1_F \mid q \in \mathbb{Q}, \ q1_X < x\}.$$

Since X is archimedean there exists $n > 0$ in $\mathbb{N}$ such that $n1_X > -x$ whence $(-n)1_X < x$ and so $L_x \neq \emptyset$. Again since X is archimedean there exists $p > 0$ in $\mathbb{N}$ such that $p1_X > x$ and so $p1_F$ is an upper bound for L_x. Since, by hypothesis, F is Dedekind complete, $\sup L_x$ exists. For every $x \in X$ define also the subset G_x of F by

$$G_x = \{q1_F \mid q \in \mathbb{Q}, \ x < q1_X\}.$$

By a similar argument to the above the set $-G_x$ has a supremum whence $\inf G_x$ exists. We now show that

$$\sup L_x = \inf G_x.$$

Now it is clear that every element of G_x is strictly greater than every element of L_x and so every element of G_x is an upper bound for L_x whence we have $\inf G_x \geqslant \sup L_x$. Here we must have equality, for otherwise, F being archimedean, there would exist $t \in \mathbb{Q}$ such that $\sup L_x < t1_F < \inf G_x$ and likewise $r, s \in \mathbb{Q}$ such that $\sup L_x < r1_F < t1_F < s1_F < \inf G_x$. We would then have $r1_X \geqslant x$ and $s1_X \leqslant x$ whence $r \geqslant s$, which gives the contradiction $r1_F \geqslant s1_F$.

We now define a mapping $h : X \to F$ by the prescription $h(x) = \sup L_x$, noting that if $\iota_1 : \mathbb{Q} \to F$ and $\iota_2 : \mathbb{Q} \to X$ are the canonical embeddings then for every $q \in \mathbb{Q}$ we obtain $(h \circ \iota_2)(q) = h(q1_X) = \sup L_{q1_X} = q1_F = \iota_1(q)$ so that $h \circ \iota_2 = \iota_1$.

We now show that h is a morphism. Suppose first that $x, y \in X$ are such that $0 < x, y$. Then if $0 < p1_F \in L_x$ and $0 < q1_F \in L_y$ we have $pq1_X = p1_X q1_X < xy$ and so $pq1_F = \sup L_{pq1_X} \leqslant \sup L_{xy} = h(xy)$. It follows that $p1_F \leqslant h(xy)(q1_F)^{-1}$ and so $h(x) \leqslant h(xy)(q1_F)^{-1}$ whence $q1_F \leqslant h(xy)[h(x)]^{-1}$ and so $h(y) \leqslant h(xy)[h(x)]^{-1}$ whence $h(x)h(y) \leqslant h(xy)$. Since we also have $h(x) = \inf G_x$ for every $x \in X$ we can argue similarly to deduce that $h(xy) \leqslant h(x)h(y)$. Thus, for $x, y > 0$ we have $h(xy) = h(x)h(y)$. Clearly, this equality also holds whenever one of x, y is 0.

A similar argument, using sums instead of products, shows that $h(x+y) = h(x) + h(y)$, irrespective of whether $x, y > 0$ or not; and in particular we have $h(-x) = -h(x)$.

To complete the proof that h is a morphism, it remains to consider the case where $x < 0$ and $y > 0$. Here we have $x = -z$ where $z > 0$ and so

$$h(x)h(y) = h(-z)h(y) = -h(z)h(y)$$
$$= -h(zy)$$
$$= -h(-xy) = -[-h(xy)] = h(xy)$$

as required.

Next, we show that h is injective. For this purpose, suppose that $x < y$ in X. Since X is archimedean there is an integer n such that $\frac{1}{n}1_X < y - x$ whence $\frac{1}{n}1_F \in L_{y-x}$. We then have $h(y) = h(y - x + x) = h(y - x) + h(x) \geqslant \frac{1}{n}1_F + h(x) > h(x)$. Thus h is injective and so is a monomorphism.

It remains to establish the uniqueness of h. For this purpose, suppose that $g : X \to F$ is also an isotone monomorphism such that $g \circ \iota_2 = \iota_1$. Suppose, by way of obtaining a contradiction, that there exists $x \in X$ such that $g(x) < h(x)$. Then since F is archimedean there exists $q \in \mathbb{Q}$ such that $g(x) < q1_F < h(x)$. It then follows from the definition of h that $q1_F \in L_x$ and so $q1_X < x$. Thus we obtain the contradiction

$$q1_F = \iota_1(q) = g[\iota_2(q)] = g(q1_X) < g(x) < q1_F.$$

It follows therefore that $(\forall x \in X) \ g(x) \geqslant h(x)$. Suppose now that there exists $x \in X$ such that $g(x) > h(x)$. Then there exists $q \in \mathbb{Q}$ such that $h(x) < q1_F < g(x)$. In this case $q1_F \notin L_x$ and so $q1_X \geqslant x$ and we have a similar contradiction. We conclude therefore that $g(x) = h(x)$ for all $x \in X$ and so $g = h$.

(1) $\Rightarrow$ (3): We note first that, by Theorem 10.14, the totally ordered field $C(\mathbb{Q})/N(\mathbb{Q})$ is archimedean and Cauchy complete. By the implications (3) $\Rightarrow$ (2) $\Rightarrow$ (1) above, this is then a universal archimedean field. Now if F is any universal archimedean field then the uniqueness of the isotone monomorphisms h, k in the diagrams

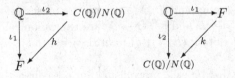

shows that $h \circ k : F \to F$ is an isotone monomorphism such that $h \circ k \circ \iota_1 = \iota_1$. But since F is univeral archimedean the only such monomorphism possible is id_F. Hence $h \circ k = \mathrm{id}_F$ and so h is also surjective. Thus we see that h is an isotone isomorphism whose inverse is the isotone isomorphism k. This then establishes the existence and uniqueness up to isomorphism of a universal archimedean field. It also establishes (1) $\Rightarrow$ (3) since then F is isomorphic to $C(\mathbb{Q})/N(\mathbb{Q})$ which satisfies (3) by Theorem 10.14. $\square$

The above result shows that there is an essentially unique universal archimedean field. As a model of this we may choose $C(\mathbb{Q})/N(\mathbb{Q})$. This essentially unique field is called the field of **real numbers** and is denoted by $\mathbb{R}$. Of course this is not the only model that can be chosen. Another useful way of defining the real numbers (to within isomorphism) involves the notion of a **Dedekind section** of the rationals, this consisting of a partition of $\mathbb{Q}$ into a down-set L and a complementary up-set R. If L has a biggest element, or R has a smallest element, then the section (L, R) is said to define a **real rational number**; on the other hand if L has no biggest element and R has no smallest element then the section (L, R) is said to define a **real irrational number**. If X, Y are subsets of $\mathbb{Q}$ and if we define $X + Y = \{x + y \mid x \in X, y \in Y\}$ and $XY = \{xy \mid x \in X, y \in Y\}$ then addition and multiplication can be defined on the set of Dedekind sections by setting $(L, R) + (L', R') = (L + L', R + R')$ and $(L, R)(L', R') = (LL', RR')$. Morover, an order can be defined by $(L, R) \leqslant (L', R')$ if and only if $L \subseteq L'$ or, equivalently, $R \supseteq R'$. It can be shown that, equipped with these operations, the set of Dedekind sections of $\mathbb{Q}$ forms a totally ordered field that is Dedekind complete and therefore serves also as a suitable model for $\mathbb{R}$.

The fact that $\mathbb{R}$ is Dedekind complete is of immense importance in analysis. We are in fact now equipped with enough machinery to embark on a rigorous course of real analysis!

EXERCISES

10.4. Use the fact that $\mathbb{Q}$ is dense in $\mathbb{R}$ to prove that for every $r \in \mathbb{R}$ there is a unique $n \in \mathbb{Z}$ such that $n \leqslant r < n + 1$.

10.5. If a is a sequence of real numbers then by the **series** defined by a we mean the sequence Σ_a of **partial sums** given by $\Sigma_a(n) = \sum_{i=0}^{n} a(i)$. If the sequence Σ_a converges then we say that a is **summable** and write $\lim \Sigma_a$ as $\sum_{n \in \mathbb{N}} a(n)$.

Prove that if the sequences a, b are summable then so is $\lambda a + \mu b$ for all $\lambda, \mu \in \mathbb{R}$ and that $\sum_{i \in \mathbb{N}} (\lambda a + \mu b) = \lambda \sum_{i \in \mathbb{N}} a(i) + \mu \sum_{i \in \mathbb{N}} b(i)$.

10.6. Let $x \in \mathbb{R}$ be such that $0 \leqslant x < 1$ and let b be a fixed integer with $b > 1$. For every $y \in \mathbb{R}$ let $[y]$ denote the biggest integer that is less than or equal to y. Show that if c is the sequence defined by $c(n) = [xb^n]/b^n$ then $\lim c = x$. Deduce that the sequence a given by $a(n) = [xb^n] - [xb^{n-1}]b$ is such that

(1) $(\forall n \in \mathbb{N})\ 0 \leqslant a(n) \leqslant b - 1$;

(2) $(\forall n \in \mathbb{N})(\exists m \geqslant n)\ a(m) < b - 1$;

(3) $x = \sum_{n \in \mathbb{N}} a(n)b^{-n}$.

The equality (3) is the **expansion of x to the base b**. With $b = 10$ we obtain the **decimal expansion** and with $b = 2$ the **binary expansion**.

10.7. If $x \in \mathbb{R}$ is such that $x > 0$ and if $n \in \mathbb{Z}$ is such that $n > 0$ show as follows that there is a unique $y \in \mathbb{R}$ such that $y^n = x$.

Let $E_x = \{t \in \mathbb{R} \mid t^n < x\}$. Show that $\dfrac{x}{1 + x} \in E_x$ so that $E_x \neq \emptyset$; and that $1 + x$ is an upper bound of E_x so that $\alpha = \sup E_x$ exists. To show that $\alpha^n = x$, suppose first that $\alpha^n < x$ and choose h such that

$$0 < h < \frac{x - \alpha^n}{(\alpha + 1)^n - \alpha^n}.$$

Use the binomial expansion of $(\alpha + h)^n$ to deduce the contradiction $\alpha + h \in E_x$. Suppose next that $x < \alpha^n$ and choose h such that

$$0 < h < \min\left\{1, \alpha, \frac{\alpha^n - x}{(\alpha + 1)^n - \alpha^n}\right\}.$$

Use the binomial expansion of $(\alpha - h)^n$ to deduce the contradiction that $\alpha - h$ is an upper bound of E_x.

10.3 Archimedean totally ordered groups

We now turn our attention to archimedean totally ordered groups.

Theorem 10.16 (Hölder [66]) *Let G be a totally ordered group. Then the following statements are equivalent:*

(1) *G is archimedean;*
(2) *G is isomorphic to a subgroup of $(\mathbb{R}; +)$;*
(3) *G has no proper non-trivial convex subgroups.*

Proof (1) $\Rightarrow$ (2): There are two cases to consider.

Suppose first that P_G contains an atom a. Since, by hypothesis, G is totally ordered, this is the only atom in P_G. It then follows by Theorem 9.24 that $G = \{a^n \mid n \in \mathbb{Z}\} \simeq (\mathbb{Z}; +)$.

Suppose now that P_G has no atoms. Then for every $g > 1_G$ there exists $h \in P_G$ such that $1_G < h < g$. Now either $h \leqslant gh^{-1}$ or $gh^{-1} \leqslant h$; i.e., either $1_G < h^2 < g$ or $1_G < (gh^{-1})^2 \leqslant g$. In each case therefore there exists $f \in P_G$ such that $1_G < f^2 \leqslant g$.

We now use this observation to show that G is commutative. Suppose that $a, b \in P_G$ and consider the commutator $[a, b]$. Without loss of generality we may suppose that $[a, b] \in P_G$. By way of obtaining a contradiction, suppose that $[a, b] \neq 1_G$. By the above observation there exists $f \in P_G$ such that $f^2 \leqslant [a, b]$. Now by the archimedean property there exist m, n such that $f^m \leqslant a < f^{m+1}$ and $f^n \leqslant b < f^{n+1}$. This gives the contradiction $[a, b] = a^{-1}b^{-1}ab < f^{-m}f^{-n}f^{m+1}f^{n+1} = f^2$. Hence we must have $[a, b] = 1_G$ for all $a, b \in P_G$. Thus P_G is commutative whence, by Theorem 9.16, so is G.

To construct an ℓ-morphism $\vartheta : G \to \mathbb{R}$ we begin by fixing an element $\beta > 1_G$.

Suppose first that $g \in G$ is rationally dependent on β, in the sense that $g^m = \beta^n$ for non-zero integers m and n. Now the equation $x^m = \beta^n$ has a unique solution. In fact, if $y \in G$ is also such that $y^m = \beta^n$ then necessarily $x^{|m|} = y^{|m|}$ whence, by Theorem 9.11, it follows that $x = y$. Denoting the unique solution of $x^m = \beta^n$ by $\beta^{n/m}$ we can therefore associate with g the rational number $\vartheta(g) = \dfrac{n}{m}$.

Suppose now that g is not rationally dependent on β. Then we can partition $\mathbb{Q}$ into the two classes L, R given by

$$\frac{m}{n} \in L \iff g^m < \beta^n, \quad \frac{m}{n} \in R \iff g^m > \beta^n.$$

We observe that neither L nor R is empty. In fact, if $g < \beta$ then $\frac{1}{1} \in L$ and, by the archimedean property, there exists k such that $g^k < \beta < g^{k+1}$ whence $R \neq \emptyset$. A similar argument holds if $g > \beta$. Now L is a down-set of $\mathbb{Q}$. In fact, if $\frac{p}{q} \in \mathbb{Q}$ is such that $\frac{p}{q} < \frac{m}{n} \in L$ then $|g|^{pn} < |g|^{qm} < \beta^{qn}$. Since G is commutative, it follows by Theorem 9.16(3) and Theorem 10.8(7) that $g^p < \beta^q$ and so $\frac{p}{q} \in L$. Dually, R is an up-set of $\mathbb{Q}$. Thus we see that (L, R) forms a Dedekind section of $\mathbb{Q}$ and therefore defines an irrational number associated with g. We denote this irrational by $\vartheta(g)$.

Consider now the mapping $\vartheta : G \to \mathbb{R}$ so defined. We observe first that if $g, h \in G$ are rationally dependent on β, say $g^m = \beta^n$ and $h^p = \beta^q$, then since g and h commute we have $(gh)^{mp} = g^{mp} h^{mp} = \beta^{np} \beta^{mq} = \beta^{np+mq}$ which shows that the real number associated with gh is the rational

$$\frac{np + mq}{mp} = \frac{n}{m} + \frac{q}{p}.$$

In the general case, suppose that $g, h \in G$ correspond to real numbers r, s respectively. Since the sum of two rationals respectively less than (or greater than) r and s belong to the lower (or upper) class of the Dedekind section that defines the real number $\vartheta(gh)$, this number is then the sum of the real numbers associated with g and h. In other words, we have $\vartheta(gh) = \vartheta(g) + \vartheta(h)$ and so ϑ is a group morphism. Clearly, ϑ is isotone and so is also an ℓ-morphism.

If now $\vartheta(g) = 0$ then only the negative powers of β are less than the positive powers of g. Consequently, for every positive integer n we have $g^n \leqslant \beta$ whence, by the archimedean property, we obtain $g = 1_G$. Thus ϑ is injective and (2) follows.

(2) $\Rightarrow$ (3): Suppose that G is a subgroup of $\mathbb{R}$ and let H be a proper non-trivial convex subgroup of G. Then if $0 < h \in H$ and $0 < g \in G$ there exists a positive integer n such that $nh \geqslant g$. The convexity of H gives $g \in H$ so that $P_G \subseteq P_H$ whence $G \subseteq H$ and we have equality.

(3) $\Rightarrow$ (1): Suppose that (3) holds. If $g, h \in G$ are such that $1_G \leqslant g \ll h$ then $G(g) \subset G(h)$. By (3) we must then have $G(g) = \{1_G\}$ whence $g = 1_G$ and consequently G is archimedean. $\qquad \square$

It is an immediate consequence of the above result that every archimedean totally ordered group is commutative. In fact, the following stepping stones will lead us to the more general result that every archimedean lattice-ordered group is commutative.

Given a lattice-ordered group G and $1_G \neq g \in G$, let V be a value of g. Then by Theorem 9.37, V is completely meet-irreducible in the lattice $\mathcal{C}(G)$. A consequence of this is that V is covered in $\mathcal{C}(G)$ by the intersection $V^\star$ of all the convex ℓ-subgroups that properly contain V.

Definition A lattice-ordered group is said to be **normal valued** if every value V in $\mathcal{C}(G)$ is normal in its cover $V^\star$.

Theorem 10.17 *Every representable lattice-ordered group is normal valued.*

Proof Let G be representable and let V be a regular subgroup of G. By the Corollaries of Theorems 9.36 and 9.37 there is a minimal prime subgroup P such that $P \subseteq V$. Moreover, by Theorem 9.39 we have $g^{-1}Pg = P$ for every $g \in G$. If now we choose $g \in V^\star$ then by Theorem 9.36(2) we see that either $V \subseteq g^{-1}Vg \subset g^{-1}V^\star g = V^\star$ or $g^{-1}Vg \subseteq V \subset V^\star = g^{-1}V^\star g$, the latter giving $V \subseteq gVg^{-1} \subset gV^\star g^{-1} = V^\star$. Since $V^\star$ covers V it follows that $V = g^{-1}Vg$ whence V is normal in $V^\star$.   □

Theorem 10.18 *Every archimedean lattice-ordered group is representable.*

Proof Let G be archimedean and let $g \in G$. Suppose that $f \in g^\perp$, so that $f \wedge |g| = 1_G$. For any $h \in G_+$ let $x = h^{-1}fh \wedge |g|$. Then $1_G \leqslant x$ and consequently $1_G \leqslant x \wedge hxh^{-1} = h^{-1}fh \wedge |g| \wedge f \wedge h|g|h^{-1} \leqslant |g| \wedge f = 1_G$. The resulting equality shows that x and hxh^{-1} are orthogonal whence, by Theorem 9.22, they commute and so, by the Corollary to Theorem 9.17, for all $n \geqslant 1$ we have $1_G = x^n \wedge (hxh^{-1})^n = x^n \wedge hx^nh^{-1} \geqslant 1_G \vee x^nh^{-1} \geqslant 1_G$. The resulting equality gives $x^nh^{-1} \leqslant 1_G$ whence $x^n \leqslant h$ for all $n \geqslant 1$. The hypothesis that G is archimedean now gives $x = 1_G$. Consequently $h^{-1}fh \in g^\perp$ for every $h \in G_+$. Using Theorem 9.14 it readily follows that $k^{-1}fk \in g^\perp$ for every $k \in G$. That G is representable now follows by Theorem 9.39(5).   □

Theorem 10.19 *Every archimedean lattice-ordered group is commutative.*

Proof Let G be an archimedean lattice-ordered group. Then, by Theorem 10.18, G is representable and so, by Theorem 10.17, G is normal valued. Now if V is a regular subgroup of G then $\mathcal{R}(V)$ is totally ordered whence so is $V^\star/V$. Since $V^\star$ covers V in $\mathcal{C}(G)$ we see that $V^\star/V$ has no non-trivial proper convex subgroups and so, by Theorem 10.16, is isomorphic to a subgroup of $\mathbb{R}$ and therefore is commutative. It follows from this that if $x, y \in V^\star$ then the commutator $[x, y] \in V$.

We use this observation to show that $[x, y] \ll |x| \vee |y|$ for all $x, y \in G$. By way of obtaining a contradiction, suppose that there exists $n \in \mathbb{N}$ such that

$$[x, y]^n \not\leqslant |x| \vee |y|. \tag{10.2}$$

Let $z = [x, y]^n(|x| \vee |y|)^{-1} \vee 1_G$. Then $z > 1_G$ and $z \in G(x) \vee G(y) = G(|x| \vee |y|)$, by the Corollary to Theorem 9.33. Now let N be a value of z. Then we have $|x| \vee |y| \notin N$; for otherwise we would have $z \in N$. There therefore exists a value M of $|x| \vee |y|$ such that $N \subseteq M$. Since $x, y \in M^\star$ it follows by the above observation that $[x, y] \in M$ whence $[x, y]^n \in M$. Now by (10.2) and the fact that $\mathcal{R}(N)$ is totally ordered we have $N[x, y]^n > N(|x| \vee |y|)$; for otherwise there exists $t \in N$ such that $t[x, y]^n \leqslant |x| \vee |y|$ which gives $1_G < z \leqslant t^{-1} \vee 1_G \in N$ whence the contradiction $z \in N$. Consequently there exists $p \in N$ such that $[x, y]^n > p(|x| \vee |y|)$. Since $N \subseteq M$ we therefore have, in the totally ordered group $\mathcal{R}(M)$, $M[x, y]^n > M(|x| \vee |y|)$. But $[x, y]^n \in M$ now gives the existence of $q \in M$ such that $1_G > q(|x| \vee |y|)$ whence $|x| \vee |y| < q^{-1} \in M$ and we have the contradiction $|x| \vee |y| \in M$. We conclude from this that for all $n \in \mathbb{N}$ we have $[x, y]^n \leqslant |x| \vee |y|$ and so $[x, y] \ll |x| \vee |y|$. Since G is archimedean it follows that $[x, y] = 1_G$ as required.   □

We now consider briefly totally ordered rings and fields that are archimedean, in the following sense.

Definition We say that an ordered ring $(R; +, \cdot, \leqslant)$ is **archimedean** if the underlying ordered abelian group $(R; +, \leqslant)$ is archimedean.

For a characterisation of such rings we require the following result concerning $\mathbb{R}$.

Theorem 10.20 (Hion [65]) *Let A and B be subgroups of $(\mathbb{R}; +)$ with $A \neq \{0\}$. If $f : A \to B$ is an isotone morphism then there exists $r \in \mathbb{R}$ with $r \geqslant 0$ such that $f(a) = ra$ for every $a \in A$.*

Proof Suppose first that there exists $a \in P_A$ such that $f(a) = 0$. By Theorem 10.16, A is archimedean and so for every $x \in P_A$ there exists $n \geqslant 1$ such that $0 < x < na$. Consequently $0 = f(0) \leqslant f(x) \leqslant f(na) = nf(a) = n0 = 0$. It follows that f is the zero morphism, whence the result holds in this case with $r = 0$.

Suppose now that $f(a) > 0$ for every $a \in P_A \backslash \{0\}$. Given $a_1, a_2 \in P_A \backslash \{0\}$, consider the quotients $\dfrac{a_1}{a_2}$ and $\dfrac{f(a_1)}{f(a_2)}$. If $\dfrac{a_1}{a_2} < \dfrac{f(a_1)}{f(a_2)}$ then since $\mathbb{Q}$ is dense in $\mathbb{R}$ there exists $\dfrac{m}{n} \in \mathbb{Q}$ such that $\dfrac{a_1}{a_2} < \dfrac{m}{n} < \dfrac{f(a_1)}{f(a_2)}$. This gives $na_1 < ma_2$ and $f(ma_2) = mf(a_2) < nf(a_1) = f(na_1)$ which contradicts the hypothesis that f is isotone. A similar contradiction results on assuming that $\dfrac{a_1}{a_2} > \dfrac{f(a_1)}{f(a_2)}$. We therefore deduce that $\dfrac{a_1}{a_2} = \dfrac{f(a_1)}{f(a_2)}$ for all $a \in P_A \backslash \{0\}$. Consequently for every $a \in P_A \backslash \{0\}$ we have $\dfrac{f(a)}{a}$ is a constant $r > 0$, from which the result follows. $\square$

Theorem 10.21 *If A is an archimedean totally ordered ring then either A is a zero ring with additive group order isomorphic to a subgroup of $\mathbb{R}$, or A is order isomorphic to a uniquely determined subring of $\mathbb{R}$.*

Proof By Theorem 10.16, $(A; +)$ is isomorphic to a subgroup of $(\mathbb{R}; +)$. For convenience we shall identify A with this subgroup. We shall also denote products in A by $\cdot$ and those in $\mathbb{R}$ by juxtaposition. If $0 \leqslant a \in A$ then the mapping $x \mapsto a \cdot x$ is an isotone endomorphism of A and so, by Theorem 10.20, there exists $r_a \in \mathbb{R}$ with $r_a \geqslant 0$ such that $a \cdot x = r_a x$ for all $x \in A$. If for $a < 0$ we let $r_a = -r_{-a}$ then it is readily verified that the assignment $\vartheta : a \mapsto r_a$ defines an isotone morphism $\vartheta : A \to \mathbb{R}$. Applying Theorem 10.20 to this, we deduce the existence of $s \geqslant 0$ such that $r_a = sa$ for every $a \in A$. If now $s = 0$ then clearly A is a zero ring. If $s > 0$ then from $a \cdot b = r_a b = sab$ we deduce that, for all $a, b \in A$, $r_{a \cdot b} = s(a \cdot b) = s^2 ab = (sa)(sb) = r_a r_b$. It follows from this that $\operatorname{Im} \vartheta$ is a subring of $\mathbb{R}$. Since $\vartheta(a) = \vartheta(b)$ gives $sa = sb$ in $\mathbb{R}$ we have that $a = b$ and so ϑ is an isotone monomorphism whence $A \overset{o}{\simeq} \operatorname{Im} \vartheta$.

To establish uniqueness, it suffices to show that if A and B are subrings of $\mathbb{R}$ with $\zeta : A \to B$ an order isomorphism then $A = B$. For this purpose we observe that by Theorem 10.16 there exists $r > 0$ in $\mathbb{R}$ such that $\zeta(a) = ra$ for every $a \in A$. Then $\zeta(a)\zeta(b) = \zeta(ab)$ gives $r^2 = r$ whence, since A and B are not zero rings, $r = 1$ and consequently $A = B$. $\square$

EXERCISES

10.8. Corresponding to the notion of a representable group, we say that a lattice-ordered ring is a **function ring**, or **f-ring**, if it is both a subring and a sublattice of a cartesian ordered cartesian product of totally ordered rings.

If A is an f-ring and $a, b, c \in A$ with $c \geqslant 0$, prove that
   (1) $|ab| = |a|\,|b|$;
   (2) $a^2 \geqslant 0$;
   (3) $b \in a^{\perp} \Rightarrow ab = 0$;
   (4) $b \in a^{\perp} \Rightarrow ac, bc \in a^{\perp}$;
   (5) $(a \vee b)c = ac \vee bc$ and $c(a \vee b) = ca \vee cb$.

10.9. If A is a lattice-ordered ring prove that the following are equivalent:
   (1) A is an f-ring;
   (2) every polar of A is an ideal of A;
   (3) every minimal prime subgroup of A is an ideal of A.

10.10. Prove that every archimedean f-ring is commutative.

Ordered semigroups; residuated semigroups

11.1 Ordered semigroups

We now turn our attention to the more general algebraic structure of an **ordered semigroup**, this being a semigroup together with a compatible order. As we shall see, the theory of such structures is vastly different from that of ordered groups. Some examples of ordered semigroups are the following. We shall use the term **monoid** to indicate a semigroup with an identity.

Example 11.1 If B is an idempotent semigroup (a **band**) then the relation

$$e \leqslant f \iff e = ef = fe$$

is a compatible order on B.

Example 11.2 If G is an ordered group then its positive cone P_G is an ordered monoid.

Example 11.3 Ordered by divisibility, $(\mathbb{N}; \cdot)$ is an ordered monoid. It has top element 0 and bottom element 1.

Example 11.4 If S is a semigroup then $(\mathbb{P}(S); \cdot, \subseteq)$ is an ordered semigroup under the multiplication $X \cdot Y = \{xy \mid x \in X,\ y \in Y\}$.

Example 11.5 If R is a commutative ring then, ordered by set inclusion, the set $I(R)$ of ideals of R is an ordered semigroup, the multiplication being the set product as defined in the previous example.

Example 11.6 If E is an ordered set then, ordered as in Example 1.7, the set of isotone mappings $f : E \to E$ forms an ordered monoid under composition. Likewise, the subset $\operatorname{Res} E$ of residuated mappings on E is an ordered monoid.

EXERCISES

11.1. Prove that, under the usual order, each of the following intervals of $\mathbb{R}$ are ordered semigroups with respect to the multiplication given:
    (1) $(0, 1]$ with $x \cdot y = \min\{x, y\}$;
    (2) $[\frac{1}{2}, 1]$ with $x \cdot y = \max\{\frac{1}{2}, xy\}$;
    (3) $[0, 1]$ with $x \cdot y = x + y - xy$;
    (4) $[0, 1]$ with $x \cdot y = (x + y)/(1 + xy)$.

11.2. If G is a group and $\mathcal{N}(G)$ is the lattice of normal subgroups of G prove that $(\mathcal{N}(G); \cdot, \subseteq)$ is an ordered semigroup where, for all $H, K \in \mathcal{N}(G)$, the product $H \cdot K$ is defined to be the commutator $[H : K]$.

Definition If $(S; \cdot, \leqslant)$ is an ordered semigroup and $x, y \in S$ then we define the right and left **quasi-residuals** of x by y to be the down-sets

$$\langle x \cdot {}^{\cdot} y \rangle = \{z \in S \mid yz \leqslant x\}, \qquad \langle x^{\cdot} . y \rangle = \{z \in S \mid zy \leqslant x\}.$$

We say that S is **quasi-residuated** if these sets are not empty for all choices of $x, y \in S$. If $x \in S$ is such that $\langle x \cdot {}^{\cdot} y \rangle = \langle x^{\cdot} . y \rangle$ for every $y \in S$ then we denote this by writing $\langle x : y \rangle$ and say that x is **equi-quasiresidual**.

For an example of a quasi-residuated semigroup we focus on a type of semigroup that plays an important role in the algebraic theory. For the reader's convenience, we include all of the relevant details.

Example 11.7 A semigroup S is said to be **regular** if, for every $a \in S$, there exists $x \in S$ such that $a = axa$, such an x being called a **pre-inverse** of a. We say that $b \in S$ is an **inverse** of $a \in S$ if $a = aba$ and $b = bab$. In this case a is also an inverse of b, and both ab and ba are idempotent. In a regular semigroup every element has at least one inverse; if $a = axa$ then for example $b = xax$ is an inverse of a. By an **inverse semigroup** we mean a regular semigroup in which every element has a unique inverse. The following conditions on a semigroup S are equivalent:

(1) S is an inverse semigroup;
(2) S is regular and its idempotents commute;
(3) every principal right [left] ideal has a unique idempotent generator.

[In fact, if (1) holds then clearly S is regular, and the set E of idempotents in S is not empty. If now $e, f \in E$ let a be the unique inverse of ef. Then $ef \cdot ae \cdot ef = efaef = ef$ and $ae \cdot ef \cdot ae = aefae = ae$ whence, by the uniqueness of inverses, $ae = a$. Similarly, we have $fa = a$. Consequently $a^2 = aefa = a$ which shows that a is an inverse of itself. By the uniqueness of inverses we then have $a = ef$ and so ef is idempotent. Likewise, fe is idempotent. Then $ef \cdot fe \cdot ef = ef$ and $fe \cdot ef \cdot fe = fe$ whence ef and fe are inverses of each other. Hence $fe = a = ef$ and therefore idempotents commute, so that (2) holds.

Suppose now that (2) holds. For every $a \in S$ there exists $x \in S$ such that $a = axa$, and the element $e = ax$ is an idempotent such that $ea = a$. Since for every $t \in S$ the principal right ideal generated by $\{t\}$ is $\{t\} \cup tS$ we see from the above that $\{a\} \cup aS = \{e\} \cup eS = eS$ and so every principal right ideal has an idempotent generator. Suppose now that e, f are idempotents that generate the same principal right ideal, so that $eS = \{e\} \cup eS = \{f\} \cup fS = fS$. Then there exists $y \in S$ such that $ey = f$ and so $ef = eey = ey = f$. Similarly we have $fe = e$. Since, by the hypothesis, idempotents commute we then have $e = f$, whence the uniqueness in (3).

Finally, suppose that (3) holds. If e is an idempotent then from $\{a\} \cup aS = \{e\} \cup eS = eS$ we have the existence of $x \in S$ such that $a = ex$, so that $ea = ex = a$; and $y \in S$ such that $ay = e$. It follows that $aya = ea = a$ and so a is regular. Since a is arbitrary, S is regular. Suppose now that b, c are inverses of a, so that $aba = a$, $bab = b$, $aca = a$, $cac = c$. Since ab and ac are idempotents, these equalities give

$$\{ab\} \cup \{ab\}S = \{ab\}S = \{a\}S = \{ac\}S = \{ac\} \cup \{ac\}S.$$

By (3), we then have $ab = ac$. In a similar way, $ba = ca$ and consequently $b = bab = bac = cac = c$ whence inverses are unique and so S is an inverse semigroup.]

In an inverse semigroup the unique inverse of a is written a^{-1}, and in this situation we have the identities $(x^{-1})^{-1} = x$ and $(xy)^{-1} = y^{-1}x^{-1}$.

[In fact, the first is clear by the uniqueness of inverses. As for the second, since idempotents commute we have $ab \cdot b^{-1}a^{-1} \cdot ab = aa^{-1}abb^{-1}b = ab$ and $b^{-1}a^{-1} \cdot ab \cdot b^{-1}a^{-1} = b^{-1}bb^{-1}a^{-1}aa^{-1} = b^{-1}a^{-1}$ whence $b^{-1}a^{-1}$ is the inverse of ab.]

We now show how an inverse semigroup can be ordered. For this purpose, we observe that in such a semigroup S the following conditions concerning $a, b \in S$ are equivalent:

$(\alpha) \quad aa^{-1} = ab^{-1};$ $\quad (\beta) \quad a^{-1}a = a^{-1}b;$ $\quad (\gamma) \quad ab^{-1}a = a;$

$(\alpha') \quad aa^{-1} = ba^{-1};$ $\quad (\beta') \quad a^{-1}a = b^{-1}a;$ $\quad (\gamma') \quad a^{-1}ba^{-1} = a^{-1}.$

[In fact, each condition (x') is equivalent to (x) as may be seen by taking inverses. We establish the result by showing that $(\alpha) \Rightarrow (\gamma)$, $(\gamma') \Rightarrow (\beta)$, and $(\beta') \Rightarrow (\alpha)$. For $(\alpha) \Rightarrow (\gamma)$ simply multiply (α) on the right by a. Now (γ') implies that $a^{-1}b$ is idempotent and so, since idempotents commute, (γ') gives $a^{-1}a = a^{-1}ba^{-1}a = a^{-1}aa^{-1}b = a^{-1}b$ which is (β). Finally, if (β') holds then $(ab^{-1})^2 = ab^{-1}ab^{-1} = aa^{-1}ab^{-1} = ab^{-1}$ so that ab^{-1} is idempotent. Then, again since idempotents commute, $aa^{-1} = aa^{-1}aa^{-1} = ab^{-1}aa^{-1} = aa^{-1}ab^{-1} = ab^{-1}$ and we have (α).]

Consider now the relation $\leqslant$ defined on S by

$$a \leqslant b \iff aa^{-1} = ab^{-1}.$$

It is clear that $\leqslant$ is reflexive on S. It is also anti-symmetric; for from $a \leqslant b$ and $b \leqslant a$ we have $a = ab^{-1}a$, $ab^{-1} = bb^{-1}$ and $b^{-1}a = b^{-1}b$, whence we obtain $a = ab^{-1}a = bb^{-1}a = bb^{-1}b = b$. Finally, $\leqslant$ is transitive; for from $a \leqslant b$ and $b \leqslant c$ we have $ac^{-1}a = aa^{-1}ac^{-1}a = aa^{-1}bc^{-1}a = aa^{-1}bb^{-1}a = aa^{-1}aa^{-1}a = a$ whence $a \leqslant c$. Thus $\leqslant$ is an order on S. To see that $\leqslant$ is compatible, we observe that if $a \leqslant b$ then, for every $x \in S$,

$$\begin{cases} xa(xb)^{-1} = xab^{-1}x^{-1} = xaa^{-1}x^{-1} = xa(xa)^{-1}; \\ (ax)^{-1}bx = x^{-1}a^{-1}bx = x^{-1}a^{-1}ax = (ax)^{-1}ax, \end{cases}$$

so that $xa \leqslant xb$ and $ax \leqslant bx$.

That an inverse semigroup, equipped with this order, is quasi-residuated can be seen as follows. Observing that $ab^{-1}b(ab^{-1}b)^{-1} = ab^{-1}bb^{-1}ba^{-1} = ab^{-1}ba^{-1} = a(ab^{-1}b)^{-1}$ we have $ab^{-1}b \leqslant a$ and consequently $ab^{-1} \in \langle a \cdot b \rangle$. Similarly, we have $b^{-1}a \in \langle a \cdot b \rangle$.

Definition If S is an inverse semigroup then the order defined in the above example is called the **natural order** on S.

In what follows we shall denote the natural order on an inverse semigroup by $\leqslant_n$.

Example 11.8 For each $p \in \mathbb{Z}$ let $H_p = \{p - \frac{1}{n} \mid n \in \mathbb{Z}, n \geqslant 2\}$ and consider the set $S = \bigcup_{p \in \mathbb{Z}} H_p$. Define the relation $\leqslant$ on S by

$$p - \tfrac{1}{n} \leqslant q - \tfrac{1}{m} \iff p = q, \; n \leqslant m.$$

Then $\leqslant$ is an order. Define a law of composition $\oplus$ on S by

$$(p - \tfrac{1}{n}) \oplus (q - \tfrac{1}{m}) = p + q - \tfrac{1}{\min\{n,m\}}.$$

Then $(S; \oplus, \leqslant)$ is an ordered (abelian) semigroup. It is convenient to consider the Hasse diagram for S, namely

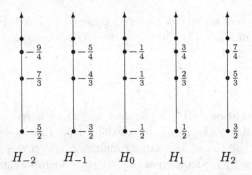

Now S is regular; for example, we have $(p - \tfrac{1}{n}) \oplus (-p - \tfrac{1}{n}) \oplus (p - \tfrac{1}{n}) = p - \tfrac{1}{n}$. The idempotents are the elements of H_0, and since these commute S is an inverse semigroup. Moreover, S is quasi-residuated; we have the formula

$$\langle p - \tfrac{1}{n} : q - \tfrac{1}{m} \rangle = \begin{cases} (p - q - \tfrac{1}{n})^{\downarrow} & \text{if } m > n; \\[2mm] H_{p-q} & \text{otherwise.} \end{cases}$$

We leave to the reader the simple task of showing that on the inverse semigroup $(S; \oplus)$ the order $\leqslant$ so defined coincides with the natural order $\leqslant_n$.

Theorem 11.1 *If S is an inverse semigroup with set of idempotents E then the natural order $\leqslant_n$ on S can be characterised as follows:*

$$x \leqslant_n y \iff (\exists e \in E) \; x = ey \iff (\exists f \in E) \; x = yf.$$

Moreover, $\leqslant_n$ has the property that

$$x \leqslant_n y \implies x^{-1} \leqslant_n y^{-1}.$$

Proof If $x \leqslant_n y$ then $x^{-1}x = x^{-1}y$ and so $x = xx^{-1}y$ where $xx^{-1} \in E$. Conversely, if $x = ey$ where $e \in E$ then $ex = ey = x$ and so $x^{-1} = y^{-1}e^{-1} = y^{-1}e$. Then $x^{-1}x = y^{-1}ex = y^{-1}x$ and hence $x \leqslant_n y$. The second characterisation is established similarly.

Finally, if $x \leqslant_n y$ then from the above $x = ey$ for some $e \in E$. Consequently $x^{-1} = y^{-1}e^{-1} = y^{-1}e$ and therefore $x^{-1} \leqslant_n y^{-1}$. $\square$

In Example 11.7 we have seen that if S is an inverse semigroup then the set E of idempotents of S forms a commutative idempotent subsemigroup. It follows from Theorem 11.1 that for $e, f \in E$ we have

$$e \leqslant_n f \iff e = ef.$$

Referring to Theorem 2.2, we thus see that $(E; \leqslant_n)$ is a meet semilattice.

EXERCISES

11.3. Let $\mathbb{R}_- = \{x \in \mathbb{R} \mid x < 0\}$. Consider the cartesian ordered set $S = \mathbb{R}_- \times \mathbb{Z}$. Define a multiplication on S by

$$(p, i)(q, j) = (-1, i + j).$$

Prove that $(S, \cdot, \leqslant)$ is a quasi-residuated semigroup and determine the quasi-residuals.

11.4. Let $T = (\mathbb{R}_- \times \mathbb{Z}) \cup \{(0, 0)\}$ and define a multiplication on T by

$$(p, i)(q, j) = (\min\{p, q\}, i + j).$$

Prove that T is a quasi-residuated regular semigroup and determine the quasi-residuals.

11.5. If S is an inverse semigroup and $x, y \in S$ are such that there exists $a \in S$ with $x, y \leqslant_n a$, prove that there exists $b \in S$ such that $b \leqslant_n x, y$.

11.6. If S is an inverse semigroup and $H \subseteq S$ define

$$H\omega = \{x \in S \mid (\exists h \in H)\ h \leqslant_n x\}.$$

Prove that the assignment $H \mapsto H\omega$ is a closure and that $H\omega$ is an inverse subsemigroup of S.

11.7. Show that the seven real matrices

$$I = \begin{bmatrix} 1 & 0 \\ 0 & 1 \end{bmatrix}, \quad A = \begin{bmatrix} 0 & 1 \\ 1 & 0 \end{bmatrix}, \quad O = \begin{bmatrix} 0 & 0 \\ 0 & 0 \end{bmatrix},$$

$$E_{11} = \begin{bmatrix} 1 & 0 \\ 0 & 0 \end{bmatrix}, \quad E_{12} = \begin{bmatrix} 0 & 1 \\ 0 & 0 \end{bmatrix}, \quad E_{21} = \begin{bmatrix} 0 & 0 \\ 1 & 0 \end{bmatrix}, \quad E_{22} = \begin{bmatrix} 0 & 0 \\ 0 & 1 \end{bmatrix}$$

form an inverse semigroup and sketch the Hasse diagram relative to the natural order.

11.2 Residuated semigroups

Definition By a **residuated semigroup** we shall mean an ordered semigroup S in which all translations $\lambda_x : y \mapsto xy$ and $\rho_x : y \mapsto yx$ are residuated.

If S is a residuated semigroup then the quasi-residuals $\langle x \cdot {}^{\cdot} y \rangle$ and $\langle x^{\cdot} \cdot y \rangle$ are principal down-sets. In what follows we shall denote their respective top elements by $x \cdot {}^{\cdot} y$ and $x^{\cdot} \cdot y$. Consequently,

$$x \cdot {}^{\cdot} y = \lambda_y^+(x) = \max\{z \in S \mid yz \leqslant x\}, \quad x^{\cdot} \cdot y = \rho_y^+(x) = \max\{z \in S \mid zy \leqslant x\}.$$

If $x \in S$ is such that $x \cdot {}^{\cdot} y = x^{\cdot} \cdot y$ for every $y \in S$ then we shall denote this by writing $x : y$ and say that x is **equiresidual**. This is precisely the notation we used in our discussion of Heyting algebras.

Example 11.9 Every ordered group G is residuated. For all $x, y \in G$ we have $x \,.^{\cdot}\, y = y^{-1}x$ and $x\,^{\cdot}.\,y = xy^{-1}$.

Example 11.10 For every semigroup S the ordered semigroup $(\mathbb{P}(S); \cdot, \subseteq)$ of Example 11.4 is residuated. We have $X \,.^{\cdot}\, Y = \{z \in S \mid (\forall y \in Y)\ yz \in X\}$ and $X\,^{\cdot}.\,Y = \{z \in S \mid (\forall y \in Y)\ zy \in X\}$.

Example 11.11 If R is a commutative ring then the ordered semigroup $I(R)$ of Example 11.5 is residuated. For all $I, J \in I(R)$ we have that $I : J$ is the ideal $\{x \in R \mid xJ \subseteq I\}$.

Example 11.12 If H is a Heyting lattice then $(H; \wedge)$ is a residuated semigroup.

Example 11.13 If B is a boolean lattice then $(B; \wedge)$ is a residuated semigroup. Here, for all $x, y \in B$, we have $x : y = x \vee y'$.

Example 11.14 Let $\mathcal{B}$ be a boolean algebra and let $\mathrm{Mat}_n\,\mathcal{B}$ be the set of $n \times n$ matrices with entries in $\mathcal{B}$. For $X = [x_{ij}]_{n \times n}$ and $Y = [y_{ij}]_{n \times n}$ in $\mathrm{Mat}_n\,\mathcal{B}$ define

$$X \leqslant Y \iff (\forall i, j)\ x_{ij} \leqslant y_{ij}.$$

Then it is clear that $(\mathrm{Mat}_n\,\mathcal{B}; \leqslant)$ becomes a boolean algebra in which $X \wedge Y = [x_{ij} \wedge y_{ij}]$, $X \vee Y = [x_{ij} \vee y_{ij}]$, and the complement of $X = [x_{ij}]$ is $X' = [x'_{ij}]$. Just as with matrices over a ring or field, we can define a multiplication on $\mathrm{Mat}_n\,\mathcal{B}$. Given $X = [x_{ij}]$ and $Y = [y_{ij}]$ we define the product XY to be the $n \times n$ matrix whose (i, j)-th element is given by

$$[XY]_{ij} = \bigvee_{k=1}^{n} (x_{ik} \wedge y_{kj}),$$

which, in engineer's notation (see Exercise 6.17), takes the form $\sum_{k=1}^{n} x_{ik}y_{kj}$. It is readily seen that in this way $\mathrm{Mat}_n\,\mathcal{B}$ becomes an ordered semigroup. Observe now that in $\mathrm{Mat}_n\,\mathcal{B}$ we have

$$
\begin{aligned}
XY \leqslant Z &\iff (\forall i, j)\ \bigvee_{k=1}^{n} (x_{ik} \wedge y_{kj}) \leqslant z_{ij} \\
&\iff (\forall i, j, k)\ x_{ik} \wedge y_{kj} \leqslant z_{ij} \\
&\iff (\forall i, j, k)\ y_{kj} \leqslant z_{ij} \vee x'_{ik} \\
&\iff (\forall j, k)\ y_{kj} \leqslant \bigwedge_{i=1}^{n} (z_{ij} \vee x'_{ik}) = \Big[\bigvee_{i=1}^{n} (z'_{ij} \wedge x_{ik}) \Big]' \\
&\iff Y \leqslant (X^t Z')'.
\end{aligned}
$$

Thus we see that $Z \,.^{\cdot}\, X$ exists and is $(X^t Z')'$. Similarly we can see that $Z\,^{\cdot}.\,X$ exists and is $(Z' X^t)'$. Hence $(\mathrm{Mat}_n\,\mathcal{B}; \cdot, \leqslant)$ is a residuated semigroup.

The principal properties of residuals are those listed in the following three theorems.

Theorem 11.2 *If S is a residuated semigroup then the following statements are equivalent:*

(1) *S is commutative;*
(2) *$(\forall x, y \in S)\ x\,^{\cdot}.\,y = x \,.^{\cdot}\, y$.*

Proof Observe that (1) is equivalent to saying that $\lambda_y = \rho_y$ for every $y \in S$. By the uniqueness of residuals, this is equivalent to $\lambda_y^+ = \rho_y^+$ which in turn is equivalent to (2). □

Theorem 11.3 *If S is a residuated semigroup then, for all $x, y, z \in S$,*

(1) $x(y\,\dot{\cdot}\,x) \leqslant y$;
(2) $(y\,\dot{\cdot}\,x)x \leqslant y$;
(3) $y \leqslant xy\,\dot{\cdot}\,x$;
(4) $y \leqslant yx\,\dot{\cdot}\,x$;
(5) $x \leqslant y\,\dot{\cdot}\,(y\,\dot{\cdot}\,x)$;
(6) $x \leqslant y\,\dot{\cdot}\,(y\,\dot{\cdot}\,x)$;
(7) $(x\,\dot{\cdot}\,y)\,\dot{\cdot}\,z = x\,\dot{\cdot}\,yz$;
(8) $(x\,\dot{\cdot}\,y)\,\dot{\cdot}\,z = x\,\dot{\cdot}\,zy$;
(9) $(x\,\dot{\cdot}\,y)\,\dot{\cdot}\,z = (x\,\dot{\cdot}\,z)\,\dot{\cdot}\,y$;
(10) $(x\,\dot{\cdot}\,y)z \leqslant xz\,\dot{\cdot}\,y$;
(11) $y(x\,\dot{\cdot}\,z) \leqslant yx\,\dot{\cdot}\,z$.

Proof Properties $(1), (2), (3), (4)$ are immediate from the definitions of $x\,\dot{\cdot}\,y$ and $x\,\dot{\cdot}\,y$. Then (5) follows from (1); and (6) from (2). Now since S is a semigroup we have $\lambda_y \circ \lambda_z = \lambda_{yz}$ which, by Theorem 1.6, gives $\lambda_z^+ \circ \lambda_y^+ = \lambda_{yz}^+$ which is none other than (7). Similarly $\rho_y \circ \rho_z = \rho_{zy}$ gives (8). Likewise, $\lambda_y \circ \rho_z = \rho_z \circ \lambda_y$ gives (9). To establish (10) and (11) we observe that if E is an ordered set and $f, g \in \operatorname{Res} E$ then

$$f \circ g \leqslant g \circ f \iff g \circ f^+ \leqslant f^+ \circ g.$$

[In fact, if $f \circ g \leqslant g \circ f$ then $g \circ f^+ \leqslant f^+ \circ f \circ g \circ f^+ \leqslant f^+ \circ g \circ f \circ f^+ \leqslant f^+ \circ g$; and conversely, if $g \circ f^+ \leqslant f^+ \circ g$ then $f \circ g \leqslant f \circ g \circ f^+ \circ f \leqslant f \circ f^+ \circ g \circ f \leqslant g \circ f$.] Writing $\lambda_y \circ \rho_z = \rho_z \circ \lambda_y$ as $\lambda_y \circ \rho_z \leqslant \rho_z \circ \lambda_y$ and $\rho_z \circ \lambda_y \leqslant \lambda_y \circ \rho_z$, we obtain from this observation the inequalities $\rho_z \circ \lambda_y^+ \leqslant \lambda_y^+ \circ \rho_z$ and $\lambda_y \circ \rho_z^+ \leqslant \rho_z^+ \circ \lambda_y$ which are (10) and (11) respectively. □

Theorem 11.4 *If S is a residuated semigroup and $x, y \in S$ then*

(1) $x(y\,\dot{\cdot}\,x) = y \iff (\exists z \in S)\ y = xz$;
(2) $(y\,\dot{\cdot}\,x)x = y \iff (\exists z \in S)\ y = zx$;
(3) $y = xy\,\dot{\cdot}\,x \iff (\exists z \in S)\ y = z\,\dot{\cdot}\,x$;
(4) $y = yx\,\dot{\cdot}\,x \iff (\exists z \in S)\ y = z\,\dot{\cdot}\,x$;
(5) $y = x\,\dot{\cdot}\,(x\,\dot{\cdot}\,y) \iff (\exists z \in S)\ y = x\,\dot{\cdot}\,z$;
(6) $y = x\,\dot{\cdot}\,(x\,\dot{\cdot}\,y) \iff (\exists z \in S)\ y = x\,\dot{\cdot}\,z$.

Proof If $f : A \to B$ is a residuated mapping then we have

$$\begin{cases} x = ff^+(x) \iff (\exists y \in A)\ x = f(y); \\ x = f^+f(x) \iff (\exists y \in B)\ x = f^+(y). \end{cases}$$

Applying these observations to λ_x and ρ_x we obtain $(1) - (4)$. As for (5) and (6), we observe that for all $x, y, z \in S$ we have

$$x\,\dot{\cdot}\,y \geqslant z \iff yz \leqslant x \iff y \leqslant x\,\dot{\cdot}\,z$$

and so the mapping $\zeta_x : S \to S^d$ given by the prescription $\zeta_x(y) = x\,\dot{\cdot}\,y$ is residuated with ζ_x^+ given by $\zeta_x^+(z) = x\,\dot{\cdot}\,z$. Applying the above observations to ζ_x we obtain (5) and (6). □

EXERCISES

11.8. Consider the following ordered set and Cayley multiplication table:

$\cdot$	a	b	c	d	e
a	b	b	c	e	e
b	c	c	c	e	e
c	c	c	c	e	e
d	d	d	e	c	c
e	e	e	e	c	c

Show that these define a residuated semigroup and compile tables of right and left residuals.

11.9. The **bicyclic semigroup** $\mathcal{C}(p,q)$ is the monoid generated by $\{p,q\}$ subject to the relation $pq = 1$. Each element of $\mathcal{C}(p,q)$ can be expressed uniquely in the form $q^m p^n$ with the convention that $q^0 = 1 = p^0$. Multiplication is then given by the formula $q^m p^n \cdot q^{m'} p^{n'} = q^r p^s$ where

$$r = m + m' - \min\{n, m'\}, \quad s = n + n' - \min\{n, m'\}.$$

(1) Prove that $\mathcal{C}(p,q)$ is an inverse semigroup in which $(q^m p^n)^{-1} = q^n p^m$.

(2) Prove that the natural order on $\mathcal{C}(p,q)$ is given by

$$q^m p^n \leqslant_n q^r p^s \iff s \leqslant n, \ m + s = n + r.$$

(3) Sketch the Hasse diagram for $(\mathcal{C}(p,q); \leqslant_n)$.

(4) Prove that left residuals exist in $(\mathcal{C}(p,q); \leqslant_n)$ and are given by

$$q^r p^s \cdot q^m p^n = \begin{cases} q^r p^{s+m-n} & \text{if } s > n; \\ q^{n+r-m-s} & \text{if } s \leqslant n \text{ and } n + r \geqslant m + s; \\ p^{m+s-n-r} & \text{if } s \leqslant n \text{ and } n + r \leqslant m + s. \end{cases}$$

(5) Obtain likewise a formula for right residuals and hence deduce that $(\mathcal{C}(p,q); \leqslant_n)$ is residuated.

11.10. Let S be the inverse semigroup of Example 11.8. Extend S to the ordered set $T = S \cup \mathbb{Z}$ where each $i \in \mathbb{Z}$ is the top element of H_i. Extend the operation $\oplus$ by defining $m \oplus n = m + n$ for all $m, n \in \mathbb{Z}$ and

$$m \oplus \left(p - \tfrac{1}{n}\right) = \left(p - \tfrac{1}{n}\right) \oplus m = m + p - \tfrac{1}{n}.$$

Show that T becomes a residuated inverse semigroup.

We now consider some equivalence relations that arise naturally on a residuated semigroup.

Definition If E is an ordered set then by the **zig-zag equivalence** we mean the relation Z defined on E by

$$(x, y) \in Z \iff (\exists x_1, \ldots, x_n \in E) \ x = x_1 \between x_2 \between \cdots \between x_{n-1} \between x_n = y.$$

It is clear that Z so defined is an equivalence relation on E and that the Z-classes are none other than the Hasse diagram components of E.

Theorem 11.5 *If S is an ordered semigroup then Z is strongly regular and is compatible with multiplication. If, moreover, S is quasi-residuated then the quotient semigroup S/Z is a group.*

Proof It is clear from the definition that every Z-crown is Z-closed and so, by Theorem 3.1, Z is regular. Now if $(a, b) \in Z$ then for any $c \nparallel b$ we have $(a, c) \in Z$. It follows that Z satisfies the link property and its dual. By Theorem 3.4 and its dual, Z is then strongly regular. To show that Z is compatible with multiplication, let $(a, b) \in Z$. Then there is a finite zig-zag chain

$$a = x_1 \nparallel x_2 \nparallel \cdots \nparallel x_{n-1} \nparallel x_n = b.$$

Since each translation in S is isotone it follows that, for every $c \in S$,

$$ca = cx_1 \nparallel cx_2 \nparallel \cdots \nparallel cx_{n-1} \nparallel cx_n = cb,$$

so that $(ca, cb) \in Z$. Similarly, $(ac, bc) \in Z$ and so Z is compatible with multiplication.

Suppose now that S is quasi-residuated. If $a, b \in S$ and $t \in \langle b . \cdot a \rangle$ then $at \leqslant b$ gives $[a]_Z [t]_Z = [at]_Z = [b]_Z$. Likewise, if $s \in \langle b . \cdot a \rangle$ then we have $[s]_Z [a]_Z = [b]_Z$. Thus quotients exist in the semigroup S/Z and consequently S/Z is a group. $\qquad\square$

We now consider consequences of the existence of extremal elements in S.

Theorem 11.6 *If S is a residuated semigroup and if $a \in S$ is maximal in S then $a = \max [a]_Z$.*

Proof Suppose that a is a maximal element of S and let $x \in S$ be such that $x^{\downarrow} \cap a^{\downarrow} \neq \emptyset$. Then necessarily $x \leqslant a$. To see this, observe that if $y \in x^{\downarrow} \cap a^{\downarrow}$ then $a \leqslant a^2 . \cdot a \leqslant a^2 . \cdot y$ and the fact that a is maximal gives $a = a^2 . \cdot a = a^2 . \cdot y$. Consequently, $a^2 . \cdot x \leqslant a^2 . \cdot y = a^2 . \cdot a$ whence $(a^2 . \cdot x)a \leqslant a^2$ and therefore $a \leqslant a^2 . \cdot (a^2 . \cdot x)$. The maximality of a now gives $a = a^2 . \cdot (a^2 . \cdot x) \geqslant x$ by Theorem 11.3(6).

Suppose now that $a^* \in [a]_Z$. By way of obtaining a contradiction, suppose that in any finite zig-zag chain joining a to a^* there exists a first element, a_k say, such that $a_k \notin a^{\downarrow}$. Then necessarily $a_k^{\downarrow} \cap a^{\downarrow} \neq \emptyset$ since the element preceding a_k in the zig-zag chain belongs to this set. The above observation then gives the contradiction $a_k \leqslant a$. All the elements in the zig-zag chain are therefore less than or equal to a. It follows from this that $a = \max [a]_Z$. $\qquad\square$

Theorem 11.7 *Let S be a residuated semigroup.*

(1) *If S contains a descending chain that is unbounded below then every Z-class contains at least one such chain.*

(2) *If S contains an ascending chain that is unbounded above then every Z-class contains at least one ascending chain that is unbounded above and at least one descending chain that is unbounded below.*

Proof Let $a_1 \geqslant a_2 \geqslant a_3 \geqslant \cdots$ be a descending chain, unbounded below, in the class $\mathcal{A}$ modulo Z and let $\mathcal{B}$ be any class modulo Z. Then there is a unique class $\mathcal{C}$ modulo Z such that $\mathcal{CB} = \mathcal{A}$ and for every $c \in \mathcal{C}$ we have in $\mathcal{B}$ the descending chain

$$a_1 . \cdot c \geqslant a_2 . \cdot c \geqslant a_3 . \cdot c \cdots .$$

Suppose, by way of obtaining a contradiction, that there exists $b \in \mathcal{B}$ such that $b \leqslant a_n . \cdot c$ for every n. Then we would have $cb \leqslant a_n$ for every n whence the chain $a_i \geqslant a_{i+1}$ would be bounded below by cb, contrary to the hypothesis. Thus $\mathcal{B}$ contains a descending chain that is unbounded below, and since $\mathcal{B}$ is arbitrary the same is true for all classes modulo Z.

If now $a_1 \leqslant a_2 \leqslant a_3 \leqslant \cdots$ is an ascending chain, unbounded above, in $\mathcal{A}$ let $\mathcal{C}$ be any class modulo Z. Then there is a unique class $\mathcal{B}$ modulo Z such that $\mathcal{BA} = \mathcal{C}$ and for every $b \in \mathcal{B}$ we have in $\mathcal{C}$ the ascending chain

$$ba_1 \leqslant ba_2 \leqslant ba_3 \leqslant \cdots .$$

Suppose, by way of obtaining a contradiction, that there exists $c \in \mathcal{C}$ such that $ba_n \leqslant c$ for every n. Then we would have $a_n \leqslant c{.}^{\cdot}b$ for every n whence the chain $a_i \leqslant a_{i+1}$ would be bounded above by $c{.}^{\cdot}b$, contrary to the hypothesis. Thus $\mathcal{C}$ contains an ascending chain that is unbounded above, and since $\mathcal{C}$ is arbitrary the same is true for all classes modulo Z.

Moreover, for every class $\mathcal{C}$ modulo Z there is a unique class $\mathcal{D}$ modulo Z such that $\mathcal{AD} = \mathcal{C}$ and for every $c \in \mathcal{C}$ we have in $\mathcal{D}$ the descending chain

$$c{.}^{\cdot}a_1 \geqslant c{.}^{\cdot}a_2 \geqslant c{.}^{\cdot}a_3 \geqslant \cdots .$$

Suppose, by way of obtaining a contradiction, that there exists $d \in \mathcal{D}$ such that $d \leqslant c{.}^{\cdot}a_n$ for every n. Then we would have $a_n \leqslant c^{\cdot}{.}d$ for every n whence the chain $a_i \leqslant a_{i+1}$ would be bounded above by $c^{\cdot}{.}d$, contrary to the hypothesis. Thus $\mathcal{D}$ contains an ascending chain that is unbounded above, and by the first part all classes modulo Z contain such a chain. □

Using the above results, we can now establish the following.

Theorem 11.8 *If a residuated semigroup S contains a maximal element then every Z-class contains a maximum element.*

Proof Suppose that a is maximal in S. Then, by Theorem 11.6, a is maximum in its Z-class $[a]_Z$. It follows that $[a]_Z$ contains no ascending chain that is unbounded above and so, by Theorem 11.7, no Z-class contains such a chain. Then, by Zorn's axiom, every Z-class contains a maximal element which, by virtue of Theorem 11.6, is maximum in its Z-class. □

We now consider the consequences of the existence of a minimal element in a residuated semigroup.

Theorem 11.9 *If a residuated semigroup S contains a minimal element then every Z-class is an upper directed set.*

Proof Let $x_\star$ be a minimal element in S and consider elements $a_1, a_2 \in S$ such that $a_1^{\downarrow} \cap a_2^{\downarrow} \neq \emptyset$. By Theorem 11.3(1) and the fact that $x_\star$ is minimal there exists $y \in S$ such that $a_1 y = x_\star$ (take for example $y = x_\star{.}^{\cdot}a_1$). Let $a \in a_1^{\downarrow} \cap a_2^{\downarrow}$ and observe that, since $x_\star$ is minimal, we have $x_\star = a_1 y = ay \leqslant a_2 y$ whence $a_1 \leqslant a_2 y^{\cdot}{.}y$. Since we know that $a_2 \leqslant a_2 y^{\cdot}{.}y$ it therefore follows that $a_1^{\uparrow} \cap a_2^{\uparrow} \neq \emptyset$.

This being the case, let $\mathcal{B}$ be any Z-class and let $\beta_1, \beta_2 \in \mathcal{B}$ with $\beta_1 \parallel \beta_2$. Choose a zig-zag chain of minimal length joining β_1 to β_2, say

$$\beta_1 = b_1 \nparallel b_2 \nparallel \cdots \nparallel b_n = \beta_2.$$

Now amongst these elements there is a finite number, N say, of elements b_k such that $b_{k-1} \leqslant b_k$ and $b_{k+1} \leqslant b_k$. Denoting such elements by $b_{\overline{k}}$, we have the finite sequence

$$b_{\overline{1}} \parallel b_{\overline{2}} \parallel \cdots \parallel b_{\overline{N}},$$

where the non-comparability results from the minimality of the length of the chain. Now from the definition of $b_{\overline{k}}$ we have $\beta_1^{\downarrow} \cap b_{\overline{1}}^{\downarrow} \neq \emptyset$, and the observation in the first paragraph gives $\beta_1^{\uparrow} \cap b_{\overline{1}}^{\uparrow} \neq \emptyset$. Consider any element $b_{1'} \in \beta_1^{\uparrow} \cap b_{\overline{1}}^{\uparrow}$. Since $b_{\overline{1}}^{\downarrow} \cap b_{\overline{2}}^{\downarrow} \neq \emptyset$ we deduce that $b_{1'}^{\downarrow} \cap b_{\overline{2}}^{\downarrow} \neq \emptyset$ so that, by the above, $b_{1'}^{\uparrow} \cap b_{\overline{2}}^{\uparrow} \neq \emptyset$ and consequently $\beta_1^{\uparrow} \cap b_{\overline{2}}^{\uparrow} \neq \emptyset$. Consider now $b_{2'} \in \beta_1^{\uparrow} \cap b_{\overline{2}}^{\uparrow}$. Since $b_{\overline{2}}^{\downarrow} \cap b_{\overline{3}}^{\downarrow} \neq \emptyset$ we have $b_{2'}^{\downarrow} \cap b_{\overline{3}}^{\downarrow} \neq \emptyset$ so that $b_{2'}^{\uparrow} \cap b_{\overline{3}}^{\uparrow} \neq \emptyset$ and consequently $\beta_1^{\uparrow} \cap b_{\overline{3}}^{\uparrow} \neq \emptyset$. Consider now any $b_{3'} \in \beta_1^{\uparrow} \cap b_{\overline{3}}^{\uparrow}$, ... and so on.

After N applications of this process we arrive at $\beta_1^{\uparrow} \cap \beta_{\overline{N}}^{\uparrow} \neq \emptyset$. Since we also have $b_{\overline{N}}^{\downarrow} \cap \beta_2^{\downarrow} \neq \emptyset$ a final application of the process yields $\beta_1^{\uparrow} \cap \beta_2^{\uparrow} \neq \emptyset$ as required. $\square$

Theorem 11.10 *If S is a residuated semigroup and if $a \in S$ is minimal in S then $a = \min [a]_Z$.*

Proof Let a be minimal in S and let $x \in [a]_Z$. Since $[a]_Z$ is upper directed by Theorem 11.9, there exists $z \in x^{\uparrow} \cap a^{\uparrow}$, and since there exists $t \in S$ such that $tz = a$ (namely $t = a^{\cdot}.z$) we have $tx = ta = a$. In other words, for every $x \in [a]_Z$ there exists an element, which we shall denote by t_x, such that $t_x x = a$ and $t_x \leqslant a^{\cdot}.a$.

Consider now any $y \in [a]_Z$. Since S is residuated there exists $x \in [a]_Z$ such that $(a^{\cdot}.a)x \leqslant y$. Consequently, $a = t_x x \leqslant (a^{\cdot}.a)x \leqslant y$ and so $a = \min[a]_Z$.
                                                                                $\square$

Theorem 11.11 *If a residuated semigroup S contains a minimal element then every Z-class contains both a top element and a bottom element.*

Proof Let a be minimal in S. By Theorem 11.10, $[a]_Z$ has a bottom element, namely a. Then $[a]_Z$ contains no descending chains that are unbounded below and so, by Theorem 11.7(2), no Z-class contains an ascending chain that is unbounded above. It follows by Zorn's axiom that S contains maximal elements whence, by Theorem 11.8, every Z-class contains a top element. Moreover, since a is the bottom element of $[a]_Z$ it follows by Theorem 11.7(1) that no Z-class can contain a descending chain that is unbounded below. Consequently every Z-class contains minimal elements and the result follows from Theorem 11.10.
                                                                                $\square$

Drawing together the above results, we obtain the following.

Theorem 11.12 (Blyth [15]) *In a residuated semigroup every Z-class contains either a top element and a bottom element, or a top element and no minimal elements, or no maximal elements and no minimal elements.* $\square$

EXERCISES

11.11. If S is a residuated semigroup prove that the zig-zag equivalence Z is either a closure, or both a closure and a dual closure, or neither.

11.12. Construct examples of each of the three types in Theorem 11.12.

11.3 Molinaro equivalences

We now introduce some further fundamental equivalence relations on a residuated semigroup. For this purpose, we recall that if S is an ordered semigroup and S^d is its order dual then for every $x \in S$ the translations $\lambda_x, \rho_x : S \to S$ given by $\lambda_x(y) = xy$, $\rho_x(y) = yx$ are residuated as are the mappings $\zeta_x, \eta_x : S \to S^d$ given by $\zeta_x(y) = x \cdot {}^{\cdot}y$, $\eta_x(y) = x^{\cdot} \cdot y$. Likewise, the mappings $\lambda_x^{\oplus}, \rho_x^{\oplus} : S^d \to S^d$ given by $\lambda_x^{\oplus}(y) = y \cdot {}^{\cdot}x$, $\rho_x^{\oplus}(y) = y^{\cdot} \cdot x$ are residuated.

Definition By the **Molinaro equivalences** we shall mean the kernels of the above six residuated mappings.

In what follows we shall use the notation

$$F_x = \ker \lambda_x, \quad {}_xF = \ker \rho_x;$$
$$A_x = \ker \zeta_x, \quad {}_xA = \ker \eta_x;$$
$$B_x = \ker \lambda_x^{\oplus}, \quad {}_xB = \ker \rho_x^{\oplus},$$

and refer to these as the Molinaro equivalences of types F, A, B respectively.

Applying Theorem 3.6 in the case where $E = F = S$ and $f = \lambda_x$ [resp. ρ_x], and using Theorem 11.4, we have the following characterisation of the Molinaro equivalences of type F.

Theorem 11.13 *For every element x of a residuated semigroup S the equivalence F_x [resp. ${}_xF$] is a closure equivalence on S and can be characterised as an equivalence relation with convex classes such that each class modulo F_x [resp. ${}_xF$] contains one and only one right [resp. left] residual by x which is the top element in its class. The top element in the class of y modulo F_x [resp. ${}_xF$] is $xy \cdot {}^{\cdot}x$ [resp. $yx^{\cdot} \cdot x$].* □

Likewise, applying the dual of Theorem 3.6 in the case where $E = S$, $F = S^d$ and $f = \zeta_x$ [resp. η_x], we have the following characterisation of the Molinaro equivalences of type A.

Theorem 11.14 *For every element x of a residuated semigroup S the equivalence A_x [resp. ${}_xA$] is a closure equivalence on S and can be characterised as an equivalence relation with convex classes such that each class modulo A_x [resp. ${}_xA$] contains one and only one left [resp. right] residual of x which is the top element in its class. The top element in the class of y modulo A_x [resp. ${}_xA$] is $x^{\cdot} \cdot (x \cdot {}^{\cdot}y)$ [resp. $x \cdot {}^{\cdot}(x^{\cdot} \cdot y)$].* □

Finally, applying Theorem 3.6 in the case where $E = F = S^d$ and $f = \lambda_x^{\oplus}$ [resp. $\rho_x^{\oplus}$], we have the following characterisation of the Molinaro equivalences of type B.

Theorem 11.15 *For every element x of a residuated semigroup S the equivalence B_x [resp. $_xB$] is a dual closure equivalence on S and can be characterised as an equivalence relation with convex classes such that each class modulo B_x [resp. $_xB$] contains one and only one right [resp. left] multiple of x which is the bottom element in its class. The bottom element in the class of y modulo B_x [resp. $_xB$] is $x(y \,\dot{} \, x)$ [resp. $(y \,\dot{} \, x)x)$].*   □

It is an immediate consequence of the above theorems that the Molinaro equivalences are all finer then the zig-zag equivalence. For example, if $(a, b) \in A_x$ then we have $a \leqslant x \,\dot{} \,.(x \,\dot{} \, a) = x \,\dot{} \,.(x \,\dot{} \, b) \geqslant b$ so that $(a, b) \in Z$. In special cases equality can occur:

Theorem 11.16 *In a residuated semigroup S the following statements are equivalent:*

(1) *S contains a maximal element;*
(2) *$(\exists x \in S)$ $A_x = Z$ [resp. $_xA = Z$].*

Proof (1) $\Rightarrow$ (2): Suppose that x^* is maximal in S. Then, by Theorem 11.6, x^* is the top element in its Z-class $\mathcal{X}$ and, by Theorem 11.8, all Z-classes have top elements. Let $\mathcal{A}$ be any class modulo Z. There is a unique Z-class $\mathcal{B}$ such that $\mathcal{AB} = \mathcal{X}$. Let b^* be the top element of $\mathcal{B}$. Since $ab^* \leqslant x^*$ for every $a \in \mathcal{A}$, the maximality of b^* gives $b^* = x^* \,\dot{} \, a$ for every $a \in \mathcal{A}$. Since $\mathcal{A}$ is arbitrary it follows that $A_{x^*} = Z$.

(2) $\Rightarrow$ (1): If there exists $x \in S$ such that $A_x = Z$ then since every class modulo A_x has a top element so does every Z-class and therefore S has a maximal element.   □

Theorem 11.17 *In a residuated semigroup S the following statements are equivalent:*

(1) *S contains a minimal element;*
(2) *$(\exists x \in S)$ $F_x = Z$ [resp. $_xF = Z$];*
(3) *$(\exists x \in S)$ $B_x = Z$ [resp. $_xB = Z$].*

Proof (1) $\Rightarrow$ (2): If S contains a minimal element then, by Theorem 11.11, every Z-class has a top element and a bottom element. Given Z-classes $\mathcal{B}$ and $\mathcal{C}$, let $\mathcal{A}$ be the Z-class such that $\mathcal{BA} = \mathcal{C}$. Let $b_\star$ be the bottom element of $\mathcal{B}$, $a^\star$ the top element of $\mathcal{A}$, and $c_\star$ the bottom element of $\mathcal{C}$. Then necessarily $b_\star a^\star = c_\star$; for otherwise every $b \in \mathcal{B}$ would be such that $ba^\star \geqslant b_\star a^\star > c_\star$ and $c_\star \,\dot{} \, a^\star$ would not exist. It now follows from $b_\star a^\star = c_\star$ and the minimality of $c_\star$ that $b_\star a = c_\star$ for every $a \in \mathcal{A}$. Since $\mathcal{B}$ and $\mathcal{C}$ are arbitrary it follows from this that $Z = F_{b_\star}$.

(2) $\Rightarrow$ (1): Suppose that there exists $x \in S$ such that $F_x = Z$. Let $\mathcal{X}$ be the Z-class of x and let $\mathcal{A}$ be any Z-class. Writing $\mathcal{XA} = \mathcal{Y}$ we have, since $F_x = Z$,

$$(\exists y \in \mathcal{Y})(\forall a \in \mathcal{A}) \quad xa = y.$$

Now if y_1 is any element of $\mathcal{Y}$ then $y_1 \,\dot{} \, x \in \mathcal{A}$ and so $y = x(y_1 \,\dot{} \, x) \leqslant y_1$. Thus y_1 is the bottom element of $\mathcal{Y}$ and so is minimal in S.

(1) $\Leftrightarrow$ (3): The proof is similar to the above.   □

EXERCISES

11.13. If a residuated semigroup contains a maximal element α prove that
$$\max S = \{\alpha \cdot\!\!\cdot x \mid x \in S\} = \{\alpha\cdot\!\cdot x \mid x \in S\}.$$

11.14. If a residuated semigroup contains a minimal element β prove that
$$\max S = \{x \cdot\!\!\cdot \beta \mid x \in S\} = \{x\cdot\!\cdot \beta \mid x \in S\};$$
$$\min S = \{x\beta \mid x \in S\} = \{\beta x \mid x \in S\}.$$

11.15. If S is a residuated semigroup and $x \in S$ prove that
$$A_x \subseteq \bigcap_{t \in S} A_{x\cdot\!\cdot t}; \quad F_x \subseteq \bigcap_{t \in S} F_{tx} \cap \bigcap_{t \in S} A_{t\cdot\!\cdot x}; \quad B_x \subseteq \bigcap_{t \in S} B_{xt}.$$

11.16. If S is a residuated semigroup prove that
$$\bigcap_{x \in S} A_x = \bigcap_{x \in S} {}_xF; \quad \bigcap_{x \in S} {}_xA = \bigcap_{x \in S} F_x; \quad \bigcap_{x \in S} B_x = \bigcap_{x \in S} {}_xB.$$

11.17. Let S be a residuated monoid. Call a Molinaro equivalence **proper** if it differs from equality and Z. Prove that if S has no proper Molinaro equivalences then every Z-class has at most two elements and the class of 1 is isomorphic to the boolean algebra **2**.

11.18. If $\mathcal{B}$ is a boolean algebra (regarded as a residuated semigroup in which multiplication is the meet operation) prove that
  (1) $(\forall x \in \mathcal{B})\ F_x = A_{x'}$;
  (2) $(y, z) \in B_x \iff (y', z') \in A_{x'}$.

11.19. In the residuated semigroup $\mathrm{Mat}_n\,\mathcal{B}$ of boolean matrices (Example 11.14) prove that
  (1) $F_{X^t} = A_{X'}$;
  (2) $(Y, Z) \in A_{X'} \iff (Y', Z') \in B_X$.

11.20. Determine the Molinaro equivalences in the bicyclic semigroup $\mathcal{C}(p, q)$ (see Exercise 11.9).

Epimorphic group images; Dubreil-Jacotin semigroups

12.1 Anticones

As we have seen in Theorem 11.5, for a quasi-residuated semigroup S the quotient semigroup S/Z is a (trivially ordered) epimorphic group image of S. It is natural to ask the general question as to just when an ordered semigroup S is such that there exist an ordered group G and an isotone epimorphism $f : S \to G$. For the purpose of obtaining an answer to this question, we recall from Example 11.10 that if S is a semigroup then $(\mathbb{P}(S); \cdot, \subseteq)$ is a residuated semigroup in which the empty subset acts as a zero element. Here residuals are given by

$$X \,\cdot\! Y = \{z \in S \mid Yz \subseteq X\}, \quad X^{\cdot}.Y = \{z \in S \mid zY \subseteq X\}.$$

Note for example that by Theorem 11.3(7) we have $X \,\cdot\! YZ = (X \,\cdot\! Y) \,\cdot\! Z$ so that in particular, for $a, b \in S$,

$$X \,\cdot\! \{ab\} = X \,\cdot\! \{a\}\{b\} = (X \,\cdot\! \{a\}) \,\cdot\! \{b\}.$$

Definition A non-empty subset H of a semigroup S is said to be **reflexive** if it is such that

$$(\forall x, y \in S) \quad xy \in H \iff yx \in H.$$

This is equivalent to saying that $H \,\cdot\! \{x\} = H^{\cdot}.\{x\}$ for every $x \in S$, in which case we shall use the notation $H : \{x\}$. We shall say that a reflexive subset H of S is **neat** if $H : \{x\} \neq \emptyset$ for every $x \in S$.

Given a neat reflexive subset H of an ordered semigroup S, consider the **Dubreil equivalence** R_H defined on S by

$$(a, b) \in R_H \iff H : \{a\} = H : \{b\}.$$

We observe first that R_H is compatible with multiplication; for if $(a, b) \in R_H$ then for every $x \in S$ we have

$$y \in H : \{xa\} \iff yx \in H : \{a\} = H : \{b\} \iff y \in H : \{xb\}$$

so that $(xa, xb) \in R_H$, and similarly $(ax, bx) \in R_H$. We can therefore form the quotient semigroup S/R_H. In what follows we shall denote the elements of S/H by $[x]_H$.

Consider the relation $\leqslant_H$ defined on S/R_H by

$$[x]_H \leqslant_H [y]_H \iff H:\{y\} \subseteq H:\{x\}.$$

Clearly, $\leqslant_H$ is an order. Moreover, if $[a]_H \leqslant_H [b]_H$ then from $H:\{b\} \subseteq H:\{a\}$ we obtain $H:\{bx\} = (H:\{b\}) \cdot \{x\} \subseteq (H:\{a\}) \cdot \{x\} = H:\{ax\}$ whence $[a]_H[x]_H = [ax]_H \leqslant_H [bx]_H = [b]_H[x]_H$. Likewise, $[x]_H[a]_H \leqslant_H [x]_H[b]_H$ and therefore $(S/R_H; \cdot, \leqslant_H)$ is an ordered semigroup.

If now we impose on H the restriction that it be a down-set of S then the natural epimorphism $\natural_H : S \to S/R_H$ becomes isotone; for if $a \leqslant b$ in S then from $xa \leqslant xb$ we deduce that if $xb \in H$ then necessarily $xa \in H$, so that $H:\{b\} \subseteq H:\{a\}$ and therefore $[a]_H \leqslant_H [b]_H$.

Thus, if H is a neat reflexive down-set of S then we can construct an ordered semigroup S/R_H with an isotone epimorphism $\natural_H : S \to S/R_H$. Let us now examine under what conditions S/R_H becomes a group.

For this purpose, suppose that H is also a subsemigroup of S. Then for every $h \in H$ we have $H \subseteq H:\{h\}$. Suppose further that H satisfies the property

$$(\exists h \in H) \quad H:\{h\} = H,$$

and consider the set

$$I_H = \{x \in S \mid H:\{x\} = H\}.$$

If $x, y \in I_H$ then $H:\{xy\} = (H:\{x\}) \cdot \{y\} = H:\{y\} = H$ so $xy \in I_H$. Also, if $y \in I_H$ then $H = H:\{y\}$ gives $yH \subseteq H$ whence $yh \in H$ for every $h \in H$ and therefore $y \in \bigcap_{h \in H} (H:\{h\}) = H$. Thus I_H is a subsemigroup of H.

Now from the definition of I_H every element of S that is equivalent modulo R_H to an element of I_H is itself an element of I_H. Moreover, any two elements of I_H are R_H-equivalent. Hence I_H forms one of the classes modulo R_H. Furthermore, from the fact that if $y \in I_H$ then

$$H:\{xy\} = H:\{x\} = H:\{yx\},$$

we see that I_H is the identity element of S/R_H.

Finally, suppose that I_H is both reflexive and neat. Then if $y \in I_H:\{x\}$ we have $xy \in I_H$ whence $[xy]_H = I_H$; and similarly $[yx]_H = I_H$. Consequently S/R_H is a group in which $[x]_H^{-1} = I_H:\{x\}$; for we have

$$y \in [x]_H^{-1} \iff [y]_H = [x]_H^{-1} \iff [yx]_H = I_H \iff yx \in I_H \iff y \in I_H:\{x\}.$$

These observations prompt the following definition.

Definition By an **anticone** of an ordered semigroup S we shall mean a subset H of S such that

(1) H is a subsemigroup;

(2) H is a down-set;

(3) H is reflexive and neat;

(4) $(\exists h \in H) \; H:\{h\} = H$;

(5) I_H is reflexive and neat.

The above discussion, coupled with the notion of an anticone, forms the substance of the first half of the following result.

Theorem 12.1 (Bigard [6]) *Let S be an ordered semigroup. If H is an anti-cone of S and R_H is the associated Dubreil equivalence then S/R_H is an isotone epimorphic group image of S. Moreover, every isotone epimorphic group image of S arises in this way for some anticone.*

Proof For the second part, suppose that G is an ordered group and that $h : S \to G$ is an isotone epimorphism. Denoting the negative cone of G as usual by N_G, consider the non-empty subset

$$H = h^{\leftarrow}(N_G) = \{x \in S \mid h(x) \leqslant 1_G\}.$$

Since h is isotone it is clear that H is both a subsemigroup and a down-set of S. That H is reflexive follows from the observation

$$xy \in H \iff h(x)h(y) \leqslant 1_G \iff h(x) \leqslant [h(y)]^{-1}$$
$$\iff h(yx) \leqslant 1_G$$
$$\iff yx \in H.$$

That H is neat follows from the fact that h is surjective; for, given any $x \in S$ there exists $y \in S$ such that $h(y) = [h(x)]^{-1}$ whence $h(xy) = h(x)h(y) = 1_G$ and then $xy \in H$ and so $H:\{x\} \neq \emptyset$.

Now for every $y \in S$ we have

$$x \in H:\{y\} \iff h(xy) \leqslant 1_G \iff h(x) \leqslant [h(y)]^{-1}.$$

Choosing y with $h(y) = 1_G$, we obtain $H:\{y\} = H$ with $y \in H$. Moreover, again since h is surjective,

$$I_H = \{y \in S \mid H:\{y\} = H\}$$
$$= \{y \in S \mid h(x) \leqslant [h(y)]^{-1} \Leftrightarrow h(x) \leqslant 1_G\}$$
$$= \{y \in S \mid h(y) = 1_G\}.$$

It follows that $I_H \subseteq H$ and is reflexive; and the argument used above to show that H is neat shows that I_H is also neat.

Thus we see that $H = h^{\leftarrow}(N_G)$ is an anticone of S.

Now the fact that h is surjective also gives

$$H:\{a\} \subseteq H:\{b\} \iff (h(xa) \leqslant 1_G \Rightarrow h(xb) \leqslant 1_G)$$
$$\iff (h(x) \leqslant [h(a)]^{-1} \Rightarrow h(x) \leqslant [h(b)]^{-1})$$
$$\iff [h(a)]^{-1} \leqslant [h(b)]^{-1}$$
$$\iff h(b) \leqslant h(a).$$

It follows from this that

$$H:\{a\} = H:\{b\} \iff h(a) = h(b)$$

and so we can define a mapping $\zeta_h : S/R_H \to G$ by setting

$$\zeta_h([x]_H) = h(x).$$

Clearly, ζ_h is a bijection such that

$$[x]_H \leqslant_H [y]_H \iff \zeta_h([x]_H) \leqslant \zeta_h([y]_H),$$

and so is an order isomorphism. Hence G is of the stated form. $\qquad\square$

Example 12.1 Consider an inverse semigroup S under its natural order $\leqslant_n$ (Example 11.7). If E is the semilattice of idempotents of S consider the set

$$E\omega = \{x \in S \mid (\exists e \in E)\ e \leqslant_n x\}.$$

Since E is a subsemigroup, so is $E\omega$. Now $E\omega$ is also a down-set. To see this, suppose that $y \leqslant_n x \in E\omega$. Since $x \in E\omega$ there exists $e \in E$ such that $e \leqslant_n x$ and so $e = ee^{-1} = ex^{-1}$. Taking inverses, we obtain $e = xe$. Thus, from $y \leqslant_n x$ we deduce that $y^{-1}ye = y^{-1}xe = y^{-1}e \leqslant_n y^{-1}$. It follows by Theorem 11.1 that $ey^{-1}y \leqslant_n y$. Since $ey^{-1}y \in E$ we deduce that $y \in E\omega$.

That $E\omega$ is reflexive follows from the fact that

$$xy \in E\omega \Rightarrow (\exists e \in E)\ e \leqslant_n xy$$
$$\Rightarrow (\exists e \in E)\ yey^{-1} \leqslant_n yxyy^{-1} \leqslant_n yx$$
$$\Rightarrow yx \in E\omega \quad (\text{since } yey^{-1} \in E).$$

Moreover, since $xx^{-1} \in E \subseteq E\omega$ for every $x \in S$ we see that $E\omega$ is neat.

Next we note that for every $f \in E$ we have $E\omega : \{f\} = E\omega$. In fact, on the one hand $f \in E \subseteq E\omega$ gives $\{f\}E\omega \subseteq E\omega$ and so $E\omega \subseteq E\omega : \{f\}$; and on the other hand

$$y \in E\omega : \{f\} \Rightarrow yf \in E\omega \Rightarrow (\exists e \in E)\ e \leqslant_n yf \leqslant_n y \Rightarrow y \in E\omega,$$

so that $E\omega : \{f\} \subseteq E\omega$ and hence we have equality.

It follows from the above that

$$E\omega = \bigcap_{x \in E\omega}(E\omega : \{x\}) = E\omega : \bigcup_{x \in E\omega}\{x\} = E\omega : E\omega$$

whence we see that

$$x \in I_{E\omega} \Rightarrow E\omega : \{x\} = E\omega \Rightarrow x \in E\omega : E\omega = E\omega$$

and therefore $I_{E\omega} \subseteq E\omega$. In fact, we have $I_{E\omega} = E\omega$. To see this, let $x \in E\omega$. Then there exists $f \in E$ such that $f \leqslant_n x$, whence $f = xf$. If $y \in E\omega : \{x\}$ then $yx \in E\omega$ and so there exists $e \in E$ such that $e \leqslant_n yx$. It follows that $ef \leqslant_n yxf = yf \leqslant_n y$ and hence, since $ef \in E$, that $y \in E\omega$. Thus we see that if $x \in E\omega$ then $E\omega : \{x\} \subseteq E\omega$ whence we have equality since $\{x\}E\omega \subseteq E\omega$.

It follows from the above observations that $E\omega$ is an anticone of S.

If S is an ordered semigroup then every non-empty quasi-residual of the form $\langle x . \cdot x \rangle$ or $\langle x \cdot . x \rangle$ is both a subsemigroup and a down-set. We call these the **principal subsemigroups** of S. By the **core** of S we shall mean the set-theoretic union of all the principal subsemigroups of S. We shall denote the core of S by $)S($.

Theorem 12.2 *If* $)S(\neq \emptyset$ *then* $)S($ *is a reflexive down-set.*

Proof If $xy \in)S($ then for some $t \in S$ either $xyt \leqslant t$ or $txy \leqslant t$, whence either $yxyt \leqslant yt$ or $txyx \leqslant tx$, so that either $yx \in \langle yt . \cdot yt \rangle$ or $yx \in \langle tx . \cdot tx \rangle$. In either case, $yx \in)S($ and so $)S($ is reflexive. Being the set-theoretic union of down-sets, $)S($ is also a down-set.   □

Theorem 12.3 *If* H *is an anticone of* S *then* $)S(\subseteq H$.

Proof For example, if $t \in \langle a^{\cdot}.a \rangle$ then from $ta \leqslant a$ we obtain $[t]_H[a]_H \leqslant_H [a]_H$ and so $[t]_H \leqslant I_H$ whence $t \in H$.   □

Example 12.2 If S is an inverse semigroup then $)S(= E\omega$. In fact, if $x \in E\omega$ then for some $e \in E$ we have $e \leqslant_n x$ whence $e = xe$ and hence $x \in \langle e^{\cdot}.e \rangle \subseteq)S($. Thus $E\omega \subseteq)S($ whence we have equality by Theorem 12.3.

Example 12.3 Let $A = \{x \in \mathbb{R} \mid -1 \leqslant x < 0\}$ and consider the cartesian ordered set $S = A \times \mathbb{Z}$ made into an ordered semigroup by the multiplication

$$(x, m)(y, n) = (-1, m + n).$$

It is readily seen that $)S(= \{(x, m) \mid m \leqslant 0\}$. By Theorem 12.2, $)S($ is a reflexive down-set of S, and is clearly a subsemigroup. Now

$$(x, m)(y, n) = (-1, m + n) \in)S(\iff n \leqslant -m$$

and therefore $)S(: \{(x, m)\} = \{(y, n) \mid n \leqslant -m\}$. This shows that $)S($ is neat, and that $I_{)S(} = \{(y, 0) \mid y \in A\}$. Since then

$$(x, m)(y, n) = (-1, m + n) \in I_{)S(} \iff n = -m$$

it is clear that $I_{)S(}$ is reflexive and neat. It follows that $)S($ is an anticone of S. The classes modulo the corresponding Dubreil equivalence are given by

$$[(x, m)]_{)S(} = \{(y, m) \mid y \in A\},$$

and $S/R_{)S(} \simeq \mathbb{Z}$. Note that in this example we have $I_{)S(} \neq)S($.

Suppose now that H is an anticone of S such that $H \neq)S($. Then $)S(\subset H$ and so H must contain the element $(-1, 1)$ whence, being a subsemigroup, it must contain the elements $(-1, m)$ for all $m \in \mathbb{Z}$. Since for all $(x, m), (y, n) \in S$ we then have $(x, m)(y, n) = (-1, m + n) \in H$ it follows that $H : \{(x, m)\} = S$ for all $(x, m) \in S$. But since H is an anticone we have $H : \{(x, m)\} = H$ for some $(x, m) \in H$. It therefore follows that $H = S$, a contradiction. Thus $)S($ is the only anticone of S.

EXERCISES

12.1. Let G be an ordered group and let E be a semilattice. Endow $G \times E$ with the cartesian order and the multiplication $(g, x)(h, y) = (gh, x)$. Show that $G \times E$ is an ordered semigroup and that $f : G \times E \to G$ given by the assignment $(g, x) \mapsto g$ is an isotone epimorphism. Determine the Dubreil equivalence associated with the anticone $f^{\leftarrow}(N_G)$.

12.2. If S is an inverse semigroup prove that each of the following equivalence relations coincides with the Dubreil equivalence that is associated with the anticone $E\omega$:

(1) $(x, y) \in M \iff (\exists e \in E) \; ex = ey$;
(2) $(x, y) \in H \iff xy^{-1} \in E\omega$;
(3) the zig-zag equivalence Z.

12.3. Let S be an ordered monoid. For each integer p let

$$[p] = \begin{cases} p & \text{if } p \geqslant 0; \\ 0 & \text{otherwise.} \end{cases}$$

and let $f_p : S \to S$ be given by

$$f_p(x, y) = \begin{cases} x & \text{if } p > 0; \\ xy & \text{if } p = 0; \\ y & \text{if } p < 0. \end{cases}$$

Verify the following identities:

(1) $[m] + [n - [-m]] = [m + [n]]$;

(2) $f_{m+[n]}\big(f_n(x, y), z\big) = f_{n-[-m]}\big(x, f_m(y, z)\big)$.

Deduce that $A = \mathbb{N} \times S \times \mathbb{N}$ is a monoid with respect to the multiplication

$$(m, x, n)(m', x', n') = (m + [m' - n], f_{n-m'}(x, x'), n' + [n - m']).$$

Define the relation $\leqslant$ on A by

$$(m, x, n) \leqslant (m', x', n') \iff m = m', x \leqslant x', n = n'.$$

Show that $(A; \leqslant)$ is an ordered monoid. Determine $)A($ and show that it is the smallest anticone.

12.2 Dubreil-Jacotin semigroups

Definition If S is an ordered semigroup and G is an ordered group then we shall say that an epimorphism $f : S \to G$ is **principal** if it is isotone and the pre-image of the negative cone of G is a principal down-set of S.

If $f : S \to G$ is a principal epimorphism then the anticone $f^{\leftarrow}(N_G)$ is a principal down-set of S. Our interest in principal anticones stems from the following observation.

Theorem 12.4 *If an ordered semigroup S has a principal anticone $h^{\downarrow}$ then this is the only principal anticone and necessarily $h^{\downarrow} =)S($.*

Proof For every anticone H of S we have

$$H = \natural_H^{\leftarrow}(N_{S/R_H}) = \{x \in S \mid [x]_H \leqslant_H I_H\} = \{x \in S \mid H \subseteq H : \{x\}\}.$$

Taking $H = h^{\downarrow}$ we obtain

$$x \in h^{\downarrow} \iff h^{\downarrow} \subseteq h^{\downarrow} : \{x\} = \langle h : x \rangle \iff h \in \langle h : x \rangle \iff x \in \langle h : h \rangle,$$

so that $h^{\downarrow} = \langle h : h \rangle \subseteq)S($. Equality now follows by Theorem 12.3. This also establishes uniqueness. $\qquad\square$

Definition By a **Dubreil-Jacotin semigroup** we shall mean an ordered semigroup which has a (necessarily unique) principal anticone.

Example 12.4 Let S be as in Example 12.3 and consider the cartesian ordered set $T = S \cup \{(0,0)\}$ with the multiplication of S extended to T by defining

$$(x, m)(0,0) = (0,0)(x, m) = (-1, m); \quad (0,0)(0,0) = (-1, 0).$$

Here the principal down-set $(0,0)^{\downarrow}$ is an anticone and so T is a Dubreil-Jacotin semigroup.

Example 12.5 Let G be an ordered group and let B be a band that is ordered as in Example 11.1. Then $G \times B$ is an ordered semigroup under the cartesian order and the multiplication $(g,e)(h,f) = (gh,e)$. The mapping $\vartheta : G \times B \to G$ given by $(g,e) \mapsto g$ is an isotone epimorphism and $\vartheta^{\leftarrow}(N_G) = \{(g,e)\mid g \leqslant 1_G\}$. Now it is readily seen that

$$\begin{cases} \langle (g,e) \cdot (g,e) \rangle = \{(h,f) \mid h \leqslant 1_G\}; \\ \langle (g,e) \cdot (g,e) \rangle = \{(h,f) \mid h \leqslant 1_G, f \leqslant e\}. \end{cases}$$

It follows that $)G \times B(= \vartheta^{\leftarrow}(N_G)$ and so is an anticone. If now B has a top element 1_B then the core of $G \times B$ is the principal anticone $(1_G, 1_B)^{\downarrow}$ whence $G \times B$ is a Dubreil-Jacotin semigroup.

We now establish the following basic characterisation.

Theorem 12.5 (Dubreil-Jacotin [47])   *An ordered semigroup is a Dubreil-Jacotin semigroup if and only if it admits a principal epimorphic image which is a group. Such a group is necessarily unique. If S is a Dubreil-Jacotin semigroup then its core $)S($ has a top element ξ which is equi-quasiresidual, and the unique principal epimorphic group image of S is given by $S/\mathcal{A}_\xi$ where*

$$(x,y) \in \mathcal{A}_\xi \iff \langle \xi : x \rangle = \langle \xi : y \rangle.$$

Proof In view of the above definitions and results it suffices to prove that the Dubreil equivalence associated with the unique principal anticone $)S($ is the equivalence $\mathcal{A}_\xi$ so described. For this purpose, let $)S(= \xi^{\downarrow}$. Since $)S(= \xi^{\downarrow}$ is reflexive we have

$$(\forall x \in S) \qquad \langle \xi \cdot x \rangle = \xi^{\downarrow} \cdot \{x\} = \xi^{\downarrow} \cdot \{x\} = \langle \xi \cdot x \rangle$$

so that ξ is equi-quasiresidual. Then the Dubreil equivalence associated with $\xi^{\downarrow}$ is precisely the relation $\mathcal{A}_\xi$ as described above. $\square$

It is natural to consider the situation where the notion of a principal epimorphism is restricted to that of a residuated epimorphism. Relative to this we have the following result.

Theorem 12.6 *Let S be an ordered semigroup. If G is an ordered group and $f : S \to G$ is a principal epimorphism with associated principal anticone $\xi^{\downarrow} = f^{\leftarrow}(N_G)$, then the following conditions are equivalent:*

(1) f *is residuated;*
(2) ξ *is residuated (in fact equiresidual).*

Proof Under the hypotheses we have $f(\xi) = 1_G$ and

$$f(x) \leqslant 1_G \iff x \leqslant \xi.$$

(1) $\Rightarrow$ (2): Suppose that (1) holds. Since f is surjective we have

$$xt \leqslant \xi \iff f(x)f(t) = f(xt) \leqslant f(\xi) = 1_G$$
$$\iff f(x) \leqslant [f(t)]^{-1}$$
$$\iff x \leqslant f^+[f(t)]^{-1},$$

and likewise

$$tx \leqslant \xi \iff x \leqslant f^+[f(t)]^{-1}.$$

Thus $\xi : t$ exists and is $f^+[f(t)]^{-1}$.

(2) $\Rightarrow$ (1): If (2) holds, let $y \in G$. Since f is surjective there exists $t \in S$ such that $f(t) = y^{-1}$. Then

$$f(x) \leqslant y \iff f(xt) = f(x)f(t) = f(x)y^{-1} \leqslant 1_G$$
$$\iff xt \leqslant \xi$$
$$\iff x \leqslant \xi^\cdot . t,$$

and similarly

$$f(x) \leqslant y \iff tx \leqslant \xi \iff x \leqslant \xi .^\cdot t.$$

Hence f is residuated with $f^+(y) = \xi^\cdot . t = \xi .^\cdot t$. □

Definition We shall say that a Dubreil-Jacotin semigroup is **strong** if the top element ξ of $)S($ is residuated.

Before obtaining a characterisation of strong Dubreil-Jacotin semigroups, we establish the following useful general result.

Theorem 12.7 *Let S be an ordered semigroup. If $t \in S$ is residuated then every right residual of t is residuated on the right, and every left residual of t is residuated on the left.*

Proof We give the proof for left residuals. Since t is residuated on the left, for any given $x, y \in S$ the set of $z \in S$ satisfying $zy \leqslant t^\cdot . x$ is not empty; for $(t^\cdot . yx)yx \leqslant t$ and so $(t^\cdot . yx)y \leqslant t^\cdot . x$. Now for any $z \in S$ satisfying $zy \leqslant t^\cdot . x$ we have $zyx \leqslant t$ and so $z \leqslant t^\cdot . yx$. This then shows that $(t^\cdot . x)^\cdot . y$ exists and is $t^\cdot . yx$. Similarly, $(t .^\cdot x) .^\cdot y$ exists and is $t .^\cdot xy$. □

Theorem 12.8 *An ordered semigroup S is a strong Dubreil-Jacotin semigroup if and only if there exists $\xi \in S$ such that*

(1) $)S(= \xi^\downarrow$;
(2) ξ *is residuated (in fact, equiresidual).*

Proof The conditions are clearly necessary. Conversely, suppose that (1) and (2) hold. Then we have $(\xi : x)x \leqslant \xi$ and so $)S(= \xi^\downarrow$ is neat. Now

$$x \in I_{)S(} \iff \xi^\downarrow : \{x\} = \xi^\downarrow \iff \xi : x \text{ exists and is } \xi.$$

That $I_{)S(} \neq \emptyset$ follows from the fact that $\xi \in I_{)S(}$. In fact, on the one hand $\xi : \xi \in \langle \xi : \xi \rangle \subseteq)S(= \xi^\downarrow$ gives $\xi : \xi \leqslant \xi$; and on the other, $)S(= \xi^\downarrow$ implies that ξ belongs to some principal subsemigroup, whence so does ξ^2 and hence $\xi^2 \in)S(= \xi^\downarrow$, so that $\xi^2 \leqslant \xi$ and then $\xi \leqslant \xi : \xi$.

Using Theorem 12.7 we see that, for all $x \in S$,

$$\xi = \xi : \xi \leqslant \xi : (\xi : x)x = (\xi : x)^\cdot . (\xi : x) \in)S(= \xi^\downarrow,$$

whence $\xi : (\xi : x)x = \xi$. Similarly we have $\xi : x(\xi : x) = \xi$. These equalities imply that the quotient semigroup $S/R_{)S(}$ is a group in which $[x]^{-1}_{)S(} = [\xi : x]_{)S(}$. Thus $)S($ is an anticone and the result follows. □

The following result is now clear.

Theorem 12.9 *An ordered semigroup is a strong Dubreil-Jacotin semigroup if and only if it admits an ordered group as an image under a residuated epimorphism. Such a group is necessarily unique. If S is a strong Dubreil-Jacotin semigroup then its core $)S($ has a top element ξ which is equiresidual. Moreover, the Molinaro closure equivalence A_ξ is a regular congruence on S and the unique residuated epimorphic group image of S is given by S/A_ξ.* ☐

We also have the following result.

Theorem 12.10 *Let S be an ordered semigroup and let ϑ be a regular congruence on S such that S/ϑ is a group. Then the following statements are equivalent:*

(1) *S is a strong Dubreil-Jacotin semigroup and $\vartheta = A_\xi$;*
(2) *one of the ϑ-classes has a top element that is residuated.*

Proof (1) $\Rightarrow$ (2): If (1) holds then ξ is the top element in its ϑ-class and is residuated.

(2) $\Rightarrow$ (1): Suppose that one of the ϑ-classes has a top element t that is residuated. Let e be any element of the identity ϑ-class. From $(et, t) \in \vartheta$ we deduce that $et \leqslant t$ and so $e \leqslant t^{\cdot}.t$, whence $et \leqslant (t^{\cdot}.t)t \leqslant t$. Since by Theorem 3.2 the ϑ-classes are convex, we then have $\big(et, (t^{\cdot}.t)t\big) \in \vartheta$ and so, S/ϑ being a group, $(e, t^{\cdot}.t) \in \vartheta$. In a similar way we can see that $(e, t.^{\cdot}t) \in \vartheta$. This then shows that the identity class modulo ϑ has a top element, namely $\xi = t^{\cdot}.t = t.^{\cdot}t$. Since t is residuated it follows by Theorem 12.7 that so is ξ.

Now let x be any element of S and let $x^\star \in [x]_\vartheta^{-1}$. From $(xx^\star, \xi) \in \vartheta$ we obtain $xx^\star \leqslant \xi$ whence $x^\star \leqslant \xi.^{\cdot}x$ and therefore $xx^\star \leqslant x(\xi.^{\cdot}x) \leqslant \xi$. The ϑ-classes being convex, we deduce that $\big(xx^\star, x(\xi.^{\cdot}x)\big) \in \vartheta$, from which it follows that $\xi.^{\cdot}x \in [x]_\vartheta^{-1}$. Similarly we can show that $\xi^{\cdot}.x \in [x]_\vartheta^{-1}$. It follows from these observations that for every $x \in S$ the element $\xi^{\cdot}.x(\xi.^{\cdot}x)$ belongs to the identity class modulo ϑ and so $\xi^{\cdot}.x(\xi.^{\cdot}x) \leqslant \xi$.

Now let $p \in \langle x^{\cdot}.x\rangle$. Then $px \leqslant x \leqslant \xi^{\cdot}.(\xi.^{\cdot}x)$ gives

$$p \leqslant [\xi^{\cdot}.(\xi.^{\cdot}x)]^{\cdot}.x = \xi^{\cdot}.x(\xi.^{\cdot}x) \leqslant \xi,$$

and consequently $\langle x^{\cdot}.x\rangle \subseteq \xi^{\downarrow}$. Now $(\xi^2, \xi) \in \vartheta$ gives $\xi^2 \leqslant \xi$ whence $\xi \leqslant \xi^{\cdot}.\xi$. It follows that $\bigcup_{x \in S}\langle x^{\cdot}.x\rangle = \xi^{\downarrow}$. In a similar way we see that $\bigcup_{x \in S}\langle x.^{\cdot}x\rangle = \xi^{\downarrow}$, from which we conclude that $)S(= \xi^{\downarrow}$. It now follows by Theorem 12.8 that S is a strong Dubreil-Jacotin semigroup.

That ϑ coincides with A_ξ is shown as follows. Since ϑ is compatible, if $(x, y) \in \vartheta$ then $\xi \overset{\vartheta}{\equiv} x(\xi{:}x) \overset{\vartheta}{\equiv} y(\xi{:}x)$ whence $y(\xi{:}x) \leqslant \xi$ and therefore $\xi{:}x \leqslant \xi{:}y$; and likewise $\xi{:}y \leqslant \xi{:}x$. Thus we see that $\vartheta \subseteq A_\xi$. But, as we observed above, we have $\xi{:}x \in [x]_\vartheta^{-1}$ and so $\xi{:}(\xi{:}x) \in [x]_\vartheta$. It follows immediately from this that if $\xi{:}x = \xi{:}y$ then $(x, y) \in \vartheta$ and so $A_\xi \subseteq \vartheta$, whence we have equality. ☐

Example 12.6 Let $S = \{(p, i) \mid p \in \mathbb{R}, p \leqslant 0, i \in \mathbb{Z}\}$. Under the cartesian order and the multiplication

$$(p, i)(q, j) = (-1, i + j),$$

S is an ordered (commutative) semigroup in which

$$\langle (r, k) : (p, i) \rangle = \begin{cases} \emptyset & \text{if } r < -1; \\ (0, k - i)^{\downarrow} & \text{if } r \geqslant -1. \end{cases}$$

It follows that for each (r, k) we have $\langle (r, k) : (r, k) \rangle$ is $\emptyset$ or $(0, 0)^{\downarrow}$. Thus $)S(= (0, 0)^{\downarrow}$. Moreover, $(0, 0)$ is residuated with $(0, 0) : (p, i) = (0, -i)$. It follows by Theorem 12.8 that S is a strong Dubreil-Jacotin semigroup, and $S/A_{(0,0)} \simeq \mathbb{Z}$.

Definition If S is a strong Dubreil-Jacotin semigroup then we shall say that $x \in S$ is *A*-nomal if it is a residual of ξ.

Theorem 12.11 *If S is a strong Dubreil-Jacotin semigroup then under the law of composition $(\xi : x) \circ (\xi : y) = \xi : xy$ the subset of A-nomal elements of S forms an ordered group which is isomorphic to S/A_{ξ}.*

Proof Let S° denote the subset of *A*-nomal elements of S. Consider the mapping $f : S^{\circ} \to S/A_{\xi}$ given by the prescription

$$f(\xi : x) = [x]_{A_{\xi}}^{-1}.$$

This mapping is clearly a bijection. Since the order on S/A_{ξ} is given by

$$[x]_{A_{\xi}} \leqslant [y]_{A_{\xi}} \iff)S(: \{y\} \subseteq)S(: \{x\} \iff \langle \xi : y \rangle \subseteq \langle \xi : x \rangle \iff \xi : y \leqslant \xi : x,$$

we see that

$$\xi : x \leqslant \xi : y \iff f(\xi : x) \leqslant f(\xi : y).$$

Moreover, we have

$$f[(\xi : x) \circ (\xi : y)] = f(\xi : yx) = [yx]_{A_{\xi}}^{-1} = [x]_{A_{\xi}}^{-1}[y]_{A_{\xi}}^{-1} = f(\xi : x)f(\xi : y)$$

and so f is also a morphism. Consequently f is an order isomorphism. $\qquad \square$

EXERCISES

12.4. With reference to Theorem 12.1, Example 12.1, and Exercise 12.2, prove that under its natural order an inverse semigroup is a Dubreil-Jacotin semigroup if and only if the identity class modulo Z contains a top element.

12.5. Prove that with respect to its natural order the bicyclic semigroup is a strong Dubreil-Jacotin semigroup.

12.6. Let $k > 1$ be a fixed integer. For every $n \in \mathbb{Z}$ let n_k be the biggest multiple of k that is less than or equal to n; i.e., $n_k = pk \leqslant n < (p+1)k$. Put another way, n_k is the integer part of n/k. Establish the properties

(1) $n_k = (n_k + k - 1)_k$;
(2) $n_k + m_k = (n + m_k)_k$;
(3) $n \leqslant n_k + k - 1$.

If the operation $\oplus$ is defined on $\mathbb{Z}$ by $m \oplus n = m + n_k$, prove that $(\mathbb{Z}; \oplus, \leqslant)$ is a totally ordered Dubreil-Jacotin semigroup and determine the top element of its core.

12.7. If $\mathbf{n}$ denotes the chain $1 < 2 < 3 < \cdots < n$, let L be the **left zero semigroup** consisting of $\mathbf{n}$ with the multiplication $xy = x$, and let R be the **right zero semigroup** consisting of $\mathbf{n}$ with the multiplication $xy = y$. Clearly, L and R are ordered semigroups. Now let G be an ordered group and let $S = L \times G \times R$ be the cartesian ordered cartesian product semigroup. Prove that $)S(= (n, 1_G, n)^{\downarrow}$. Show that $(n, 1_G, n)$ is residuated and deduce that S is a strong Dubreil-Jacotin semigroup.

12.8. Consider the set $D = \{a_{i,-p}, b_{i,-p} \mid i \in \mathbb{Z}, p \in \mathbb{N}\}$, ordered by
$$\begin{cases} a_{i,-p} \leqslant a_{j,-q} \iff b_{i,-p} \leqslant b_{j,-q} \iff i \leqslant j, -p \leqslant -q; \\ (\forall i, j, p, q) \quad a_{i,-p} \parallel b_{j,-q}. \end{cases}$$
Endow D with the multiplication
$$\begin{cases} a_{i,-p} a_{j,-q} = b_{i,-p} b_{j,-q} = a_{i+j,\min\{-p,-q\}}; \\ a_{i,-p} b_{j,-q} = b_{j,-q} a_{i,-p} = b_{i+j,\min\{-p,-q\}}. \end{cases}$$
Prove that D is a strong Dubreil-Jacotin inverse semigroup and determine its natural order.

12.3 Residuated Dubreil-Jacotin semigroups

We now consider necessary and sufficient conditions for a residuated semigroup S to be a (strong) Dubreil-Jacotin semigroup. For this purpose, consider the subsets
$$S^{\cdot\cdot} = \{x \cdot x \mid x \in S\}, \quad S^{\cdot\cdot} = \{x \cdot x \mid x \in S\}.$$
For all $x, y \in S$ we have, using Theorem 11.3,
$$(y \cdot x)(x \cdot x) \leqslant (y \cdot x) x \cdot x \leqslant y \cdot x$$
and therefore $x \cdot x \leqslant (y \cdot x) \cdot (y \cdot x)$. It follows from this that every upper bound of $S^{\cdot\cdot}$ is also an upper bound of $S^{\cdot\cdot}$. In a dual manner we can establish the converse. Thus $S^{\cdot\cdot}$ has a top element if and only if $S^{\cdot\cdot}$ has a top element, in which case these elements are the same. When this occurs we shall denote the common top element by ξ and call it the **bimaximum element** of S.

Example 12.7 Consider the cartesian ordered set
$$S = \{(x, y) \in \mathbb{Z} \times \mathbb{Z} \mid x \leqslant 0\}.$$
Let $k > 1$ be a fixed integer and for every $n \in \mathbb{Z}$ let n_k denote the biggest multiple of k that is less than or equal to n; see Exercise 12.6. Endow S with the multiplication
$$(x, y)(u, v) = (\min\{x, u\}, y + v_k).$$
Then S is residuated with residuals given by
$$\begin{cases} (c, d) \cdot (a, b) = \begin{cases} (0, d - b_k) & \text{if } a \leqslant c; \\ (c, d - b_k) & \text{if } a > c, \end{cases} \\ (c, d) \cdot (a, b) = \begin{cases} (0, (d - b)_k + k - 1) & \text{if } a \leqslant c; \\ (c, (d - b)_k + k - 1) & \text{if } a > c. \end{cases} \end{cases}$$
In this residuated semigroup we have
$$(c, d) \cdot (c, d) = (0, d - d_k), \qquad (c, d) \cdot (c, d) = (0, k - 1)$$
and so S has a bimaximum element, namely $(0, k - 1)$.

Theorem 12.12 *In a residuated semigroup S the following conditions are equivalent*:

(1) S *is a (strong) Dubreil-Jacotin semigroup*;

(2) S *has a bimaximum element*;

(3) $\max\{x \in S \mid x^2 \leqslant x\}$ *exists.*

Moreover, if S is a monoid then each of the above conditions is equivalent to

(4) S *has a biggest idempotent.*

Proof (1) $\Rightarrow$ (2): If (1) holds then $)S(= \xi^{\downarrow}$. Consequently, for every $x \in S$ we have $x \cdot\, \dot{} x \leqslant \xi$ and $x\dot{}\, . x \leqslant \xi$. Since, as seen in the proof of Theorem 12.8, we have $\xi \leqslant \xi \cdot\, \dot{} \xi$ and $\xi \leqslant \xi\dot{}\, . \xi$, it follows that S has a bimaximum element, namely ξ.

(2) $\Rightarrow$ (1): If a bimaximum element ξ exists then clearly $)S(= \xi^{\downarrow}$ and (1) follows by Theorem 12.8.

(2) $\Rightarrow$ (3): Let $J = \{x \in S \mid x^2 \leqslant x\}$. If ξ is the bimaximum element of S then $\xi = \xi\dot{}\, . \xi$ gives $\xi \in J$. Since for every $x \in J$ we have $x \leqslant x \cdot\, \dot{} x \leqslant \max S^{\cdot\cdot} = \xi$ it follows that $\max J$ exists and is ξ.

(3) $\Rightarrow$ (2): With J as above, observe that for every $x \in S$ the inequalities

$$(x\dot{}\, . x)(x \cdot\, \dot{} x) \leqslant (x\dot{}\, . x)x \cdot\, \dot{} x \leqslant x\dot{}\, . x$$

give $x\dot{}\, . x \in J$ and consequently $S^{\cdot\cdot} \subseteq J$; and likewise $S^{\cdot\cdot} \subseteq J$. If now (3) holds let $e = \max J$. Then from $e^2 \leqslant e$ we obtain $e \leqslant e \cdot\, \dot{} e \in S^{\cdot\cdot} \subseteq J$ whence $e = e \cdot\, \dot{} e \in S^{\cdot\cdot}$; and likewise $e = e\dot{}\, . e \in S^{\cdot\cdot}$. It follows immediately that e is the bimaximum element of S.

Finally, suppose that S is a monoid and let E be the set of idempotents. Then $1x = x$ gives $1 \leqslant x\dot{}\, . x$ and hence $x\dot{}\, . x \leqslant (x\dot{}\, . x)^2$. But as we have seen above, $x\dot{}\, . x \in J$. Hence $x\dot{}\, . x$ is idempotent. Thus $S^{\cdot\cdot} \subseteq E$; and likewise $S^{\cdot\cdot} \subseteq E$.

Now if S has a biggest idempotent, say e, then $e \leqslant e\dot{}\, . e \in S^{\cdot\cdot} \subseteq E$ gives $e = e\dot{}\, . e$; and likewise $e = e\dot{}\, . e$. Consequently e is the bimaximum element of S. Conversely, if ξ is the bimaximum element of S then $\xi \in S^{\cdot\cdot} \cap S^{\cdot\cdot} \subseteq E$ and so ξ is idempotent. Since ξ is the biggest element of J and since $E \subseteq J$ it follows that ξ is also the biggest element of E. $\qquad\square$

EXERCISES

12.9. Let X and D be non-empty subsets of a semigroup S. Then we say that X is D-**transportable** if $D \cdot\, \dot{} X$ and $D\dot{}\, . X$ are non-empty; and that X is D-**neat** if $X \cdot\, \dot{} D$, $X\dot{}\, . D$ and $(X \cdot\, \dot{} D)\dot{}\, . D$ are non-empty. If X is both D-transportable and D-neat then we say that X is a D-**complex**.

Prove that if D is a subsemigroup of S that is equiresidual in $\mathbb{P}(S)$ then for every D-transportable subset X the subsets XD, DX and DXD are D-complexes.

12.10. Let D be a subsemigroup that is equiresidual in $\mathbb{P}(S)$ with $D{:}S = \emptyset$. Prove that the set $C(D)$ of D-complexes is a residuated subsemigroup of $\mathbb{P}(S)$ that contains neither $\emptyset$ nor S.

12.11. A D-transportable subset of the form $\{x^n \mid x \geqslant 1\}$ is said to be **principal**. Let D be a subsemigroup that is equiresidual in $\mathbb{P}(S)$ with $D{:}S = \emptyset$. Prove that the following statements are equivalent:

(1) $C(D)$ is a Dubreil-Jacotin semigroup with bimaximum element D;
(2) every principal D-transportable subset is contained in D.

In a residuated Dubreil-Jacotin semigroup the Molinaro closure equivalence A_ξ is a congruence and is cancellative. We now take a closer look at closure equivalences with these properties.

Theorem 12.13 *If S is an ordered semigroup then a closure equivalence R on S is compatible on the left with multiplication if and only if*

$$(\forall x, y \in S) \quad xf(y) \leqslant f(xy)$$

where f is the associated closure mapping.

Proof $\Rightarrow$: If R is compatible on the left then $y \stackrel{R}{\equiv} f(y)$ gives $xy \stackrel{R}{\equiv} xf(y)$. Since $f(xy)$ is the top element in the R-class of xy it follows that $xf(y) \leqslant f(xy)$.

$\Leftarrow$: Suppose that the condition holds. Then since f is a closure, we have $f[xf(y)] \leqslant f^2(xy) = f(xy)$; and from $y \leqslant f(y)$ we obtain the converse inequality. Hence $f[xf(y)] = f(xy)$. It follows from this that if $f(y) = f(z)$ then $f(xy) = f(xz)$ and so R is compatible on the left. $\square$

Theorem 12.14 *Let S be a residuated semigroup and let R be a closure equivalence on S with associated closure mapping f. Then the following statements are equivalent:*

(1) R *is compatible on the left with multiplication;*
(2) $x \in \operatorname{Im} f \Rightarrow (\forall y \in S)\; x{\cdot}{\cdot}y \in \operatorname{Im} f$;
(3) $R \subseteq \bigcap\limits_{x \in \operatorname{Im} f} {}_x A$;
(4) $(\forall x, y \in S)\; f(x){\cdot}{\cdot}y = f(x){\cdot}{\cdot}f(y)$.

Proof (1) $\Rightarrow$ (2): Suppose that (1) holds and let $x \in \operatorname{Im} f$. If $a = x{\cdot}{\cdot}y$ then by Theorem 12.13 we have $yf(a) \leqslant f(ya) \leqslant f(x) = x$, so that $f(a) \leqslant x{\cdot}{\cdot}y = a$. Consequently $f(a) = a$ and (2) follows.

(2) $\Rightarrow$ (1): If (2) holds then from $y \leqslant xy{\cdot}{\cdot}x \leqslant f(xy){\cdot}{\cdot}x \in \operatorname{Im} f$ we deduce that $f(y) \leqslant f[f(xy){\cdot}{\cdot}x] = f(xy){\cdot}{\cdot}x$ whence $xf(y) \leqslant f(xy)$ and so, by Theorem 12.13, (1) holds.

(1) $\Rightarrow$ (3) Let $(a,b) \in R$ and let $x \in \operatorname{Im} f$. By (1) we have $(x{\cdot}{\cdot}b)a \stackrel{R}{\equiv} (x{\cdot}{\cdot}b)b$. Now $x \in \operatorname{Im} f$ is the top element in its R-class, so that $[x]_R \subseteq x^{\downarrow}$. Since $(x{\cdot}{\cdot}b)b \leqslant x$ it follows from the above that $[(x{\cdot}{\cdot}b)a]_R \cap x^{\downarrow} \neq \emptyset$. Since R is strongly upper regular we deduce that $[(x{\cdot}{\cdot}b)a]_R \subseteq x^{\downarrow}$ whence $(x{\cdot}{\cdot}b)a \leqslant x$ and so $x{\cdot}{\cdot}b \leqslant x{\cdot}{\cdot}a$. Similarly, we have $x{\cdot}{\cdot}a \leqslant x{\cdot}{\cdot}b$, whence $(a,b) \in {}_x A$.

(3) $\Rightarrow$ (4): This clearly follows from $y \stackrel{R}{\equiv} f(y)$ and $f(x) \in \operatorname{Im} f$.

(4) $\Rightarrow$ (1): If (4) holds then for all $a, b \in S$ we have

$$a \leqslant ab{\cdot}{\cdot}b \leqslant f(ab){\cdot}{\cdot}b = f(ab){\cdot}{\cdot}f(b)$$

from which we obtain $af(b) \leqslant f(ab)$, and (1) follows by Theorem 12.13. $\square$

Theorem 12.15 *If S is a residuated semigroup then a closure equivalence R on S is left cancellative if and only if*

$$(\forall x, y \in S) \quad y \stackrel{R}{\equiv} f(xy) \,.\!\cdot\, x$$

where f is the associated closure mapping.

Proof $\Rightarrow$: From $xy \leqslant f(xy)$ we have $y \leqslant f(yx) \,.\!\cdot\, x$ and so $xy \leqslant x[f(xy) \,.\!\cdot\, x] \leqslant f(xy)$. The R-classes being convex, it follows that $xy \stackrel{R}{\equiv} x[f(xy) \,.\!\cdot\, x]$ whence the result follows by left cancellation.

$\Leftarrow$: Suppose that $xy \stackrel{R}{\equiv} xz$. Then $f(xy) = f(xz)$ and so $y \stackrel{R}{\equiv} f(xy) \,.\!\cdot\, x = f(xz) \,.\!\cdot\, x \stackrel{R}{\equiv} z$ whence R is left cancellative. $\square$

The above result provides the following characterisation.

Theorem 12.16 *A residuated semigroup S is a Dubreil-Jacotin semigroup if and only if there is a closure equivalence on S that is left or right cancellative.*

Proof $\Rightarrow$: If S is a Dubreil-Jacotin semigroup let ξ be its bimaximum element. Then by Theorem 12.9 the Molinaro equivalence A_ξ is a closure such that S/A_ξ is a group. Hence A_ξ is cancellative.

$\Leftarrow$: Suppose that R is a left cancellative closure equivalence on S. Let f be the associated closure mapping. Then from Theorem 12.15 we have $y \stackrel{R}{\equiv} f(xy) \,.\!\cdot\, x$. Since $y \leqslant xy \,.\!\cdot\, x \leqslant f(xy) \,.\!\cdot\, x$ we have, again by the convexity of the R-classes, that $y \stackrel{R}{\equiv} xy \,.\!\cdot\, x$. Thus, for every $z \in S$, we have $f(z) \stackrel{R}{\equiv} xf(z) \,.\!\cdot\, x$. Since $f(z)$ is the top element in its R-class we deduce that

$$(\forall x, z \in S) \qquad f(z) = xf(z) \,.\!\cdot\, x.$$

It then follows that

$$f(z) \,.\!\cdot\, f(z) = [xf(z) \,.\!\cdot\, x] \,.\!\cdot\, f(z) = [xf(z) \,.\!\cdot\, f(z)] \,.\!\cdot\, x \geqslant x \,.\!\cdot\, x.$$

Thus $S^{\,.\!\cdot}$ is bounded above, whence so is $S^{\,.\!\cdot}$. Now since

$$[f(z) \,.\!\cdot\, f(z)] \,.\!\cdot\, [f(z) \,.\!\cdot\, f(z)] \geqslant f(z) \,.\!\cdot\, f(z)$$

it follows that a bimaximum element exists, namely $[f(z) \,.\!\cdot\, f(z)] \,.\!\cdot\, [f(z) \,.\!\cdot\, f(z)]$. Thus, by Theorem 12.12, S is a Dubreil-Jacotin semigroup. A similar conclusion results from the dual hypothesis that R is right cancellative. $\square$

Theorem 12.17 *If S is a residuated semigroup then a closure equivalence R on S is both left compatible and left cancellative if and only if*

$$(\forall x, y \in S) \quad f(y) = f(xy) \,.\!\cdot\, x$$

where f is the associated closure mapping.

Proof $\Rightarrow$: Since R is left cancellative, Theorem 12.15 gives $y \stackrel{R}{\equiv} f(xy) \,.\!\cdot\, x$; and since R is left compatible, Theorem 12.14 gives $f(xy) \,.\!\cdot\, x \in \operatorname{Im} f$. Consequently, $f(y) = f[f(xy) \,.\!\cdot\, x] = f(xy) \,.\!\cdot\, x$.

$\Leftarrow$: The equality $f(y) = f(xy) \,.\!\cdot\, x$ implies that $xf(y) \leqslant f(xy)$ so that, by Theorem 12.13, R is left compatible; and clearly, by Theorem 12.15, R is left cancellative. $\square$

We now investigate closures of the form A_x and $_xA$ in relation to the above. In this connection, we note that by Theorem 11.3(7) the equivalence A_x is compatible on the right; and dually $_xA$ is compatible on the left.

Theorem 12.18 *If S is a residuated semigroup then for every $x \in S$ the following conditions are equivalent*:

(1) A_x *is a congruence that is left cancellative*;
(2) $(\forall t \in S)\ \ A_x = A_{x.\cdot t}$.

Proof Taking $R = A_x$ in Theorem 12.17, we see that (1) holds if and only if

$$(\forall t, y \in S)\ \ x^{\cdot}.(x.^{\cdot}y) = [x^{\cdot}.(x.^{\cdot}ty)].^{\cdot}t = (x.^{\cdot}t)^{\cdot}.[(x.^{\cdot}t).^{\cdot}y]$$

which is equivalent to (2). $\qquad\square$

We can now determine precisely when a Molinaro closure equivalence A_x is a group congruence.

Theorem 12.19 (Molinaro–Querré [84],[92]) *If S is a residuated semigroup then the following conditions concerning $x \in S$ are equivalent*:

(1) A_x *is a congruence and S/A_x is a group*;
(2) S *is a Dubreil-Jacotin semigroup and $A_x = A_\xi$*;
(3) $(\forall t \in S)\ \ A_x = A_{x.\cdot t} = A_{x^{\cdot}.t}$.

Proof (1) $\Leftrightarrow$ (2): This is immediate from Theorem 12.10.

(2) $\Rightarrow$ (3): If (2) holds then for every $t \in S$ there exists $p \in S$ such that $x^{\cdot}.t = \xi{:}p$ so, by Theorem 12.18 and the hypothesis, we have $A_{x.\cdot t} = A_x = A_\xi = A_{\xi{:}p} = A_{x^{\cdot}.t}$.

(3) $\Rightarrow$ (2): If (3) holds then, by Theorem 12.18, A_x is a left cancellative congruence and so, by Theorem 12.16, S is a Dubreil-Jacotin semigroup. It follows by Theorem 12.12 that S has a bimaximum element ξ. Then $A_{x^{\cdot}.\xi} = A_x$ is a left cancellative congruence and so, by Theorem 12.18, $A_{x^{\cdot}.\xi} = A_{(x^{\cdot}.\xi).^{\cdot}(x^{\cdot}.\xi)}$. But $(x^{\cdot}.\xi).^{\cdot}(x^{\cdot}.\xi) = [x.^{\cdot}(x^{\cdot}.\xi)]^{\cdot}.\xi \geqslant \xi^{\cdot}.\xi = \xi$ whence $(x^{\cdot}.\xi).^{\cdot}(x^{\cdot}.\xi) = \xi$. Combining these observations, we obtain $A_x = A_\xi$ as required. $\qquad\square$

It is immediate from Theorem 12.19 and its dual that if S is a residuated semigroup and $x \in S$ then the following statements are equivalent:

(1) A_x is a congruence and x is the top element in its A_x-class;
(2) $_xA$ is a congruence and x is the top element in its $_xA$-class;
(3) S is a Dubreil-Jacotin semigroup and x is A-nomal.

Example 12.8 Let D be a commutative integral domain and let F be its field of quotients. By a **fractionary ideal** of D we mean a D-submodule $\mathbf{a}$ of the D-module F such that the elements of $\mathbf{a}$ have a common denominator $d \neq 0$ in D; more precisely, if there exists $d \neq 0$ such that $d\mathbf{a} \subseteq D$. Thus, if $\mathbf{a}$ is a fractionary ideal then $\mathbf{a} = d^{-1}\mathbf{b}$ where $\mathbf{b}$ is an ordinary ideal of D. For example, every ordinary ideal is a fractionary ideal (take $d = 1$). Every finitely generated D-submodule $\mathbf{a}$ of F is a fractionary ideal; for if $\{a_1, \ldots, a_n\}$ is a set of generators of $\mathbf{a}$ then each a_i can be written in the form $a_i = b_i d_i^{-1}$ with

$d_i \neq 0$, and if we define $d = \prod\limits_{i=1}^{n} d_i$ then $d \neq 0$ and $\mathbf{a} \subseteq d^{-1}F$. In particular, if $\mathbf{a}$ has a single generator, say $x = ad^{-1}$, then xD is the **principal fractionary ideal** generated by x. It is clear that the set of fractionary ideals of F, ordered by set inclusion, forms a lattice-ordered semigroup under the laws

$$(\mathbf{a}, \mathbf{b}) \mapsto \mathbf{a} + \mathbf{b}, \quad (\mathbf{a}, \mathbf{b}) \mapsto \mathbf{a} \cap \mathbf{b}, \quad (\mathbf{a}, \mathbf{b}) \mapsto \mathbf{ab}.$$

In particular, the set $\mathrm{PF}^{\star}(D)$ of all non-zero principal fractionary ideals forms a lattice-ordered group, the identity element of which is $1D = D$ and the inverse of xD is $x^{-1}D$. This ordered group is dually isomorphic to the quotient group $F^{\star}/U$ where $F^{\star} = F\backslash\{0\}$ and U is the group of units of D.

[To be more explicit, if we let $D^{\star} = D\backslash\{0\}$ and define the relation $|$ by

$$x|y \iff (\exists p \in D^{\star}) \; xp = y$$

then the relation given by

$$x \equiv y \iff x|y \text{ and } y|x$$

is an equivalence relation which is none other that the relation $xy^{-1} \in U$. We can therefore order $F^{\star}/U$ by setting

$$[x]_U \leqslant [y]_U \iff x|y.$$

Since $x|y$ holds if and only if $xD \supseteq yD$, the dual isomorphism follows.]

Now let $F^{\star}(D)$ be the set of non-zero fractionary ideals of D. Then $F^{\star}(D)$ is a residuated semigroup.

[It suffices to observe that if $\mathbf{a}, \mathbf{b} \in F^{\star}(D)$ then $\mathbf{a}:\mathbf{b} = \{x \in F \mid x\mathbf{b} \subseteq \mathbf{a}\}$ is also an element of $F^{\star}(D)$. For this purpose, let let $d \neq 0$ be such that $\mathbf{a} \subseteq d^{-1}D$ and let $b \in \mathbf{b}\backslash\{0\}$; then $db(\mathbf{a}:\mathbf{b}) \subseteq da \subseteq D$. Moreover, if $a \in \mathbf{a}\backslash\{0\}$ and $d'\mathbf{b} \subseteq D$ then $ad'\mathbf{b} \subseteq \mathbf{a}$, so that $\mathbf{a}:\mathbf{b} \neq \emptyset$.]

Since for each $\mathbf{a} \in F^{\star}(D)$ there exists $d \neq 0$ such that $\mathbf{a} \subseteq d^{-1}D$, the set of principal fractionary ideals that contain $\mathbf{a}$ is not empty. Let $\mathrm{Pr}(\mathbf{a})$ be the set of principal fractionary ideals of D that contain $\mathbf{a}$ and define a relation R on $F^{\star}(D)$ by

$$(\mathbf{a}, \mathbf{b}) \in R \iff \mathrm{Pr}(\mathbf{a}) = \mathrm{Pr}(\mathbf{b}).$$

Since $F^{\star}(D)$ is residuated we have

$$xD \in \mathrm{Pr}(\mathbf{a}) \iff x^{-1}\mathbf{a} \subseteq D \iff x^{-1} \in D:\mathbf{a}$$

whence we see that R coincides with the Molinaro closure A_D on $F^{\star}(D)$.

We say that $\mathbf{a} \in F^{\star}(D)$ is a **divisorial ideal** if $\mathbf{a} = \mathbf{a}^{\circ}$ where $\mathbf{a}^{\circ}$ denotes the intersection of all the principal fractionary ideals that contain $\mathbf{a}$. It is clear that for every $\mathbf{a} \in F^{\star}(D)$ we have $\mathbf{a} \subseteq \mathbf{a}^{\circ}$ and $(\mathbf{a}, \mathbf{a}^{\circ}) \in R = A_D$. It follows that the divisorial ideals are none other than the top elements in the A_D-classes.

We say that D is **completely integrally closed** if, whenever $x \in F$ is such that all the powers x^n $(n \geqslant 0)$ are contained in a finitely generated D-submodule of F then $x \in D$.

We now show that the following statements are equivalent:

(1) *$F^{\star}(D)$ is a residuated Dubreil-Jacotin semigroup in which the A-nomal elements are the divisorial ideals*;

(2) D *is completely integrally closed.*

Suppose first that (1) holds. Let $x \in F$ be such that $D[x]$ is contained in a finitely generated D-submodule M of F. Since M is an element of $F^\star(D)$, so also is $D[x]$. Let $D[x] = \mathbf{a}$. Then we have $x\mathbf{a} \subseteq \mathbf{a}$ and $Dx\mathbf{a} \subseteq \mathbf{a}$ so that, $F^\star(D)/A_D$ being a group with identity element D/A_D, we have $Dx/A_D \leqslant D/A_D$. It follows that $Dx = (Dx)^\circ \subseteq D^\circ = (1D)^\circ = 1D = D$ and hence that $x = 1x \in D$. Thus D is completely integrally closed.

Conversely, suppose that (2) holds. To show that $F^\star(D)/A_D$ is a group, it suffices to establish the following identity

$$\mathbf{a}/A_D \cdot (D\!:\!\mathbf{a})/A_D = D/A_D$$

where $\mathbf{a}$ is divisorial; and since $\mathbf{a}(D\!:\!\mathbf{a}) \subseteq D$ it is sufficient, in view of the fact that $A_D = R$, to show that every principal fractionary ideal which contains $\mathbf{a}(D\!:\!\mathbf{a})$ also contains D.

Suppose then that $\mathbf{a}(D\!:\!\mathbf{a}) \subseteq x^{-1}D$ and let $y \in F^\star$ be such that $\mathbf{a} \subseteq yD$. Then $y^{-1}\mathbf{a} \subseteq (D\!:\!\mathbf{a})\mathbf{a} \subseteq x^{-1}D$ and so $x\mathbf{a} \subseteq yD$. Thus every principal fractionary ideal that contains $\mathbf{a}$ also contains $x\mathbf{a}$. Since, by hypothesis, $\mathbf{a}$ is divisorial, we deduce that $x\mathbf{a} \subseteq \mathbf{a}$. Consequently, for every positive integer n we have $x^n\mathbf{a} \subseteq \mathbf{a}$. Now let $x_1, x_2 \in F^\star$ be such that $x_1 \in \mathbf{a} \subseteq x_2D$. Then for every positive integer n we have $x^n x_1 \in x_2D$ which gives $x^n \in x_1^{-1}x_2D$. Applying the standing hypothesis, we deduce that $x \in D$, so that $xD \subseteq D$ and $D \subseteq x^{-1}D$. This then establishes the required property.

EXERCISES

12.12. Prove that every residuated semigroup with a minimal element (in particular, every finite residuated semigroup) is a Dubreil-Jacotin semigroup in which $A_\xi = Z$.

12.13. Let S be a residuated Dubreil-Jacotin semigroup. Prove that, for every $x \in S$, the following conditions are equivalent:

(1) $A_x \leqslant A_\xi$;
(2) $\xi = x^\cdot.(x.^\cdot\xi)$.

In Example 12.7 show that $A_{(c,d)} \leqslant A_\xi$ if and only if $d \equiv k - 1 \pmod{k}$, whereas $_{(c,d)}A \leqslant {}_\xi A$ for all $(c,d) \in S$.

12.14. If S is a residuated Dubreil-Jacotin semigroup and $E = [\xi]_{A_\xi}$, prove that

$$(a,b) \in A_\xi \iff (\exists x, y \in E) \ ax = yb.$$

12.15. Let G be an ordered group and let B be a boolean algebra. Show that the cartesian ordered cartesian product semigroup $G \times B$ is a residuated Dubreil-Jacotin semigroup. Identify the A-nomal elements and deduce that $(G \times B)/A_\xi \simeq G$.

12.16. Prove that in a residuated Dubreil-Jacotin semigroup S an A-nomal element is equiresidual if and only if its A_ξ-class belongs to the centre of S/A_ξ.

12.17. Consider the ordered semigroup S of Example 12.7. Extend the cartesian order of S to include the set $T = \{(0, n - \frac{1}{2}) \mid n \in \mathbb{Z}\}$ and let $V = S \cup T$ be the inflation of S obtained by defining multiplication by $(0, n - \frac{1}{2})$ to be the same as multiplication by $(0, n)$.

Prove that V is a residuated Dubreil-Jacotin semigroup with bimaximum element $\xi = (0, k - 1)$. Taking $k = 3$, describe diagrammatically the A_ξ-classes. Do likewise for the classes modulo $F_x, {}_xF, B_x, {}_xB$ where $x = (-4, 2)$.

12.18. Let G be an ordered group. Prove that
$$(\forall x \in G) \quad x^{-1}N_G = N_Gx^{-1}$$
and deduce that N_G is equiresidual in $\mathbb{P}(G)$. Show also that $N_G:G = \emptyset$. With terminology as defined in Exercise 12.9, prove that for every subset X of G the following conditions are equivalent:

(1) X is bounded above in G;

(2) X is N_G-transportable.

An ordered group G is said to be **completely integrally closed** if whenever $\{x^n \mid n \geqslant 1\}$ admits an upper bound in G then it admits 1_G as an upper bound. Prove that the following conditions are equivalent:

(3) G is completely integrally closed;

(4) the set $C(N_G)$ of N_G-complexes is a residuated Dubreil-Jacotin semigroup with bimaximum element N_G.

12.19. By a **down-ideal** of an ordered semigroup S we mean an ideal of S that is a down-set. By a **base** of S we mean a reflexive down-ideal H such that

(1) $(\forall x \in S) \ H:\{x\} = H:\{x^2\}$;

(2) $(\forall x \in S)(\exists x^\circ \in S) \ H:(H:\{x\}) = H:\{x^\circ\}$.

Prove that if H is a base of S and if R_H is the associated Dubreil equivalence then S/R_H is an epimorphic boolean image of S. Show moreover that every such image arises in this way.

12.20. Let S be an ordered semigroup, B a boolean algebra, and $f : S \to B$ an isotone epimorphism. Prove that f is residuated if and only if Ker f has a top element that is residuated.

12.21. Prove that an ordered semigroup S admits a boolean algebra as an image under a residuated epimorphism if and only if it admits an equiresidual top element t such that
$$(\forall x \in S) \quad t:x = t:x^2.$$

Ordered regular semigroups

13.1 Regular Dubreil-Jacotin semigroups

In the algebraic theory of semigroups an important part is played by those that are regular. In what follows we shall concentrate on this class. If S is a regular semigroup and $x \in S$ then we shall denote the set of inverses of x by $V(x)$, and the subset of idempotents of S by $E(S)$.

We begin by describing the anticones in an ordered regular semigroup. For this purpose, we require the following terminology.

Definition If S is an ordered regular semigroup then a subsemigroup H of S is said to be

    (a) **full** if $E(S) \subseteq H$;
    (b) **idempotent-closed** if $(\forall e \in E(S))\ H = H:\{e\}$;
    (c) **self-conjugate** if $(\forall x \in S)(\forall x' \in V(x))\ x'Hx \subseteq H$.

Theorem 13.1 *Let S be an ordered regular semigroup. Then a subset H of S is an anticone if and only if H is a down-set that is a full idempotent-closed self-conjugate subsemigroup.*

Proof $\Rightarrow$: Suppose that H is an anticone of S. In the group quotient S/R_H let the R_H-class of $x \in S$ be denoted by $[x]$, and let the identity class be denoted by I_H. If $e \in E(S)$ then clearly $[e] = I_H$ whence $e \in \natural_H^\leftarrow(N_{S/R_H}) = H$ and so H is a full subsemigroup. Since H is full, it is clear that $H \subseteq H:\{e\}$ for every $e \in E(S)$. Conversely, if $x \in H:\{e\}$ for some $e \in E(S)$ then $ex = h \in H$ and on passing to quotients we obtain $[x] = I_H[x] = [e][x] = [ex] = [h] \leqslant_H I_H$ whence $x \in \natural_H^\leftarrow(N_{S/R_H}) = H$. Thus H is idempotent-closed. Finally, if $h \in H$ then for all $x \in S$ and all $x' \in V(x)$ we have

$$[x'hx] = [x]^{-1}[h][x] \leqslant_H [x]^{-1}I_H[x] = I_H$$

so that $x'hx \in H$ and consequently H is self-conjugate.

$\Leftarrow$: Conversely, suppose that H is a down-set of S that is a full idempotent-closed self-conjugate subsemigroup. If $xy \in H$ then since H is self-conjugate we have $x'xyx \in H$ with $x'x \in E(S)$. Since H is idempotent-closed, it follows that $yx \in H$. Thus H is reflexive. Since $x'x \in E(S) \subseteq H$ we have $H : \{x\} \neq \emptyset$ and so H is also neat.

It remains to show that I_H is reflexive and neat. Since H is idempotent-closed we have $E(S) \subseteq I_H$. Thus for every $x \in S$ we have $x'x, xx' \in I_H$ whence I_H is neat. Finally, suppose that $xy \in I_H$, so that $H = H : \{xy\}$. Then

$$H : \{y'\} = H : \{xyy'\} = (H : \{yy'\})^{\cdot} . \{x\} = H : \{x\},$$

whence we obtain

$$H : \{yx\} = (H : \{x\})^{\cdot} . \{y\} = (H : \{y'\})^{\cdot} . \{y\} = H : \{yy'\} = H.$$

Thus $yx \in I_H$ and consequently I_H is reflexive. $\qquad\square$

Using the above description of the anticones, we can deduce for ordered regular semigroups the following version of Theorem 12.1.

Theorem 13.2 *Let S be an ordered regular semigroup. If H is a down-set of S that is a full idempotent-closed self-conjugate subsemigroup then the relation defined by*

$$(x, y) \in \beta_H \iff (\exists y' \in V(y)) \; xy' \in I_H$$

is a congruence on S and S/β_H is an isotone epimorphic group image of S. Moreover, every isotone epimorphic group image of S arises in this way.

Proof In view of Theorem 13.1, it suffices to show that the Dubreil equivalence R_H coincides with the relation β_H.

Now, on the one hand, we have

$$(x, y) \in R_H \Rightarrow H : \{x\} = H : \{y\}$$
$$\Rightarrow (\forall y' \in V(y)) \; H : \{xy'\} = H : \{yy'\} = H$$
$$\Rightarrow (\forall y' \in V(y)) \; xy' \in I_H.$$

On the other hand, if there exists $y' \in V(y)$ such that $xy' \in I_H$ then, on passing to quotients, we obtain $[x][y]^{-1} = I_H$ whence $[x] = [y]$ and consequently $(x, y) \in R_H$. $\qquad\square$

Remark From the proof of Theorem 13.2 we see that the quantifier $\exists$ can be replaced by $\forall$.

We can now characterise the regular Dubreil-Jacotin semigroups.

Theorem 13.3 *In a regular Dubreil-Jacotin semigroup ξ is the biggest idempotent.*

Proof Let $\xi' \in V(\xi)$. Now ξ is the top element of the identity class in the quotient S/A_ξ and so $\xi^2 \in [\xi]$ and $\xi' \in [\xi]$ give $\xi^2 \leqslant \xi$ and $\xi' \leqslant \xi$. Consequently $\xi = \xi\xi'\xi \leqslant \xi^3 \leqslant \xi^2 \leqslant \xi$ from which it follows that $\xi^2 = \xi$. $\qquad\square$

Theorem 13.4 *An ordered regular semigroup is a Dubreil-Jacotin semigroup if and only if*

(1) *there exists $\alpha = \max\{x \in S \mid x^2 \leqslant x\}$ and α is idempotent;*
(2) *α is equi-quasiresidual;*
(3) *for every $e \in E(S)$ the residual $\alpha : e$ exists.*

Proof $\Rightarrow$: (1) If $x \in S$ is such that $x^2 \leqslant x$ then $x \in \langle x .\,{}^{\cdot} x \rangle \subseteq)S(= \xi^{\downarrow}$. It follows by Theorem 13.3 that $\xi = \max\{x \in S \mid x^2 \leqslant x\}$.

(2) That ξ is equi-quasiresidual follows from Theorem 12.5.

(3) Since ξ is idempotent we have $\xi \in \langle \xi : \xi \rangle \subseteq)S(= \xi^{\downarrow}$ whence $\langle \xi : \xi \rangle = \xi^{\downarrow}$ and hence $\xi : \xi$ exists and is ξ. Since for every $e \in E(S)$ we have $\langle \xi : e \rangle = \langle \xi : \xi \rangle$ it follows that $\xi : e$ exists and is ξ.

$\Leftarrow$: Conversely, suppose that S satisfies properties $(1), (2), (3)$. Then we note first that $\alpha : e = \alpha$ for every $e \in E(S)$. In fact, using (3) we have $\alpha(\alpha : e)e \leqslant \alpha^2 = \alpha$ so $\alpha(\alpha : e) \leqslant \alpha : e$ whence $e(\alpha : e)^2 \leqslant \alpha(\alpha : e) \leqslant \alpha : e$ which gives $e(\alpha : e)^2 = e^2(\alpha : e)^2 \leqslant \alpha$ whence $(\alpha : e)^2 \leqslant \alpha : e$ and therefore, by (1), $\alpha : e \leqslant \alpha$. But by (1) we have $e \leqslant \alpha$ which gives $e\alpha \leqslant \alpha^2 = \alpha$ so that $\alpha \leqslant \alpha : e$. Hence we have equality.

Using (2) we can consider the Dubreil equivalence $R_{\alpha^{\downarrow}} = \mathcal{A}_{\alpha}$ given by

$$(x, y) \in \mathcal{A}_{\alpha} \iff \langle \alpha : x \rangle = \langle \alpha : y \rangle.$$

Observing that

$$t \in \langle \alpha : xz \rangle \iff xzt \leqslant \alpha \iff zt \in \langle \alpha : x \rangle,$$

we see immediately that if $(x, y) \in \mathcal{A}_{\alpha}$ then $(xz, yz) \in \mathcal{A}_{\alpha}$ for all $z \in S$; and similarly $(zx, zy) \in \mathcal{A}_{\alpha}$. Thus $\mathcal{A}_{\alpha}$ is a congruence on S.

Since, as seen above, $\alpha : e = \alpha$ for every $e \in E(S)$ it follows that $S/\mathcal{A}_{\alpha}$ is a monoid, the identity element of which is

$$I = [\alpha]_{\mathcal{A}_{\alpha}} = \{x \in S \mid \langle \alpha : x \rangle = \langle \alpha : \alpha \rangle\}$$
$$= \{x \in S \mid \alpha : x \text{ exists and is } \alpha\}.$$

Consider now the principal down-set $\alpha^{\downarrow}$. Since for $x \in I$ we have $\alpha = \alpha : x$, so that $x \leqslant \alpha : \alpha = \alpha$, we see that α is the top element of I. Then since every idempotent belongs to I we have that $\alpha^{\downarrow}$ is a full subsemigroup of S. It is also idempotent-closed since

$$x \in \alpha^{\downarrow} : \{e\} \iff x \in \langle \alpha : e \rangle \iff x \leqslant \alpha.$$

Finally to see that $\alpha^{\downarrow}$ is self-conjugate, let $x \leqslant \alpha$ and let $y \in S$. Observing that in $S/\mathcal{A}_{\alpha}$ we have

$$[x] \leqslant [\alpha] \iff \alpha \in \langle \alpha : \alpha \rangle \subseteq \langle \alpha : x \rangle \iff x \leqslant \alpha : \alpha = \alpha,$$

we have, for every $y' \in V(y)$,

$$[y'xy] = [y'][x][y] \leqslant [y'][\alpha][y] = [y']I[y] = [y'y] = I = [\alpha]$$

whence $y'xy \leqslant \alpha$. It now follows by Theorem 13.1 that $\alpha^{\downarrow}$ is an anticone of S. Since it is principal it follows that S is a Dubreil-Jacotin semigroup. $\square$

Corollary *An ordered regular semigroup is a strong Dubreil-Jacotin semigroup if and only if*

(1) *there exists $\alpha = \max\{x \in S \mid x^2 \leqslant x\}$ and α is idempotent;*

(2) *α is equiresidual.* $\square$

EXERCISES

13.1. Let L be a semilattice with a top element 1_L, let B be an ordered rectangular band, and let G be an ordered group. Consider the cartesian ordered semigroup $S = L \times B \times G$. Prove that S is a strong Dubreil-Jacotin regular semigroup.

13.2. Let G be a discretely ordered group. Adjoin to G an element z and add the single relation $z < 1_G$. Extend the multiplication of G to $S = G \cup \{z\}$ by defining $z^2 = z$ and $(\forall x \in G)\ xz = x = zx$. Show that S is a strong Dubreil-Jacotin inverse semigroup.

13.2 The Nambooripad order

In the algebraic theory of semigroups a fundamental role is played by equivalences that are known as **Green's relations**. In the case of a regular semigroup S these are defined on by

$$(x, y) \in \mathcal{L} \iff Sx = Sy, \quad (x, y) \in \mathcal{R} \iff xS = yS.$$

For every $x \in S$ and every $x' \in V(x)$, the idempotent $x'x$ belongs to the $\mathcal{L}$-class L_x of x; and the idempotent xx' belongs to the $\mathcal{R}$-class R_x of x. Every idempotent in L_x acts as a right identity in L_x; and every idempotent in R_x acts as a left identity in R_x.

We now use these relations in establishing an important order relation that can be defined on every regular semigroup.

Theorem 13.5 *In a regular semigroup S the relations*

(1) $x \in yS$ *and* $x = ey$ *for some* $e \in E(S) \cap R_x$;

(1′) $x \in Sy$ *and* $x = yf$ *for some* $f \in E(S) \cap L_x$;

(2) $x \in yS$ *and* $x = ey$ *for some* $e \in E(S)$;

(2′) $x \in Sy$ *and* $x = yf$ *for some* $f \in E(S)$;

(3) $x \in yS \cap Sy$ *and* $x = xy'x$ *for all* $y' \in V(y)$

are equivalent and define an order on S.

Proof (1) $\Rightarrow$ (2): This is clear.

(2) $\Rightarrow$ (3): If (2) holds then $x \in yS \cap Sy$. Let $z \in S$ be such that $x = yz$. Then if $y' \in V(y)$ we have $xy'x = ey \cdot y' \cdot yz = eyz = ex = ey = x$.

(3) $\Rightarrow$ (1): If (3) holds then clearly $x \in yS$ and $xy' \in E(S)$. Then $x = xy'x \in xy'S$ whence $xS \subseteq xy'S \subseteq xS$. Thus $xy' \in R_x$ and we have (1).

Thus we see that (1), (2), (3) are equivalent; and dually (1′), (2′), (3) are equivalent.

Now let $\leqslant$ denote the relation (1). We show as follows that $\leqslant$ is an order on S. That $\leqslant$ is reflexive follows from the identity $x = xx'x$. That it is anti-symmetric follows from the fact that if $x \leqslant y$ and $y \leqslant x$ then $x \in yS$ gives $xS \subseteq yS$, and $y \in xS$ gives $yS \subseteq xS$, so that $R_x = R_y$. Since every idempotent in R_x acts as a left identity element in R_x, we then have $x = ey = y$. Finally, $\leqslant$ is transitive. To see this, let $x \leqslant y$ and $y \leqslant z$. Then $xS \subseteq yS$ and there is an idempotent e such that $x = ey$ and $eS = xS$. Likewise $yS \subseteq zS$ and there is an idempotent f such that $y = fz$ and $fS = yS$. Thus $eS \subseteq fS$ and so $e = fe = efe$ which gives ef idempotent with $ef \in R_e = R_x$. Since $x = ey = efz$ it follows that $x \leqslant z$ as required.   □

Definition The order described by any of the above equivalent relations is called the **Nambooripad order** (or the **natural order**) on the regular semigroup S and will henceforth be denoted by $\leqslant_n$.

Because (2) and (2′) in the above are equivalent, they are each equivalent to their conjunction. Consequently we see that the Nambooripad order on a regular semigroup S can be described in the following useful way:

$$x \leqslant_n y \iff (\exists e, f \in E(S)) \quad x = ey = yf.$$

We note here that when S is an inverse semigroup condition (3) of Theorem 13.5 is condition (γ) of Example 11.7, so that $\leqslant_n$ coincides with the order defined there, and justifies the use of the same notation. Note also that, when S is inverse, $\leqslant_n$ takes on the simpler form obtained in Theorem 11.1.

We shall require the following property of the Nambooripad order.

Theorem 13.6 *Let S be a regular semigroup and let $a, b \in S$ be such that $a \leqslant_n b$. Then*

(1) $(\forall b' \in V(b))(\exists a' \in V(a))$ $ab' = aa' \leqslant_n bb'$ *and* $a = aa'b$;
(2) $(\forall b' \in V(b))(\exists a' \in V(a))$ $b'a = a'a \leqslant_n b'b$ *and* $a = ba'a$.

Proof (1) Let $e, f \in E(S)$ be such that $a = eb = bf$. Then $ab' \cdot ab' = ebb' \cdot bfb' = ebfb' = afb' = ab'$ so that $ab' \in E \cap R_a$. Thus $b'ab' \in V(a)$ and if we define $a' = b'ab'$ then we have $aa' = ab'$. Now since $ab' \cdot bb' = ab'$ and $bb' \cdot ab' = bb'a \cdot b' = ab'$ we also have $ab' = aa' \leqslant_n bb'$. Finally, $aa'b = ab'b = a$.
(2) is established similarly.   □

As the following example shows, the Nambooripad order on a regular semigroup is not in general compatible with multiplication.

Example 13.1 Consider the semigroup $\mathrm{Mat}_n\,\mathcal{B}$ of $n \times n$ matrices with entries in the boolean algebra $\mathcal{B}$. Here we suppose that $n \geqslant 2$. That $\mathrm{Mat}_2\,\mathcal{B}$ is regular follows from the easily verified observation that if $A = \begin{bmatrix} a & b \\ c & d \end{bmatrix}$ then $A = AXA$ where $X = \begin{bmatrix} ad + b' + c' & bc + a' + d' \\ bc + a' + d' & ad + b' + c' \end{bmatrix}$. In contrast, in $\mathrm{Mat}_3\,\mathcal{B}$ it is readily seen that for the matrix $I_3' = \begin{bmatrix} 0 & 1 & 1 \\ 1 & 0 & 1 \\ 1 & 1 & 0 \end{bmatrix}$ there is no $X \in \mathrm{Mat}_3\,\mathcal{B}$ such that $I_3'XI_3' = I_3'$; and more generally, for $n > 3$, the same is true for the matrix $M = \begin{bmatrix} I_3' & 0 \\ 0 & I_{n-3} \end{bmatrix}$. Hence we see that $\mathrm{Mat}_n\,\mathcal{B}$ is regular if and only if $n = 2$.

Consider now the idempotent matrices $X = \begin{bmatrix} 0 & 0 \\ 1 & 1 \end{bmatrix}$ and $Y = \begin{bmatrix} 0 & 0 \\ 0 & 1 \end{bmatrix}$. Clearly, $X \leqslant_n I_2$. Now $YX = X$ and $XY = Y$ so that $X \nleqslant_n Y$. Thus $YX \nleqslant_n YI_2$ and so $\leqslant_n$ is not compatible on the left with multiplication.

Our objective now is to determine precisely when the Nambooripad order on a regular semigroup is compatible with multiplication. For this purpose we require the following facts (see any standard text on semigroups).

If S is a regular semigroup and $e, f \in E(S)$ then the **sandwich set** of the ordered pair (e, f) is defined by

$$S(e, f) = \{g \in E(S) \mid ge = g = fg, \; egf = ef\} = f \cdot V(ef) \cdot e.$$

A **rectangular band** is the cartesian product semigroup $L \times R$ where L is a left zero semigroup $[xy = x]$ and R is a right zero semigroup $[xy = y]$. A rectangular band is characterised by the identity $xyx = x$.

The principal properties of sandwich sets are the following, all of which are easily derived from the above definition.

Theorem 13.7 *If S is a regular semigroup and $e, f \in E(S)$ then*

(1) $S(e, f)$ *is a rectangular band;*

(2) $\begin{cases} (e, f) \in \mathcal{L} \implies (\forall g \in E(S)) \; S(e, g) = S(f, g), \\ (e, f) \in \mathcal{R} \implies (\forall g \in E(S)) \; S(g, e) = S(g, f); \end{cases}$

(3) *if $a' \in V(a)$, $b' \in V(b)$ and $g \in S(a'a, bb')$ then $b'ga' \in V(ab)$.* $\qquad\square$

If S is a regular semigroup and $e \in E(S)$ then eSe is a subsemigroup that is also regular [for if $x \in eSe$ and $x' \in V(x)$ then $ex'e \in V(x) \cap eSe$], and has e as an identity. We call eSe a **local submonoid** of S.

A regular semigroup S is said to be **$\mathcal{L}$-unipotent** if every $\mathcal{L}$-class of S contains a unique idempotent; equivalently, if $E(S)$ is a **right regular band** $[efe = fe]$.

We say that S is **locally $\mathcal{L}$-unipotent** if every local submonoid of S is $\mathcal{L}$-unipotent.

Theorem 13.8 (Blyth–Gomes [22]) *Let S be a regular semigroup. Then the following statements are equivalent:*

(1) $\leqslant_n$ *is compatible on the right with multiplication;*

(2) $(\forall e, f \in E(S)) \; S(e, f)$ *is a right zero semigroup;*

(3) S *is locally $\mathcal{L}$-unipotent.*

Proof (1) $\Rightarrow$ (2): If $g \in S(e, f)$ then $gf \in E(S)$ with $gf \leqslant_n f$. If also $h \in S(e, f)$ then by (1) we have $gh = gfh \leqslant_n fh = h$ whence, $S(e, f)$ being a rectangular band, $gh = hgh = h$. Consequently $S(e, f)$ is a right zero semigroup.

(2) $\Rightarrow$ (3): Given $e \in E(S)$, let $a \in eSe$ and $a' \in V(a) \cap eSe$. Then clearly $a'a \in S(a'a, e)$. If also $a'' \in V(a) \cap eSe$ then, by Theorem 13.7(2), $a'a$ and $a''a$ belong to $S(a'a, e) = S(a''a, e)$ which by (2) is a right zero semigroup. Thus we have $a'a = a''a \cdot a'a = a'' \cdot aa'a = a''a$ and consequently the local submonoid eSe is $\mathcal{L}$-unipotent.

(3) $\Rightarrow$ (1): Let $a \leqslant_n b$ and let $c \in S$. Choose $a' \in V(a)$ such that $a = aa'b$ and let $c' \in V(c)$ and $g \in S(a'a, cc')$. Then we have

$$ac \cdot c'ga' \cdot bc = aga'bc = aga'aa'bc = aga'ac = agc = ac$$

and so $ac = e \cdot bc$ where $e = ac \cdot c'ga' \in E(S)$ since, by Theorem 13.7(3), we have $c'ga' \in V(ac)$.

Now let $b' \in V(b)$ and, using Theorem 13.6, choose $a'' \in V(a)$ such that $a''a \leqslant_n b'b$ and $a = ba''a$. Let $c'' \in V(c)$ and $h \in S(a''a, cc'')$. Then we note first that $a''ah \in E(S)$; and that

$$b'bh \cdot b'bh = b'bha''ab'bh = b'bha''ah = b'bh,$$

so that $b'bh \in E(S)$. Now

$$a''ah \cdot b'bh = a''aha''ab'bh = a''aha''ah = a''ah, \quad b'bh \cdot a''ah = b'bh,$$

whence $(a''ah, b'bh) \in \mathcal{L}$. Since $h = ha''a$ and $a''a \leqslant_n b'b$ we have that $a''ah$ and $b'bh$ are $\mathcal{L}$-related idempotents in the local submonoid $b'bSb'b$. It follows by (3) that $a''ah = b'bh$. Consequently,

$$bc \cdot c''ha'' \cdot ac = bhc = bb'bhc = ba''ahc = ahc = ac$$

so that $ac = bc \cdot f$ where $f = c''ha'' \cdot ac \in E(S)$ since $c''ha'' \in V(ac)$.

The above observations give

$$ac = e \cdot bc = bc \cdot f \quad \text{where} \quad e, f \in E(S)$$

whence $ac \leqslant_n bc$ and we have (1).   □

If a regular semigroup S is locally $\mathcal{L}$-unipotent and locally $\mathcal{R}$-unipotent then every local submonoid eSe is such that every $\mathcal{L}$-class and every $\mathcal{R}$-class contains a unique idempotent. Thus every local submonoid eSe is an inverse semigroup. In this case we say that S is **locally inverse**.

A combination of Theorem 13.8 and its dual gives the following fundamental result, which provides an answer to the question of precisely when a regular semigroup S is such that $(S; \leqslant_n)$ is an ordered semigroup.

Theorem 13.9 (Nambooripad [86]) *Let S be a regular semigroup. Then the following statements are equivalent*:

(1) $\leqslant_n$ *is compatible with multiplication*;
(2) $(\forall e, f \in E(S))$ $S(e, f)$ *is a singleton*;
(3) S *is locally inverse*.   □

EXERCISES

13.3. A band is **right quasi-normal** if it satisfies the identity $yxa = yax$ (i.e. if $x \mapsto xa$ is a morphism). Prove that if S is a regular semigroup then the following are equivalent:

(1) $\leqslant_n$ is compatible on the left with multiplication;
(2) $(\forall e \in E(S))$ $E(S) \cap eS$ is a right regular band and $E(S) \cap Se$ is a right quasi-normal band;
(3) $(\forall e, f \in E(S))$ $E(S) \cap eSf$ is a right regular band.

13.4. A band is **right normal** if it satisfies the identity $yxa = xya$, and **left normal** if it satisfies the identity $axy = ayx$. Prove that if S is a regular semigroup then the following are equivalent:

(1) $\leqslant_n$ is compatible with multiplication;
(2) $(\forall e \in E(S))$ $E(S) \cap eS$ is a right normal band and $E(S) \cap Se$ is a left normal band;
(3) $(\forall e, f \in E(S))$ $E(S) \cap eSf$ is a semilattice.

13.3 Natural orders on regular semigroups

Definition An ordered regular semigroup $(S; \leqslant)$ is said to be **naturally ordered** if the order $\leqslant$ extends the Nambooripad order $\leqslant_n$ on the set $E(S)$ of idempotents of S, in the sense that

$$(\forall e, f \in E(S)) \quad e \leqslant_n f \Rightarrow e \leqslant f.$$

Example 13.2 The regular semigroup $\mathrm{Mat}_2\, \mathcal{B}$ of Example 13.1 is not naturally ordered since, for the idempotent X, we have $X \leqslant_n I_2$ but $X \nleqslant I_2$.

Example 13.3 The cartesian ordered semigroup S of Example 12.7 is regular. To see this it suffices to observe that $(x, y)(x, -y)(x, y) = (x, y)$. The idempotents of S are the pairs (x, y) where $y_k = 0$, i.e. $0 \leqslant y \leqslant k - 1$. If (x, y) and (a, b) are idempotents then it is readily seen that $(x, y) \leqslant_n (a, b)$ if and only if $x \leqslant a$ and $y = b$, whence $(x, y) \leqslant (a, b)$. Thus $(S; \leqslant)$ is naturally ordered.

Theorem 13.10 (McAlister [81]) *A regular semigroup can be naturally ordered if and only if it is locally inverse.*

Proof Suppose that $\leqslant$ is a compatible order on S such that $(S; \leqslant)$ is naturally ordered. Let $e \in E(S)$ and consider the subsemigroup eSe. Let u, v be $\mathcal{R}$-equivalent idempotents in eSe. Then $u = eue \leqslant_n e$ and $v = eve \leqslant_n e$ so that, by the hypothesis, $u \leqslant e$ and $v \leqslant e$. But $(u, v) \in \mathcal{R}$ implies that $u = vu \leqslant ve = v$ and $v = uv \leqslant ue = u$ so that $u = v$. Hence eSe is $\mathcal{R}$-unipotent. Similarly, it is $\mathcal{L}$-unipotent. Hence S is locally inverse.

The converse is immediate from Theorem 13.9.   □

A simple but useful property of ordered regular semigroups is the following.

Theorem 13.11 *Let $(S; \leqslant)$ be an ordered regular semigroup. If e, f are idempotents in S such that $e \leqslant f$ then the products ef, fe, efe, fef are all idempotent. If, moreover, $(S; \leqslant)$ is naturally ordered then $e = efe$.*

Proof That ef is idempotent follows from the fact that

$$efef \leqslant efff = ef = eeef \leqslant efef.$$

Similarly, fe is idempotent. Since $e = ee \leqslant ef$ it follows that efe and fef are also idempotent.

As for the final statement, we have $efe \leqslant_n e$ and so, since $\leqslant$ extends $\leqslant_n$ on the idempotents, $efe \leqslant e = eee \leqslant efe$ whence we obtain $e = efe$.   □

Our next objective is to investigate the naturally ordered Dubreil-Jacotin regular semigroups. Now we have seen in Theorem 13.3 that such a semigroup contains a biggest idempotent ξ. The following result highlights a particular situation in which natural ordering occurs.

Theorem 13.12 *Let S be a regular Dubreil-Jacotin semigroup. Then the following conditions are equivalent:*

(1) *ξ is an identity element;*
(2) *S is an inverse semigroup and on the semilattice of idempotents the order coincides with the natural order.*

Proof (1) $\Rightarrow$ (2): Suppose that ξ is the identity element 1. Let $x \in [1]_{A_1}$ and let $x' \in V(x)$. Then $x' \in [1]_{A_1}$ and therefore $x = xx'x \leqslant x1x = x^2 \leqslant x1 = x$ and so x is idempotent. Thus the set of idempotents is the subsemigroup $[1]_{A_1}$. Now if e, f are idempotents then $ef \leqslant e1 = e$ and $ef \leqslant 1f = f$, while if x is an idempotent such that $x \leqslant e$ and $x \leqslant f$ then $x = x^2 \leqslant ef$. Thus ef is the infimum of e and f in $[1]_{A_1}$. Similarly, so is fe and therefore $ef = fe$. Since idempotents commute, S is then an inverse semigroup.

If now e and f are idempotents then $e \leqslant f$ gives $e = e^2 \leqslant ef \leqslant e1 = e$ and $e = e^2 \leqslant fe \leqslant 1e = e$, whence $e = ef = fe$ and so $e \leqslant_n f$. Conversely, if $e \leqslant_n f$ then $e = ef \leqslant 1f = f$. Thus the order of S, restricted to the semilattice of idempotents, coincides with the natural order.

(2) $\Rightarrow$ (1): Every idempotent of S belongs to the identity class $[\xi]_{A_\xi}$ and so if (2) holds then $x^{-1}x \leqslant_n \xi$ for every $x \in S$. Thus $x^{-1}x = x^{-1}x\xi$ whence $x = xx^{-1}x = xx^{-1}x\xi = x\xi$. Similarly, from $xx^{-1} \leqslant_n \xi$ we obtain $x = \xi x$. Thus ξ is the identity element.   □

Definition By an **integral** Dubreil-Jacotin semigroup we shall mean a strong Dubreil-Jacotin semigroup in which ξ is an identity element.

It is immediate from the above that if S is an integral Dubreil-Jacotin inverse semigroup then on the semilattice of idempotents of S the order coincides with the natural order.

Theorem 13.13 *If S is an integral Dubreil-Jacotin inverse semigroup then for every $x \in S$ we have*

(1) $x = xx^{-1}[1:(1:x)]$;
(2) $x = x(1:x)x$;
(3) $x^{-1} = (1:x)x(1:x)$;
(4) $xx^{-1} = x(1:x)$, $x^{-1}x = (1:x)x$;
(5) $x = x(1:x)(1:x)^{-1}$.

Proof (1) Since xx^{-1} is idempotent we have $xx^{-1} \leqslant 1$ and so $x^{-1} \leqslant 1:x$ whence $1:x^{-1} \geqslant 1:(1:x)$. Then

$$x = xx^{-1}x \leqslant xx^{-1}[1:(1:x)] \leqslant xx^{-1}(1:x^{-1}) \leqslant x1 = x$$

whence we have (1).

(2) We have $x = xx^{-1}x \leqslant x(1:x)x \leqslant x1 \leqslant x$, whence equality.

(3) By (2), $x(1:x)$ is idempotent whence $(1:x)x(1:x) \in V(x)$ and (3) follows by the uniqueness of inverses.

(4) This follows from (2) and (3).

(5) $x \overset{(1)}{=} xx^{-1}[1:(1:x)] \overset{(4)}{=} x(1:x)[1:(1:x)] \overset{(4)}{=} x(1:x)(1:x)^{-1}$.   □

Corollary $E(S) = [1]_{A_1}$.

Proof Clearly, $E \subseteq [1]_{A_1}$. Conversely, if $x \in [1]_{A_1}$ then $1:x = 1:1 = 1$ and so, by (1) above, we have $x = xx^{-1} \in E(S)$.   □

Theorem 13.14 *If S is an integral Dubreil-Jacotin inverse semigroup and if G is the set of A-nomal elements then*

$$x \in G \iff x^{-1} \in G \iff x = 1:x^{-1} \iff x^{-1} = 1:x.$$

Proof If $x \in G$ then $x = 1:(1:x)$ and so

$$\begin{aligned} x &= x(1:x)(1:x)^{-1} \quad \text{by Theorem 13.13(5)} \\ &= [1:(1:x)](1:x)(1:x)^{-1} \\ &= (1:x)^{-1}(1:x)(1:x)^{-1} \quad \text{by Theorem 13.13(4)} \\ &= (1:x)^{-1} \end{aligned}$$

whence $x^{-1} = 1:x \in G$. For the converse, replace x by x^{-1}.   □

Theorem 13.15 *If S is an integral Dubreil-Jacotin inverse semigroup then the natural order on S is given by*

$$x \leqslant_n y \iff x \leqslant y, \quad (x, y) \in A_1.$$

Moreover, for every $x \in S$,

$$[x]_{A_1} = E(S)\,[1:(1:x)].$$

Proof Suppose first that $x \leqslant y$ with $(x, y) \in A_1$. Then $1:x = 1:y$ and, by Theorem 13.13(4), $xx^{-1} = x(1:x) \leqslant y(1:y) = yy^{-1}$ so that $xx^{-1} \leqslant_n yy^{-1}$. Thus $xx^{-1} = xx^{-1}yy^{-1}$ and it follows by Theorem 13.13(1) that

$$x = xx^{-1}[1:(1:x)] = xx^{-1}yy^{-1}[1:(1:y)] = xx^{-1}y$$

whence we have $x \leqslant_n y$. Conversely, if $x \leqslant_n y$ then $x = xx^{-1}y$ from which we deduce on the one hand that $x \leqslant 1y = y$ and on the other, on passing to quotients, that $(x, y) \in A_1$.

Note that we have shown that

$$x \in [y]_{A_1} \cap y^\downarrow \Rightarrow x = xx^{-1}y.$$

Consequently, $[y]_{A_1} \cap y^\downarrow \subseteq E(S)\,y$. But since 1 is the top element of $E(S)$ the reverse inclusion holds. Hence we have

$$(\forall x \in S) \quad E(S)\,x = [x]_{A_1} \cap x^\downarrow.$$

It follows that $E(S)\,[1:(1:x)] = [1:(1:x)]_{A_1} = [x]_{A_1}$.   □

Corollary *If Z_n denotes the zig-zag equivalence relative to the natural order on S then $Z_n = A_1$.*

Proof By Theorem 13.15 we have $Z_n = Z \cap A_1 = A_1$.   □

If S is an ordered inverse semigroup then for every $x \in S$ we can define an order on the quotient set S/R_x by

$$[x]_{R_x} \leqslant [y]_{R_x} \iff xx^{-1} \leqslant yy^{-1},$$

and similarly on R/L_x. It follows that Green's relations $\mathcal{R}$ and $\mathcal{L}$ on S are regular if and only if

$$x \leqslant y \Rightarrow xx^{-1} \leqslant yy^{-1}, \quad x^{-1}x \leqslant y^{-1}y.$$

In the final chapter we shall obtain a structure theorem for integral Dubreil-Jacotin inverse semigroups on which $\mathcal{R}$ and $\mathcal{L}$ are regular.

EXERCISES

13.5. Let S be an integral Dubreil-Jacotin inverse semigroup. Prove that the following statements are equivalent:

    (1) the order coincides with the natural order;
    (2) the set of A-nomal elements is discretely ordered;
    (3) $(\forall x \in S)\ x \leqslant y \Rightarrow x^{-1} \leqslant y^{-1}$.

13.6. Let S be an integral Dubreil-Jacotin inverse semigroup. Prove that $\mathcal{R}$ and $\mathcal{L}$ are regular on S if and only if they have convex classes and the set of A-nomal elements in the identity class modulo Z is a subgroup of S.

13.7. In the integral Dubreil-Jacotin inverse semigroup D of Exercise 12.8 show that Green's relations are not regular, but that their restrictions to the subsemigroup $D^{*} = \{a_{i,-p} \mid i \in \mathbb{Z}, p \in \mathbb{N}\}$ are regular.

In order to investigate the more general situation in which ξ need not be an identity element, we shall first consider the wider class of ordered regular semigroups that have a biggest idempotent. For this purpose we require to identify certain particular types of idempotent. In so doing we shall denote by $\langle E(S) \rangle$, or simply $\overline{E}$, the subsemigroup generated by the idempotents of S (so that $x \in \overline{E}$ if and only if $x = e_1 e_2 \cdots e_n$ where each $e_i \in E(S)$).

Definition Let S be a regular semigroup. Then $\alpha \in E(S)$ is said to be

    (1) a **weak middle unit** if $(\forall x \in S)(\forall x' \in V(x))\ x\alpha x' = xx'$;
    (2) **medial** if $(\forall \overline{e} \in \overline{E})\ \overline{e}\alpha\overline{e} = \overline{e}$;
    (3) **normal** if $\alpha \overline{E} \alpha$ is a semilattice.

Definition By an **inverse transversal** of a regular semigroup S we mean an inverse subsemigroup T that contains exactly one inverse of each element of S; i.e. $(\forall x \in S)\ |T \cap V(x)| = 1$.

The following results show how the above concepts are related in the context of an ordered regular semigroup that has a biggest idempotent.

Theorem 13.16 (Blyth–Santos [30], Blyth–McFadden [25], McAlister [81]) *Let S be an ordered regular semigroup with a biggest idempotent α. Then the following statements are equivalent:*

    (1) *S is naturally ordered;*
    (2) *α is a normal medial idempotent;*
    (3) *α is a weak middle unit;*
    (4) *$\alpha S\alpha$ is an inverse transversal of S;*
    (5) *every $x \in S$ has a biggest inverse x° and $\alpha x^{\circ} = x^{\circ} = x^{\circ}\alpha$;*
    (6) *every $e \in E(S)$ has a biggest inverse e° and $\alpha e^{\circ} = e^{\circ} = e^{\circ}\alpha$;*
    (7) *$(\forall e \in E(S))\ \inf_{E(S)}\{e\alpha, \alpha e\}$ exists and is e.*

Proof We establish the result by showing that

$$(5) \Rightarrow (6) \Rightarrow (3) \Rightarrow (7) \Rightarrow (1) \Rightarrow (4) \Rightarrow (5) \text{ and } (5) \Rightarrow (2) \Rightarrow (1).$$

$(5) \Rightarrow (6)$: This is clear.

$(6) \Rightarrow (3)$: If (6) holds then for every $e \in E(S)$ we have

$$e^\circ = e^\circ ee^\circ = e^\circ eeee^\circ \leqslant \alpha e\alpha \leqslant \alpha e^\circ \alpha = e^\circ$$

and so $e^\circ = \alpha e\alpha$. Then $e = ee^\circ e = e\alpha e\alpha e$. But

$$e\alpha e = e\alpha eeee \leqslant e\alpha e \cdot e\alpha e \leqslant e\alpha\alpha\alpha\alpha = e\alpha e$$

and so $e\alpha e \in E(S)$. Hence $e = e\alpha e$. Taking $e = x'x$ we obtain $x'x = x'x\alpha x'x$ whence, pre-multiplying by x and post-multiplying by x', we obtain $xx' = x\alpha x'$. Thus α is a weak middle unit.

$(3) \Rightarrow (7)$: If (3) holds then taking $x = e \in E(S)$ and using the fact that $e \in V(e)$ we obtain $e = e\alpha e$. If now $f \in E(S)$ is such that $f \leqslant e\alpha$ and $f \leqslant \alpha e$ then $f = f^2 \leqslant e\alpha\alpha e = e$. Since $e = e^2 \leqslant e\alpha$ and $e = e^2 \leqslant \alpha e$, (7) follows.

$(7) \Rightarrow (1)$: If (7) holds and $e \leqslant_n f$ then $e = ef = fe \leqslant \alpha f \wedge f\alpha = f$. Thus S is naturally ordered.

$(1) \Rightarrow (4)$: If S is naturally ordered then, by Theorem 13.10, S is locally inverse and hence $\alpha S\alpha$ is an inverse subsemigroup. Now if $x' \in V(x)$ then $x'x \leqslant \alpha$ and Theorem 13.11 gives $x'x = x'x\alpha x'x$ whence $x = x\alpha x'x$ and then $xx' = x\alpha x'$. Likewise $x'x = x'\alpha x$, and so $\alpha x'\alpha \in \alpha S\alpha \cap V(x)$. Suppose now that $y, z \in \alpha S\alpha \cap V(x)$. Then $y = \alpha y\alpha \in V(x)$ and $z = \alpha z\alpha \in V(x)$ whence

$$y = \alpha y\alpha = \alpha yxy\alpha = \alpha yxzxy\alpha \leqslant \alpha\alpha z\alpha\alpha = z.$$

Similarly, $z \leqslant y$ and consequently we see that $|\alpha S\alpha \cap V(x)| = 1$ whence $\alpha S\alpha$ is an inverse transversal of S.

$(4) \Rightarrow (5)$: If (4) holds define x° by $\alpha S\alpha \cap V(x) = \{x^\circ\}$. Then clearly $\alpha x^\circ = x^\circ = x^\circ \alpha$. If now $x' \in V(x)$ then since $x^\circ \in V(x)$ we have

$$x' = x'xx' = x'xx^\circ xx' \leqslant \alpha x^\circ \alpha = x^\circ.$$

Hence x° is the biggest inverse of x.

$(5) \Rightarrow (2)$: If (5) holds then for every $\bar e \in \overline E$ we have, by a result of Fitz-Gerald [48], $\bar e^\circ \in \overline E$ whence $\bar e \bar e^\circ \leqslant \bar e\alpha = \bar e \bar e^\circ \bar e\alpha \leqslant \bar e \bar e^\circ \alpha\alpha = \bar e \bar e^\circ$ and therefore $\bar e \bar e^\circ = \bar e\alpha$. Consequently $\bar e = \bar e\alpha \bar e$ and so α is medial. It follows that $\alpha \bar e\alpha$ is idempotent. Thus $\alpha \overline E\alpha \subseteq E(S)$ and consequently $\alpha \overline E\alpha = \alpha E(S)\alpha$. Now since we have seen above that (5) implies (1) we have that S is locally inverse. Then $\alpha S\alpha$ is an inverse semigroup with semilattice of idempotents $\alpha E(S)\alpha = \alpha \overline E\alpha$. Hence α is also normal.

$(2) \Rightarrow (1)$: If (2) holds let $e, f, g \in E(S)$ with $f, g \in eSe$. Then we have

$$
\begin{aligned}
fg &= efeege = e\alpha efe\alpha ege\alpha e & &\alpha \text{ medial} \\
&= e \cdot \alpha ege\alpha \cdot \alpha efe\alpha \cdot e & &\alpha \text{ normal} \\
&= gf.
\end{aligned}
$$

Thus every local submonoid eSe is inverse and so S is locally inverse. □

From Theorem 13.16 we see that if S is a naturally ordered regular semigroup that has a biggest idempotent then every $x \in S$ has a biggest inverse x° which is the unique inverse of x that belongs to the local submonoid $\alpha S\alpha$. Suppose now that $x' \in V(x)$. Then since α is a weak middle unit we have $\alpha x'\alpha \in \alpha S\alpha \cap V(x)$. Thus we have

$$(\forall x' \in V(x)) \quad x^\circ = \alpha x'\alpha.$$

Writing $(x^\circ)^\circ$ as $x^{\circ\circ}$, we deduce from $x \in V(x^\circ)$ that $x^{\circ\circ} = \alpha x\alpha$. In particular, for every $e \in E(S)$ we have $e^\circ = \alpha e\alpha = e^{\circ\circ}$.

Theorem 13.17 *If S is a naturally ordered regular semigroup with a biggest idempotent α then*

(1) $(\forall x, y \in S)$ $(x\alpha y)^\circ = y^\circ x^\circ$;

(2) $(\forall x, y \in S)$ $xy = xyy^\circ x^\circ xy$;

(3) $(\forall x, y \in S)$ $(xy)^\circ = y^\circ x^\circ xyy^\circ x^\circ$.

Proof (1) Using the fact that $\alpha \overline{E} \alpha$ is a semilattice, we have

$$y^\circ x^\circ \cdot x\alpha y \cdot y^\circ x^\circ = y^\circ x^\circ \alpha x\alpha \cdot \alpha y\alpha y^\circ x^\circ$$
$$= y^\circ x^\circ x^{\circ\circ} y^{\circ\circ} y^\circ x^\circ$$
$$= y^\circ y^{\circ\circ} y^\circ x^\circ x^\circ x^{\circ\circ} x^\circ = y^\circ x^\circ.$$

Likewise, using the fact that α is a weak middle unit, we have

$$x\alpha y \cdot y^\circ x^\circ \cdot x\alpha y = x\alpha y\alpha y^\circ x^\circ \alpha x\alpha y$$
$$= xy^{\circ\circ} y^\circ x^\circ x^{\circ\circ} y$$
$$= xx^\circ x^{\circ\circ} y^{\circ\circ} y^\circ y$$
$$= xx^\circ \alpha x\alpha \alpha y\alpha y^\circ y = x\alpha y.$$

Thus we see that $y^\circ x^\circ \in V(x\alpha y)$ whence $(x\alpha y)^\circ = \alpha y^\circ x^\circ \alpha = y^\circ x^\circ$.

(2) As seen above, we have $\alpha \overline{E} \alpha \subseteq E(S)$ and so $x^\circ xyy^\circ = \alpha x^\circ xyy^\circ \alpha \in E(S)$ whence (2) follows.

(3) It follows from (2) that $y^\circ x^\circ xyy^\circ x^\circ \in V(xy)$. But if $z \in V(xy)$ then, by (2) again,

$$z = zxyz = zxyy^\circ x^\circ xyz \leqslant \alpha y^\circ x^\circ xyy^\circ x^\circ \alpha = y^\circ x^\circ xyy^\circ x^\circ$$

whence (3) follows.   □

Definition Let S be a regular semigroup. Then $\alpha \in E(S)$ is said to be

(4) a **middle unit** if $(\forall x, y \in S)$ $x\alpha y = xy$.

A regular semigroup is said to be **orthodox** if $E(S)$ is a subsemigroup of S; i.e. if $\overline{E} = E(S)$. From the algebraic theory, a regular semigroup S is orthodox if and only if it satisfies the property

$$(\forall x, y \in S) \quad V(x) \cap V(y) \neq \emptyset \;\Rightarrow\; V(x) = V(y).$$

The connection between orthodox semigroups and middle units is highlighted in the following result.

Theorem 13.18 *Let S be an ordered regular semigroup with a biggest idempotent α. Then the following statements are equivalent:*

(1) S *is naturally ordered and orthodox;*

(2) α *is a middle unit;*

(3) S *is naturally ordered and* $(\forall x, y \in S)$ $(xy)^\circ = y^\circ x^\circ$;

(4) S *is naturally ordered and* $(\forall e, f \in E(S))$ $\alpha e f \alpha = \alpha f e \alpha$.

Proof (1) $\Rightarrow$ (2): Suppose that S is naturally ordered and orthodox. By Theorem 13.16, α is a weak middle unit and so $e\alpha e = e$ for every $e \in E(S)$; and the product of any two idempotents is idempotent. Using these facts, we see that

$$\begin{cases} fe \cdot \alpha e \alpha f \alpha \cdot fe = fe\alpha fe = fe; \\ \alpha e \alpha f \alpha \cdot fe \cdot \alpha e \alpha f \alpha = \alpha e \alpha f \alpha f e \alpha f \alpha = \alpha e \alpha f e \alpha f \alpha = \alpha e \alpha f \alpha, \end{cases}$$

so that $fe \in V(\alpha e \alpha f \alpha)$. Likewise,

$$\begin{cases} fe \cdot \alpha e f \alpha \cdot fe = fe f e = fe; \\ \alpha e f \alpha \cdot fe \cdot \alpha e f \alpha = \alpha e f e f \alpha = \alpha e f \alpha, \end{cases}$$

so that also $fe \in V(\alpha e f \alpha)$. It therefore follows that $V(\alpha e \alpha f \alpha) = V(\alpha e f \alpha)$ and so, equating biggest inverses, we have $(\alpha e \alpha f \alpha)^\circ = (\alpha e f \alpha)^\circ$. Now since S is orthodox we have $\overline{E} = E(S)$ and so $z^\circ = \alpha z \alpha$ for every $z \in \overline{E}$. Thus the above equality reduces to $\alpha e \alpha f \alpha = \alpha e f \alpha$. Pre-multiplying this by e and post-multiplying by f, we obtain $e\alpha f = ef$. Since this holds for all $e, f \in E(S)$ we can take $e = x'x$ and $f = yy'$ to obtain

$$x\alpha y = xx'x\alpha yy'y = xx'xyy'y = xy,$$

whence α is a middle unit.

(2) $\Rightarrow$ (3): Suppose that α is a middle unit. Then α is a weak middle unit and so, by Theorem 13.16, S is naturally ordered. Moreover, by Theorem 13.17, we have the identity $(xy)^\circ = y^\circ x^\circ$.

(3) $\Rightarrow$ (4): If (3) holds then for all $e, f \in E(S)$ we have

$$(ef)^\circ = f^\circ e^\circ = \alpha f \alpha \cdot \alpha e \alpha = \alpha e \alpha \cdot \alpha f \alpha = e^\circ f^\circ = (fe)^\circ$$

whence $\alpha e f \alpha = (ef)^{\circ\circ} = (fe)^{\circ\circ} = \alpha f e \alpha$.

(4) $\Rightarrow$ (1): If (4) holds then for all $x, y \in S$ we have

$$\alpha yy^\circ x^\circ x\alpha = \alpha x^\circ xyy^\circ \alpha.$$

Now the right-hand side of this is $x^\circ xyy^\circ$; and the left-hand side can be written

$$\alpha yy^\circ \alpha \cdot \alpha x^\circ x\alpha = \alpha x^\circ x\alpha \cdot \alpha yy^\circ \alpha = x^\circ x\alpha yy^\circ.$$

Thus $x^\circ x\alpha yy^\circ = x^\circ xyy^\circ$. Pre-multiplying this by x and post-multiplying by y, we obtain $x\alpha y = xy$ and so α is a middle unit. Now by the standing hypothesis we have the identity $\alpha e f \alpha = \alpha f e \alpha$. Pre-multiplying this by e and post-multiplying by f, we then obtain $ef = (ef)^2$. Hence S is orthodox.   □

Corollary (McAlister [81]) *A naturally ordered regular semigroup with a biggest idempotent α is orthodox if and only if α is a middle unit.*   □

Theorem 13.19 (Blyth–McFadden [25]) *The smallest naturally ordered non-orthodox regular semigroup with a biggest idempotent is isomorphic to the ordered semigroup N_5 that is described by the following Hasse diagram and Cayley table:*

	u	e	f	f	b
u	u	u	f	f	b
e	e	e	a	a	b
f	u	b	f	b	b
a	e	b	a	b	b
b	b	b	b	b	b

Proof Let S be a naturally ordered non-orthodox regular semigroup with a biggest idempotent α. Since S is not orthodox there exist idempotents e and f such that $ef \neq (ef)^2$. Let $x = ef \in \overline{E}$. From the proof of Theorem 13.16 we have $x\alpha = xx^\circ \in E(S)$ and $\alpha x = x^\circ x \in E(S)$. These give $x = x\alpha x$ and $\alpha x\alpha = x^{\circ\circ} \in E(S)$. We therefore have the Hasse diagram

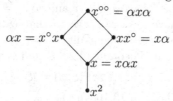

$$x^{\circ\circ} = \alpha x\alpha$$
$$\alpha x = x^\circ x \qquad\qquad xx^\circ = x\alpha$$
$$x = x\alpha x$$
$$x^2$$

in which the lines of positive gradient indicate $\mathcal{R}$-equivalence and those of negative gradient indicate $\mathcal{L}$-equivalence. It is now a routine matter to verify that this ordered set is closed under multiplication and that consequently S contains an ordered regular semigroup isomorphic to N_5. Thus, to within isomorphism, N_5 is the smallest such ordered regular semigroup.   □

EXERCISES

13.8. Let S be a naturally ordered regular semigroup with a biggest idempotent α. Prove that the following are equivalent:

(1) the mapping $x \mapsto x^\circ$ is isotone;
(2) on the inverse transversal $\alpha S\alpha$ the order coincides with the Namboripad order.

13.9. In a naturally ordered regular semigroup S with a biggest idempotent show that all conjugates of $a \in S$ (i.e. elements of the form xax° and $x^\circ ax$) are idempotent.

13.10. Prove that, in a naturally ordered regular semigroup with a biggest idempotent, the biggest idempotent-separating congruence is given by

$$(x, y) \in \mu \iff (\forall e \in E(S)) \quad x^\circ ex = y^\circ ey \text{ and } xex^\circ = yey^\circ.$$

Theorem 13.20 *If S is a strong Dubreil-Jacotin regular semigroup then the following statements are equivalent*:

(1) S *is naturally ordered*;
(2) ξ *is medial*;
(3) ξ *is a weak middle unit*;
(4) $(\forall x \in S)$ $x = x(\xi\!:\!x)x$.

Proof (1) $\Rightarrow$ (2): This is clear from Theorem 13.16.

(2) $\Rightarrow$ (3): If (2) holds then $\overline{e}\xi\overline{e} = \overline{e}$ gives $e\xi e = e$ whence $x'x\xi x'x = x'x$ and consequently $x\xi x' = xx'$.

(3) $\Rightarrow$ (4): If (3) holds then every $x \in S$ is such that

$$x(\xi\!:\!x)x = xx'x(\xi\!:\!x)x \leqslant xx'\xi x = xx'x = x.$$

But $xx' \leqslant \xi$ gives $x' \leqslant \xi\!:\!x$ whence $x = xx'x \leqslant x(\xi\!:\!x)x$. Hence we have (4).

(4) $\Rightarrow$ (1): If (4) holds then $e = e\xi e$ for every $e \in E(S)$. If now $e, f \in E(S)$ are such that $e \leqslant_n f$ then $e = e\xi e = fe\xi ef = fef \leqslant f\xi f = f$. Hence S is naturally ordered.   □

Corollary *If S is a naturally ordered strong Dubreil-Jacotin regular semigroup then, for every $x \in S$,*

(a) $x^\circ = (\xi\!:\!x)x(\xi\!:\!x)$ *is the biggest inverse of x;*

(b) $\xi(\xi\!:\!x) = \xi\!:\!x = (\xi\!:\!x)\xi$.

Proof (a) Since $xx^\circ \leqslant \xi$ we have $x \leqslant \xi\!:\!x$ and so $x^\circ = x^\circ xx^\circ \leqslant (\xi\!:\!x)x(\xi\!:\!x)$. But by the above we have $x = x(\xi\!:\!x)x$ so that $(\xi\!:\!x)x(\xi\!:\!x) \in V(x)$ and consequently we have the reverse inequality.

(b) Writing $\xi\!:\!x$ for x in Theorem 13.20(4) and using (a), we obtain

$$\xi\!:\!x = (\xi\!:\!x)[\xi\!:\!(\xi\!:\!x)](\xi\!:\!x) = [\xi\!:\!(\xi\!:\!x)]^\circ$$

whence (b) follows by Theorem 13.16(5). □

Using the above, we can characterise naturally ordered strong Dubreil-Jacotin regular semigroups as follows.

Theorem 13.21 (Blyth [17]) *An ordered regular semigroup is a naturally ordered strong Dubreil-Jacotin semigroup if and only if*

(α) $\xi = \max\{x \in S \mid x^2 \leqslant x\}$ *exists*;

(β) $(\forall x \in S)$ $x^\star = \max\{y \in S \mid xyx \leqslant x\}$ *exists*;

(γ) $(\forall x \in S)$ $\xi x^\star = x^\star = x^\star \xi$.

Proof $\Rightarrow$: If S is a naturally ordered strong Dubreil-Jacotin semigroup then its bimaximum element ξ satisfies (α) by Theorem 13.4. Now for every $x \in S$ the set $\{y \in S \mid xyx \leqslant x\}$ is not empty since it contains every $x' \in V(x)$. But if $xyx \leqslant x$ then $xy \in \langle x^\cdot.x\rangle \subseteq)S(= \xi^\downarrow$ and so $y \leqslant \xi\!:\!x$. Since S is naturally ordered, it follows by Theorem 13.20 that (β) is satisfied with $x^\star = \xi\!:\!x$. As for (γ), we observe that $\xi(\xi\!:\!x)x \leqslant \xi^2 \leqslant \xi$ and so $\xi(\xi\!:\!x) \leqslant \xi\!:\!x$. But since $x = xx'x \leqslant \xi x$ for every $x \in S$ we have $\xi\!:\!x \leqslant \xi(\xi\!:\!x)$. Thus we see that $\xi x^\star = \xi(\xi\!:\!x) = \xi\!:\!x = x^\star$. Likewise we have $x^\star \xi = x^\star$, whence we have (γ).

$\Leftarrow$: Conversely, suppose that S is an ordered regular semigroup that satisfies the conditions (α), (β), (γ). By (β) we have $xx^\star x \leqslant x$ so that $xx^\star xx^\star x \leqslant xx^\star x \leqslant x$ and consequently $x^\star xx^\star \leqslant x^\star$ whence

(1) $x \leqslant x^{\star\star}$.

Now for every $x' \in V(x)$ we have $xx'x = x$ and so, by (β), $x' \leqslant x^\star$. Thus $x = xx'x \leqslant xx^\star x \leqslant x$ whence

(2) $x = xx^\star x$.

It follows from (2) that $xx^\star$ and $x^\star x$ are idempotents and so, by (α), we have

(3) $xx^\star \leqslant \xi, \quad x^\star x \leqslant \xi$.

Let us now establish the property

(4) $xax \leqslant \xi x \Rightarrow a \leqslant x^\star$.

In fact, we have

$$xax \leqslant \xi x \Rightarrow xax = xx^\star xax \leqslant xx^\star \xi x = xx^\star x = x \quad \text{by (2), (γ)}$$
$$\Rightarrow a \leqslant x^\star \quad \text{by (β)}.$$

Using (4) we can now show that

(5) $x \leqslant y \Rightarrow y^* \leqslant x^*$.

In fact, if $x \leqslant y$ then by (3) we have $xy^*x \leqslant yy^*x \leqslant \xi x$ whence (5) follows from (4).

It is an immediate consequence of (1) and (5) that $x^* = x^{***}$ for every $x \in S$ and that $x \mapsto x^{**}$ is a closure mapping on S. Let $A_\star$ be the associated closure equivalence, so that

$$(x, y) \in A_\star \iff x^{**} = y^{**} \iff x^* = y^*.$$

We show that $A_\star$ is a congruence on S. For this purpose, we require the following observations.

First we note that, by (3) and (α), $x\xi yy^*x^*x\xi y \leqslant x\xi y$ so, by (β) and (5), we have

(6) $y^*x^* \leqslant (x\xi y)^* \leqslant (xy)^*$.

Now since, by (3), $xy(xy)^*x \leqslant \xi x$ it follows from (4) that

(7) $y(xy)^* \leqslant x^*$.

From the above we see that

$$x^{**}y(xy)^*x^{**}y \leqslant x^{**}x^*x^{**}y \qquad \text{by (7)}$$
$$= x^{**}x^{***}x^{**}y$$
$$= x^{**}y \qquad \text{by (2)}$$

and consequently $(xy)^* \leqslant (x^{**}y)^*$. But it is clear from (1) and (5) that the converse inequality holds. Hence we see that $(xy, x^{**}y) \in A_\star$ and so $A_\star$ is compatible on the right with multiplication. Similarly, $A_\star$ is compatible on the left and is therefore a congruence on S.

We next note that

(8) $\xi^* = \xi$.

In fact, from (2) and (3) we have $x \leqslant x\xi$ and $x \leqslant \xi x$ so that $x\xi \leqslant \xi x\xi$. Taking $x = \xi^*$ we obtain $\xi^* \leqslant \xi\xi^*\xi = \xi$. But by ($\alpha$) we have $\xi^2 \leqslant \xi$ and so $\xi^3 \leqslant \xi^2 \leqslant \xi$ and therefore, by (β), $\xi \leqslant \xi^*$.

It follows from this that if $e \in E(S)$ then

(9) $e^* = \xi$.

In fact, by (α), we have $e \leqslant \xi$ and so, by (5) and (8), $e^* \geqslant \xi^* = \xi$. On the other hand, by (6), $e^*e^* \leqslant (ee)^* = e^*$; and since $eee = e$ we have $e \leqslant e^*$ so that $e = ee \leqslant ee^*$ whence, by (5) and (6), we have $e^* \geqslant (ee^*)^* \geqslant e^{**}e^*$ whence, by (2), $e^*e^* \geqslant e^*e^{**}e^* = e^*$. Thus we see that e^* is idempotent. Then, by (α), $e^* \leqslant \xi$ whence (9) follows.

We now show that the quotient semigroup $S/A_\star$ has an identity. For this purpose, we note by (2) and (3) that $x = xx^*x \leqslant x\xi$ and so by (5), $(x\xi)^* \leqslant x^*$. On the other hand, by (6), (8) and (γ), we have $(x\xi)^* \geqslant \xi^*x^* = \xi x^* = x^*$. Thus $(x\xi)^* = x^*$ and so $(x\xi, x) \in A_\star$. Similarly, we can show that $(\xi x, x) \in A_\star$. Consequently $[\xi]_{A_\star}$ is the identity element of $S/A_\star$.

Since by (9) we have $xx^* \equiv \xi \equiv x^*x\,(A_\star)$ it follows that $S/A_\star$ is a group in which the inverse of $[x]_{A_\star}$ is $[x^*]_{A_\star}$. Now since $A_\star$ is a closure we can order the group quotient by

$$[x]_{A_\star} \leqslant [y]_{A_\star} \iff x^{**} \leqslant y^{**}.$$

The natural epimorphism $\natural : S \to S/A_\star$ is then residuated; for the mapping $\natural^+ : S/A_\star \to S$ given by $\natural^+([x]_{A_\star}) = x^{\star\star}$ is isotone and such that $\natural^+ \circ \natural \geqslant \mathrm{id}_S$ and $\natural \circ \natural^+ = \mathrm{id}_{S/A_\star}$.

Thus we see that S is a strong Dubreil-Jacotin semigroup. Now the pre-image under $\natural$ of the negative cone of $S/A_\star$ is the principal down-set $\xi^\downarrow$. Thus, by Theorem 12.6, ξ is residuated and so, by Theorem 12.10, $A_\star$ coincides with the Molinaro equivalence A_ξ. It follows from this that $x^\star = \xi : x$; for $x^\star$ is the biggest element of $[x]_{A_\star}^{-1}$ and $\xi : x$ is that of $[x]_{A_\xi}^{-1}$. Consequently $x = xx^\star x = x(\xi : x)x$ whence, by Theorem 13.20, S is naturally ordered. $\quad\square$

EXERCISES

13.11. Consider the subset of $\mathbb{R} \times \mathbb{R}$ given by $S = \{(0,0)\} \cup \{(x,y) \mid x < 0\}$. Show that, under the cartesian order and the multiplication
$$(p, i)(q, j) = (\min\{p, q\}, i + j),$$
S becomes an ordered regular semigroup in which property (α) of Theorem 13.21 holds but property (β) fails.

13.12. Let $S = \{(x, y, z) \in \mathbb{Z} \times \mathbb{Z} \times \mathbb{Z} \mid 0 \leqslant y \leqslant x\}$. Show that, under the cartesian order and the multiplication
$$(x, y, z)(a, b, c) = (\max\{x, a\}, y, z + c),$$
S is an ordered regular semigroup in which property (β) of Theorem 13.21 holds but property (α) fails.

13.13. Let k be a fixed positive integer and consider the subset S_k of the semigroup S of the previous exercise that is obtained by imposing the restriction that $x \leqslant k$. Prove that S_k is an ordered regular semigroup in which properties (α) and (β) of Theorem 13.21 hold but property (γ) fails.

13.14. In a naturally ordered strong Dubreil-Jacotin regular semigroup establish the identities $x\xi = x(\xi : x)[\xi : (\xi : x)]$ and $\xi x = [\xi : (\xi : x)](\xi : x)x$.

13.15. If S is a strong Dubreil-Jacotin regular semigroup and $B = [\xi]_{A_\xi}$ prove that $\xi B \xi$ is a semilattice. Deduce that a naturally ordered strong Dubreil-Jacotin regular semigroup is orthodox if and only if
$$(\forall x, y \in S)(\forall e, f \in E(S)) \quad xefy = xfey.$$

13.16. Prove that, in a naturally ordered Dubreil-Jacotin regular semigroup, Green's relations are given by
$$(x, y) \in \mathcal{R} \iff xx^\circ = yy^\circ, \quad (x, y) \in \mathcal{L} \iff x^\circ x = y^\circ y.$$

13.17. Prove that, in a strong Dubreil-Jacotin regular semigroup S, the following statements are equivalent:
(1) S is naturally ordered and $\mathcal{R}$-unipotent;
(2) ξ is a right identity element;
(3) $(\forall x \in S) \ x = x(\xi : x)[\xi : (\xi : x)]$.

13.18. Let $\mathbb{N}^\infty$ consist of the chain of natural numbers with a new top element ∞ adjoined, and let $S = \mathbb{Z} \times \mathbb{N}^\infty$. Show that S, under the order given by
$$(m, x) \leqslant (n, y) \iff m = n, \ x \leqslant y$$
and the multiplication given by $(m, x)(n, y) = (m + n, x)$, is a naturally ordered strong Dubreil-Jacotin semigroup that is $\mathcal{R}$-unipotent and has the bimaximum element as a right identity. Describe Green's relations on S.

13.19. In a strong Dubreil-Jacotin regular semigroup S an element x is said to be **perfect** if $x = x(\xi : x)x$. Prove that in a strong Dubreil-Jacotin regular semigroup S the set $P(S)$ of perfect elements of S is a regular subsemigroup and an ideal of S.

13.20. Let L be a bounded lattice and let G be an ordered group. Show that the cartesian ordered set $S = L \times L \times G$ with the multiplication

$$(x, y, g)(a, b, h) = (x \vee a, y, gh)$$

is a strong Dubreil-Jacotin regular semigroup. Determine the set of perfect elements of S.

13.4 Biggest inverses

There are two wider situations that arise from the above results. The first of these, which we shall now consider, concerns ordered regular semigroups in which every element x has a biggest inverse x°.

Example 13.4 Let $k > 1$ be a fixed integer and consider the cartesian ordered set

$$S = \{(x, y, z) \in \mathbb{Z} \times \mathbb{Z} \times \mathbb{Z} \mid 0 \leqslant y \leqslant k\}$$

together with the multiplication

$$(x, y, z)(a, b, c) = (\min\{x, a\}, \ y, \ z + c).$$

Then S is an ordered regular semigroup in which every element (x, y, z) has a biggest inverse, namely $(x, y, z)^\circ = (x, k, -z)$. Note that here the idempotents are the elements of the form $(x, y, 0)$, so no biggest idempotent exists.

If S is a regular semigroup with biggest inverses then for every $x \in S$ we have $(x^\circ)^{\circ\circ} = ((x^\circ)^\circ)^\circ = (x^{\circ\circ})^\circ$ which we can write unambiguously as $x^{\circ\circ\circ}$. The principal properties of the unary operation $x \mapsto x^\circ$ are the following.

Theorem 13.22 (Saito [99]) *Let S be an ordered regular semigroup in which every $x \in S$ has a biggest inverse x°. Then the following properties hold:*

(1) $x^\circ = (x^\circ x)^\circ x^\circ = x^\circ (xx^\circ)^\circ$;
(2) xx° *is the biggest idempotent in* R_x;
(3) $x^\circ x$ *is the biggest idempotent in* L_x;
(4) $(x, y) \in \mathcal{R} \iff xx^\circ = yy^\circ, \quad (x, y) \in \mathcal{L} \iff x^\circ x = y^\circ y$;
(5) $(xx^\circ)^\circ = x^{\circ\circ}x^\circ, \quad (x^\circ x)^\circ = x^\circ x^{\circ\circ}$;
(6) $(\forall e \in E(S)) \ e = ee^{\circ\circ}e^\circ e = ee^\circ e^{\circ\circ}e$;
(7) $(\forall e \in E(S)) \ e^{\circ\circ} \in E(S) \cap V(e)$;
(8) $(\forall e \in E(S)) \ e^\circ \leqslant e^\circ e^\circ$, *and* $e^\circ \in E(S) \iff e^\circ = e^{\circ\circ}$;
(9) $x^\circ = x^{\circ\circ\circ}$.

Proof (1) Since $x^\circ x \in E(S)$ we have $x^\circ x \leqslant (x^\circ x)^\circ$ whence $x^\circ = x^\circ xx^\circ \leqslant (x^\circ x)^\circ x^\circ$. To obtain the reverse inequality, we observe that since

$$x \cdot (x^\circ x)^\circ x^\circ \cdot x = xx^\circ x(x^\circ x)^\circ x^\circ x = xx^\circ x = x$$

and $(x^\circ x)^\circ x^\circ \cdot x \cdot (x^\circ x)^\circ x^\circ = (x^\circ x)^\circ x^\circ$ we have $(x^\circ x)^\circ x^\circ \in V(x)$ and therefore $(x^\circ x)^\circ x^\circ \leqslant x^\circ$. The dual equality is established similarly.

(2) Since $xS = xx^\circ S$ we have $(x, xx^\circ) \in \mathcal{R}$. If now $e, f \in E(S)$ are such that $(e, f) \in \mathcal{R}$ then since $f \in V(e)$ we have $f \leqslant e^\circ$. Then $f = ef \leqslant ee^\circ$ and so ee° is the biggest idempotent in R_e. But by (1) we have $xx^\circ = xx^\circ(xx^\circ)^\circ$ whence it follows that xx° is the biggest idempotent in $R_{xx^\circ} = R_x$.

(3) This is dual to (2).

(4) This is clear from (2) and (3).

(5) Since $x^\circ S = x^\circ xS$ we have $(x^\circ, x^\circ x) \in \mathcal{R}$ and so, by (2), $x^\circ x^{\circ\circ} = x^\circ x(x^\circ x)^\circ$. But by (1) we have $x^\circ x(x^\circ x)^\circ = (x^\circ x)^\circ x^\circ x(x^\circ x)^\circ = (x^\circ x)^\circ$. Hence $x^\circ x^{\circ\circ} = (x^\circ x)^\circ$. The dual equality is established similarly.

(6) If $e \in E(S)$ then $e \in V(e)$ and $e \in V(e^\circ)$, so $e \leqslant e^\circ$ and $e \leqslant e^{\circ\circ}$. Consequently $e = eeee \leqslant ee^{\circ\circ}e^\circ e = eee^{\circ\circ}e^\circ e \leqslant ee^\circ e^{\circ\circ}e^\circ e = ee^\circ e = e$ which gives the first equality. Likewise we obtain the second.

(7) Using (6), we have $ee^{\circ\circ}e = e \cdot e^{\circ\circ}e^\circ eee^\circ e^{\circ\circ} \cdot e = ee = e$ and $e^{\circ\circ}ee^{\circ\circ} = e^{\circ\circ}e^\circ ee^\circ e^{\circ\circ} \cdot ee \cdot e^{\circ\circ}e^\circ ee^\circ e^{\circ\circ} = e^{\circ\circ}e^\circ ee^\circ e^{\circ\circ} = e^{\circ\circ}$, so that $e^{\circ\circ} \in V(e)$.

The above gives $e^{\circ\circ} \leqslant e^\circ$. But $e^\circ ee^{\circ\circ}e^\circ = e^\circ ee^{\circ\circ}e^\circ ee^\circ = e^\circ ee^\circ = e^\circ$ and $ee^{\circ\circ}e^\circ ee^{\circ\circ} = ee^{\circ\circ}$, whence $ee^{\circ\circ} \in V(e^\circ)$ so that $ee^{\circ\circ} \leqslant e^{\circ\circ}$. Thus we have

$$e^{\circ\circ}e^{\circ\circ} = e^{\circ\circ}ee^{\circ\circ}e^{\circ\circ} \leqslant e^{\circ\circ}e^\circ e^{\circ\circ} = e^{\circ\circ} = e^{\circ\circ}ee^{\circ\circ} \leqslant e^{\circ\circ}e^{\circ\circ}$$

and consequently $e^{\circ\circ} \in E(S)$.

(8) Using (6) we have $e^\circ e^{\circ\circ} \cdot e \cdot e^\circ e^{\circ\circ} = e^\circ ee^{\circ\circ}e^{\circ\circ}ee^\circ e^{\circ\circ} = e^\circ ee^\circ e^{\circ\circ} = e^\circ e^{\circ\circ}$ which, together with $e = ee^\circ e^{\circ\circ}e$, gives $e^\circ e^{\circ\circ} \in V(e)$. Consequently $e^\circ e^{\circ\circ} \leqslant e^\circ$ and hence $e^\circ = e^\circ e^{\circ\circ}e^\circ \leqslant e^\circ e^\circ$. Now since, by (7), $e^{\circ\circ} \in V(e)$ we have $e^{\circ\circ} \leqslant e^\circ$. If then $e^\circ \in E(S)$ then from $e^\circ \in V(e^\circ)$ we obtain $e^\circ \leqslant e^{\circ\circ}$, whence $e^\circ = e^{\circ\circ}$. Conversely, if $e^\circ = e^{\circ\circ}$ then (7) gives $e^\circ \in E(S)$.

(9) For every $x \in S$ we have from (5) that $(xx^\circ)^\circ$ is an idempotent and so, by (8), $(xx^\circ)^\circ = (xx^\circ)^{\circ\circ}$. Thus, using (5) again, we have $x^{\circ\circ}x^\circ = (xx^\circ)^\circ = (xx^\circ)^{\circ\circ} = x^{\circ\circ}x^{\circ\circ\circ}$. Similarly, $x^\circ x^{\circ\circ} = x^{\circ\circ\circ}x^{\circ\circ}$ whence $x^{\circ\circ\circ} = x^{\circ\circ\circ}x^{\circ\circ}x^{\circ\circ\circ} = x^\circ x^{\circ\circ}x^\circ = x^\circ$. □

If now S is an ordered regular semigroup in which every element has a biggest inverse it is clear from Theorem 13.22(4) that for every $x \in S$ we can define an order on the quotient set S/R_x by

$$[x]_{R_x} \leqslant [y]_{R_y} \iff xx^\circ \leqslant yy^\circ$$

and similarly on R/L_x. It follows therefore that Green's relations $\mathcal{R}$ and $\mathcal{L}$ on S are regular if and only if in S we have

$$x \leqslant y \Rightarrow xx^\circ \leqslant yy^\circ, \quad x^\circ x \leqslant y^\circ y.$$

In what follows we shall use a weaker version of this which concentrates on the idempotents.

Definition If S is an ordered regular semigroup with biggest inverses then $\mathcal{R}$ and $\mathcal{L}$ are said to be **weakly regular** if

$$(\forall e, f \in E(S)) \quad e \leqslant f \Rightarrow ee^\circ \leqslant ff^\circ, \quad e^\circ e \leqslant f^\circ f.$$

Example 13.5 Let $k > 1$ be a fixed integer. For every $n \in \mathbb{Z}$ let n_k be the biggest multiple of k that is less than or equal to n. On the cartesian ordered set $S = \mathbb{Z} \times -\mathbb{N} \times \mathbb{Z}$ define a multiplication by

$$(x, -p, m)(y, -q, n) = (\min\{x, y\}, -q, m + n_k).$$

Then S is an ordered semigroup in which the idempotents are $(x, -p, m)$ where $m_k = 0$, i.e where $0 \leqslant m \leqslant k - 1$. Every element $(x, -p, m)$ has a biggest inverse, namely $(x, -p, m)^\circ = (x, 0, -m_k + k - 1)$. Since $(x, -p, m)^\circ(x, -p, m) = (x, -p, k - 1)$ we see that $\mathcal{L}$ is regular, hence weakly regular. On the other hand, $(x, -p, m)(x, -p, m)^\circ = (x, 0, m - m_k)$ shows that $\mathcal{R}$ is not regular but is weakly regular.

Weak regularity is characterised in the following result which generalises a property of the natural order on an inverse semigroup. This is due to M. H. Almeida Santos and appears in [28].

Theorem 13.23 *If S is an ordered regular semigroup with biggest inverses then the following statements are equivalent:*

(1) *$\mathcal{L}$ and $\mathcal{R}$ are weakly regular;*
(2) *$(\forall e, f \in E(S))$ $e \leqslant f \Rightarrow e^\circ \leqslant f^\circ$.*

Proof (1) $\Rightarrow$ (2): If $\mathcal{L}$ and $\mathcal{R}$ are weakly regular and if $e \leqslant f$ then $e^\circ e \leqslant f^\circ f$ and $ee^\circ \leqslant ff^\circ$ and therefore $e^\circ = e^\circ e \cdot ee^\circ \leqslant f^\circ f \cdot ff^\circ = f^\circ$.

(2) $\Rightarrow$ (1): If (2) holds then clearly $e \leqslant f$ implies both $e^\circ e \leqslant f^\circ f$ and $ee^\circ \leqslant ff^\circ$.   □

Corollary *If $\mathcal{L}$ and $\mathcal{R}$ are weakly regular then*

$$(\forall e \in E(S))\ e^\circ = e^{\circ\circ} \in E(S).$$

Proof Applying (2) above to the inequality $e \leqslant ee^\circ$ and using Theorem 13.22(5) we obtain $e^\circ \leqslant (ee^\circ)^\circ = e^{\circ\circ}e^\circ$. It follows that $e^\circ e^\circ \leqslant e^\circ e^{\circ\circ}e^\circ = e^\circ$ whence, by Theorem 13.22(8), $e^\circ = e^{\circ\circ} \in E(S)$.   □

We now observe that if S is an ordered regular semigroup and e, f are idempotents such that $e \leqslant f$ then, by Theorem 13.11, the subsemigroup $\langle e, f \rangle$ is the band $B_{e,f}$ with Hasse diagram

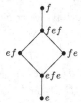

In particular, if $f \in V(e)$ then $B_{e,f}$ reduces to

In the case where $\mathcal{L}$ and $\mathcal{R}$ are weakly regular and e, f are such that $e \leqslant f$, we can describe the subsemigroup $\langle e, f, e^\circ, f^\circ \rangle$ that is generated by $\{e, f, e^\circ, f^\circ\}$.

Theorem 13.24 (Blyth–Pinto [28]) *Let S be an ordered regular semigroup with biggest inverses on which $\mathcal{L}$ and $\mathcal{R}$ are weakly regular. Let e, f be idempotents of S such that $e \leqslant f$. Then $\langle e, f, e^\circ, f^\circ \rangle$ is a band with at most 24 elements. If $\langle e, f, e^\circ, f^\circ \rangle$ has precisely 24 elements then it is order isomorphic to the cartesian ordered cartesian product semigroup $B_{e,f} \times B_{f,f^\circ}$.*

Proof Let $B = \langle e, f, e^\circ, f^\circ \rangle$ and consider the following observations.

(1)   *If $a \in \{e, e^\circ\}$ and $b \in \{f, f^\circ\}$ then $(\forall x \in B)$ $axb = ab$, $bxa = ba$.*
There are four cases to consider:

(α) $a = e$, $b = f^\circ$.
Here $ef^\circ = eef^\circ \leqslant exf^\circ \leqslant ef^\circ f^\circ = ef^\circ$ so $exf^\circ = ef^\circ$. Likewise $f^\circ xe = f^\circ e$.

(β) $a = e$, $b = f$.
Using (α) twice, we have $e^\circ xf = e^\circ ee^\circ xff^\circ f = e^\circ ef^\circ f = e^\circ ee^\circ ff^\circ f = e^\circ f$, and similarly $fxe^\circ = fe^\circ$.

(γ) $a = e^\circ$, $b = f^\circ$.
In fact, $e^\circ xf^\circ = e^\circ ee^\circ xf^\circ = e^\circ ef^\circ = e^\circ ee^\circ f^\circ = e^\circ f^\circ$; likewise $f^\circ xe^\circ = f^\circ e^\circ$.

(δ) $a = e$, $b = f$.
In fact, $exf = exff^\circ f = ef^\circ f = eff^\circ f = ef$, and likewise $fxe = fe$.

(2)   *If $a, b \in \{e, e^\circ\}$ then $(\forall x \in B)$ $axb \in \{ab, afb\}$.*
First observe that if $x \in \langle e, e^\circ \rangle$ then clearly $axb = ab$. If now $x \notin \langle e, e^\circ \rangle$ then $x = x_1 x_2 \cdots x_r$ where at least one x_i is either f or f°. In this case we have

$$
\begin{aligned}
axb &= ax_1 \cdots x_i \cdots x_r b \\
&= ax_i \cdots x_r b \quad \text{by (1)} \\
&= ax_i b \quad \text{by (1)} \\
&= ax_i fx_i b \quad \text{since } x_i \in \{f, f^\circ\} \\
&= afb \quad \text{by (1).}
\end{aligned}
$$

(3)   *If $a, b \in \{f, f^\circ\}$ then $(\forall x \in B)$ $axb \in \{ab, aeb\}$.*
The proof is similar to that of (2).

We conclude from the above that every element of $B = \langle e, f, e^\circ, f^\circ \rangle$ can be written in one of the following 24 ways:

$$
\begin{array}{ll}
ef, \ fe, \ e^\circ f, \ f^\circ e, \ ef^\circ, \ fe^\circ, \ e^\circ f^\circ, \ f^\circ e^\circ & \text{from (1);} \\
e, \ ee^\circ, \ e^\circ e, \ e^\circ, \ efe, \ e^\circ fe, \ efe^\circ, \ e^\circ fe^\circ & \text{from (2);} \\
f, \ ff^\circ, \ f^\circ f, \ f^\circ, \ fef, \ f^\circ ef, \ fef^\circ, \ f^\circ ef^\circ & \text{from (3).}
\end{array}
$$

It follows that every element of B can be written (non-uniquely) in the form acb where $a, b \in \{e, f, e^\circ, f^\circ\}$ and $c \in \{e, f\}$, and $|B| \leqslant 24$. Moreover, by $(1), (2), (3)$ it is readily seen that $x^2 = x$ for every $x \in B$ and so B is a band.

When the above 24 elements are distinct, the Hasse diagram for $(B; \leqslant)$ is

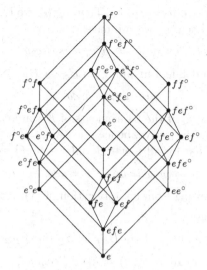

Observe that in this diagram elements that are joined by lines of negative gradient are $\mathcal{L}$-equivalent, elements that are joined by lines of positive gradient are $\mathcal{R}$-equivalent, and vertical lines indicate the natural order $\leqslant_n$ or its dual. Since the bottom 'pillar' of this is $B_{e,f}$ and the top 'face' is B_{f,f° it is clear that this set is order isomorphic to the cartesian ordered set $B_{e,f} \times B_{f,f^\circ}$.

Our objective now is to show that if $|B| = 24$ then B and $B_{e,f} \times B_{f,f^\circ}$ are isomorphic as ordered bands. For this purpose, note that every element of $B_{e,f} = \langle e, f \rangle$ can be written in the form acb where $a, c, b \in \{e, f\}$, and that in $B_{e,f}$ we clearly have

$$(\epsilon) \qquad a_1 c_1 b_1 \leqslant a_2 c_2 b_2 \;\Rightarrow\; a_1 \leqslant a_2, \; b_1 \leqslant b_2.$$

Now given $acb \in B_{e,f}$ and $x \in B_{f,f^\circ}$, define $a^x cb^y \in B$ by setting

$$a^x = \begin{cases} a^\circ & \text{if } x \in \{f^\circ f, f^\circ\}, \\ a & \text{if } x \in \{f, f f^\circ\}; \end{cases}$$

$$b^y = \begin{cases} b^\circ & \text{if } y \in \{f f^\circ, f^\circ\}, \\ b & \text{if } y \in \{f, f^\circ f\}. \end{cases}$$

Then we have the following properties in B:

$(\zeta) \quad a^x cb^y = a^{xy} cb^{xy}.$

This can readily be verified by considering the various cases. For example, if $a^x = a^\circ$ and $b^y = b$ then $x \in \{f^\circ f, f^\circ\}$ and $y \in \{f, f^\circ f\}$ whence $xy = f^\circ f$ and so $a^{xy} = a^\circ$ and $b^{xy} = b$.

$(\eta) \quad$ If $x \leqslant y$ then $a^x \leqslant a^y$ and $b^x \leqslant b^y$.

$(\vartheta) \quad$ If $a_1 \leqslant a_2$ then $a_1^x \leqslant a_2^x$; and if $b_1 \leqslant b_2$ then $b_1^x \leqslant b_2^x$.

$(\iota) \quad a^x cb^x = a^x acbb^x.$

Again this can be verified by considering the various cases. For example, if $a^x = a$ and $b^x = b$ then $a^x acbb^x = aacbb = acb = a^x cb^x$; and if $a^x = a$ and $b^x = b^\circ$ then $a^x acbb^x = acbb^\circ$ which by $(1), (2), (3)$ reduces to $acb^\circ = a^x cb^x$. The remaining cases are similar.

Suppose now that $a_1c_1b_1 = a_2c_2b_2$. Then by (ϵ) we have $a_1 = a_2 = a$ say, and $b_1 = b_2 = b$ say. Then, by (ι), we have

$$ac_1b = ac_2b \;\Rightarrow\; a^x c_1 b^x = a^x c_2 b^x.$$

Consequently, we can define a mapping $\mu : B_{e,f} \times B_{f,f^\circ} \to B$ by setting

$$\mu(acb, x) = a^x cb^x.$$

Observing now that each of the 24 elements of B can be written in the form $a^x cb^x$, we see that μ is surjective, hence bijective.

Next we observe that for all $u, v \in B$ and $c \in \{e, f\}$ we have

$$uc^\circ v = uc^\circ cc^\circ v = ucv,$$

the final equality resulting from an application of (1). It follows that

$(\kappa) \quad ub^x a^y v = ubav.$

Now given $a_1c_1b_1$ and $a_2c_2b_2$ in $B_{e,f}$ let $a_1c_1b_1 \cdot a_2c_2b_2 = a_1c_3b_2$. Then

$$
\begin{aligned}
\mu(a_1c_1b_1, x)\,\mu(a_2c_2b_2, y) &= a_1^x c_1 b_1^x \cdot a_2^y c_2 b_2^y \\
&= a_1^x c_1 b_1 a_2 c_2 b_2^y \qquad \text{by } (\kappa) \\
&= a_1^x c_3 b_2^y \\
&= a_1^{xy} c_3 b_2^{xy} \qquad \text{by } (\zeta) \\
&= \mu(a_1c_3b_2, xy) \\
&= \mu((a_1c_1b_1, x)(a_2c_2b_2, y)),
\end{aligned}
$$

and so μ is a morphism.

Finally, μ is isotone. To see this, suppose that $(a_1c_1b_1, x) \leqslant (a_2c_2b_2, y)$. Then we have $a_1c_1b_1 \leqslant a_2c_2b_2$ and $x \leqslant y$ so that, by (ϵ), $a_1 \leqslant a_2$ and $b_1 \leqslant b_2$ and hence, by (η) and (ϑ), $a_1^x \leqslant a_1^y \leqslant a_2^y$ and $b_1^x \leqslant b_1^y \leqslant b_2^y$. Consequently, using (ι), we obtain

$$\mu(a_1c_1b_1, x) = a_1^x c_1 b_1^x = a_1^x a_1 c_1 b_1 b_1^x \leqslant a_2^y a_2 c_2 b_2 b_2^y = a_2^y c_2 b_2^y = \mu(a_2c_2b_2, y),$$

as required. $\qquad\qquad\qquad\qquad\qquad\qquad\qquad\qquad\qquad\qquad\qquad\square$

Corollary *If $|B| = 24$ then $B^\circ = \{e^\circ, f^\circ ef^\circ, f^\circ\}$.*

Proof If C and D are ordered regular semigroups with biggest inverses then so also is the cartesian ordered cartesian product semigroup $C \times D$; clearly, we have $(c, d)^\circ = (c^\circ, d^\circ)$. Now $B_{e,f}$ has biggest inverses; we have

$$
x^\circ = \begin{cases}
fef & \text{if } x \neq e, f; \\
e & \text{if } x = e; \\
f & \text{if } x = f.
\end{cases}
$$

Likewise, so does B_{f, f°; here we have $x^\circ = f^\circ$ for every x. It follows that

$$(B_{e,f} \times B_{f, f^\circ})^\circ = \{(e, f^\circ), (fef, f^\circ), (f, f^\circ)\}.$$

Now, from the isomorphism of Theorem 13.24, we have $\mu(e, f^\circ) = \mu(eee, f^\circ) = e^\circ ee^\circ = e^\circ$, $\mu(fef, f^\circ) = f^\circ ef^\circ$, and $\mu(f, f^\circ) = \mu(fff, f^\circ) = f^\circ ff^\circ = f^\circ$. So the elements of B that correspond to the pairs (e, f°), (fef, f°), (f, f°) are e°, $f^\circ ef^\circ$, f° respectively. We therefore deduce that $B^\circ = \{e^\circ, f^\circ ef^\circ, f^\circ\}$. $\quad\square$

We note in particular from the above Corollary that S° is not in general a subsemigroup of S.

If S is a naturally ordered regular semigroup with a biggest idempotent α and if $S^\circ = \{x^\circ \mid x \in S\}$ is the set of biggest inverses in S then by Theorem 13.16(4)(5) we have $S^\circ = \alpha S \alpha$ and is an inverse transversal of S. Moreover, for all $x, y \in S$ we have $x^\circ xyy^\circ \in E(S)$ by Theorem 13.17(2); and $x^\circ xyy^\circ = x^\circ \alpha xy \alpha y^\circ = x^\circ (xy)^{\circ\circ} y^\circ \in S^\circ$. Consequently $x^\circ xyy^\circ \in E(S^\circ)$, a property that we refer to by saying that the inverse transversal S° is **multiplicative** [26]. We now show that a corresponding result holds in the case of a naturally ordered regular semigroup with biggest inverses in which Green's relations are weakly regular.

Theorem 13.25 (Saito [99]) *Let S be a naturally ordered regular semigroup with biggest inverses on which $\mathcal{R}$ and $\mathcal{L}$ are weakly regular. Then*

$$(\forall x, y \in S) \quad (xy)^\circ = (x^\circ xy)^\circ x^\circ = y^\circ (xyy^\circ)^\circ.$$

Moreover, the set S° of biggest inverses in S is a multiplicative inverse transversal of S.

Proof Observe first that $xy(x^\circ xy)^\circ x^\circ \in E(S)$ with $xy(x^\circ xy)^\circ x^\circ \leqslant_n xx^\circ$ and so $xy(x^\circ xy)^\circ x^\circ \leqslant xx^\circ$. Since $\mathcal{R}$ is weakly regular we deduce that

$$xy(x^\circ xy)^\circ x^\circ [xy(x^\circ xy)^\circ x^\circ]^\circ \leqslant xx^\circ (xx^\circ)^\circ = xx^\circ.$$

But $xy \,\mathcal{R}\, xy(xy)^\circ \,\mathcal{R}\, xy(x^\circ xy)^\circ x^\circ$ and so the left-hand side of the above reduces to $xy(xy)^\circ$. It follows that $x^\circ xy(xy)^\circ \leqslant x^\circ$.

Now $(x^\circ xy)^\circ x^\circ \in V(xy)$ gives $(x^\circ xy)^\circ x^\circ \leqslant (xy)^\circ$ and consequently

$$xy(xy)^\circ = xx^\circ xy(x^\circ xy)^\circ x^\circ xy(xy)^\circ \leqslant xy(x^\circ xy)^\circ x^\circ \leqslant xy(xy)^\circ$$

whence we have $xy(xy)^\circ = xy(x^\circ xy)^\circ x^\circ$. Using the fact that $xy \,\mathcal{L}\, x^\circ xy$ we also have $(xy)^\circ xy = (x^\circ xy)^\circ x^\circ xy$. Combining these observations, we obtain

$$\begin{aligned}
(xy)^\circ &= (xy)^\circ xy(xy)^\circ \\
&= (xy)^\circ xy(x^\circ xy)^\circ x^\circ \\
&= (x^\circ xy)^\circ x^\circ xy(x^\circ xy)^\circ x^\circ \\
&= (x^\circ xy)^\circ x^\circ.
\end{aligned}$$

Similarly, we can show that $(xy)^\circ = y^\circ (xyy^\circ)^\circ$.

We now proceed to show that S° is a multiplicative inverse transversal of S. Clearly, $x^\circ \in S^\circ \cap V(x)$ for every $x \in S$. Suppose now that $a \in S^\circ \cap V(x)$. Then $a \leqslant x^\circ$, $x \leqslant a^\circ$, $a = a^{\circ\circ}$, and $xa \in E(S)$. Observing that $a^\circ a \in V(xa)$ we then have $a^\circ a \leqslant (xa)^\circ$ and so, by Theorem 13.23 and Theorem 13.22(5),

$$a^\circ a \leqslant (xa)^\circ \leqslant (xx^\circ)^\circ = x^{\circ\circ} x^\circ.$$

Likewise, observing that $xx^\circ \in V(xa)$, we have $xx^\circ \leqslant (xa)^\circ$ and so

$$x^{\circ\circ} x^\circ = (xx^\circ)^\circ \leqslant (xa)^{\circ\circ} \leqslant (a^\circ a)^{\circ\circ} = a^\circ a^{\circ\circ} = a^\circ a.$$

Thus we see that $a^\circ a = x^{\circ\circ} x^\circ$; and similarly $aa^\circ = x^\circ x^{\circ\circ}$. Consequently,

$$a^\circ = a^\circ aa^\circ = a^\circ axaa^\circ = x^{\circ\circ} x^\circ xx^\circ x^{\circ\circ} = x^{\circ\circ}$$

and therefore $a = a^{\circ\circ} = x^{\circ\circ\circ} = x^\circ$. Thus we see that $S^\circ \cap V(x) = \{x^\circ\}$.

We show next that $S^\circ SS^\circ \subseteq S^\circ$. For this purpose, let $a, b \in S^\circ$ and $x \in S$. Then $a = a^{\circ\circ}$ and $b = b^{\circ\circ}$. Using the formulae obtained above, we have

$$axb(axb)^\circ aa^\circ = axb(a^\circ axb)^\circ a^\circ aa^\circ = axb(a^\circ axb)^\circ a^\circ = axb(axb)^\circ$$

from which we see that $axb(axb)^\circ \leqslant_n aa^\circ$ and hence $axb(axb)^\circ \leqslant aa^\circ$; and in a similar way $(axb)^\circ axb \leqslant b^\circ b$. We therefore have

$$
\begin{aligned}
(axb)^{\circ\circ}(axb)^\circ axb &= [axb(axb)^\circ]^\circ axb(axb)^\circ axb \\
&\leqslant (aa^\circ)^\circ aa^\circ axb \\
&= axb \\
&= axb(axb)^\circ axb \\
&\leqslant (axb)^{\circ\circ}(axb)^\circ axb,
\end{aligned}
$$

whence $axb = (axb)^{\circ\circ}(axb)^\circ axb$; and likewise $axb = axb(axb)^\circ(axb)^{\circ\circ}$. Combining these, we have

$$axb = (axb)^{\circ\circ}(axb)^\circ axb(axb)^\circ(axb)^{\circ\circ} = (axb)^{\circ\circ} \in S^\circ.$$

Thus we see that $S^\circ SS^\circ \subseteq S^\circ$. Also S° is a subsemigroup; for if $x^\circ, y^\circ \in S^\circ$ then $x^\circ y^\circ = x^\circ x^{\circ\circ} x^\circ y^\circ \in S^\circ SS^\circ \subseteq S^\circ$.

Thus S° is an inverse transversal of S. For the purpose of showing that it is multiplicative, we observe that

$$x' \in V(x) \;\Rightarrow\; x'^\circ = x^{\circ\circ}. \tag{13.1}$$

In fact, from $(xx', x) \in \mathcal{R}$ we have $xx'(xx')^\circ = xx^\circ$ whence, by the Corollary to Theorem 13.23, $xx^\circ = xx^\circ(xx')^\circ$. Also, $(xx')^\circ xx^\circ = (xx')^\circ$ and therefore $(xx^\circ, (xx')^\circ) \in \mathcal{L}$. Then $x^{\circ\circ}x^\circ = (xx^\circ)^\circ xx^\circ = (xx')^{\circ\circ}(xx')^\circ = (xx')^\circ$. Similarly, we have $x^\circ x^{\circ\circ} = (x'x)^\circ$. Consequently, on the one hand, $x^{\circ\circ}x'x^{\circ\circ} = x^{\circ\circ}x'xx'x^{\circ\circ}x^\circ x^{\circ\circ} = x^{\circ\circ}x'xx'(xx')^\circ x^{\circ\circ} = x^{\circ\circ}x'xx^\circ x^{\circ\circ} = x^{\circ\circ}(x'x)^{\circ\circ}x^\circ x^{\circ\circ} = x^{\circ\circ}x^\circ x^{\circ\circ}x^\circ x^{\circ\circ} = x^{\circ\circ}$; and on the other $x'x^{\circ\circ}x' = x'xx'x^{\circ\circ}x^\circ x^{\circ\circ}x' = x'xx'(xx')^\circ x^{\circ\circ}x' = x'xx^\circ x^{\circ\circ}x' = x'x(x'x)^\circ x' = x'$. Hence $x^{\circ\circ} \in V(x') \cap S^\circ = \{x'^\circ\}$ and consequently $x'^\circ = x^{\circ\circ}$.

Consider now idempotents $e = xx^\circ$ and $f = y^\circ y$. We have $fe = y^\circ yxx^\circ \in S^\circ SS^\circ \subseteq S^\circ$ and so $fe = (fe)^{\circ\circ}$. Also, since $e(fe)^\circ f \in E(S) \cap V(fe)$ we have, by (12.1), $(fe)^{\circ\circ} = [e(fe)^\circ f]^\circ \in E(S^\circ)$. Hence $fe \in E(S^\circ)$ and so the inverse transversal S° is multiplicative. $\square$

EXERCISES

13.21. Let S be a naturally ordered regular semigroup with biggest inverses on which $\mathcal{L}$ and $\mathcal{R}$ are weakly regular. Prove that if $e, f \in E(S)$ are such that $e \leqslant f$ then $\langle e, f, e^\circ, f^\circ \rangle$ has the Hasse diagram

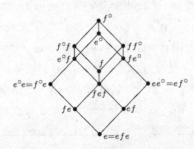

13.22. In the semigroup of Example 13.5 determine the band $\langle e, f, e^\circ, f^\circ \rangle$ for idempotents $e = (x, -p, m)$ and $f = (y, -q, n)$ with $e < f$.

13.5 Principally ordered regular semigroups

The second wider situation that we now consider is the following which is prompted by property (β) of Theorem 13.21.

Definition An ordered regular semigroup is said to be **principally ordered** if, for every $x \in S$ there exists $x^\star = \max\{y \in S \mid xyx \leqslant x\}$.

Example 13.6 Let $S = \mathbb{Z} \times \mathbb{Z}$ with the cartesian ordering. Let $k > 1$ be a fixed integer and for every $n \in \mathbb{Z}$ let n_k be the biggest multiple of k that is less than or equal to n. Endow S with the multiplication

$$(m, n)(p, q) = (\max\{m, p\}, n + q_k).$$

Then S is a regular semigroup; for example $(m, -n) \in V(m, n)$. Now

$$(m, n)(p, q)(m, n) = (\max\{m, p\}, n + q_k + n_k)$$
$$\leqslant (m, n) \iff p \leqslant m, \quad q \leqslant -n_k + k - 1.$$

Thus S is principally ordered with $(m, n)^\star = (m, -n_k + k - 1)$.

Example 13.7 Consider the ordered band $B = \langle e, f, e^\circ, f^\circ \rangle$ of Theorem 13.24. If $|B| = 24$ then B is principally ordered. In fact, it is an easy matter to verify that $B_{e, f}$ is principally ordered, with

$$x^\star = \begin{cases} f & \text{if } x \neq e; \\ e & \text{if } x = e, \end{cases}$$

and so is B_{f, f° with $x^\star = f^\circ$ for every x. But if C and D are principally ordered regular semigroups then so also is the cartesian ordered cartesian product semigroup $C \times D$, with $(c, d)^\star = (c^\star, d^\star)$. It therefore follows that $B \simeq B_{e, f} \times B_{f, f^\circ}$ is principally ordered.

Writing $(x^\star)^\star$ as $x^{\star\star}$ we have $(x^{\star\star})^\star = ((x^\star)^\star)^\star = (x^\star)^{\star\star}$ which we can write unambiguously as $x^{\star\star\star}$.

Theorem 13.26 *In a principally ordered regular semigroup the unary operation $x \mapsto x^\star$ has the following properties:*

(1) $x' \leqslant x^\star$ *for every* $x' \in V(x)$;
(2) $x = xx^\star x$;
(3) $x^\circ = x^\star x x^\star$ *is the biggest inverse of* x;
(4) $xx^\circ = xx^\star$ *is the biggest idempotent in* R_x;
(5) $x^\circ x = x^\star x$ *is the biggest idempotent in* L_x;
(6) $x \leqslant x^{\star\star}$;
(7) $x^{\circ\star} = x^{\star\star}$;
(8) $x^\star = x^{\star\star\star}$;
(9) $x^{\circ\circ} \leqslant x^{\star\star}$;
(10) $x^{\star\circ} = x^{\star\star}$.

Proof (1) This is immediate from the equality $xx'x = x$.

(2) By (1) we have $x = xx'x \leqslant xx^\star x \leqslant x$ whence (2) follows.

(3) It follows from (2) that $x^\star xx^\star \in V(x)$. But if $x' \in V(x)$ then, by (1), $x' = x'xx' \leqslant x^\star xx^\star$. Hence we have (3).

(4) It is clear from (3) that $xx^\circ = xx^\star$. Now since $xS = xx^\star S$ we have $(x, xx^\star) \in \mathcal{R}$. Suppose now that $e \in E(S)$ is such that $(e, x) \in \mathcal{R}$. Then there exist $y, z \in S$ such that $e = xy$ and $x = ez$. It follows that $xyx = ex = ez = x$ whence $y \leqslant x^\star$ and consequently $e = xy \leqslant xx^\star$.

(5) is dual to (4).

(6) By (3) and (1) we have $x^\star xx^\star = x^\circ \leqslant x^\star$ whence $x \leqslant x^{\star\star}$.

(7) Using (4) and (5), we have

$$
\begin{aligned}
y \leqslant x^{\circ\star} &\iff x^\circ yx^\circ \leqslant x^\circ \\
&\iff xx^\circ yx^\circ x \leqslant xx^\circ x = x \\
&\iff xx^\star yx^\star x \leqslant x \\
&\iff x^\star yx^\star \leqslant x^\star \\
&\iff y \leqslant x^{\star\star}.
\end{aligned}
$$

(8) It is clear from (6) that $x^\star \leqslant x^{\star\star\star}$. But, by (2), we have

$$(\forall y \in S) \quad xyx \leqslant x^{\star\star} \Rightarrow x^\star xyxx^\star \leqslant x^\star \Rightarrow xyx \leqslant x$$

whence, by (6), we see that

$$xyx \leqslant x^{\star\star} \iff xyx \leqslant x.$$

But by (6) and (2) we have $xx^{\star\star\star}x \leqslant x^{\star\star}x^{\star\star\star}x^{\star\star} = x^{\star\star}$. Using the above observation we then have $x^{\star\star\star} \leqslant x^\star$. This produces (8).

(9) We have

$$x^{\circ\circ} \overset{(3)}{=} x^{\circ\star}x^\circ x^{\circ\star} \overset{(7)}{=} x^{\star\star}x^\circ x^{\star\star} \overset{(1)}{\leqslant} x^{\star\star}x^\star x^{\star\star} \overset{(8)}{=} x^{\star\star}x^{\star\star\star}x^{\star\star} \overset{(2)}{=} x^{\star\star}.$$

(10) Using (2), (3), (8) we obtain $x^{\star\circ} = x^{\star\star}x^\star x^{\star\star} = x^{\star\star}x^{\star\star\star}x^{\star\star} = x^{\star\star}$.   □

Theorem 13.27 *If S is a principally ordered regular semigroup then the following statements are equivalent:*

(1) *S is naturally ordered;*

(2) *$(\forall x, y \in S)$ $xy(xy)^\star \leqslant xx^\star$;*

(3) *$(\forall x, y \in S)$ $(xy)^\star xy \leqslant y^\star y$;*

(4) *$(\forall x \in S)$ $x \mapsto x^\star$ is antitone.*

Proof (1) $\Rightarrow$ (2): Suppose that (1) holds. Then since $x^\star xy(xy)^\star x \in E(S)$ with $x^\star xy(xy)^\star x \leqslant_n x^\star x$ we have $x^\star xy(xy)^\star x \leqslant x^\star x$ whence $xy(xy)^\star x \leqslant x$ and therefore $y(xy)^\star \leqslant x^\star$ from which (2) follows.

(2) $\Rightarrow$ (1): Suppose that (2) holds and let $e, f \in E(S)$ be such that $e \leqslant_n f$. Then $fe(fe)^\star f \leqslant ff^\star f = f$ and $e = ef = fe$. These give $ee^\star f \leqslant f$ whence $e = eef \leqslant ee^\star f \leqslant f$ and so S is naturally ordered.

(1) $\Leftrightarrow$ (3): This is dual to the above.

(2) $\Rightarrow$ (4): Suppose that $x \leqslant y$. Then

$$xy^\star \cdot y^{\star\star}y^\star \cdot xy^\star = xy^\star xy^\star \leqslant xy^\star yy^\star = xy^\circ \leqslant xy^\star$$

and so $y^{\star\star}y^\star \leqslant (xy^\star)^\star$. Using (2) we then have

$$xy^*x = xy^*y^{**}y^*x \leqslant xy^*(xy^*)^*x \leqslant xx^*x = x$$

and so (4) holds.

(4) $\Rightarrow$ (2): Observe first that since $xy(xy)^* \cdot xx^* \cdot xy(xy)^* = xy(xy)^*$ we have $xx^* \leqslant [xy(xy)^*]^*$. So if (4) holds then

$$[xy(xy)^*]^{**} \leqslant (xx^*)^*.$$

It follows from this that

$$\begin{aligned}
xy(xy)^* &= xx^*xy(xy)^* \\
&\leqslant xx^*[xy(xy)^*]^{**} \quad \text{by Theorem 13.26(6)} \\
&\leqslant xx^*(xx^*)^* = xx^* \quad \text{by Theorem 13.26(4),}
\end{aligned}$$

and so we have (2). □

Theorem 13.28 *A naturally ordered regular semigroup is a strong Dubreil-Jacotin semigroup if and only if*

(1) *S is principally ordered;*

(2) *S has a biggest idempotent.*

Proof $\Rightarrow$: If S is a naturally ordered Dubreil-Jacotin semigroup then S is principally ordered by Theorem 13.21, and has a biggest idempotent by Theorem 13.3.

$\Leftarrow$: Suppose conversely that S satisfies properties (1) and (2). Let the biggest idempotent of S be α. Then if $x \in S$ is such that $x^2 \leqslant x$ we have $xxx^*x \leqslant xx^*x = x$ and so $xx^* \leqslant x^*$ whence $x = xx^*x \leqslant x^*x \leqslant \alpha$. It follows that the first condition of Theorem 13.21 holds.

Now from Theorem 13.16 we know that $\alpha \overline{E} \alpha \subseteq E(S)$ and consequently, by Theorems 13.26 and 13.16,

$$x^*xyy^* = x^\circ xyy^\circ = \alpha x^\circ xyy^\circ \alpha \in E(S)$$

and therefore $x^*xyy^* \cdot x^*xyy^* = x^*xyy^*$ which gives $xyy^*x^*xy = xy$. There follows $y^*x^* \leqslant (xy)^*$. Writing xx^* for both x and y in this, we obtain $(xx^*)^*(xx^*)^* \leqslant (xx^*)^*$ whence, by the first observation, we have $(xx^*)^* \leqslant \alpha$. But it is clear from Theorem 13.27 that $\alpha \leqslant \alpha^* \leqslant (xx^*)^*$. Thus we see that $(xx^*)^* = \alpha$ for every $x \in S$.

This being so, we now have $xx^* = xx^*(xx^*)^*xx^* = xx^*\alpha xx^*$ which gives $x = xx^*\alpha x$ whence $x^*\alpha \leqslant x^*$. But clearly $x^* = x^*x^{**}x^* \leqslant x^*\alpha$ and so we have the equality $x^*\alpha = x^*$. Likewise, $\alpha x^* = x^*$ and so the third condition of Theorem 13.21 is satisfied. Since, by hypothesis, S is principally ordered, the second condition of Theorem 13.21 holds. It therefore follows that S is a strong Dubreil-Jacotin semigroup. □

EXERCISES

13.23. In the principally ordered regular semigroup S of Exercise 13.12 determine, for each $x \in S$, the elements $x^*, x^\circ, x^{**}, x^{\circ\circ}$. Do likewise for the cartesian ordered cartesian product semigroup $N_5 \times S$ where N_5 is as in Theorem 13.19.

13.24. If S is a principally ordered regular semigroup and $e \in E(S)$ prove that

(1) $e^* \leqslant e^{**}$;
(2) $e = ee^{**}e$;
(3) $e^{**} \leqslant e^*$;
(4) $e^* \in E(S) \iff e^* = e^{**} \iff e^\circ \in E(S)$;
(5) $e^{**}e^{**} \leqslant e^{**}$;
(6) $(ee^*)^* = (e^{**}e^*)^*$;
(7) $(e^*e^{**}e^{**}e^*)^* = e^{**}$;
(8) $e^{**}e^{**}e^{**} = e^{**}e^{**}$;
(9) $e^{\circ\circ} \in E(S)$ and $e^{\circ\circ} \leqslant_n e^{**}e^{**}$.

13.25. Prove that a principally ordered regular semigroup S is naturally ordered if and only if

$$(\forall e, f \in E(S)) \quad e \leqslant_n f \implies e^* = f^*.$$

13.26. Let S be a principally ordered regular semigroup in which the unary operation $x \mapsto x^*$ is **weakly isotone** in the sense that

$$(\forall e, f \in E(S)) \quad e \leqslant f \implies e^* \leqslant f^*.$$

If $e, f \in E(S)$ are such that $e < f$ let T be the $*$-subsemigroup that is generated by $\{e, f\}$; i.e. the smallest subsemigroup that contains e and f and is closed under the unary operation $*$. Prove that if T is naturally ordered then it is a lattice-ordered band with at most 14 elements and has Hasse diagram

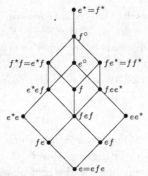

in which elements joined by lines of positive gradient are $\mathcal{R}$-related, those joined by lines of negative gradient are $\mathcal{L}$-related, and vertical lines also indicate $\leqslant_n$.

13.27. Prove that every residuated regular semigroup is principally ordered in which $x^* = (x^{\cdot}.x)^{\cdot}.x = (x.^{\cdot}x)^{\cdot}.x$.

13.28. If S is a residuated regular semigroup prove that, for every $x \in S$,

(1) $(x^{\cdot}.x)x = x = x(x.^{\cdot}x)$;
(2) $x^{\cdot}.x = (x^{\cdot}.x)^{\cdot}.(x^{\cdot}.x)$, $x.^{\cdot}x = (x.^{\cdot}x).^{\cdot}(x.^{\cdot}x)$;
(3) $(x^{\cdot}.x)^* = (x^{\cdot}.x)^{\cdot}.(x^{\cdot}.x)$, $(x.^{\cdot}x)^* = (x.^{\cdot}x).^{\cdot}(x.^{\cdot}x)$;
(4) $(x^{\cdot}.x)^\circ = (x^{\cdot}.x)^*(x^{\cdot}.x)$, $(x.^{\cdot}x)^\circ = (x.^{\cdot}x)(x.^{\cdot}x)^*$;
(5) $x^{\cdot}.x \in E(S)$, $x.^{\cdot}x \in E(S)$.

13.29. If S is a residuated regular semigroup prove that, associated with every $x \in S$, there are the chains of idempotents

$$xx^* \leqslant x^{\cdot}.x \leqslant (x^{\cdot}.x)^\circ \leqslant (x^{\cdot}.x)^* \leqslant x^*.^{\cdot}x^* = x^{**}.^{\cdot}x = (xx^*)^*;$$

$$x^*x \leqslant x.^{\cdot}x \leqslant (x.^{\cdot}x)^\circ \leqslant (x.^{\cdot}x)^* \leqslant x^*.^{\cdot}x^* = x^{**}.^{\cdot}x = (x^*x)^*.$$

13.30. Prove that a residuated regular semigroup is a strong Dubreil-Jacotin semi-group if and only if it has a biggest idempotent.

13.31. If S is a residuated regular semigroup prove that

$$(\forall e \in E(S)) \quad e^\circ = (e.\check{}e)e(e\check{}.e).$$

Deduce that S is

(1) $\mathcal{L}$-unipotent if and only if $(\forall e \in E(S))$ $e.\check{}e \leqslant e\check{}.e$;
(2) $\mathcal{R}$-unipotent if and only if $(\forall e \in E(S))$ $e\check{}.e \leqslant e.\check{}e$.

Hence show that the following statements are equivalent:

(a) S is an inverse semigroup;
(b) $(\forall e \in E(S))$ $e.\check{}e = e\check{}.e$;
(c) $(\forall e \in E(S))$ $e = (e.\check{}e)e(e\check{}.e)$.

13.6 Principally and naturally ordered semigroups

It is clear from Theorem 13.28 that if S is an ordered regular semigroup that is both naturally and principally ordered then S is a strong Dubreil-Jacotin semigroup if and only if it has a biggest idempotent. We now investigate the situation where such a semigroup S does not have a biggest idempotent. For this purpose, we shall use the following result, in which the term **maximal idempotent** is taken to mean an idempotent that is maximal in the set of idempotents.

Theorem 13.29 *Let S be a regular semigroup that is both principally and naturally ordered. Then*

(1) $e^\circ \in E(S) \Rightarrow e^\star \in E(S)$;
(2) $e \in E(S)$ *is a maximal idempotent if and only if* $e = e^\star$;
(3) $(\forall x \in S)$ $(xx^\star)^\star$ *and* $(x^\star x)^\star$ *are maximal idempotents.*

Proof (1) Observe first that if $e \in E(S)$ then we have

$$
\begin{aligned}
e &= ee^{\circ\circ}e^\circ e && \text{by Theorem 13.22(6)} \\
&= ee^{\circ\star}e^\circ e && \text{by Theorem 13.26(5)} \\
&= ee^{\star\star}e^\circ e && \text{by Theorem 13.26(7)} \\
&= ee^{\star\star}e^\star e && \text{by Theorem 13.26(5)}
\end{aligned}
$$

whence we obtain $e^{\star\star}e^\star \leqslant e^\star$ and therefore $e^\star = e^\star e^{\star\star}e^\star \leqslant e^\star e^\star$. If now $e^\circ \in E(S)$ then $e^\circ e^\circ = e^\circ$ gives $ee^\star e^\star e = e$ whence $e^\star e^\star \leqslant e^\star$ and therefore $e^\star \in E(S)$.

(2) If e is a maximal idempotent then from $e \leqslant e^\star$ and $e \leqslant e^{\star\star}$ we have $e \leqslant e^\star e^{\star\star}$ and $e \leqslant ee^\star$ whence, by the maximality, $e = e^\star e^{\star\star} = ee^\star$ and consequently $e = ee^\star = e^\star e^{\star\star}e^\star = e^\star$.

Conversely, suppose that $e \in E(S)$ is such that $e = e^\star$. If $f \in E(S)$ is such that $e \leqslant f$ then by Theorem 13.11 we have $e = efe$ whence $f \leqslant e^\star = e$. Hence $f = e$ and so e is a maximal idempotent.

(3) Let $\alpha_x = x^{\circ\circ}x^\circ$ and observe that $\alpha_x^2 = \alpha_x \in E(S)$ whence, by (1), $\alpha_x^\star \in E(S)$. But

$$\alpha_x^\star = \alpha_x^{\circ\star} = (x^{\circ\circ}x^\circ)^{\circ\star} = (xx^\circ)^{\circ\circ\star} = (xx^\circ)^{\star\star\star} = (xx^\circ)^\star = (xx^\star)^\star.$$

Hence $(xx^\star)^\star \in E(S)$. Now since

$$(xx^\star)^{\star\star} = (xx^\circ)^{\circ\star} = (x^{\circ\circ}x^\circ)^\star = \alpha_x^\star = (xx^\star)^\star$$

we see by (2) that $(xx^\star)^\star$ is a maximal idempotent. Likewise so is $(x^\star x)^\star$.   □

Our next observation is that in a regular semigroup that is both principally and naturally ordered the Saito formulae of Theorem 13.25 hold.

Theorem 13.30 *If S is a regular semigroup that is both principally and naturally ordered then*

$$(\forall x, y \in S) \quad (xy)^\circ = (x^\circ xy)^\circ x^\circ = y^\circ (xyy^\circ)^\circ.$$

Proof Observing that $(xy, x^\circ xy) \in \mathcal{L}$, we have $(xy)^\circ xy = (x^\circ xy)^\circ x^\circ xy$. Consequently

$$
\begin{aligned}
(xy)^\circ &= (xy)^\circ xy(xy)^\circ \\
&= (x^\circ xy)^\circ x^\circ xy(xy)^\circ \\
&\leqslant (x^\circ xy)^\circ x^\circ xx^\circ \qquad \text{by Theorem 13.27} \\
&= (x^\circ xy)^\circ x^\circ.
\end{aligned}
$$

But, as is readily seen, $(x^\circ xy)^\circ x^\circ \in V(xy)$ and therefore $(x^\circ xy)^\circ x^\circ \leqslant (xy)^\circ$. The first equality thus holds. The second is established similarly.   □

If S is a regular semigroup that is both principally and naturally ordered then we recall from Theorems 13.9 and 13.10 that sandwich sets in S are singletons. Since $S(e, f) = fV(ef)e$, in this situation we therefore clearly have

$$S(e, f) = \{f(ef)^\circ e\}.$$

Theorem 13.31 *Let S be a regular semigroup that is both principally and naturally ordered. If e and f are maximal idempotents then $S(e, f) = \{(ef)^\circ\}$.*

Proof For every idempotent e we have $e \leqslant e^\circ \leqslant e^\star$. Thus, if e and f are maximal idempotents we have, by Theorem 13.29(2), that $e = e^\circ$ and $f = f^\circ$. Using the formulae of Theorem 13.30 we obtain

$$(ef)^\circ = f^\circ (eff^\circ)^\circ = f^\circ (e^\circ eff^\circ)^\circ e^\circ = f(ef)^\circ e.$$

Thus $S(e, f) = \{(ef)^\circ\}$.   □

Theorem 13.32 (Blyth–Pinto [27]) *Let S be a regular semigroup that is both principally and naturally ordered. If S is not a Dubreil-Jacotin semigroup then S contains the crown*

in which e and f are maximal idempotents.

Proof If S is not a Dubreil-Jacotin semigroup then by Theorem 13.29 there exist at least two maximal idempotents e, f. Let $S(e, f) = g$ and $S(f, e) = h$. Since $eg \leqslant_n e$ we have $eg \leqslant e$ and so $ege \leqslant e$ whence $g \leqslant e^*$. Then $g \leqslant e$ by Theorem 13.29. Likewise, from $gf \leqslant_n f$ we obtain $g \leqslant f$. In fact, $g < e$ and $g < f$ since equality in either gives the contradiction $e \not{|} f$. To complete the proof, it suffices to show that $g \parallel h$. Suppose, by way of obtaining a contradiction, that we had $g \leqslant h$ so that, by Theorem 13.31, $(ef)^\circ \leqslant (fe)^\circ$. Then by Theorem 13.27 we would have $fe(ef)^\circ \leqslant fe(fe)^\circ \leqslant ff^\circ = f$ whence $fef = fegf = fe(ef)^\circ f \leqslant ff = f$ which gives $e \leqslant f^* = f$, a contradiction. Similarly, $h \leqslant g$ leads to the contradiction $f \leqslant e$. Thus we have the crown of idempotents as described. $\square$

EXERCISES

13.32. Let S be a regular semigroup that is both principally and naturally ordered and is not a Dubreil-Jacotin semigroup. If the idempotents in the crown of Theorem 13.32 are the only idempotents in S, prove that they generate the subsemigroup described by the Hasse diagram

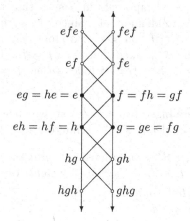

In this semigroup describe Green's relations $\mathcal{L}$ and $\mathcal{R}$.

13.33. Let S be a regular semigroup that is both principally and naturally ordered. Prove that S is a Dubreil-jacotin semigroup if and only if $\overline{E} = \langle E(S) \rangle$ is periodic (in particular, finite).

13.34. Let S be a regular semigroup that is both principally and naturally ordered. Prove that the following statements are equivalent:
   (1) S is a Dubreil-Jacotin semigroup;
   (2) $\mathcal{R}$ is regular on $\overline{E}$;
   (3) $\mathcal{L}$ is regular on $\overline{E}$.

13.35. A principally ordered regular semigroup is said to be **compact** if $x^\circ = x^*$ for every $x \in S$. Prove that if S is a principally ordered regular semigroup then the following statements are equivalent:
   (1) S is naturally ordered and compact;
   (2) $x \mapsto x^\circ$ is antitone;
   (3) S is completely simple [in the sense that the natural order $\leqslant_n$ reduces to equality on $E(S)$].

13.36. A principally ordered regular semigroup is said to be **weakly compact** if
$$(\forall e \in E(S)) \quad e^{**} \in E(S), \quad (ee^*)^* = e^{**}e^*, \quad (e^*e)^* = e^*e^{**}.$$

Let S be a regular semigroup that is both principally and naturally ordered. Prove that if S is weakly compact then any two maximal idempotents generate the semigroup described in Exercise 13.32.

13.7 Ordered completely simple semigroups

We recall (see, for example, Grillet [59], Higgins [64], or Howie [67]) that a semigroup S is **completely simple** if every idempotent is minimal with respect to $\leqslant_n$; in other words, if $\leqslant_n$ reduces to equality on $E(S)$.

Theorem 13.33 *An ordered regular semigroup S is completely simple if and only if it is (dually) naturally ordered and any two comparable idempotents are mutually inverse.*

Proof $\Rightarrow$: If S is completely simple then trivially S is (dually) naturally ordered. If now e, f are idempotents in S such that $e \leqslant f$ then, by Theorem 13.11, efe and fef are idempotents. Since $efe \leqslant_n e$ and $fef \leqslant_n f$ it follows that $efe = e$ and $fef = f$ and so e, f are mutually inverse.

$\Leftarrow$: Suppose that $e \leqslant_n f$, so that $e = ef = fe$. Then the hypotheses give $e \leqslant f$ (or $f \leqslant e$) and $f = fef$. It follows that $f = e$ and so $\leqslant_n$ reduces to equality. $\qquad \square$

In the case where biggest inverses exist, we have the following characterisation of ordered completely simple semigroups.

Theorem 13.34 *Let S be an ordered semigroup in which every element has a biggest inverse. Then the following statements are equivalent:*

(1) *S is completely simple;*

(2) *S is naturally ordered and*
$$(\forall e, f \in E(S)) \quad e \leqslant f \Rightarrow f^\circ \leqslant e^\circ;$$

(3) *S is dually naturally ordered and every idempotent of the form $x^{\circ\circ}x^\circ$ is maximal.*

Proof (1) $\Rightarrow$ (2): Suppose that (1) holds and that $e, f \in E(S)$ with $e \leqslant f$. By Theorem 13.33 we have $f \in V(e)$ and therefore
$$\begin{cases} e = efe \leqslant ef^\circ e = eef^\circ ee \leqslant eff^\circ fe = efe = e; \\ f^\circ = f^\circ ff^\circ = f^\circ ffefff^\circ \leqslant f^\circ ff^\circ ef^\circ ff^\circ = f^\circ ef^\circ \leqslant f^\circ ff^\circ = f^\circ, \end{cases}$$
whence we see that $ef^\circ e = e$ and $f^\circ ef^\circ = f^\circ$. Thus we also have $f^\circ \in V(e)$ and so $f^\circ \leqslant e^\circ$. That S is naturally ordered follows from Theorem 13.33.

(2) $\Rightarrow$ (1): Suppose that (2) holds and that $e, f \in E(S)$ are such that $e \leqslant_n f$. Then $e \leqslant f$ and $f^\circ \leqslant e^\circ$. Now since $ef = e$ we have that $fe^\circ e$ is idempotent with $fe^\circ e \leqslant_n f$ and so $fe^\circ e \leqslant f$. It follows that
$$e = ee^\circ e \leqslant fe^\circ e = fe^\circ ee \leqslant fe = e$$

and so $fe^\circ e = e$. Similarly, $ee^\circ f = e$. Consequently

$$f = ff^\circ f \leqslant fe^\circ f = fe^\circ eee^\circ f = ee = e$$

and therefore $e = f$. Thus $\leqslant_n$ reduces to equality on $E(S)$.

(1) $\Rightarrow$ (3): Suppose that (1) holds and consider the idempotent $\alpha_x = x^{\circ\circ}x^\circ$. If g is an idempotent such that $\alpha_x \leqslant g$ then by Theorem 13.33 we have $g \in V(\alpha_x)$. Hence $g \leqslant \alpha_x^\circ = \alpha_x$ by Theorem 13.22(5). Thus α_x is a maximal idempotent. That S is dually naturally ordered follows from Theorem 13.33.

(3) $\Rightarrow$ (1): Suppose that (3) holds and that $e, f \in E(S)$ are such that $e \leqslant_n f$, so that $f \leqslant e$. Observe that $ef^\circ e = eff^\circ fe = efe = e$ and so ef° is idempotent. Then $f^{\circ\circ}f^\circ ef^\circ$ is idempotent. But $f^{\circ\circ}f^\circ ef^\circ \leqslant_n f^{\circ\circ}f^\circ$ so it follows by the hypothesis that $f^{\circ\circ}f^\circ = f^{\circ\circ}f^\circ ef^\circ$ whence $f^\circ = f^\circ ef^\circ$. Consequently

$$f = ff^\circ f = ff^\circ ef^\circ f = ff^\circ feff^\circ f = fef = e$$

and so S is completely simple.   □

More generally, we also have the following result.

Theorem 13.35 (Blyth–Santos [30]) *Let S be a naturally ordered regular semigroup with a biggest idempotent α. Then the following statements are equivalent:*

(1) *$(\forall x \in S)$ $x \mapsto x^\circ$ is antitone;*

(2) *the inverse transversal $\alpha S\alpha$ is a group;*

(3) *S is orthodox strong Dubreil-Jacotin with a rectangular band of idempotents;*

(4) *S is strong Dubreil-Jacotin and $(\forall x \in S)$ $x^\circ = \alpha : x$;*

(5) *S is completely simple.*

Proof (1) $\Rightarrow$ (2): If (1) holds then by Theorem 13.16(5) we have

$$(\forall \overline{e} \in \overline{E}) \quad \alpha = \alpha^\circ \leqslant \overline{e}^\circ = \alpha\overline{e}\alpha.$$

But clearly $\alpha\overline{e}\alpha \leqslant \alpha$ and therefore $\alpha\overline{e}\alpha = \alpha$. But, as seen in the proof of Theorem 13.16, the inverse semigroup $\alpha S\alpha$ has semilattice of idempotents $\alpha\overline{E}\alpha$. Hence $\alpha S\alpha$ has only one idempotent and is therefore a group.

(2) $\Rightarrow$ (3): If (2) holds then we have $\alpha\overline{e}\alpha = \alpha$ for all $\overline{e} \in \overline{E}$. In particular, $\alpha ef\alpha = \alpha fe\alpha$ for all $e, f \in E(S)$. Then, by Theorem 13.18, S is orthodox and α is a middle unit. Consequently

$$(\forall e, f \in E(S)) \quad fef = f\alpha ef\alpha f = f\alpha f = f$$

and so $E(S)$ is a rectangular band. Consider now the mapping $\vartheta : S \to \alpha S\alpha$ given by $\vartheta(x) = \alpha x\alpha$. Since α is a middle unit, ϑ is a morphism which is clearly isotone. Define $\vartheta^+ : \alpha S\alpha \to S$ to act as the identity on $\alpha S\alpha$. Then we have $\vartheta\vartheta^+ = \mathrm{id}_{\alpha S\alpha}$, and $\vartheta^+\vartheta(x) = \alpha x\alpha = x^{\circ\circ} \geqslant x$ gives $\vartheta^+\vartheta \geqslant \mathrm{id}_S$. Since ϑ^+ is also isotone it follows that ϑ is residuated with residual ϑ^+. Hence S is a strong Dubreil-Jacotin semigroup with bimaximum element $\vartheta^+(\alpha) = \alpha$.

(3) $\Rightarrow$ (4): If (3) holds then $\xi e\xi = \xi$ for every $e \in E(S)$. In particular, $\xi x(\xi : x)\xi = \xi$ for every $x \in S$. Then, by the Corollary to Theorem 13.20,

$$x^\circ = (\xi:x)x(\xi:x) = (\xi:x)\xi x(\xi:x)\xi = (\xi:x)\xi = \xi:x.$$

$(4) \Rightarrow (1)$: If (4) holds then (1) follows by Theorem 13.34.

$(3) \Rightarrow (5)$: This is clear.

$(5) \Rightarrow (2)$: By Theorem 13.11, for every $e \in E(S)$ we have $\alpha e\alpha \in E(S)$. Then $\alpha e\alpha \leqslant_n \alpha$ and (5) give $\alpha e\alpha = \alpha$. Consequently

$$(\forall x \in S) \quad \alpha x\alpha \cdot \alpha x^\circ \alpha = \alpha x x^\circ \alpha = \alpha = \alpha x^\circ x\alpha = \alpha x^\circ \alpha \cdot \alpha x\alpha.$$

Since α is the identity of $\alpha S\alpha$ it follows that $\alpha S\alpha$ is a group. $\square$

EXERCISES

13.37. Let S be an ordered completely simple semigroup in which every element has a biggest inverse. Prove that S° is a subsemigroup of S. Show also that the Saito formulae $(xy)^\circ = (x^\circ xy)^\circ x^\circ = y^\circ(xyy^\circ)^\circ$ hold.

13.38. Let S be an ordered completely simple semigroup in which every element has a biggest inverse. Prove that $e \in E(S)$ is a maximal idempotent if and only if $e = e^\circ$.

13.39. If S is an ordered completely simple semigroup prove that the following statements are equivalent:

(1) S is orthodox and every $x \in S$ has a biggest inverse;

(2) S has a biggest idempotent.

13.40. Prove that the semigroup described in the Hasse diagram of Exercise 13.32 is completely simple and is residuated.

A fundamental result in semigroup theory is that a completely simple semigroup can be represented by a **Rees matrix semigroup** $M(G; I, \Lambda; P)$ which consists of a cartesian product $I \times G \times \Lambda$ where G is a group, I and Λ are sets, and $P = [p_{ij}]_{\Lambda \times I}$ is a **sandwich matrix** that controls the multiplication which is given by

$$(i, g, \lambda)(j, h, \mu) = (i, g\, p_{\lambda j}\, h, \mu).$$

In what follows we shall consider various orders that can be defined on a completely simple semigroup. For this purpose we shall suppose that G, I, Λ are ordered sets and in each we shall write the local order as $\leqslant$.

Definition On the cartesian product set $S = I \times G \times \Lambda$ the **lexicographic order** $\leqslant_{lex}$ is given by

$$(i, x, \lambda) \leqslant_{lex} (j, y, \mu) \iff \begin{cases} x < y, \\ \text{or} \quad x = y, i \leqslant j, \lambda \leqslant \mu. \end{cases}$$

In relation to this we can also define

(a) the **left lexicographic order** $\leqslant_l$ by

$$(i, x, \lambda) \leqslant_l (j, y, \mu) \iff \lambda \leqslant \mu \quad \text{and} \quad \begin{cases} x < y, & (1) \\ \text{or} \quad x = y, i \leqslant j; & (2) \end{cases}$$

(b) the **right lexicographic order** $\leqslant_r$ by

$$(j, y, \mu) \leqslant_r (k, z, \nu) \iff j \leqslant k \quad \text{and} \quad \begin{cases} y < z, & (3) \\ \text{or} \quad y = z, \mu \leqslant \nu. & (4) \end{cases}$$

It is readily seen that $\leqslant_l \cap \leqslant_r = \leqslant_c$, the cartesian order on S. In contrast, the relation $\leqslant_l \cup \leqslant_r$ is not in general transitive. However, its transitive closure (i.e. the smallest transitive relation that contains $\leqslant_l \cup \leqslant_r$) is an order on S which we call the **bootlace order** and write as $\leqslant_b$. We can obtain a useful description of $\leqslant_b$ from the following result.

Theorem 13.36 *The relations $\leqslant_l$ and $\leqslant_r$ commute.*

Proof Suppose that, for given (i, x, λ) and (k, z, ν), there exists (j, y, μ) such that

$$(i, x, \lambda) \leqslant_l (j, y, \mu) \leqslant_r (k, z, \nu).$$

To show that the relations $\leqslant_l$ and $\leqslant_r$ commute, it suffices to establish the existence of (j', y', μ') such that

$$(i, x, \lambda) \leqslant_r (j', y', \mu') \leqslant_l (k, z, \nu).$$

For this, it is readily seen that

$$\text{if conditions} \quad \begin{cases} (2), (4) \\ (2), (3) \\ (1), (4) \\ (1), (3) \end{cases} \quad \text{hold then} \quad (j', y', \mu') = \begin{cases} (j, y, \mu) \\ (k, z, \nu) \\ (i, x, \mu) \\ (i, y, \nu) \end{cases} \quad \text{suffices.} \qquad \square$$

Corollary $\leqslant_b = \leqslant_l \circ \leqslant_r = \leqslant_r \circ \leqslant_l = \leqslant_l \vee \leqslant_r.$ $\qquad \square$

Theorem 13.37 *The bootlace order $\leqslant_b$ is given by*

$$(i, x, \lambda) \leqslant_b (j, y, \mu) \iff \begin{cases} x = y, \ i \leqslant j, \ \lambda \leqslant \mu; & (5) \\ or \ x < y, \ i \leqslant j; & (6) \\ or \ x < y, \ \lambda \leqslant \mu. & (7) \end{cases}$$

Proof Observe that

(α) if (5) holds then $(i, x, \lambda) \leqslant_c (j, y, \mu)$;
(β) if (6) holds then $(i, x, \lambda) \leqslant_r (j, y, \mu)$;
(γ) if (7) holds then $(i, x, \lambda) \leqslant_l (j, y, \mu)$.

Since each of $\leqslant_c, \leqslant_l, \leqslant_r$ is contained in $\leqslant_l \vee \leqslant_r = \leqslant_b$ by Theorem 13.33, we see that if (5) or (6) or (7) holds then we have $(i, x, \lambda) \leqslant_b (j, y, \mu)$.

Conversely, it is readily seen from the definitions of $\leqslant_l$ and $\leqslant_r$ that if $(i, x, \lambda) \leqslant_b (j, y, \mu)$ then (5) or (6) or (7) holds. $\qquad \square$

It is useful to bear in mind the following lattice diagram which describes the hierarchy of the various orders that are considered above:

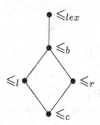

The above types of order arise naturally in connection with a completely simple semigroup $S = M(\langle x \rangle; I, \Lambda; P)$ over an ordered non-trivial cyclic group $\langle x \rangle$ with $|I| \geqslant 2$ and $|\Lambda| \geqslant 2$. Here we assume that I and Λ are also ordered, each having a bottom element which we shall denote by 1. In this connection, we say that the sandwich matrix P is **isotone** if $p_{\lambda i} \leqslant p_{\mu j}$ whenever $\lambda \leqslant \mu$ and $i \leqslant j$. There are two situations to consider, namely when $\langle x \rangle$ is totally ordered and when it is discretely ordered.

Suppose first that $\langle x \rangle$ is totally ordered with $x > 1$. If $\leqslant$ is one of the five orders described above then precisely when $(S; \leqslant)$ is an ordered semigroup is recorded in the following result, the proof of which is routine and left to the reader.

Theorem 13.38 *Compatibility with multiplication occurs if and only if, for*

(1) $\leqslant_c$: P *is isotone;*
(2) $\leqslant_l$: P *is isotone and* $(\forall i, \lambda)$ $p_{\lambda i} \leqslant p_{\lambda 1} x$;
(3) $\leqslant_r$: P *is isotone and* $(\forall i, \lambda)$ $p_{\lambda i} \leqslant x p_{1i}$;
(4) $\leqslant_b$: P *is isotone and* $(\forall i, \lambda)$ $p_{\lambda i} \leqslant \inf\{p_{\lambda 1} x, x p_{1i}\}$;
(5) $\leqslant_{lex}$: $(\forall i, \lambda)$ $p_{\lambda i} = p_{11}$.   □

Suppose now that $\langle x \rangle$ is discretely ordered. In this case we consider the **discrete lexicographic order** $\leqslant_{dl}$ defined on $I \times \langle x \rangle \times \Lambda$ by

$$(i, a, \lambda) \leqslant_{dl} (j, b, \mu) \iff i \leqslant j, \ a = b, \ \lambda \leqslant \mu.$$

As is readily seen, $\leqslant_{dl}$ is compatible if and only if $p_{\lambda i} = p_{11}$ for all i and λ.

In illustration of the above, consider the case where $I = \Lambda = \mathbf{2}$. In the completely simple semigroup $S_x = M(\langle x \rangle; \mathbf{2}, \mathbf{2}; P)$ we shall assume that P is **normalised** [59] with $p_{11} = x^{-1}$. If $\leqslant$ denotes any of the six orders defined above we shall determine precisely when $(S_x; \leqslant)$ is an ordered semigroup. For this purpose, we observe by Theorem 13.38 that this is the case if and only if P is of the form

$$P = \begin{bmatrix} x^{-1} & x^{-1} \\ x^{-1} & z \end{bmatrix}$$

where $z \geqslant x^{-1}$. There are then two cases to consider.

(a) $z = x^{-1}$.

In all cases there exists a biggest idempotent, namely $\alpha = (2, x, 2)$. Then by Theorem 13.16 every element β has a biggest inverse, namely $\beta^\circ = \alpha \beta' \alpha$ for any $\beta' \in V(\beta)$. Moreover, by Theorem 13.35, S_x is an orthodox strong Dubreil-Jacotin semigroup and, by Theorem 13.18, α is a middle unit. The band of idempotents is then $B_4 = \langle e, \alpha \rangle$ where e and α are mutually inverse idempotents with $e < \alpha$, the corresponding Hasse diagram being

In this situation, the compatible order $\leqslant$ can be any one of the six orders considered above. It is readily seen that the ordered semigroups arising from

each of these orders are described by the following Hasse diagrams, with multiplication achieved by using the sandwich matrix

$$P = \begin{bmatrix} x^{-1} & x^{-1} \\ x^{-1} & x^{-1} \end{bmatrix}.$$

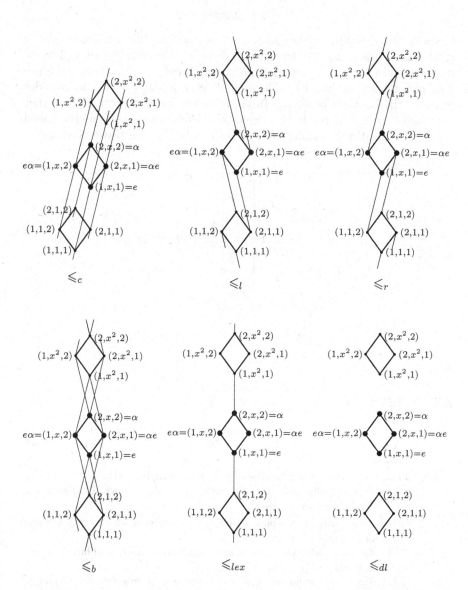

Observe that of the above only $(S_x; \leqslant_{dl})$ can be finite, and that this occurs precisely when x is of finite order.

(b) $z \neq x^{-1}$.

In this case we necessarily have $x^{-1} < z$. Consider the idempotents $e = (1, x, 2)$ and $f = (2, x, 1)$. We have $fe = (2, x, 2)$ which is not idempotent. Now

$(i, y, \lambda) \in V(2, x, 2)$ if and only if $p_{2i} y p_{\lambda 2} = x^{-1}$, and the only possibilities for this are

$$(i, y, \lambda) = \begin{cases} (1, x, 1) \\ (1, z^{-1}, 2) \\ (2, z^{-1}, 1) \\ (2, x^{-1} z^{-2}, 2). \end{cases}$$

If now S_x is ordered by $\leqslant_c$ then since $z^{-1} < x$ it follows that no biggest inverse of $(2, x, 2)$ exists. If S_x is ordered by $\leqslant_l$ or by $\leqslant_r$ then by Theorem 13.38 we must have $z^{-1} = p_{22} = 1$ and again no biggest inverse of $(2, x, 2)$ exists. Since $\leqslant_{lex}$ and $\leqslant_{dl}$ are excluded by Theorem 13.38, the only case left to consider concerns the bootlace order $\leqslant_b$. In this case we see by Theorem 13.38 that necessarily $z = 1$. The resulting ordered completely simple semigroup is called the **crown bootlace semigroup** and is described by the Hasse diagram

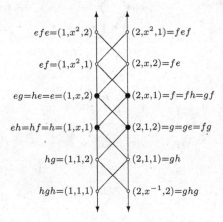

EXERCISES

13.41. Prove that the crown bootlace semigroup is residuated and determine formulae for residuals. Determine also formulae that give x^* and x° for each element x.

13.42. Prove that the crown bootlace semigroup can be characterised as the smallest principally and naturally ordered regular semigroup that is not a Dubreil-Jacotin semigroup.

13.43. Let S be an ordered completely simple semigroup in which every element has a biggest inverse. Prove that

(1) if S is orthodox then S contains a subsemigroup isomorphic to $(S_x; \leqslant)$ where $\leqslant$ is one of $\leqslant_c, \leqslant_l, \leqslant_r, \leqslant_b, \leqslant_{lex}, \leqslant_{dl}$;

(2) if S is not orthodox then S contains a subsemigroup isomorphic to the crown bootlace semigroup.

Structure theorems

In the classification of semigroups an important part is played by various structure theorems, some of which are very complicated. Our purpose here is to obtain structure theorems for some of the ordered semigroups discussed previously. Of necessity, these too are rather complicated in general, though there are special cases that have a relatively simple structure.

14.1 Naturally ordered regular semigroups

14.1.1 Inverse transversals

In the investigation of naturally ordered regular semigroups we have met with the notion of an *inverse transversal*. This has its roots in a paper by McAlister and Blyth [80], the terminology being introduced by Blyth and McFadden [26]. If T is an inverse transversal of S then we shall denote by x° the unique element of $T \cap V(x)$ and write T as $S^\circ = \{x^\circ \mid x \in S\}$. Then in the inverse semigroup S° we have $(x^\circ)^{-1} = x^{\circ\circ}$, so that $x^\circ = x^{\circ\circ\circ}$ for every $x \in S$. As shown by Blyth and Chen [21] all inverse transversals of S are isomorphic.

We list below the basic properties of an inverse transversal on a regular semigroup S. These are drawn variously from Blyth and Santos [31], McAlister and McFadden [82], Saito [97, 98], and Tang [108]. For a survey article on the classification of inverse transversals and examples of the many various types we refer the reader to Blyth and Santos [32].

Theorem 14.1 $(xy)^\circ = (x^\circ xy)^\circ x^\circ = y^\circ(xyy^\circ)^\circ = y^\circ(x^\circ xyy^\circ)^\circ x^\circ.$

Proof For example, $(x^\circ xy)^\circ x^\circ \in S^\circ \cap V(xy)$ gives the first equality. $\qquad\square$

Two particularly important subsets of $E(S)$ are

$$I = \{x \in S \mid x = xx^\circ\} = \{x \in S \mid x^\circ x = x^\circ\};$$
$$\Lambda = \{x \in S \mid x = x^\circ x\} = \{x \in S \mid xx^\circ = x^\circ\}.$$

Equivalent definitions are

$$I = \{xx^\circ \mid x \in S\}, \quad \Lambda = \{x^\circ x \mid x \in S\}.$$

For example, we have $xx^\circ(xx^\circ)^\circ = xx^\circ(x^\circ xx^\circ)^\circ x^\circ = xx^\circ x^{\circ\circ}x^\circ = xx^\circ.$

Theorem 14.2 $I \cap \Lambda = E(S^\circ)$.

Proof If $x \in I \cap \Lambda$ then $x^\circ x = x^\circ$ and $x = x^\circ x$. Thus $x = x^\circ \in S^\circ \cap E(S) = E(S^\circ)$. Conversely, if $x \in E(S^\circ)$ then $x^2 = x = x^\circ$ whence $x = xx^\circ = x^\circ x \in I \cap \Lambda$. $\square$

Theorem 14.3   *I and Λ are sub-bands with I left regular and Λ right regular.*

Proof Suppose that $i, j \in I$ and consider the sandwich element $g \in S(i,j)$ given by $g = j(ij)^\circ i$. Using Theorem 14.1, we have $g = j(ij)^\circ i = j(i^\circ ij)^\circ i^\circ i = j(i^\circ ij)^\circ i^\circ = j(ij)^\circ$ whence we see that $ig = ij(ij)^\circ \in I$. We also have, by Theorem 14.1 again, $g^\circ = [j^\circ j(ij)^\circ i]^\circ j^\circ$. This, together with the fact that $j^\circ = j^\circ j$, gives $g^\circ = g^\circ j$. Using these facts, we see that $g \mathcal{L} ig \mathcal{L} (ig)^\circ$ whence $(ig)^\circ \in S^\circ \cap V(g)$ and therefore $g^\circ = (ig)^\circ$. Thus $g^\circ \mathcal{L} g$ and so we have $g = gg^\circ = gg^\circ j = gj$ which gives $ij = igj = ig \in I$. Hence I is a subsemigroup which is clearly a sub-band. To see that I is left regular, observe from the above that $ij = ig$ gives $ij = iji$. Dually, Λ is a right regular sub-band. $\square$

There are two important consequences of Theorem 14.3, each of which provides other useful formulae.

Theorem 14.4 $(xy^\circ)^\circ = y^{\circ\circ} x^\circ$, $(x^\circ y)^\circ = y^\circ x^{\circ\circ}$.

Proof Since I is left regular, we have
$$y^\circ x^{\circ\circ} \cdot x^\circ y \cdot y^\circ x^{\circ\circ} = y^\circ yy^\circ x^{\circ\circ} x^\circ yy^\circ x^{\circ\circ} = y^\circ yy^\circ x^{\circ\circ} x^\circ x^{\circ\circ} = y^\circ x^{\circ\circ}$$
and similarly $x^\circ y \cdot y^\circ x^{\circ\circ} \cdot x^\circ y = x^\circ y$. Thus $y^\circ x^{\circ\circ} \in S^\circ \cap V(x^\circ y)$ and so $y^\circ x^{\circ\circ} = (x^\circ y)^\circ$. Dually, we have the second identity. $\square$

Theorem 14.5 *Green's relations $\mathcal{L}$ and $\mathcal{R}$ on S are given by*
$$(x, y) \in \mathcal{L} \iff x^\circ x = y^\circ y, \quad (x, y) \in \mathcal{R} \iff xx^\circ = yy^\circ.$$

Proof Suppose that $(x, y) \in \mathcal{R}$. Then $xS = yS$ gives $xx^\circ = yz$ for some $z \in S$ whence $xx^\circ = yy^\circ xx^\circ$. Similarly, $yy^\circ = xx^\circ yy^\circ$. Since I is left regular we have $xx^\circ = yy^\circ xx^\circ = xx^\circ yy^\circ xx^\circ = xx^\circ yy^\circ = yy^\circ$. Conversely, if $xx^\circ = yy^\circ$ then $x = xx^\circ x = yy^\circ x \in yS$ and similarly $y \in xS$ whence $(x, y) \in \mathcal{R}$. $\square$

Other significant subsets of S are
$$L = \{x \in S \mid x = xx^\circ x^{\circ\circ}\} = \{x \in S \mid x^\circ x = x^\circ x^{\circ\circ}\};$$
$$R = \{x \in S \mid x = x^{\circ\circ} x^\circ x\} = \{x \in S \mid xx^\circ = x^{\circ\circ} x^\circ\}.$$

Theorem 14.6 $L \cap R = S^\circ$.

Proof If $x \in L \cap R$ then $x = xx^\circ x^{\circ\circ} = x^{\circ\circ} x^\circ xx^\circ x^{\circ\circ} = x^{\circ\circ} \in S^\circ$; conversely if $x \in S^\circ$ then $x = x^{\circ\circ}$ gives $x = xx^\circ x = xx^\circ x^{\circ\circ} \in L$ and similarly $x \in R$. $\square$

Theorem 14.7 *If $x \in L$ or $y \in R$ then $(xy)^\circ = y^\circ x^\circ$.*

Proof Suppose, for example, that $x \in L$. Then $(xy)^\circ = y^\circ (x^\circ xyy^\circ)^\circ x^\circ = y^\circ (x^\circ x^{\circ\circ} yy^\circ)^\circ x^\circ = y^\circ (yy^\circ)^\circ (x^\circ x^{\circ\circ})^\circ x^\circ = y^\circ y^{\circ\circ} y^\circ x^\circ x^{\circ\circ} x^\circ = y^\circ x^\circ$. $\square$

Theorem 14.8 *L and R are subsemigroups of S, with L left inverse and R right inverse. Moreover, $E(L) = I$ and $E(R) = \Lambda$.*

Proof If $x, y \in L$ then by Theorem 14.7 we have $(xy)^\circ = y^\circ x^\circ$. It follows that

$$(xy)^\circ xy = y^\circ x^\circ xy = y^\circ y^{\circ\circ} y^\circ x^\circ x^{\circ\circ} yy^\circ y^{\circ\circ} = y^\circ x^\circ x^{\circ\circ} y^{\circ\circ} \in S^\circ$$

and so $(xy)^\circ xy = [(xy)^\circ xy]^\circ = (xy)^\circ (xy)^{\circ\circ}$ whence $xy \in L$. Thus L is a subsemigroup. We now observe that $E(L) = I$. In fact, if $i \in I$ then $i^\circ \in E(S^\circ)$ and so $i^\circ = i^{\circ\circ}$ whence $i = ii^\circ = ii^\circ i^\circ = ii^\circ i^{\circ\circ} \in L$; and conversely if $g \in E(L)$ then $g^\circ g = g^\circ g^{\circ\circ} \in S^\circ$ and $g^\circ g \in V(g)$ give $g^\circ g = g^\circ$ whence $g = gg^\circ \in I$. It follows that L is orthodox with a left regular band of idempotents and so is left inverse. Dually, R is right inverse. □

The following Venn diagram provides a useful summary:

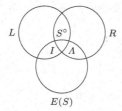

Note from the above that S° is an inverse transversal of both L and R.

Theorem 14.9 *S is left [right] inverse if and only if $\Lambda = E(S^\circ)$ $[I = E(S^\circ)]$.*

Proof Suppose, for example, that S is left inverse. Then $E(S)$ is a left regular band. If $l \in \Lambda$ we then have $l = ll^\circ l = ll^\circ = l^\circ$ and so $\Lambda = E(S^\circ)$. Conversely, if $\Lambda = E(S^\circ)$ then for every $x \in S$ we have $x^\circ x = x^\circ x^{\circ\circ}$ whence $S = L$ and so S is left inverse. □

From the definition of R we clearly have $R \subseteq S^\circ \Lambda$. The reverse inclusion follows from the fact that R is a subsemigroup. Hence $R = S^\circ \Lambda$, and similarly $L = IS^\circ$. In view of this it is natural to consider also the set product $I\Lambda$.

Theorem 14.10 $I\Lambda = \{x \in S \mid x^\circ \in E(S^\circ)\}$.

Proof If $i \in I$ and $l \in \Lambda$ we have $(il)^\circ = l^\circ (i^\circ ill^\circ)^\circ i^\circ = l^\circ (i^\circ l^\circ)^\circ i^\circ = l^\circ i^\circ \in E(S^\circ)$. Conversely, if $x \in S$ is such that $x^\circ \in E(S^\circ)$ then $x = xx^\circ x = xx^\circ x^\circ x \in I\Lambda$. □

Corollary $S^\circ \cap I\Lambda = E(S^\circ)$.

Proof $x^\circ \in I\Lambda \iff x^{\circ\circ} \in E(S^\circ) \iff x^\circ = (x^{\circ\circ})^{-1} \in E(S^\circ)$. □

Since $i^\circ \in E(S^\circ)$ for every $i \in I$ we have $I \subseteq I\Lambda$, and similarly $\Lambda \subseteq I\Lambda$.

Theorem 14.11 *The following statements are equivalent:*
  (1) *$I\Lambda$ is a subsemigroup of S;*
  (2) *$I\Lambda = \langle E(S) \rangle$;*
  (3) *$e \in E(S) \Rightarrow e^\circ \in E(S^\circ)$ [i.e., $E(S) \subseteq I\Lambda$];*
  (4) *$(\Lambda I)^\circ \subseteq E(S^\circ)$.*

Proof (1) $\Rightarrow$ (3): If (1) holds then $\Lambda I \subseteq I\Lambda$, so if $x \in E(S)$ then we have $x^\circ = x^\circ x x^\circ = x^\circ x x x^\circ \in I\Lambda$ whence, by the above Corollary, $x^\circ \in E(S^\circ)$.

(3) $\Rightarrow$ (4): For $i \in I$ and $l \in \Lambda$ we have $i(li)^\circ l \in E(S)$. Observe that

$$[i(li)^\circ l]^\circ = l^\circ [i^\circ i(li)^\circ l l^\circ]^\circ i^\circ = l^\circ [i^\circ (li)^\circ l^\circ]^\circ i^\circ = l^\circ (li)^{\circ\circ} i^\circ = (li)^{\circ\circ}.$$

Consequently, if (3) holds then $(li)^{\circ\circ} \in E(S^\circ)$ whence $(li)^\circ \in E(S^\circ)$.

(4) $\Rightarrow$ (1): Suppose that (4) holds and that $x, y \in I\Lambda$. Then $(x^\circ x y y^\circ)^\circ \in E(S^\circ)$ and $x^\circ, y^\circ \in E(S^\circ)$. It follows that $(xy)^\circ = y^\circ (x^\circ x y y^\circ)^\circ x^\circ \in E(S^\circ)$ and therefore $xy \in I\Lambda$. Hence $I\Lambda$ is a subsemigroup.

(3) $\Rightarrow$ (2): If (3) holds then $E(S) \subseteq I\Lambda$. But clearly $I\Lambda \subseteq \langle E(S) \rangle$ and, since (3) $\Rightarrow$ (1) from the above, $I\Lambda$ is a subsemigroup. It follows that $I\Lambda = \langle E(S) \rangle$.

(2) $\Rightarrow$ (1): This is clear. $\qquad\square$

Theorem 14.12 *The following statements are equivalent*:

(1) S *is orthodox*;

(2) $(\forall x, y \in S)$ $(xy)^\circ = y^\circ x^\circ$;

(3) $(\forall i \in I)(\forall l \in \Lambda)$ $(li)^\circ = i^\circ l^\circ$;

(4) $e \in E(S) \Longleftrightarrow e^\circ \in E(S^\circ)$ [i.e., $E(S) = I\Lambda$].

Proof (1) $\Rightarrow$ (2): If S is orthodox then $y^\circ x^\circ \in S^\circ \cap V(xy)$ whence (2) follows.

(2) $\Rightarrow$ (3): This is clear.

(3) $\Rightarrow$ (2): If (3) holds then we have the identity $(y^\circ y x x^\circ)^\circ = x^{\circ\circ} x^\circ y^\circ y^{\circ\circ}$. On pre-multiplying by x° and post-multiplying by y°, we obtain $(yx)^\circ = x^\circ y^\circ$.

(2) $\Rightarrow$ (4): Suppose now that (2) holds. If $x \in E(S)$ then $x^\circ = (xx)^\circ = x^\circ x^\circ$ whence $x^\circ \in E(S^\circ)$; and conversely if $x^\circ \in E(S^\circ)$ then $x = xx^\circ x = x(x^\circ)^2 x = x(x^2)^\circ x = x(x^2)^\circ x^2 (x^2)^\circ x = x(x^\circ)^2 x^2 (x^\circ)^2 x = xx^\circ xxx^\circ x = xx$ whence $x \in E(S)$.

(4) $\Rightarrow$ (1): If (4) holds then for all $e, f \in E(S)$ we have $e^\circ, f^\circ \in E(S^\circ)$ whence $(e^\circ e f f^\circ)^\circ = f^{\circ\circ}(ef)^\circ e^{\circ\circ} = f^\circ(ef)^\circ e^\circ = (ef)^\circ$. But (4) of Theorem 14.11 also holds, and therefore $(e^\circ e f f^\circ)^\circ \in E(S^\circ)$. Thus $(ef)^\circ \in E(S^\circ)$ and so, by (4), $ef \in E(S)$. Hence S is orthodox. $\qquad\square$

We recall that an inverse transversal S° is said to be **multiplicative** if $\Lambda I \subseteq E(S^\circ)$. As shown in [31], an inverse transversal S° is multiplicative if and only if it is a **quasi-ideal** $[S^\circ S S^\circ \subseteq S^\circ]$ and is **weakly multiplicative** $[(\Lambda I)^\circ \subseteq E(S^\circ)]$. The algebraic structure of a regular semigroup with a multiplicative inverse transversal is known and can be summarised as follows.

Theorem 14.13 (Blyth–McFadden [26]) *Let S be a regular semigroup with an inverse transversal S° that is multiplicative. Then $I = \{e \in S \mid e = ee^\circ\}$ is a left normal band and $\Lambda = \{f \in S \mid f = f^\circ f\}$ is a right normal band. For every $a \in S^\circ$ let $\mu_a : E(S^\circ) \to E(S^\circ)$ be given by $\mu_a(e) = aea^{-1}$. Then*

$$W(I, S^\circ, \Lambda) = \{(e, a, f) \mid e \mathcal{L} \, a^{-1} \mathcal{R} \, f\}$$

equipped with the multiplication

$$(e, a, f)(g, b, h) = \big(e\mu_a(fg), \, afgb, \, \mu_{b^{-1}}(fg)h\big)$$

is a regular semigroup. Moreover, S is isomorphic to $W(I, S^\circ, \Lambda)$ under the assignment $x \mapsto (xx^\circ, x^{\circ\circ}, x^\circ x)$ whose inverse is $(e, a, f) \mapsto eaf$. $\qquad\square$

In the proof of the above result it is shown that

$$(e, a, f)^\circ = (a^{-1}a, a^{-1}, aa^{-1}) = (f^\circ, a^{-1}, e^\circ) \in V(e, a, f)$$

so that, under the isomorphism, $(eaf)^\circ = a^{-1}$. It is also shown that

$$(e, a, f)(e, a, f)^\circ = (e, aa^{-1}, aa^{-1})$$

whence $eaf(eaf)^\circ = eaa^{-1} = ee^\circ = e$. Finally, the idempotents are the elements (e, a, f) where $a \in E(S^\circ)$ and e, f are mutually inverse.

14.1.2 Biggest idempotent

If S is a naturally ordered regular semigroup that has a biggest idempotent α then, as observed in Theorem 13.16, $S^\circ = \alpha S \alpha$ is an inverse transversal of S and, as observed prior to Theorem 13.25, S° is multiplicative. We may therefore seek to obtain an order-theoretic version of Theorem 14.13 in this case. The above algebraic isomorphism shows that in order to achieve this it will be necessary to suppose that $\mathcal{R}$ and $\mathcal{L}$ are regular on S. We shall denote by $W_c(I, S^\circ, \Lambda)$ the set $W(I, S^\circ, \Lambda)$ under the cartesian order.

Theorem 14.14 *Let S be a naturally ordered regular semigroup with a biggest idempotent α and suppose that $\mathcal{R}$ and $\mathcal{L}$ are regular on S. Then there is an ordered semigroup isomorphism*

$$S \simeq W_c\big(E(S)\alpha, \, \alpha S \alpha, \, \alpha E(S)\big).$$

Proof We have seen in the proof of Theorem 13.16 that for every $\overline{e} \in \overline{E}$ we have $\overline{e}\,\overline{e}^\circ = \overline{e}\alpha$. In particular therefore $ee^\circ = e\alpha$ for every $e \in E(S)$. Consequently, $I = E(S)\alpha$; and similarly $\Lambda = \alpha E(S)$. Moreover, since $x^{\circ\circ} = \alpha x \alpha$ we have $S^\circ = \alpha S \alpha$.

Now if $a, c \in S$ are such that $a \leqslant c$ then for every $e \in E(S^\circ)$ we have $ea \leqslant ec$ and so, by the regularity of $\mathcal{L}$, $(ea)^\circ ea \leqslant (ec)^\circ ec$. But since $e = e^\circ$ by Theorem 14.2 we have, by Theorem 14.4, $(ea)^\circ ea = a^\circ e^\circ ea = a^\circ ea = \mu_a(e)$. Consequently $\mu_a \leqslant \mu_c$. Likewise, if $b, d \in S^\circ$ are such that $b \leqslant d$ then, since $\mathcal{R}$ is regular, for every $e \in E(S^\circ)$ we have $be \leqslant de$ whence $be(be)^\circ \leqslant de(de)^\circ$. But $be(be)^\circ = bee^\circ b^\circ = beb^\circ = \mu_{b^{-1}}(e)$. Consequently $\mu_{b^{-1}} \leqslant \mu_{d^{-1}}$.

It follows from these observations that $W_c\big(E(S)\alpha, \, \alpha S \alpha, \, \alpha E(S)\big)$ is an ordered semigroup and that the isomorphism of Theorem 14.13 is an order isomorphism. $\qquad\square$

14.1.3 Biggest inverses

The algebraic isomorphism of Theorem 14.13 may also be applied in the case where S has biggest inverses. We recall that by Theorem 13.25 that if S is a naturally ordered regular semigroup with biggest inverses on which $\mathcal{R}$ and $\mathcal{L}$ are (weakly) regular then the set S° of biggest inverses is a multiplicative inverse transversal of S. In this connection we have the following interesting result.

Theorem 14.15 (Saito [99]) *Every regular semigroup S with a multiplicative inverse transversal $S°$ can be naturally ordered in such a way that $x°$ is the biggest inverse of x and both $\mathcal{R}$ and $\mathcal{L}$ are regular.*

Proof The strategy is to represent S by $W(I, S°, \Lambda)$ as in Theorem 14.13, and to define orders on the components $I, S°, \Lambda$ such that $W_c(I, S°, \Lambda)$ has the stated properties. We begin by defining relations $\leqslant_I$ on I and $\leqslant_\Lambda$ on Λ by

$$e \leqslant_I g \iff ge \in \{e, e°\}, \quad f \leqslant_\Lambda h \iff fh \in \{f, f°\}.$$

A routine analysis of the cases arising shows that $\leqslant_I$ and $\leqslant_\Lambda$ are orders.

These orders are compatible with multiplication. For example, suppose that $e, g \in I$ with $e \leqslant_I g$ and let $k \in I$. If $ge = e$ then $kgke = kkge = ke$ whence $ke \leqslant_I kg$; and $gkek = gekk = ek$ whence $ek \leqslant_I gk$. On the other hand, if $ge = e°$ then $kgke = kkge = ke°$ so $kgke = ke°e = ke$ and $ke \leqslant_I kg$; and $gkek = gekk = e°k = e°kk° = e°k°°k° = e°k° = k°e° = (ek)°$ and so $ek \leqslant_I gk$.

We observe that $e°$ is the biggest inverse of $e \in I$. In fact, if $g \in V(e) \cap I$ then $g = geg = ge$ and $e = ege = eg$ which gives $g° = (ge)° = e°g° = g°e° = (eg)° = e°$ so that $e°g = g°g = g°$ and therefore $g \leqslant_I e°$.

We next order the inverse semigroup $S°$ by its natural order $\leqslant_n$. Now if $e, g \in I$ with $e \leqslant_I g$, and if $f, h \in \Lambda$ with $f \leqslant_\Lambda h$, then a routine analysis of the possible cases reveals that $fe \leqslant_n hg$ in $E(S°)$. Further simple calculations now show that $W_c(I, S°, \Lambda)$ is an ordered semigroup.

If now (e, a, f) and (g, b, h) are idempotents such that $(e, a, f) \leqslant_n (g, b, h)$ then we have $gbheb^{-1} = e$, $afgb = a$, and $b^{-1}fgbh = f$. Consequently $ge = e$ and $fh = f$, so that $e \leqslant_I g$ and $f \leqslant_\Lambda h$. Moreover, $ba = ab = afgbb = afbg = a$ and so $a \leqslant_n b$. It follows that $W_c(I, S°, \Lambda)$ is naturally ordered.

To see that each (e, a, f) has a biggest inverse, let $(g, b, h) \in V(e, a, f)$. Then, using the property (1) established in the proof of Theorem 13.25, we have $(e, a, f)° = (g, b, h)°°$ so that $a^{-1} = b$. Consequently $g \leqslant g° = bb^{-1} = a^{-1}a$; and similarly $h \leqslant e° = aa^{-1}$. Thus $(g, b, h) \leqslant (a^{-1}a, a^{-1}, aa^{-1}) = (e, a, f)°$.

To see that $\mathcal{R}$ is regular, let $(e, a, f) \leqslant (g, b, f)$. Then $a \leqslant_n b$ and so $aa^{-1} \leqslant_n bb^{-1}$. Consequently

$$(e, a, f)(e, a, f)° = (e, aa^{-1}, aa^{-1}) \leqslant (g, bb^{-1}, bb^{-1}) = (g, b, h)(g, b, h)°.$$

Similarly, $\mathcal{L}$ is regular.

Finally, we define an order $\sqsubseteq$ on S by setting $x \sqsubseteq y$ if and only if

(1) $xx° \leqslant_I yy°$ in I;
(2) $x°° \leqslant y°°$ in $S°$;
(3) $x°x \leqslant_\Lambda y°y$ in Λ.

Then $(e, a, f) \leqslant (g, b, h)$ implies

(4) $eaf(eaf)° = e \leqslant g = (g, b, h)(g, b, h)°$ in I;
(5) $(eaf)°° = a \leqslant b = (gbh)°°$ in $S°$;
(6) $(eaf)°(eaf) = f \leqslant h = (gbh)°gbh$ in Λ.

Hence the isomorphism $(e, a, f) \mapsto eaf$ is isotone. □

14.2 Integral Dubreil-Jacotin inverse semigroups

In an integral Dubreil-Jacotin inverse semigroup we have, by Theorem 13.13(1) and its dual,

$$xx^{-1}[1:(1:x)] = x = [1:(1:x)]x^{-1}x.$$

It is therefore natural to seek coordinatisations of the form

$$\left(xx^{-1},\, 1:(1:x)\right) \sim x \sim \left(1:(1:x),\, x^{-1}x\right).$$

For order isomorphisms to result from this it is clearly necessary that Green's relations $\mathcal{R}$ or $\mathcal{L}$ be regular. Our first objective therefore is to describe the structure of an integral Dubreil-Jacotin inverse semigroup S on which both $\mathcal{R}$ and $\mathcal{L}$ are regular. This we shall accomplish in relation to the $\wedge$-semilattice $E(S)$ of idempotents of S, the set of A-nomal elements (equivalently, the group S/A_ξ), and the set $\operatorname{End} E$ of $\wedge$-morphisms on $E(S)$. For this purpose, we introduce a special order on the set $\operatorname{End} E$ of $\wedge$-morphisms of a $\wedge$-semilattice, as described in the following simple result.

Theorem 14.16 *Let E be a $\wedge$-semilattice with a top element 1 and let $\operatorname{End} E$ be the set of $\wedge$-morphisms on E. Then the relation $\leqslant^*$ defined on $\operatorname{End} E$ by*

$$f \leqslant^* g \iff (\forall x \in E)\ f(x) = f(1) \wedge g(x)$$

is an order and $(\operatorname{End} E; \leqslant^*, \circ)$ is an ordered semigroup.

Proof It is clear that $\leqslant^*$ so defined is an order. If now $f \leqslant^* g$ then for every $h \in \operatorname{End} E$ we have $hf(x) = hf(1) \wedge hg(x)$, so that $hf \leqslant^* hg$; and $fh(x) = f(1) \wedge gh(x) = f(1) \wedge gh(1) \wedge gh(x) = fh(1) \wedge gh(x)$, so that $fh \leqslant^* gh$. Thus $(\operatorname{End} E; \leqslant^*, \circ)$ is an ordered semigroup. $\qquad\square$

A structure theorem for integral Dubreil-Jacotin inverse semigroups on which $\mathcal{R}$ and $\mathcal{L}$ are regular is then the following, in which 'semilattice' means $\wedge$-semilattice and $\wedge$ is written as multiplication.

Theorem 14.17 (Blyth [16]) *Let E be a semilattice with a top element 1 and let G be an ordered group. Let $\zeta : G \to \operatorname{End} E$, described by $g \mapsto \zeta_g$, be an isotone mapping such that*

    (1) $(\forall g, h \in G)\ \zeta_g \circ \zeta_h \leqslant^* \zeta_{gh}$;
    (2) $(\forall g \in G)\ \zeta_g \circ \zeta_{g^{-1}} \circ \zeta_g = \zeta_g$;
    (3) $\zeta_1 = \operatorname{id}_E$;
    (4) $(\forall g, h \in G)\ g \leqslant h \implies \zeta_g = \zeta_h$.

On the cartesian ordered set

$$S_\zeta = \{(x, g) \in E \times G \mid x \leqslant \zeta_g(1)\}$$

define a multiplication by

$$(x, g)(y, h) = \left(x\zeta_g(y),\, gh\right).$$

Then S_ζ is an integral Dubreil-Jacotin inverse semigroup on which $\mathcal{R}$ and $\mathcal{L}$ are regular.

Moreover, every such semigroup arises in this manner. More precisely, let S be an integral Dubreil-Jacotin inverse semigroup on which $\mathcal{R}$ and $\mathcal{L}$ are regular. Then the set G of A-nomal elements is a group under the law of composition $g \star h = 1:(1:gh)$, the class $[1]_{A_1}$ is a semilattice with top element 1, the mapping $\zeta : G \to \mathrm{End}[1]_{A_1}$ given by $g \mapsto \zeta_g$, where $\zeta_g(x) = gxg^{-1}$, satisfies conditions $(1) - (4)$ above, and $S \simeq S_\zeta$.

Proof Observe that the multiplication in S_ζ is well defined since if $x \leqslant \zeta_g(1)$ and $y \leqslant \zeta_h(1)$ then $x\zeta_g(y) \leqslant \zeta_g(y) \leqslant \zeta_g\zeta_h(1) \leqslant \zeta_{gh}(1)$. The multiplication is also associative since

$$\begin{cases} [(x,g)(y,h)](z,k) = \big(x\zeta_g(y)\zeta_{gh}(z), ghk\big); \\ (x,g)[(y,h)(z,k)] = \big(x\zeta_g[y\zeta_h(z)], ghk\big), \end{cases}$$

and these are equal since, by (1),

$$\zeta_g[y\zeta_h(z)] = \zeta_g(y)\zeta_g\zeta_h(z) = \zeta_g(y)\zeta_g\zeta_h(1)\zeta_{gh}(z) = \zeta_g(y)\zeta_{gh}(z),$$

the last equality following from the fact that $y \leqslant \zeta_h(1)$.

It is clear that, since $g \mapsto \zeta_g$ is isotone, this multiplication is compatible with the cartesian order, so that S_ζ is an ordered semigroup. That S_ζ is regular follows from the observation that, by (2) and (1),

$$\zeta_g(1) = (\zeta_g \circ \zeta_{g^{-1}})[\zeta_g(1)] = \zeta_g\zeta_{g^{-1}}(1)\zeta_g(1) = \zeta_g\zeta_{g^{-1}}(1)$$

so that consequently

$$\begin{aligned} (x,g)\big(\zeta_{g^{-1}}(1), g^{-1}\big)(x,g) &= \big(x\zeta_g\zeta_{g^{-1}}(1)\zeta_1(x), gg^{-1}g\big) \\ &= \big(x\zeta_g\zeta_{g^{-1}}(1)x, g\big) \quad \text{by (3)} \\ &= \big(x\zeta_g(1)x, g\big) \\ &= (x,g). \end{aligned}$$

It is readily seen that $)S_\zeta(= (1,1)^{\downarrow}$ and that $(1,1)$ is equiresidual with

$$(1,1):(x,g) = \big(\zeta_{g^{-1}}(1), g^{-1}\big).$$

Clearly, $(1,1)$ is the identity of S_ζ and so, by Theorem 13.12, S_ζ is an integral Dubreil-Jacotin inverse semigroup. Now, using Theorem 13.13(3), we have

$$\begin{aligned} (x,g)^{-1} &= \big(\zeta_{g^{-1}}(1), g^{-1}\big)(x,g)\big(\zeta_{g^{-1}}(1), g^{-1}\big) \\ &= \big(\zeta_{g^{-1}}(x), 1\big)\big(\zeta_{g^{-1}}(1), g^{-1}\big) \\ &= \big(\zeta_{g^{-1}}(x), g^{-1}\big), \end{aligned}$$

and therefore

$$\begin{aligned} (x,g)(x,g)^{-1} = (x,g)\big(\zeta_{g^{-1}}(x), g^{-1}\big) &= \big(x\zeta_g\zeta_{g^{-1}}(x), 1\big) \\ &= \big(x\zeta_g\zeta_{g^{-1}}(1)x, 1\big) \\ &= (x,1). \end{aligned}$$

We deduce from this that $\mathcal{R}$ is regular. On the other hand, we have

$$(x,g)^{-1}(x,g) = \big(\zeta_{g^{-1}}(x), g^{-1}\big)(x,g) = \big(\zeta_{g^{-1}}(x), 1\big)$$

from which, by (4), it follows that $\mathcal{L}$ is also regular.

Conversely, suppose that S is an integral Dubreil-Jacotin inverse semigroup on which $\mathcal{R}$ and $\mathcal{L}$ are regular. Let G be the set of A-nomal elements.

Recall from Theorem 12.11 that G is a group under the law $g \star h = 1:(1:gh)$. Consider now the mapping $\zeta : G \to \mathrm{End}[1]_{A_1}$ given by $g \mapsto \zeta_g$ where

$$(\forall x \in [1]_{A_1}) \quad \zeta_g(x) = gxg^{-1}.$$

If $g, h \in G$ are such that $g \leqslant h$ and $x \in [1]_{A_1}$ then from $gx \leqslant hx$ we obtain, by the regularity of $\mathcal{R}$,

$$gxg^{-1} = gxx^{-1}g^{-1} = gx(gx)^{-1} \leqslant hx(hx)^{-1} = hxx^{-1}h^{-1} = hxh^{-1}.$$

Since $h^{-1} \leqslant g^{-1}$ we also have $xh^{-1} \leqslant xg^{-1}$ and so, by the regularity of $\mathcal{L}$,

$$hxh^{-1} = (xh^{-1})^{-1}xh^{-1} \leqslant (xg^{-1})^{-1}xg^{-1} = gxg^{-1}.$$

It follows that $gxg^{-1} = hxh^{-1}$ and therefore ζ satisfies property (4) above.

It is clear that each ζ_g is an endomorphism on $E(S)$. Now in G we have $g^{-1} = 1:g$ and so $\zeta_g\zeta_h(1) = gh(1:h)(1:g)$ and $\zeta_{g\star h}(x) = [1:(1:gh)]x(1:gh)$. Now $g \star h = 1:(1:gh) \geqslant gh$ and so

$$gh(gh)^{-1}(g \star h) \geqslant gh(gh)^{-1}gh = gh.$$

But $gh(1:h)(1:g) \leqslant g(1:g) \leqslant 1$ gives $(1:h)(1:g) \leqslant 1:gh$ and therefore

$$gh(gh)^{-1}(g \star h) = ghh^{-1}g^{-1}[1:(1:gh)]$$
$$= gh(1:h)(1:g)[1:(1:gh)]$$
$$\leqslant gh(1:gh)[1:(1:gh)]$$
$$= gh \quad \text{by Theorem 13.13.}$$

Thus we see that $gh = gh(1:h)(1:g)[1:(1:gh)]$ from which it follows that

$$\zeta_g\zeta_h(1)\zeta_{g\star h}(x) = ghx(1:gh) = ghx(gh)^{-1} = ghxh^{-1}g^{-1} = \zeta_g\zeta_h(x)$$

and so condition (1) is satisfied. Since conditions (2) and (3) are clearly satisfied, we may therefore form the semigroup S_ζ.

Observe that with $g = 1:(1:x)$ we have

$$\zeta_g(1) = gg^{-1} = [1:(1:x)](1:x) \geqslant x(1:x)$$

so we can define a mapping $\vartheta : S \to S_\zeta$ by

$$\vartheta(x) = \big(x(1:x),\, 1:(1:x)\big).$$

Now ϑ is surjective since, for every $(x, g) \in S_\zeta$,

$$\vartheta(xg) = \big(xg(1:xg),\, 1:(1:xg)\big) = \big(xg(1:g),\, 1:(1:g)\big) = \big(x\zeta_g(1),\, g\big) = (x, g).$$

By Theorem 13.13 and the regularity of $\mathcal{R}$ we have

$$x \leqslant y \iff \vartheta(x) \leqslant \vartheta(y)$$

and so ϑ is an order isomorphism. Finally, it is a semigroup morphism; for

$$\vartheta(x)\vartheta(y) = \big(x(1:x), 1:(1:x)\big)\big(y(1:y), 1:(1:y)\big)$$
$$= \big(x(1:x)\zeta_{1:(1:x)}[y(1:y)],\, [1:(1:x)] \star [1:(1:y)]\big)$$
$$= \big(x(1:x)[1:(1:x)]y(1:y)(1:x),\, 1:(1:xy)\big),$$

and the left-hand component reduces to

$$xy(1:y)(1:x) = xx^{-1}xyy^{-1}(1:x)$$
$$= xyy^{-1}x^{-1}x(1:x)$$
$$= xyy^{-1}x^{-1}xx^{-1}$$
$$= xy(xy)^{-1},$$

whence we have $\vartheta(x)\vartheta(y) = \vartheta(xy)$. $\qquad\square$

Theorem 14.17 is greatly simplified in certain special cases. In particular, if we impose on the morphisms ζ_g the restriction $\zeta_g(1) = 1$ then condition (1) of Theorem 14.17 becomes $\zeta_g \circ \zeta_h = \zeta_{gh}$. Thus G acts on E by order automorphisms.

Now if A, B are semigroups and $\zeta : B \to \operatorname{End} A$, denoted by $b \mapsto \zeta_b$, is a morphism such that each ζ_b is an automorphism on A then the cartesian product set $A \times B$ equipped with the multiplication

$$(a,b)(a',b') = \big(a\zeta_b(a'), bb'\big)$$

is a semigroup, called the **semidirect product of A and B relative to ζ** and is denoted by $A \times_\zeta B$. If ζ is also isotone then $A \times_\zeta B$, equipped with the cartesian order, is an ordered semigroup. We shall denote this cartesian ordered semidirect product also by $A \times_\zeta B$.

We can now deduce from Theorem 14.17 the following special case.

Theorem 14.18 *Let E be a $\wedge$-semilattice with a top element and let G be an ordered group with an action $\zeta : g \mapsto \zeta_g$ on E by order automorphisms in such a way that $\zeta_g = \zeta_h$ whenever g, h are in the same Z-class of G. Then the cartesian ordered semidirect product $E \times_\zeta G$ is an integral Dubreil-Jacotin inverse semigroup on which $\mathcal{R}$ and $\mathcal{L}$ are regular and the set of A-nomal elements is a subgroup. Moreover, every such semigroup arises in this way.*

Proof It suffices to note from Theorem 14.17 that in this case $(1,1):(x,g) = (1,g^{-1})$ and so the A-nomal elements form a subgroup. $\qquad\square$

In the even more restrictive situation where the group G acts trivially on E (for example, when G consists of a single Z-class) we can take each ζ_g to be id_E. Then $E \times_\zeta G$ reduces to the cartesian product $E \times G$.

Remark It is readily seen from the proof of Theorem 14.17 that the structure of an integral Dubreil-Jacotin inverse semigroup on which only $\mathcal{R}$ is regular can be obtained by omitting condition (4).

14.3 Orthodox Dubreil-Jacotin semigroups

Whereas there remains open the difficult question of describing the general structure of strong Dubreil-Jacotin regular semigroups, it is possible to obtain structure theorems for a variety of those that are naturally ordered and orthodox; equivalently, by Theorem 13.18, those in which ξ is a middle unit. For the purpose of so doing, we require certain facts from the algebraic theory that we shall now incorporate.

Let Y be a semilattice and let $(S_\alpha)_{\alpha \in Y}$ be a family of pairwise disjoint semigroups. For $\alpha, \beta \in Y$ with $\alpha \geq \beta$ let $\varphi_{\beta,\alpha} : S_\alpha \to S_\beta$ be a morphism such that

(1) $(\forall \alpha \in Y) \ \varphi_{\alpha,\alpha} = \mathrm{id}_{S_\alpha}$;

(2) if $\alpha \geq \beta \geq \gamma$ then $\varphi_{\gamma,\beta} \circ \varphi_{\beta.\alpha} = \varphi_{\gamma,\alpha}$.

Then the set $\bigcup_{\alpha \in Y} S_\alpha$ together with the multiplication

$$(\forall x \in S_\alpha)(\forall y \in S_\beta) \quad xy = \varphi_{\alpha\beta,\alpha}(x)\varphi_{\alpha\beta,\beta}(y)$$

is a semigroup, called a **strong semilattice of semigroups**.

Since we shall be dealing with ordered semigroups it is necessary to incorporate an order in this. It is in fact a routine matter to establish the following result.

Theorem 14.19 *Let* $S = \bigcup_{\alpha \in Y} S_\alpha$ *be a strong semilattice of semigroups. Suppose that each* S_α *is an ordered semigroup and that each of the structure mappings* $\varphi_{\beta,\alpha}$ *is isotone. Then the relation* $\sqsubseteq$ *defined on* S *by*

$$(\forall x \in S_\alpha)(\forall y \in S_\beta) \quad x \sqsubseteq y \iff \alpha \leqslant \beta, \; x \leqslant \varphi_{\alpha,\beta}(y)$$

is an order and $(S; \sqsubseteq)$ *is an ordered semigroup.*   □

An ordered semigroup that arises as in Theorem 14.19 will be called a **strong semilattice of ordered semigroups**.

Since in an orthodox semigroup the idempotents form a band we shall require the following facts concerning bands.

A band B is said to be **normal** if $efgh = egfh$ for all $e, f, g, h \in B$.

In a regular semigroup Green's relation $\mathcal{D}$ is defined by $\mathcal{D} = \mathcal{R} \vee \mathcal{L}$. On a band we have

$$(e, f) \in \mathcal{D} \iff V(e) = V(f).$$

Theorem 14.20 (Yamada–Kimura [115]) *A band* B *is normal if and only if it is a strong semilattice of rectangular bands, these being the* $\mathcal{D}$-classes of B.   □

Definition An ordered set is **pointed** if it has a top element. An ordered semigroup S is a **pointed semilattice of pointed semigroups** if $S = \bigcup_{\alpha \in Y} S_\alpha$ is a strong semilattice of semigroups in which Y and each S_α is pointed.

Theorem 14.21 *Let* S *be a naturally ordered orthodox Dubreil-Jacotin semigroup. Then the band of idempotents of* S *is a pointed semilattice of pointed rectangular bands on which the order* $\sqsubseteq$ *coincides with that of* S.

Proof By Theorem 13.16, $\xi S \xi$ is an inverse semigroup and, by Theorem 13.18, ξ is a middle unit. The band $E(S)$ is then normal since

$$efgh = e \cdot \xi f \xi \cdot \xi g \xi \cdot h = e \cdot \xi g \xi \cdot \xi f \xi \cdot h = egfh.$$

It follows by Theorem 14.20 that $E(S)$ is a strong semilattice of rectangular bands, these being the $\mathcal{D}$-classes of $E(S)$. In $E(S)$ the relation $\mathcal{D}$ is given by

$$(e, f) \in \mathcal{D} \iff e^\circ = f^\circ.$$

Note that each such $\mathcal{D}$-class then has a top element, that of $[e]_\mathcal{D}$ being e°.

The structure semilattice of $E(S)$ is then $Y = \xi E(S) \xi = \{e^\circ \mid e \in E(S)\}$ and it too has a top element, namely ξ.

Now it can readily be verified that for $e^\circ, f^\circ \in Y$ with $e^\circ \geqslant f^\circ$ the isotone mappings $\varphi_{f^\circ, e^\circ} : [e^\circ]_\mathcal{D} \to [f^\circ]_\mathcal{D}$ given by the prescription

$$\varphi_{f^\circ, e^\circ}(x) = x f^\circ x$$

serve as structure mappings. Hence we see that $E(S)$ is a pointed semilattice of pointed rectangular bands.

Finally, we observe that the order $\sqsubseteq$ as defined in Theorem 14.19 coincides with that of S. In fact, suppose that $x \in [e^\circ]_\mathcal{D}$ and $y \in [f^\circ]_\mathcal{D}$ are such that $x \sqsubseteq y$. Then $e^\circ \leqslant f^\circ$ and $x \leqslant \varphi_{e^\circ, f^\circ}(y) = y e^\circ y \leqslant y f^\circ y = y$ since $[f^\circ]_\mathcal{D}$ is a rectangular band. Conversely, if $x \leqslant y$ then we have

$$x = x^3 = x \cdot \xi x \xi \cdot x \leqslant y \cdot \xi x \xi \cdot y = \varphi_{\xi x \xi, \xi y \xi}(y) = \varphi_{e^\circ, f^\circ}(y)$$

whence the result follows. $\square$

As we have just observed, the band of idempotents of a naturally ordered orthodox Dubreil-Jacotin semigroup is normal. Now the structure of orthodox semigroups with a normal band of idempotents is known. They are characterised as follows.

Theorem 14.22 (Yamada [114]) *Let S be an inverse semigroup with semilattice E of idempotents. Let L and R be respectively a left normal band and a right normal band with structure decompositions $L = \bigcup\limits_{\alpha \in E} L_\alpha$ and $R = \bigcup\limits_{\alpha \in E} R_\alpha$. Then on the set*

$$L \otimes S \otimes R = \{(e, x, f) \mid x \in S, \ e \in L_{xx^{-1}}, \ f \in R_{x^{-1}x}\}$$

the multiplication given by

$$(e, x, f)(g, y, h) = (eu, xy, vh),$$

where $u \in L_{xy(xy)^{-1}}$ and $v \in R_{(xy)^{-1}xy}$, is well defined and $L \otimes S \otimes R$ is an orthodox semigroup with a normal band of idempotents.

Moreover, every such semigroup can be constructed in this way. $\square$

The problem at hand is therefore to find suitable conditions to place on the Yamada construction in order to produce a naturally ordered strong Dubreil-Jacotin semigroup. This we can do in several different ways which we now elucidate. In each of these we take S to be an integral Dubreil-Jacotin inverse semigroup.

14.3.1 The cartesian order

It is natural to first consider imposing a cartesian order on the set $L \otimes S \otimes R$ and to choose the central building brick S to be an integral Dubreil-Jacotin inverse semigroup on which $\mathcal{R}$ and $\mathcal{L}$ are regular, the structure of such having been determined above. This produces the following characterisation.

Theorem 14.23 (Blyth–Santos [29]) *Let S be an integral Dubreil-Jacotin inverse semigroup on which $\mathcal{R}$ and $\mathcal{L}$ are regular. Let L be a pointed left normal band in which the top element is a right identity, and let R be a pointed right normal band in which the top element is a left identity. Then $L = \bigcup\limits_{\alpha \in A} L_\alpha$*

is a pointed semilattice of pointed left zero semigroups, and $R = \bigcup_{\beta \in B} R_\beta$ *is a pointed semilattice of pointed right zero semigroups. Suppose that* $A = B = E$, *the semilattice of idempotents of* S. *Let* $(L \otimes S \otimes R)_c$ *denote the set*

$$L \otimes S \otimes R = \{(e, x, f) \mid x \in S, \ e \in L_{xx^{-1}}, \ f \in R_{x^{-1}x}\}$$

together with the cartesian order and the multiplication

$$(e, x, f)(g, y, h) = (eL^\star_{xy(xy)^{-1}}, \ xy, \ R^\star_{(xy)^{-1}xy}h),$$

where $L^\star_\alpha$ *is the top element of* L_α *and* $R^\star_\alpha$ *is that of* R_α. *Then* $(L \otimes S \otimes R)_c$ *is a naturally ordered orthodox strong Dubreil-Jacotin semigroup on which* $\mathcal{R}$ *and* $\mathcal{L}$ *are regular.*

Proof Since 1_L is a right identity for L we have, for $e, f \in L$,

$$e \leqslant_n f \Rightarrow e = fe \leqslant f1_L = f$$

and so L is naturally ordered. Since 1_L is the greatest element of L, it is clear that L is a strong Dubreil-Jacotin semigroup. So we can apply Theorem 14.21 with $S = E = L$ to deduce that L is a pointed semilattice of pointed rectangular bands. These $\mathcal{D}$-class rectangular bands are in fact left zero semigroups; for, if $(e, f) \in \mathcal{D}$ then $1_L e1_L = 1_L f1_L$ so $1_L e = 1_L f$ and consequently $e = e1_L e = e1_L f = ef$. In a similar way, R is a pointed semilattice of pointed right zero semigroups.

Suppose that L and R have the structure decompositions $L = \bigcup_{\alpha \in E} L_\alpha$ and $R = \bigcup_{\alpha \in E} R_\alpha$. Observe first that for $\alpha, \beta \in E$ we have

$$\beta \leqslant \alpha \Rightarrow L^\star_\beta \leqslant L^\star_\alpha, \ R^\star_\beta \leqslant R^\star_\alpha.$$

In fact, if $\beta \leqslant \alpha$ then since the structure maps in L take top elements to top elements we have $\varphi_{\beta,\alpha}(L^\star_\alpha) = L^\star_\beta$. Consequently, $L^\star_\beta \sqsubseteq L^\star_\alpha$ and hence, by Theorem 14.21, we have $L^\star_\beta \leqslant L^\star_\alpha$; similarly, $R^\star_\beta \leqslant R^\star_\alpha$.

Now by the Yamada construction, $L \otimes S \otimes R$ is an orthodox semigroup. To see that, under the cartesian order and the regularity of $\mathcal{R}, \mathcal{L}$ on S, it is an ordered semigroup let $(e, x, f) \leqslant (e_1, x_1, f_1)$. Then $x \leqslant x_1$ and so $xy \leqslant x_1 y$ for every $y \in S$; and since $\mathcal{R}$ is regular on S we have $xy(xy)^{-1} \leqslant x_1 y(x_1 y)^{-1}$ so that, by the above observation, $L^\star_{xy(xy)^{-1}} \leqslant L^\star_{x_1 y(x_1 y)^{-1}}$. Using the regularity of $\mathcal{L}$, we see similarly that $R^\star_{(xy)^{-1}xy} \leqslant R^\star_{(x_1 y)^{-1}x_1 y}$ and so

$$\begin{aligned}(e, x, f)(g, y, h) &= (eL^\star_{xy(xy)^{-1}}, xy, R^\star_{(xy)^{-1}xy}h)\\ &\leqslant (e_1 L^\star_{x_1 y(x_1 y)^{-1}}, x_1 y, R^\star_{(x_1 y)^{-1}x_1 y}h) = (e_1, x_1, f_1)(g, y, h).\end{aligned}$$

Likewise, we have $(g, y, h)(e, x, f) \leqslant (g, y, h)(e_1, x_1, f_1)$ and so $(L \otimes S \otimes R)_c$ is indeed an ordered semigroup.

Since each L_α is a left zero semigroup and each R_α is a right zero semigroup, it is readily seen that the idempotents of $L \otimes S \otimes R$ are the elements of the form (e, x, f) where $x \in E$. Suppose then that $(e, x, f), (g, y, h)$ are idempotents such that $(e, x, f) \leqslant_n (g, y, h)$. Then $(e, x, f) = (e, x, f)(g, y, h) = (g, y, h)(e, x, f)$ which gives $x = xy = yx$ so that $x \leqslant_n y$ in E. Since S is naturally ordered we then have $x \leqslant y$. Also, $e = gL^\star_{yx(yx)^{-1}} \leqslant g1_L = g$ and similarly $f \leqslant h$. It follows that $(L \otimes S \otimes R)_c$ is naturally ordered.

To see that $(L \otimes S \otimes R)_c$ is strong Dubreil-Jacotin note that, since S is so, there is an ordered group G and a residuated epimorphism $\vartheta : S \to G$. Define the mapping $\psi : (L \otimes S \otimes R)_c \to G$ by the prescription $\psi(e, x, f) = \vartheta(x)$. Then it is readily seen that ψ is an isotone epimorphism. It is also residuated; in fact, if ϑ^+ denotes the residual of ϑ consider the map $\psi^+ : G \to (L \otimes S \otimes R)_c$ given by

$$\psi^+(g) = (L^\star_{aa^{-1}}, a, R^\star_{a^{-1}a}) \qquad \text{where} \quad a = \vartheta^+(g).$$

Since ϑ^+ is isotone and $\mathcal{R}, \mathcal{L}$ are regular on S, the map ψ^+ is also isotone. Moreover, $\psi^+ \circ \psi \geq \mathrm{id}$ and $\psi \circ \psi^+ \leq \mathrm{id}$, so ψ is residuated with residual ψ^+. It follows that $(L \otimes S \otimes R)_c$ is strong Dubreil-Jacotin.

Finally, consider Green's relation $\mathcal{R}$ on $(L \otimes S \otimes R)_c$. In order to show that it is regular, we need to identify $(e, x, f)^\circ$ and for this we require to compute $\xi : (e, x, f)$. Since the bimaximum element of S is an identity element 1, that of $(L \otimes S \otimes R)_c$ is $\xi = \psi^+(1) = (1_L, 1, 1_R)$. A routine calculation now gives

$$(1_L, 1, 1_R) : (e, x, f) = (L^\star_{(1:x)(1:x)^{-1}}, 1 : x, R^\star_{(1:x)^{-1}(1:x)}).$$

Observe now that for $\alpha, \beta \in E$ we have, since the structure maps preserve top elements,

$$L^\star_\alpha L^\star_\beta = \varphi_{\alpha\beta,\alpha}(L^\star_\alpha)\varphi_{\alpha\beta,\beta}(L^\star_\beta) = L^\star_{\alpha\beta}L^\star_{\alpha\beta} = L^\star_{\alpha\beta}.$$

A more involved calculation that uses Theorem 13.13 now gives

$$
\begin{aligned}
& (e, x, f)^\circ \\
&= (L^\star_{(1:x)(1:x)^{-1}}, 1:x, R^\star_{(1:x)^{-1}(1:x)})(e, x, f)(L^\star_{(1:x)(1:x)^{-1}}, 1:x, R^\star_{(1:x)^{-1}(1:x)}) \\
&= (L^\star_{(1:x)[1:(1:x)]}L^\star_{(1:x)x}, (1:x)x, R^\star_{(1:x)x}f)(L^\star_{(1:x)(1:x)^{-1}}, 1:x, R^\star_{(1:x)^{-1}(1:x)}) \\
&= (L^\star_{(1:x)x}, (1:x)x, f)(L^\star_{(1:x)(1:x)^{-1}}, 1:x, R^\star_{(1:x)^{-1}(1:x)}) \\
&= (L^\star_{(1:x)x}, x^{-1}, R^\star_{x(1:x)}).
\end{aligned}
$$

Consequently,

$$
\begin{aligned}
(e, x, f)(e, x, f)^\circ &= (e, x, f)(L^\star_{(1:x)x}, x^{-1}, R^\star_{x(1:x)}) \\
&= (eL^\star_{xx^{-1}}, xx^{-1}, R^\star_{xx^{-1}}R^\star_{x(1:x)}) \\
&= (e, xx^{-1}, R^\star_{xx^{-1}}).
\end{aligned}
$$

It follows immediately from this that $\mathcal{R}$ is regular on $(L \otimes S \otimes R)_c$, and similarly so is $\mathcal{L}$. □

Corollary *If $Y = (L \otimes S \otimes R)_c$ has band of idempotents $E(Y)$ and bimaximum element ξ then there are ordered semigroup isomorphisms*

$$\xi Y \xi \simeq S, \quad E(Y)\xi \simeq L, \quad \xi E(Y) \simeq R.$$

Proof Since $\xi = (1_L, 1, 1_R) = (L^\star_1, 1, R^\star_1)$ it is readily seen that

$$\xi(e, x, f)\xi = (L^\star_{xx^{-1}}, x, R^\star_{x^{-1}x}).$$

The mapping $\vartheta : \xi Y \xi \to S$ given by $\vartheta(L^\star_{xx^{-1}}, x, R^\star_{x^{-1}x}) = x$ is then a semigroup isomorphism. Since $\mathcal{R}, \mathcal{L}$ are regular on S, ϑ is also an order isomorphism.

Now $E(Y) = \{(e, \alpha, f) \mid \alpha \in E\}$ and

$$(e, \alpha, f)\xi = (e, \alpha, f)(L^\star_1, 1, R^\star_1) = (eL^\star_\alpha, \alpha, R^\star_\alpha R^\star_1) = (e, \alpha, R^\star_\alpha).$$

Consider the mapping $\psi : E(Y)\,\xi \to L$ given by $\psi(e, \alpha, R_\alpha^\star) = e$. Since

$$(e, \alpha, R_\alpha^\star)(g, \beta, R_\beta^\star) = (eL_{\alpha\beta}^\star, \alpha\beta, R_{\alpha\beta}^\star R_\beta^\star) = (eL_{\alpha\beta}^\star, \alpha\beta, R_{\alpha\beta}^\star),$$

and, $L_{\alpha\beta}$ being a left zero semigroup,

$$\begin{cases} eL_{\alpha\beta}^\star = \varphi_{\alpha\beta,\alpha}(e)\varphi_{\alpha\beta,\alpha\beta}(L_{\alpha\beta}^\star) = \varphi_{\alpha\beta,\alpha}(e)L_{\alpha\beta}^\star = \varphi_{\alpha\beta,\alpha}(e); \\ eg = \varphi_{\alpha\beta,\alpha}(e)\varphi_{\alpha\beta,\beta}(g) = \varphi_{\alpha\beta,\alpha}(e), \end{cases}$$

it follows that ψ is a semigroup morphism. Clearly, ψ is surjective; and since for $e \in L_\alpha$ and $g \in L_\beta$ the equality $e = g$ implies $\alpha = \beta$, ψ is also injective. It is clear that ψ is isotone. Finally, if $e \leqslant g$ with $e \in L_\alpha$ and $g \in L_\beta$ then, by Theorem 14.21, $e \sqsubseteq g$ and therefore $\alpha \leqslant \beta$, whence it follows that ψ is an order isomorphism.

That $\xi\,E(Y) \simeq R$ is established similarly. $\square$

We now establish the converse of Theorem 14.23. The above Corollary gives an indication as to how this may be achieved.

Theorem 14.24 (Blyth–Santos [29]) *Every naturally ordered orthodox strong Dubreil-Jacotin semigroup on which $\mathcal{R}$ and $\mathcal{L}$ are regular arises in the above way. More precisely, if T is such a semigroup then $E\xi$ is a pointed left normal band in which the top element is a right identity, and ξE is a pointed right normal band in which the top element is a left identity. Moreover, $\xi T\xi$ is an integral Dubreil-Jacotin inverse semigroup whose semilattice of idempotents $\xi E\xi$ is the structure semilattice of $E\xi$ and ξE. Finally, there is an ordered semigroup isomorphism*

$$T \simeq (E\xi \otimes \xi T\xi \otimes \xi E)_c.$$

Proof It follows from Theorem 13.18 that ξ is a middle unit and that E is normal, so $efgh = egfh$ for all $e, f, g, h \in E$. Taking $e, f, g \in E\xi$ and $h = \xi$ we deduce, using the fact that ξ is a right identity for $E\xi$, that $efg = egf$ for all $e, f, g \in E\xi$. Thus $E\xi$, which is a band since ξ is a middle unit, is left normal; and clearly ξ is the top element of $E\xi$. Similarly, ξE is a right normal band with top element ξ which is a left identity. As for $\xi T\xi$, this is clearly a subsemigroup which is regular since T is regular and ξ is a middle unit. If now $\xi x\xi$ is an idempotent of $\xi T\xi$ then $\xi x\xi = \xi x\xi \cdot \xi x\xi = \xi x^2\xi$ and so, on passing to quotients modulo A_ξ and using the fact that T/A_ξ is a group, we have that $x \in [\xi]_{A_\xi}$. Then $\xi{:}x = \xi$ and so, using Theorem 13.20, we have $x = x(\xi{:}x)x = x\xi x = x^2$. It follows that $[\xi]_{A_\xi}$ is the band of idempotents of T. Consequently the set of idempotents of $\xi T\xi$ is $\xi E\xi$ which is a semilattice since, E being normal, $\xi e\xi \cdot \xi f\xi = \xi ef\xi = \xi fe\xi = \xi f\xi \cdot \xi e\xi$. It follows that $\xi T\xi$ is an inverse subsemigroup of T. To see that it is strong Dubreil-Jacotin, let $\varphi : T \to G$ be a residuated epimorphism from T onto an ordered group G. Then $\varphi(\xi) = 1_G$ and so $\varphi|_{\xi T\xi} : \xi T\xi \to G$ is also an isotone epimorphism. Now

$$\varphi|_{\xi T\xi}(\xi x\xi) \leqslant g \;\Rightarrow\; \varphi(x) \leqslant g \;\Rightarrow\; x \leqslant \varphi^+(g) \;\Rightarrow\; \xi x\xi \leqslant \xi\varphi^+(g)\xi.$$

Since we also have $\varphi|_{\xi T\xi}\big(\xi\varphi^+(g)\xi\big) = \varphi\varphi^+(g) \leqslant g$, it follows that $\varphi|_{\xi T\xi}$ is residuated with residual $g \mapsto \xi\varphi^+(g)\xi$. Hence $\xi T\xi$ is strong Dubreil-Jacotin. Since T is naturally ordered so is $\xi T\xi$. That $\mathcal{R}$ and $\mathcal{L}$ are regular on $\xi T\xi$ follows from the fact that $(\xi x\xi)^{-1} = \xi x^\circ\xi$.

The structure semilattice of $E\xi$ is $\xi E\xi$. This follows from the proof of Theorem 14.21 (on taking $S = E\xi$ there). Likewise, that of ξE is $\xi E\xi$. Using Theorem 14.23 we can construct the orthodox semigroup $(E\xi \otimes \xi T\xi \otimes \xi E)_c$. Consider the mapping

$$\vartheta : T \to (E\xi \otimes \xi T\xi \otimes \xi E)_c$$

given by the prescription

$$\vartheta(x) = (xx^\circ, \xi x\xi, x^\circ x).$$

Note that ϑ is well defined. For example, $xx^\circ = xx^\circ \xi \in E\xi$ and

$$xx^\circ \in L_{\xi xx^\circ \xi} = L_{\xi x\xi(\xi x\xi)^\circ} = L_{\xi x\xi(\xi x\xi)^{-1}}.$$

Now since ξ is a middle unit, we have $xx^\circ . \xi x\xi . x^\circ x = xx^\circ xx^\circ x = x$, from which it follows that ϑ is injective. To see that it is also surjective, suppose that $(e\xi, \xi x\xi, \xi f) \in E\xi \otimes \xi T\xi \otimes \xi E$. Then we have

$$e\xi \in L_{\xi x\xi(\xi x\xi)^{-1}} = L_{\xi x\xi(\xi x\xi)^\circ} = L_{\xi xx^\circ \xi} = L_{\xi xx^\circ}$$

whence $(e\xi, \xi xx^\circ) \in \mathcal{D}$ and therefore, since T is orthodox,

$$\xi e\xi = \xi e^\circ = (e\xi)^\circ = (\xi xx^\circ)^\circ = x^{\circ\circ} x^\circ \xi = \xi xx^\circ.$$

Similarly, $\xi f \in R_{x^\circ x\xi}$ so $(\xi f, x^\circ x\xi) \in \mathcal{D}$ and therefore

$$\xi f\xi = f^\circ \xi = (\xi f)^\circ = (x^\circ x\xi)^\circ = \xi x^\circ x^{\circ\circ} = x^\circ x\xi.$$

Consider now

$$\vartheta(exf) = \big(exf(exf)^\circ, \xi exf\xi, (exf)^\circ exf\big).$$

By the above observations, we have

$$\begin{aligned}
exf(exf)^\circ = exff^\circ x^\circ e^\circ &= exf\xi x^\circ e^\circ \\
&= ex.\xi f\xi . x^\circ e^\circ \\
&= ex.x^\circ x\xi . x^\circ e^\circ \\
&= exx^\circ e^\circ \\
&= e.\xi xx^\circ . e^\circ \\
&= e.\xi e\xi . \xi e\xi \\
&= e\xi.
\end{aligned}$$

Similarly, $(exf)^\circ exf = \xi f$; and

$$\xi exf\xi = \xi e\xi . x . \xi f\xi = \xi xx^\circ . x . x^\circ x\xi = \xi x\xi.$$

Thus $\vartheta(exf) = (e\xi, \xi x\xi, \xi f)$ and so ϑ is surjective.

We now observe that

$$\begin{aligned}
\vartheta(x)\vartheta(y) &= (xx^\circ, \xi x\xi, x^\circ x)(yy^\circ, \xi y\xi, y^\circ y) \\
&= (xx^\circ L^\star_{\xi xy\xi(\xi xy\xi)^{-1}}, \ \xi xy\xi, \ R^\star_{(\xi xy\xi)^{-1}\xi xy\xi} y^\circ y) \\
&= (xx^\circ L^\star_{\xi xyy^\circ x^\circ}, \ \xi xy\xi, \ R^\star_{y^\circ x^\circ xy\xi} y^\circ y) \\
&= (xx^\circ . \xi xyy^\circ x^\circ \xi, \ \xi xy\xi, \ \xi y^\circ x^\circ xy\xi . y^\circ y) \\
&= (xyy^\circ x^\circ, \ \xi xy\xi, \ y^\circ x^\circ xy) \\
&= \big(xy(xy)^\circ, \ \xi xy\xi, \ (xy)^\circ xy\big) \\
&= \vartheta(xy).
\end{aligned}$$

Thus ϑ is a semigroup isomorphism.

Since $\mathcal{R}, \mathcal{L}$ are regular on T, it is clear that ϑ is isotone; and since

$$\vartheta(x) \leqslant \vartheta(y) \Rightarrow xx^\circ \leqslant yy^\circ,\ \xi x\xi \leqslant \xi y\xi,\ x^\circ x \leqslant y^\circ y$$
$$\Rightarrow x = xx^\circ \cdot \xi x\xi \cdot x^\circ x \leqslant yy^\circ \cdot \xi y\xi \cdot y^\circ y = y,$$

it follows that ϑ is also an order isomorphism. Hence we have an ordered semigroup isomorphism $T \simeq (E\xi \otimes \xi T\xi \otimes \xi E)_c$. □

Corollary *If E is a naturally ordered band with a greatest element ξ then*

$$E \simeq (E\xi \otimes \xi E\xi \otimes \xi E)_c.$$

Proof For every $e \in E$ we have $\xi : e = \xi$ and so $e^\circ = (\xi : e)e(\xi : e) = \xi e\xi$. It follows that if $e \leqslant f$ then $ee^\circ = e\xi e\xi \leqslant f\xi f\xi = ff^\circ$ and similarly $e^\circ e \leqslant f^\circ f$. Thus $\mathcal{R}$ and $\mathcal{L}$ are regular. The result now follows from the above on taking $T = E$. □

A particular case of the above is the following.

Theorem 14.25 *An ordered semigroup S is a completely simple orthodox Dubreil-Jacotin semigroup on which $\mathcal{R}$ and $\mathcal{L}$ are regular if and only if S is a cartesian ordered cartesian product of a pointed rectangular band and an ordered group.*

Proof $\Rightarrow$: By Theorem 14.24 we have $S \simeq (E\xi \otimes \xi S\xi \otimes \xi E)_c$. Since, by Theorem 13.35, $\xi S\xi$ is a subgroup we also have $\xi E\xi = \{\xi\}$. The structure semilattice of $E\xi$ and ξE is therefore a singleton, so $E\xi$ is a left zero semigroup and ξE is a right zero semigroup. Moreover, for every $x \in \xi S\xi$ we have $L_{xx^{-1}} = E\xi$ and $R_{x^{-1}x} = \xi E$. The multiplication in $(E\xi \otimes \xi S\xi \otimes \xi E)_c$ is then given by

$$(e\xi, \xi x\xi, \xi f)(g\xi, \xi y\xi, \xi h) = (e\xi L^\star_{\xi x y\xi(\xi x y\xi)^{-1}},\ \xi x\xi \cdot \xi y\xi,\ R^\star_{(\xi x y\xi)^{-1}\xi x y\xi}\xi h)$$
$$= (e\xi \cdot \xi,\ \xi x\xi \cdot \xi y\xi,\ \xi \cdot \xi h)$$
$$= (e\xi g\xi,\ \xi x\xi \cdot \xi y\xi,\ \xi f\xi h)$$

i.e., S is the cartesian ordered cartesian product $E\xi \times \xi S\xi \times \xi E$. It suffices now to observe that $E\xi \times \xi E$ is a pointed rectangular band.

$\Leftarrow$: Let B be a pointed rectangular band and G an ordered group. Consider the cartesian ordered cartesian product $T = B \times G$. It is clear that T is orthodox, and is completely simple since B and G are trivially naturally ordered. Now $\vartheta : T \to G$ given by $\vartheta(b, g) = g$ is a residuated epimorphism, so T is strong Dubreil-Jacotin. If ξ_B denotes the top element of B then the bimaximum element of T is $(\xi_B, 1_G)$. It is readily seen that $(\xi_B, 1_G) : (b, g) = (\xi_B, g^{-1})$ and so

$$(b, g)^\circ = (\xi_B, g^{-1})(b, g)(\xi_B, g^{-1}) = (\xi_B b\xi_B, g^{-1}) = (\xi_B, g^{-1}).$$

Consequently,

$$(b, g)(b, g)^\circ = (b, g)(\xi_B, g^{-1}) = (b\xi_B, 1_G)$$

and so $\mathcal{R}$ is regular on T. Similarly, so is $\mathcal{L}$. □

Referring back to the completely simple semigroups $(S_x; \leqslant)$ considered at the end of Chapter 13, we see by Theorem 14.25 that there are only two on which $\mathcal{R}$ and $\mathcal{L}$ are regular, namely $(S_x; \leqslant_c)$ and $(S_x; \leqslant_{dl})$. Here the pointed rectangular band is B_4 and in the former case the ordered group is $(\mathbb{Z}; \leqslant)$ whereas in the latter it is $(\mathbb{Z}; =)$.

14.3.2 Unilateral lexicographic orders

We now turn to a consideration of orders other than the cartesian order on the semigroup $L \otimes S \otimes R$. In this connection, all of the structure theorems that follow are also drawn from [29].

As we shall see first, by keeping only one of $\mathcal{R}, \mathcal{L}$ regular on S and by imposing a simple condition on the A_ξ-classes, we are led to consider a new type of order that has a lexicographic component. For this purpose, we note that the A_ξ-classes can be ordered by

$$[x]_{A_\xi} \leq [y]_{A_\xi} \iff \xi\!:\!(\xi\!:\!x) \leq \xi\!:\!(\xi\!:\!y),$$

and that the $\mathcal{L}$-classes can be ordered by

$$[x]_\mathcal{L} \leq [y]_\mathcal{L} \iff x^\circ x \leq y^\circ y.$$

Definition We shall say that A_ξ is $\mathcal{L}$-**linear** if, for all $x, y \in S$,

$$\left.\begin{array}{c} [x]_{A_\xi} < [y]_{A_\xi} \\ [x]_\mathcal{L} \leq [y]_\mathcal{L} \end{array}\right\} \Rightarrow x < y.$$

Example 14.1 Let $k > 1$ be a fixed integer. For every $n \in \mathbb{Z}$ let n_k be the greatest multiple of k that is less than or equal to n. On the cartesian ordered set $S = -\mathbb{N} \times -\mathbb{N} \times \mathbb{Z}$ define a multiplication by

$$(-p, -m, x)(-q, -n, y) = (\min\{-p, -q\}, -n, x + y_k).$$

Then S is an ordered semigroup which is regular since, for example,

$$(-p, -m, x)(-p, -m, -x_k)(-p, -m, x) = (-p, -m, x).$$

The idempotents are of the form $(-p, -m, x)$ where $x_k = 0$, i.e. $0 \leq x \leq k-1$. It follows that S is orthodox. If $(-p, -m, x), (-q, -n, y)$ are idempotents such that $(-p, -m, x) \leqslant_n (-q, -n, y)$ then

$$(-p, -m, x) = (\min\{-p, -q\}, -n, x) = (\min\{-p, -q\}, -m, y)$$

and so $-p \leq -q, -m = -n, x = y$ which gives $(-p, -m, x) \leq (-q, -n, y)$. Hence S is naturally ordered. Since

$$(0, 0, nk + k - 1)(0, 0, mk + k - 1) = (0, 0, (n + m)k + k - 1)$$

we see that $G = \{(0, 0, nk + k - 1) \mid n \in \mathbb{Z}\}$ is a group ($\simeq \mathbb{Z}$). The mapping $\vartheta : S \to G$ given by $\vartheta(-p, -m, x) = (0, 0, x_k + k - 1)$ is readily seen to be a residuated epimorphism. Hence S is strong Dubreil-Jacotin. The bimaximum element is $\xi = (0, 0, k - 1)$, and a simple calculation gives

$$\xi : (-p, -m, x) = (0, 0, -x_k + k - 1).$$

It follows that $(-p, -m, x)^\circ = (-p, 0, -x_k + k - 1)$ whence

$$(-p, -m, x)^\circ(-p, -m, x) = (-p, -m, k - 1),$$

which shows that $\mathcal{L}$ is regular on S. In contrast,

$$(-p, -m, x)(-p, -m, x)^\circ = (-p, 0, x - x_k),$$

and $\mathcal{R}$ is not regular. Also, A_ξ is $\mathcal{L}$-linear. In fact, the conditions

$$\begin{cases} \xi : (-q, -n, y) < \xi : (-p, -m, x), \\ (-p, -m, x)^\circ(-p, -m, x) \leq (-q, -n, y)^\circ(-q, -n, y) \end{cases}$$

give $x_k < y_k, -p \leq -q, -m \leq -n$ whence $(-p, -m, x) < (-q, -n, y)$.

Theorem 14.26 *In an integral Dubreil-Jacotin inverse semigroup A_1 is $\mathcal{L}$-linear.*

Proof The conditions $1\!:\!(1\!:\!x) < 1\!:\!(1\!:\!y)$ and $(1\!:\!x)x \leq (1\!:\!y)y$ give, by Theorem 13.13, $x = [1\!:\!(1\!:\!x)](1\!:\!x)x \leq [1\!:\!(1\!:\!y)](1\!:\!y)y = y$, and equality is excluded since otherwise $[x]_{A_\xi} = [y]_{A_\xi}$. $\qquad\qquad\qquad\qquad\square$

Our objective now is to describe the structure of a naturally ordered orthodox strong Dubreil-Jacotin semigroup on which $\mathcal{L}$ is regular and A_ξ is $\mathcal{L}$-linear. For this purpose, consider the relation $\leq_{lq}$ defined on $L \otimes S \otimes R$ by

$$(e, x, f) \leq_{lq} (g, y, h) \iff x \leq y, \ f \leq h \text{ and } \begin{cases} \text{either } 1\!:\!y < 1\!:\!x, \\ \text{or } \ 1\!:\!y = 1\!:\!x, \ e \leq g. \end{cases}$$

It is readily seen that $\leq_{lq}$ is an order on $L \otimes S \otimes R$ which we shall call the **left quasi-lexicographic order**. We shall denote by $(L \otimes S \otimes R)_{lq}$ the set $L \otimes S \otimes R$ equipped with this order and the multiplication in Theorem 14.23.

Theorem 14.27 *Let S be an integral Dubreil-Jacotin inverse semigroup on which $\mathcal{L}$ is regular, let L be an ordered left normal band with a top element 1_L that is a right identity, and let R be an ordered right normal band with a top element 1_R that is a left identity. Then $(L \otimes S \otimes R)_{lq}$ is a naturally ordered orthodox strong Dubreil-Jacotin semigroup on which $\mathcal{L}$ is regular and A_ξ is $\mathcal{L}$-linear.*

Moreover, every such semigroup arises in this way.

Proof To see that $(L \otimes S \otimes R)_{lq}$ is an ordered semigroup, let $(e, x, f) \leq_{lq} (g, y, h)$. Then for $(i, z, k) \in L \otimes S \otimes R$ we have to compare the products

$$(e, x, f)(i, z, k) = (eL^{\star}_{xz(xz)^{-1}}, \ xz, \ R^{\star}_{(xz)^{-1}xz}j),$$
$$(g, y, h)(i, z, k) = (gL^{\star}_{yz(yz)^{-1}}, \ yz, \ R^{\star}_{(yz)^{-1}yz}j).$$

Since $x \leq y$ in S we have $xz \leq yz$. Since $\mathcal{L}$ is regular on S by hypothesis, it follows that $(xz)^{-1}xz \leq (yz)^{-1}yz$ and hence that $R^{\star}_{(xz)^{-1}xz} \leq R^{\star}_{(yz)^{-1}yz}$. If now $1\!:\!xz = 1\!:\!yz$ then $xz(xz)^{-1} = xz(1\!:\!xz) \leq yz(1\!:\!yz) = yz(yz)^{-1}$ whence $L^{\star}_{xz(xz)^{-1}} \leq L^{\star}_{yz(yz)^{-1}}$. Also, since S/A_ξ is a group, if $1\!:\!xz = 1\!:\!yz$ then $[x]_{A_\xi} = [y]_{A_\xi}$ and so $1\!:\!x = 1\!:\!y$, so by the hypothesis that $(e, x, f) \leq_{lq} (g, y, h)$ it follows that $e \leq g$. Thus $eL^{\star}_{xz(xz)^{-1}} \leq gL^{\star}_{yz(yz)^{-1}}$. It follows from these observations that $(e, x, f)(i, z, j) \leq_{lq} (g, y, h)(i, z, j)$. In a similar way, we have $(i, z, j)(e, x, f) \leq_{lq} (i, z, j)(g, y, h)$ and so $(L \otimes S \otimes R)_{lq}$ is indeed an ordered semigroup.

To see that $(L \otimes S \otimes R)_{lq}$ is naturally ordered, recall that the idempotents are of the form (e, x, f) where $x \in E$. If then $(e, x, f) \leqslant_n (g, y, h)$ we have

$$(e, x, f) = (e, x, f)(g, y, h) = (g, y, h)(e, x, f).$$

These equations give $x \leqslant_n y$ in E, whence $x \leq y$ by Theorem 13.15; and as $x, y \in E$ we have $1\!:\!x = 1 = 1\!:\!y$. Since the above equations reduce to

$$(e, x, f) = (e, x, R^{\star}_x h) = (gL^{\star}_x, x, f),$$

we also have $e = gL^{\star}_x \leq g1_L = g$ and $f = R^{\star}_x h \leq 1_L h = h$. Thus we see that $(e, x, f) \leq_{lq} (g, y, h)$ and so $(L \otimes S \otimes R)_{lq}$ is naturally ordered.

To see that $(L \otimes S \otimes R)_{lq}$ is strong Dubreil-Jacotin note that, since S is so, there is an ordered group G and a residuated epimorphism $\vartheta : S \to G$. Define $\psi : (L \otimes S \otimes R)_{lq} \to G$ by the prescription $\psi(e, x, f) = \vartheta(x)$. Then it is readily seen that ψ is an isotone epimorphism. It is also residuated; in fact, if ϑ^+ denotes the residual of ϑ consider the map $\psi^+ : G \to (L \otimes S \otimes R)_{lq}$ given by

$$\psi^+(g) = (L_{aa^{-1}}^\star, a, R_{a^{-1}a}^\star) \quad \text{where} \quad a = \vartheta^+(g).$$

To see that ψ^+ is isotone, observe that if $g \leq h$ then, since ϑ^+ is isotone, we have $a = \vartheta^+(g) \leq \vartheta^+(h) = b$ whence, $\mathcal{L}$ being regular on S, $a^{-1}a \leq b^{-1}b$ and so $R_{a^{-1}a}^\star \leq R_{b^{-1}b}^\star$; and if $1 : a = 1 : b$ then $aa^{-1} = a(1 : a) \leq b(1 : b) = bb^{-1}$ gives $L_{aa^{-1}}^\star \leq L_{bb^{-1}}^\star$. It follows that $\psi^+(g) \leq_{lq} \psi^+(h)$ as required. Now, with the above notation, we have on the one hand

$$\psi\psi^+(g) = \psi(L_{aa^{-1}}^\star, a, R_{a^{-1}a}^\star) = \vartheta(a) = \vartheta\vartheta^+(g) \leq g.$$

On the other hand,

$$\psi^+\psi(e, x, f) = \psi^+\vartheta(x) = (L_{aa^{-1}}^\star, a, R_{a^{-1}a}^\star)$$

where $a = \vartheta^+\vartheta(x) \geq x$. Since $\mathcal{L}$ is regular on S we have $x^{-1}x \leq a^{-1}a$ whence $f \leq R_{x^{-1}x}^\star \leq R_{a^{-1}a}^\star$; and if $1 : a = 1 : x$ then $xx^{-1} = x(1 : x) \leq a(1 : a) = aa^{-1}$ gives $L_{xx^{-1}}^\star \leq L_{aa^{-1}}^\star$ and so, since $e \in L_{xx^{-1}}$, we have $e \leq L_{aa^{-1}}^\star$. Consequently, $(e, x, f) \leq_{lq} \psi^+\psi(e, x, f)$. It follows from these inequalities that ψ^+ is the residual of ψ. Therefore $(L \otimes S \otimes R)_{lq}$ is strong Dubreil-Jacotin.

The bimaximum element is

$$\xi = \psi^+(1) = (L_1^\star, 1, R_1^\star) = (1_L, 1, 1_R)$$

and a routine calculation shows that

$$\xi : (e, x, f) = (L_{(1:x)(1:x)^{-1}}^\star, 1 : x, R_{(1:x)^{-1}(1:x)}^\star).$$

We therefore have the same formulae as in Theorem 14.23. As in the proof of that result, we have

$$(e, x, f)^\circ(e, x, f) = (L_{x^{-1}x}^\star, x^{-1}x, f).$$

It now follows from the hypothesis that $\mathcal{L}$ is regular on S that $\mathcal{L}$ is regular on $(L \otimes S \otimes R)_{lq}$. To show that A_ξ is $\mathcal{L}$-linear on $(L \otimes S \otimes R)_{lq}$ we must show that the inequalities

(1) $(L_{(1:y)(1:y)^{-1}}^\star, 1 : y, R_{(1:y)^{-1}(1:y)}^\star) <_{lq} (L_{(1:x)(1:x)^{-1}}^\star, 1 : x, R_{(1:x)^{-1}(1:x)}^\star),$

(2) $(L_{x^{-1}x}^\star, x^{-1}x, f) \leq_{lq} (L_{y^{-1}y}^\star, y^{-1}y, h)$

imply that $(e, x, f) <_{lq} (g, y, h)$. Now from (1) we must have $1 : y < 1 : x$. From (2) we obtain $x^{-1}x \leq y^{-1}y$ and $f \leq h$, the former giving

$$x = [1 : (1 : x)](1 : x)x \leq [1 : (1 : y)](1 : y)y = y$$

with, in fact, $x < y$ since $1 : y < 1 : x$. It follows immediately that we have $(e, x, f) <_{lq} (g, y, h)$. Thus we see that $(L \otimes S \otimes R)_{lq}$ is an orthodox strong Dubreil-Jacotin semigroup on which $\mathcal{L}$ is regular and A_ξ is $\mathcal{L}$-linear.

We now show, conversely, that every such semigroup arises in this way. If T is such a semigroup then, precisely as in Theorem 14.24, we have an algebraic isomorphism $\vartheta : T \to (E\xi \otimes \xi T\xi \otimes \xi E)_{lq}$ given by $\vartheta(x) = (xx^\circ, \xi x\xi, x^\circ x)$. It suffices, therefore, to show that ϑ is an order isomorphism.

Now if $x \leq y$ then $\xi x \xi \leq \xi y \xi$, and since by hypothesis $\mathcal{L}$ is regular on T, $x^\circ x \leq y^\circ y$. If $\xi : \xi x \xi = \xi : \xi y \xi$ then $\xi : x = \xi : y$ and therefore $xx^\circ = x(\xi : x) \leq y(\xi : y) = yy^\circ$. Thus $\vartheta(x) \leq_{lq} \vartheta(y)$.

Conversely, if $\vartheta(x) \leq_{lq} \vartheta(y)$ then

$$\xi x \xi \leq \xi y \xi, \; x^\circ x \leq y^\circ y \text{ and } \begin{cases} \text{either } \xi : y < \xi : x, \\ \text{or} \quad \xi : y = \xi : x, xx^\circ \leq yy^\circ. \end{cases}$$

Now the conditions $\xi : y = \xi : x, xx^\circ \leq yy^\circ, x^\circ x \leq y^\circ y$ give $x \leq y$; for then

$$x = xx^\circ x \leq yy^\circ x = y(\xi : y)x = y(\xi : x)x = yx^\circ x \leq yy^\circ y = y.$$

Also, since A_ξ is $\mathcal{L}$-linear, the conditions $x^\circ x \leq y^\circ y, \xi : y < \xi : x$ give $x < y$. It follows therefore that if $\vartheta(x) \leq_{lq} \vartheta(y)$ then $x \leq y$. Hence ϑ is also an order isomorphism and $T \simeq (E\xi \otimes \xi T\xi \otimes \xi E)_{lq}$. $\qquad\square$

Now if, in $L \otimes S \otimes R$, the central component S is a group then $1 : x$ is the inverse of $x \in S$, in which case we see that the left quasi-lexicographic order reduces to the left lexicographic order. Hence a particular case of the above is the following analogue of Theorem 14.25.

Theorem 14.28 *An ordered semigroup S is a completely simple orthodox Dubreil-Jacotin semigroup on which $\mathcal{L}$ is regular and A_ξ is $\mathcal{L}$-linear if and only if S is a left lexicographically ordered cartesian product of a pointed rectangular band and an ordered group.* $\qquad\square$

Referring back to the completely simple semigroups $(S_x; \leqslant)$ considered at the end of Chapter 13, we see by Theorem 14.27 that there is only one on which $\mathcal{L}$ is regular and A_ξ is $\mathcal{L}$-linear, namely $(S_x; \leqslant_l)$.

There are of course dual results to the above involving the notions of a **right quasi-lexicographic order** and A_ξ being **R-linear**. Here the corresponding example is $(S_x; \leqslant_r)$.

14.3.3 Bootlace orders

It is readily seen from their definitions that $\leqslant_{lq} \cap \leqslant_{rq} = \leqslant_c$. In contrast, the relation $\leqslant_{lq} \cup \leqslant_{rq}$ is not in general an order since it fails to be transitive.

Definition By the **quasi-bootlace order** $\leqslant_{qb}$ on $L \otimes S \otimes R$ we shall mean the transitive closure of the relation $\leqslant_{lq} \cup \leqslant_{rq}$.

Thus $\leqslant_{qb}$ is the smallest order that contains both the orders $\leqslant_{lq}$ and $\leqslant_{rq}$, so $(e, x, f) \leqslant_{qb} (g, y, h)$ if and only if there exist $(i_1, z_1, k_1), \ldots, (i_n, z_n, k_n)$ such that

$$(e, x, f) \ll (i_1, z_1, k_1) \ll \cdots \ll (i_n, z_n, k_n) \ll (g, y, h)$$

where $\ll$ indicates either $\leqslant_{lq}$ or $\leqslant_{rq}$.

An important property of $\leqslant_{lq}$ and $\leqslant_{rq}$ is the following.

Theorem 14.29 *If $\mathcal{L}$ and $\mathcal{R}$ are regular on S then the orders $\leqslant_{lq}$ and $\leqslant_{rq}$ commute.*

Proof We show that if, in $L \otimes S \otimes R$,

$$(e, x, f) \leqslant_{lq} (i, z, k) \leqslant_{rq} (g, y, h)$$

then there exists (i', z', k') such that

$$(e, x, f) \leqslant_{rq} (i', z', k') \leqslant_{lq} (g, y, h).$$

It follows from this property and its dual that $\leqslant_{lq}$ and $\leqslant_{rq}$ commute.

Now from $(e, x, f) \leqslant_{lq} (i, z, k)$ we have

$$x \leqslant z, f \leqslant k \text{ and } \begin{cases} \text{either} & \xi : z < \xi : x, \quad (1) \\ \text{or} & \xi : z = \xi : x, e \leqslant i; \quad (2) \end{cases}$$

and from $(i, z, k) \leqslant_{rq} (g, y, h)$ we have

$$z \leqslant y, i \leqslant g \text{ and } \begin{cases} \text{either} & \xi : y < \xi : z, \quad (3) \\ \text{or} & \xi : y = \xi : z, k \leqslant h. \quad (4) \end{cases}$$

If (2) and (4) hold then we have $(e, x, f) \leqslant_c (i, z, k)$ and $(i, z, k) \leqslant_c (g, y, h)$, whence the required inequalities hold with $(i', z', k') = (i, z, k)$; if (2) and (3) hold then $(e, x, f) \leqslant_c (i, z, k)$ and $(i, z, k) \leqslant_{rq} (g, y, h)$ whence, since $\leqslant_c$ implies $\leqslant_{rq}$, we have $(e, x, f) \leqslant_{rq} (g, y, h)$ and so the required inequalities hold with $(i', z', k') = (g, y, h)$; similarly, if (1) and (4) hold then the required inequalities hold with $(i', z', k') = (e, x, f)$. Suppose finally that (1) and (3) hold. Since $z \leqslant y$ and $\mathcal{L}$ is regular on S, we have $z^{-1}z \leqslant y^{-1}y$. Let $\varphi_{z^{-1}z, y^{-1}y}$ be the associated structure map in the semilattice decomposition of R. Since $h \in R_{y^{-1}y}$ we have $\varphi_{z^{-1}z, y^{-1}y}(h) \in R_{z^{-1}z}$. Consider now the element

$$\left(L^\star_{zz^{-1}}, z, \varphi_{z^{-1}z, y^{-1}y}(h) \right).$$

Since $\mathcal{R}$ is also regular on S we have $xx^{-1} \leqslant zz^{-1}$ and so $e \leqslant L^\star_{xx^{-1}} \leqslant L^\star_{zz^{-1}}$. It follows from (1) that

$$(e, x, f) \leqslant_{rq} \left(L^\star_{zz^{-1}}, z, \varphi_{z^{-1}z, y^{-1}y}(h) \right).$$

Observing now by Theorem 14.9 that

$$\varphi_{z^{-1}z, y^{-1}y}(h) \sqsubseteq h,$$

and using the fact that in R the orders $\sqsubseteq$ and $\leqslant$ coincide, we deduce from (3) that

$$\left(L^\star_{zz^{-1}}, z, \varphi_{z^{-1}z, y^{-1}y}(h) \right) \leqslant_{lq} (g, y, h).$$

Thus in this case we may choose $(i', z', k') = \left(L^\star_{zz^{-1}}, z, \varphi_{z^{-1}z, y^{-1}y}(h) \right)$. $\square$

Corollary *If $\mathcal{L}$ and $\mathcal{R}$ are regular then $\leqslant_{qb} = \leqslant_{lq} \circ \leqslant_{rq} = \leqslant_{rq} \circ \leqslant_{lq}$.* $\square$

In order to consider the structure under the quasi-bootlace order, we require the following notion which is weaker than the regularity of $\mathcal{L}$ or $\mathcal{R}$.

Definition If S is an orthodox strong Dubreil-Jacotin semigroup then we shall say that $\mathcal{L}$ and $\mathcal{R}$ are **interlaced** on S if, for all $x, y \in S$,

$$x \leqslant y \Rightarrow (\exists z \in S) \quad x \leqslant z \leqslant y \text{ and } \begin{cases} \text{either} & xx^\circ \leqslant zz^\circ, z^\circ z \leqslant y^\circ y; \\ \text{or} & x^\circ x \leqslant z^\circ z, zz^\circ \leqslant yy^\circ. \end{cases}$$

Theorem 14.30 *Let S be a naturally ordered integral Dubreil-Jacotin inverse semigroup on which $\mathcal{L}$ and $\mathcal{R}$ are regular. Then $\mathcal{L}$ and $\mathcal{R}$ are interlaced.*

Proof Suppose that $x, y \in S$ are such that $x \leqslant y$. Then the element $z = xx^{-1}y$ is such that

$$x = xx^{-1}x \leqslant xx^{-1}y = z \leqslant 1y = y.$$

Since $\mathcal{R}$ is regular on S, we have $xx^{-1} \leqslant zz^{-1}$. Also,

$$z^{-1}z = (xx^{-1}y)^{-1}xx^{-1}y = y^{-1}xx^{-1}y \leqslant y^{-1}1y = y^{-1}y.$$

The other possibility follows on considering the element $z = yx^{-1}x$. $\qquad\square$

Definition If S is a naturally ordered orthodox strong Dubreil-Jacotin semigroup then we shall say that $\mathcal{L}, \mathcal{R}$ are **knotted** if they are interlaced on S and are regular on $\xi S \xi$.

Theorem 14.31 *Let S be an integral Dubreil-Jacotin inverse semigroup on which $\mathcal{L}$ and $\mathcal{R}$ are regular, let L be an ordered left normal band with a top element 1_L that is a right identity, and let R be an ordered right normal band with a top element 1_R that is a left identity. Then $(L \otimes S \otimes R)_{qb}$ is a naturally ordered orthodox strong Dubreil-Jacotin semigroup on which $\mathcal{L}$ and $\mathcal{R}$ are knotted and A_ξ is both $\mathcal{L}$-linear and $\mathcal{R}$-linear.*

Moreover, every such semigroup arises in this way.

Proof Since $\mathcal{L}$ and $\mathcal{R}$ are regular on S, the orders $\leqslant_{lq}$ and $\leqslant_{rq}$ are compatible with multiplication. It follows that so also is $\leqslant_{qb}$, whence $(L \otimes S \otimes R)_{qb}$ is an ordered orthodox semigroup.

Since $\leqslant_c$ implies $\leqslant_{qb}$, it follows exactly as in the proof of Theorem 14.23 that $(L \otimes S \otimes R)_{qb}$ is naturally ordered.

To show that $(L \otimes S \otimes R)_{qb}$ is strong Dubreil-Jacotin, let $\vartheta : S \to G$ be a residuated epimorphism from S to an ordered group G and consider the map $\psi : (L \otimes S \otimes R)_{qb} \to G$ given by $\psi(e, x, f) = \vartheta(x)$. Then from the definition of $\leqslant_{qb}$ it follows that ψ is an isotone epimorphism. As in the proof of Theorem 14.23, consider $\psi^+ : G \to (L \otimes S \otimes R)_{qb}$ given by

$$\psi^+(g) = \left(L^*_{aa^{-1}}, a, R^*_{a^{-1}a}\right) \qquad \text{where} \quad a = \vartheta^+(g).$$

Since ϑ^+ is isotone, it follows from $g \leqslant h$ that $\psi^+(g) \leqslant_c \psi^+(h)$ whence $\psi^+(g) \leqslant_{qb} \psi^+(h)$. Hence ψ^+ is isotone. It is readily seen that ψ^+ is the residual of ψ. Hence $(L \otimes S \otimes R)_{qb}$ is strong Dubreil-Jacotin. Moreover, the bimaximum element is $(1_L, 1, 1_R)$ and we have the same formulae as in Theorem 14.23. Thus $(L \otimes S \otimes R)_{qb}$ is an orthodox strong Dubreil-Jacotin semigroup.

As in the Corollary to Theorem 14.23, the algebraic isomorphism described by $(L^*_{xx^{-1}}, x, R^*_{x^{-1}x}) \mapsto x$ is isotone and has an isotone inverse; consequently $\left(\xi(L \otimes S \otimes R)\xi\right)_{qb}$ and S are order isomorphic. It follows that $\mathcal{L}, \mathcal{R}$ are regular on $\left(\xi(L \otimes S \otimes R)\xi\right)_{qb}$.

To show that $\mathcal{L}$ and $\mathcal{R}$ are also interlaced on $(L \otimes S \otimes R)_{qb}$, suppose that $(e, x, f) \leqslant_{qb} (g, y, h)$. Then, by the Corollary to Theorem 14.29, there exists (i, z, k) such that

$$(e, x, f) \leqslant_{lq} (i, z, k) \leqslant_{rq} (g, y, h).$$

The first of these equalities gives $x \leqslant z$ and $f \leqslant k$, whereas the second gives $z \leqslant y$ and $i \leqslant g$. Let

$$\alpha = (e, x, f), \quad \beta = (g, y, h), \quad \gamma = (i, z, k).$$

Then since $\leqslant_{lq}$ and $\leqslant_{rq}$ imply $\leqslant_{qb}$ we have $\alpha \leqslant_{qb} \gamma \leqslant_{qb} \beta$. Also, since

$$\alpha^\circ \alpha = (L^\star_{x^{-1}x}, x^{-1}x, f), \quad \alpha\alpha^\circ = (e, xx^{-1}, R^\star_{xx^{-1}})$$

and since $\mathcal{L}, \mathcal{R}$ are regular on S, we have

$$\alpha^\circ \alpha \leqslant_c \gamma^\circ \gamma, \quad \gamma\gamma^\circ \leqslant_c \beta\beta^\circ.$$

Since $\leqslant_c$ implies $\leqslant_{qb}$, it follows that $\mathcal{L}, \mathcal{R}$ are interlaced on $(L \otimes S \otimes R)_{qb}$.

To see that A_ξ is $\mathcal{L}$-linear, we have to show that the conditions

(1) $(L^\star_{(1:y)(1:y)^{-1}}, 1{:}y, R^\star_{(1:y)^{-1}(1:y)}) <_{qb} (L^\star_{(1:x)(1:x)^{-1}}, 1{:}x, R^\star_{(1:x)^{-1}(1:x)})$

(2) $(L^\star_{x^{-1}x}, x^{-1}x, f) \leqslant_{qb} (L^\star_{y^{-1}y}, y^{-1}y, h)$

imply that $(e, x, f) <_{qb} (g, y, h)$. Now (1) gives $1{:}y < 1{:}x$, whereas (2) gives $x^{-1}x \leqslant y^{-1}y$, whence $L^\star_{x^{-1}x} \leqslant L^\star_{y^{-1}y}$. We thus have

$$x = xx^{-1}x \leqslant xy^{-1}y = x(1{:}y)y \leqslant x(1{:}x)y \leqslant 1y = y.$$

Since $1{:}x^{-1}x = 1{:}y^{-1}y$, it follows from (2) and the definition of $\leqslant_{qb}$ that $f \leqslant h$. Consequently (1) and (2) imply that $(e, x, f) \leqslant_{lq} (g, y, h)$ whence $(e, x, f) \leqslant_{qb} (g, y, h)$, with in fact strict inequality since $1{:}y < 1{:}x$ implies that $y \neq x$. Hence A_ξ is $\mathcal{L}$-linear; and similarly it is $\mathcal{R}$-linear.

For the converse, suppose that T is an orthodox strong Dubreil-Jacotin semigroup on which $\mathcal{L}, \mathcal{R}$ are knotted and A_ξ is both $\mathcal{L}$-linear and $\mathcal{R}$-linear. It suffices to show that the algebraic isomorphism $\vartheta : T \to (E\xi \otimes \xi T\xi \otimes \xi E)_{qb}$ given by $\vartheta(x) = (xx^\circ, \xi x\xi, x^\circ x)$ is an order isomorphism. Since by hypothesis $\mathcal{L}, \mathcal{R}$ are interlaced we have, for $x, y \in T$,

$$x \leqslant y \Rightarrow (\exists z \in T) \quad x \leqslant z \leqslant y \text{ and } \begin{cases} \text{either} & xx^\circ \leqslant zz^\circ, z^\circ z \leqslant y^\circ y; \quad (1) \\ \text{or} & x^\circ x \leqslant z^\circ z, zz^\circ \leqslant yy^\circ. \quad (2) \end{cases}$$

There are several cases to consider:

(α) Suppose that $\xi{:}y = \xi{:}z = \xi{:}x$. Then $xx^\circ = x(\xi{:}x) \leqslant z(\xi{:}z) = zz^\circ$ and similarly the three other inequalities of (1), (2) hold. Hence in this case we have

$$(xx^\circ, \xi x\xi, x^\circ x) \leqslant_c (zz^\circ, \xi z\xi, z^\circ z) \leqslant_c (yy^\circ, \xi y\xi, y^\circ y)$$

whence the result follows from the fact that $\leqslant_c$ implies $\leqslant_{qb}$.

(β) Suppose now that $\xi{:}y < \xi{:}z = \xi{:}x$. Then $xx^\circ \leqslant zz^\circ$ and $x^\circ x \leqslant z^\circ z$. If (1) holds we then have

$$(xx^\circ, \xi x\xi, x^\circ x) \leqslant_c (zz^\circ, \xi z\xi, z^\circ z) \leqslant_{lq} (yy^\circ, \xi y\xi, y^\circ y);$$

and if (2) holds we have

$$(xx^\circ, \xi x\xi, x^\circ x) \leqslant_c (zz^\circ, \xi z\xi, z^\circ z) \leqslant_{rq} (yy^\circ, \xi y\xi, y^\circ y).$$

In each case, $\vartheta(x) \leqslant_{qb} \vartheta(y)$.

(γ) The case $\xi{:}y = \xi{:}z < \xi{:}x$ is similar to (β).

(δ) Suppose now that $\xi{:}y < \xi{:}z < \xi{:}x$. If (1) holds then we have

$$(xx^\circ, \xi x\xi, x^\circ x) \leqslant_{rq} (zz^\circ, \xi z\xi, z^\circ z) \leqslant_{lq} (yy^\circ, \xi y\xi, y^\circ y);$$

and if (2) holds we have

$$(xx°, \xi x\xi, x°x) \leqslant_{lq} (zz°, \xi z\xi, z°z) \leqslant_{rq} (yy°, \xi y\xi, y°y).$$

In each case, therefore, $\vartheta(x) \leqslant_{qb} \vartheta(y)$.

It follows from the above observations that ϑ is isotone. To see that ϑ^{-1} is also isotone, suppose that $(xx°, \xi x\xi, x°x) \leqslant_{qb} (yy°, \xi y\xi, y°y)$. Then there exists $(zz°, \xi z\xi, z°z)$ such that

$$(xx°, \xi x\xi, x°x) \leqslant_{lq} (zz°, \xi z\xi, z°z) \leqslant_{rq} (yy°, \xi y\xi, y°y),$$

so that, in particular,

$$\xi x\xi \leqslant \xi z\xi \leqslant \xi y\xi, \quad x°x \leqslant z°z, \quad zz° \leqslant yy°.$$

Again, there are several cases to consider:

(a) Suppose that $\xi\!:\!y = \xi\!:\!z = \xi\!:\!x$. Then from the inequality involving $\leqslant_{lq}$ we also have $xx° \leqslant zz°$, and from that involving $\leqslant_{rq}$ we also have $z°z \leqslant y°y$. Consequently,

$$\xi x\xi \leqslant \xi y\xi, \quad x°x \leqslant y°y, \quad xx° \leqslant yy°,$$

from which it follows that $x = xx°\xi x\xi x°x \leqslant yy°\xi y\xi y°y = y$.

(b) Suppose now that $\xi\!:\!y < \xi\!:\!z = \xi\!:\!x$. Then we also have $xx° \leqslant zz°$ and so $xx° \leqslant yy°$. It now follows from the fact that A_ξ is $\mathcal{R}$-linear that $x \leqslant y$.

(c) The case $\xi\!:\!y = \xi\!:\!z < \xi\!:\!x$ is similar to (b).

(d) Suppose now that $\xi\!:\!y < \xi\!:\!z < \xi\!:\!x$. Then $\xi\!:\!y < \xi\!:\!z$ and $zz° \leqslant yy°$ together with the fact that A_ξ is $\mathcal{R}$-linear give $z \leqslant y$. Similarly, $\xi\!:\!z < \xi\!:\!x$ and $x°x \leqslant z°z$ and the fact that A_ξ is $\mathcal{L}$-linear give $x \leqslant z$. Hence, in this case also, we have $x \leqslant y$.

It follows from these observations that ϑ^{-1} is isotone. Consequently, we have that $T \simeq (E\xi \otimes \xi T\xi \otimes \xi E)_{qb}$. □

In the case where S is a group, the left quasi-lexicographic order $\leqslant_{lq}$ reduces to the left lexicographic order $\leqslant_l$, and correspondingly $\leqslant_{rq}$ reduces to $\leqslant_r$. In this case, $\mathcal{L}$ and $\mathcal{R}$ are automatically regular on S and so, by Theorem 14.29, the quasi-bootlace order $\leqslant_{qb}$ reduces to the bootlace order $\leqslant_b$. From the above we may therefore deduce the following analogue of Theorems 14.25 and 14.28.

Theorem 14.32 *An ordered semigroup S is a completely simple orthodox Dubreil-Jacotin semigroup on which $\mathcal{L}$ and $\mathcal{R}$ are knotted and A_ξ is both $\mathcal{L}$-linear and $\mathcal{R}$-linear if and only if S is a bootlace ordered cartesian product of a pointed rectangular band and an ordered group.* □

Referring back to the completely simple semigroups $(S_x; \leqslant)$ considered at the end of Chapter 13, we see by Theorem 14.32 that there is only one on which $\mathcal{L}$ and $\mathcal{R}$ are regular and A_ξ is both $\mathcal{L}$-linear and $\mathcal{R}$-linear, namely $(S_x; \leqslant_b)$.

14.3.4 Lexicographic orders

We shall now consider a situation in which neither $\mathcal{L}$ nor $\mathcal{R}$ is regular. This will involve another lexicographic-type order. For this purpose, let S be an orthodox strong Dubreil-Jacotin semigroup and consider the relation $\mathcal{C}$ defined variously on S by

$$(x, y) \in \mathcal{C} \iff \xi x \xi = \xi y \xi \iff x^{\circ\circ} = y^{\circ\circ} \iff x^{\circ} = y^{\circ}.$$

Since $x, x^{\circ\circ} \in V(x^{\circ})$ and $x^{\circ\circ}$ is the greatest inverse of x°, we have $x \leq x^{\circ\circ}$ from which it follows that $\mathcal{C}$ is a closure. Since ξ is a middle unit, $\mathcal{C}$ is also a congruence on S. It is clear that $\mathcal{C} \subseteq A_\xi$. We can order the $\mathcal{C}$-classes by

$$[x]_\mathcal{C} \leq [y]_\mathcal{C} \iff \xi x \xi \leq \xi y \xi.$$

Definition We shall say that A_ξ is $\mathcal{C}$-**linear** if, for all $x, y \in S$,

$$\left.\begin{array}{c} [x]_{A_\xi} < [y]_{A_\xi} \\ [x]_\mathcal{C} < [y]_\mathcal{C} \end{array}\right\} \Rightarrow x < y.$$

Example 14.2 Let $T = \mathbb{Z} \times -\mathbb{N} \times \mathbb{Z}$ with the cartesian order. Let $k > 1$ be a fixed integer and for every $n \in \mathbb{Z}$ let n_k be the biggest multiple of k that is less than or equal to n. Define a multiplication on T by

$$(x, -p, a)(y, -q, b) = (x_k + y, \min\{-p, -q\}, a + b_k).$$

Then it is readily verified that T is a naturally ordered orthodox strong Dubreil-Jacotin semigroup in which the idempotents are the elements of the form $(x, -p, a)$ with $x_k = a_k = 0$. Moreover, $\xi = (k - 1, 0, k - 1)$ and

$$\left.\begin{array}{l} \xi\colon(x, -p, a) = (-x_k + k - 1, 0, -a_k + k - 1); \\ \xi(x, -p, a)\xi = (x_k + k - 1, -p, a_k + k - 1). \end{array}\right\} \tag{14.1}$$

Consider the subset $T^\star = \{(x, -p, a) \in T \; ; \; x_k = a_k\}$. Then $T^\star$ is the same type of semigroup and the same formulae as in (14.1) hold. To show that A_ξ is $\mathcal{C}$-linear on $T^\star$, we have to show that the conditions

$$\xi\colon[\xi\colon(x, -p, a)] < \xi\colon[\xi\colon(y, -q, b)],$$
$$\xi(x, -p, a)\xi < \xi(y, -q, b)\xi$$

imply that $(x, -p, a) < (y, -q, b)$. Now by (14.1) the first of these gives $a_k = x_k < y_k = b_k$, and the second gives $-p \leqslant -q$. Hence $a < b, x < y, -p \leq -q$ and therefore $(x, -p, a) < (y, -q, b)$.

Theorem 14.33 *In an integral Dubreil-Jacotin inverse semigroup A_1 is $\mathcal{C}$-linear.*

Proof In such a semigroup $\xi = 1$ and $\mathcal{C}$ reduces to equality. □

We can now describe the structure of orthodox strong Dubreil-Jacotin semigroups on which A_ξ is $\mathcal{C}$-linear. For this, let S be an integral Dubreil-Jacotin inverse semigroup, let L be an ordered left normal band with a top element that is a right identity, and let R be an ordered right normal band with a top element that is a left identity. Let $L \otimes S \otimes R$ be as in Theorem 14.23. Consider the relation $\leq_q$ defined on $L \otimes S \otimes R$ by

$$(e, x, f) \leq_q (g, y, h) \iff x \leqslant y \text{ and } \begin{cases} \text{either} & 1\!:\!y < 1\!:\!x, \\ \text{or} & 1\!:\!y = 1\!:\!x, e \leqslant g, f \leqslant h. \end{cases}$$

It is readily seen that $\leq_q$ is an order which we call the **quasi-lexicographic order**. We shall denote by $(L \otimes S \otimes R)_q$ the set $L \otimes S \otimes R$ equipped with this order and the multiplication of Theorem 14.23.

We then have the following structure theorem, the proof of which we leave as an exercise.

Theorem 14.34 *Let S be an integral Dubreil-Jacotin inverse semigroup, let L be an ordered left normal band with a greatest element 1_L that is a right identity, and let R be an ordered right normal band with a greatest element 1_R that is a left identity. Then $(L \otimes S \otimes R)_q$ is a naturally ordered orthodox strong Dubreil-Jacotin semigroup on which A_ξ is C-linear.*

Moreover, every such semigroup arises in this way. □

In the case where S is a group, the above quasi-lexicographic order on $L \otimes S \otimes R$ reduces to the lexicographic order $\leqslant_{lex}$. From the above we may therefore deduce the following analogue of Theorems 14.25, 14.28 ansd 14.32.

Theorem 14.35 *An ordered semigroup S is a completely simple orthodox Dubreil-Jacotin semigroup on which A_ξ is C-linear if and only if S is a lexicographically ordered cartesian product of a pointed rectangular band and an ordered group.* □

Referring back to the completely simple semigroups $(S_x; \leqslant)$ considered at the end of Chapter 13, we see by Theorem 14.35 that there is only one on which A_ξ is C-linear, namely $(S_x; \leqslant_{lex})$.

14.4 Lattices for which Res L is regular

We end by returning to semigroups of residuated mappings. If E is a bounded ordered set then the semigroup of isotone mappings on E is an ordered semigroup under the order defined in Example 1.7. In particular, this is true of Res L for any bounded lattice L. The natural question of precisely when the ordered monoid Res L of residuated mappings on a bounded lattice L is regular, and the structure of such lattices L, has been considered by Janowitz [70].

Consider first the case where L is a finite chain. It is easy to see that Res L is regular (see Exercise 1.20). For more general ordered sets this is not always the case. Indeed, we observe that in a general bounded ordered set E a residuated mapping f is a regular element of the semigroup Res E if and only if there exist idempotents $a, b \in$ Res E such that $\operatorname{Im} a = \operatorname{Im} f$ and $\operatorname{Im} b^+ = \operatorname{Im} f^+$ (see Exercise 1.19).

In fact, the smallest lattice L for which the semigroup Res L is not regular has been identified by computer and turns out to be the six-element lattice $\mathbf{2} \times \mathbf{3}$. It is interesting to note that Res $\mathbf{2} \times \mathbf{3}$ has 108 elements, only one of which is not regular!

An important consequence of this is the fact that if a lattice L contains the fence

then the semigroup $\operatorname{Res} L$ is not regular.

In [70] there are given necessary and sufficient conditions for a lattice L of finite length to be such that $\operatorname{Res} L$ is a regular semigroup. Particularly interesting cases in which the structure of L can be described explicitly are upper semi-modular lattices and distributive lattices. If we denote by M_k the lattice formed by adding top and bottom elements to the discretely ordered set $\{1, 2, \ldots, k\}$ then we have the following result.

Theorem 14.36 (Janowitz [70]) (1) *If L is an upper semi-modular lattice of finite length then $\operatorname{Res} L$ is regular if and only if L is a vertical sum of lattices of the form M_k. Such a lattice is necessarily modular.*

(2) *If L is a distributive lattice of finite length then $\operatorname{Res} L$ is regular if and only if L is a vertical sum of lattices of the form M_k with $k \leqslant 2$.* $\quad\square$

References

1. J. C. Abbott, *Sets, lattices and boolean algebras*, Allyn & Bacon, Boston, 1969.

2. R. Baer, *Linear algebra and projective geometry*, Academic Press, New York, 1952.

3. R. Balbes and P. Dwinger, *Distributive lattices*, University of Missouri Press, Columbia, 1974.

4. J. Berman and P. Köhler, Cardinalities of finite distributive lattices, Mitt. Math. Sem. Giessen, **121**, 1976, 103–124.

5. F. Bernstein, Untersuchungen aus der Mengenlehre, Math. Annal., **61**, 1905, 117-155.

6. A. Bigard, Sur les images homomorphes d'un demi-groupe ordonné, C. R. Acad. Sc. Paris, **260**, 1965, 5987-5988.

7. A. Bigard, K. Keimel and S. Wolfenstein, *Groupes et anneaux réticulés*, Lecture Notes in Mathematics, 608, Springer, Berlin, 1971.

8. G. Birkhoff, On the combination of subalgebras, Proc. Camb. Phil. Soc., **29**, 1933, 441–464.

9. G. Birkhoff, Applications of lattice algebra, Proc. Camb. Phil. Soc., **30**, 1934, 115-122.

10. G. Birkhoff, On the structure of abstract algebras, Proc. Cambridge Philos. Soc., **31**, 1935, 433–454.

11. G. Birkhoff, Lattice-ordered groups, Annals of Math., **43**, 1942, 298–331.

12. G. Birkhoff, Subdirect unions in universal algebra, Bull. Amer. Math. Soc., **50**, 1944, 764–768.

13. G. Birkhoff, *Lattice theory*, American Mathematical Society, Colloquium Publications, XXV, 1940; third edition 1967.

14. G. Birkhoff and M. Ward, A characterisation of boolean algebras, Ann. Math., **40**, 609–610.

15. T. S. Blyth, The general form of residuated algebraic structures, Bull. Soc. Math. France, **93**, 1965, 109–127.

16. T. S. Blyth, Dubreil-Jacotin inverse semigroups, Proc. Roy. Soc. Edinburgh, **A71**, 1973, 345–360.

17. T. S. Blyth, Perfect Dubreil-Jacotin semigroups, Proc. Roy. Soc. Edinburgh, **78A**, 1977, 101–104.

294   References

18. T. S. Blyth, Baer semigroup coordinatisations of modular lattices, Proc. Roy. Soc. Edinburgh, **81A**, 1978, 49–56.

19. T. S. Blyth, A coordinatisation of implicative (semi)lattices by Baer semigroups, Proc. Roy. Soc. Edinburgh, **82A**, 1978, 111–113.

20. T. S. Blyth, Baer semigroup coordinatisations of distributive lattices, Proc. Roy. Soc. Edinburgh, **85A**, 1980, 307–312.

21. T. S. Blyth and J. F. Chen, Inverse transversals are mutually isomorphic, Communications in Algebra, **29**, 2001, 799–804.

22. T. S. Blyth and G. M. S. Gomes, On the compatibility of the natural order on a regular semigroup, Proc. Roy. Soc. Edinburgh, **94A**, 1983, 79–84.

23. T. S. Blyth and M. F. Janowitz, On decreasing Baer semigroups, Bull. Soc. Roy. Sc. Liège, **38**, 1969, 414–423.

24. T. S. Blyth and M. F. Janowitz, *Residuation theory*, Pergamon Press, 1972.

25. T. S. Blyth and R. McFadden, Naturally ordered regular semigroups with a greatest idempotent, Proc. Roy. Soc. Edinburgh, **91A**, 1981, 107–122.

26. T. S. Blyth and R. McFadden, Regular semigroups with a multiplicative inverse transversal, Proc. Roy. Soc. Edinburgh, **92A**, 1982, 253–270.

27. T. S. Blyth and G. A. Pinto, Principally ordered regular semigroups, Glasgow Math. J., **32**, 1990, 349–364.

28. T. S. Blyth and G. A. Pinto, On ordered regular semigroups with biggest inverses, Semigroup Forum, **54**, 1997, 154–165.

29. T. S. Blyth and M. H. Almeida Santos, Naturally ordered orthodox Dubreil-Jacotin semigroups, Communications in Algebra, **20**, 1992, 1167–1199.

30. T. S. Blyth and M. H. Almeida Santos, On naturally ordered regular semigroups with biggest idempotents, Communications in Algebra, **21**, 1993, 1761–1771.

31. T. S. Blyth and M. H. Almeida Santos, A simplistic approach to inverse transversals, Proc, Edinburgh Math. Soc., **39**, 1996, 57–69.

32. T. S. Blyth and M. H. Almeida Santos, A classification of inverse transversals, Communications in Algebra, **29**, 2001, 611–624.

33. T. S. Blyth and J. C. Varlet, Sur la construction de certaines MS-algèbres, Portugaliae Math., **39**, 1983, 489–496; corrigendum, ibid., **42**, 1985, 469–471.

34. T. S. Blyth and J. C. Varlet, *Ockham algebras*, Oxford University Press, Oxford, 1994.

35. G. Boole, *The mathematical analysis of logic*, Cambridge, 1847.

36. G. Boole, *An investigation into the laws of thought*, London, 1854.

37. J. Certaine, Lattice-ordered groupoids and some related problems, Harvard doctoral thesis, 1943.

38. C. C. Chen and G. Grätzer, Stone lattices I. Construction theorems, Canad. J. Math., **21**, 1969, 884–894; Stone lattices II. Structure theorems, ibid., 895–903.

39. P. Conrad, *Lattice-ordered groups*, Tulane Lecture Notes, Tulane University, 1970.

40. P. Crawley and R. P. Dilworth, *Algebraic theory of lattices*, Prentice–Hall, Englewood Cliffs, 1973.

41. M. R. Darnell, *Theory of lattice-ordered groups*, Marcel Dekker, New York, 1995.

42. B. A. Davey and H. A. Priestley, *Lattices and order*, second edition, Cambridge University Press, 2002.

43. R. P. Dilworth, Non-commutative residuated lattices, Trans. Amer. Math. Soc., **46**, 1939, 426–444.

44. R. P. Dilworth, Lattices with unique complements, Trans. Amer. Math. Soc., **57**, 1945, 123–154.

45. R. P. Dilworth, The structure of relatively complemented lattices, Annals of Math., **51**, 1950, 348–359.

46. M. L. Dubreil-Jacotin, L. Lesieur and R. Croisot, *Leçons sur la théorie des treillis, des structures algébriques ordonnées et des treillis géométriques*, Gauthiers-Villars, Paris, 1953.

47. M. L. Dubreil-Jacotin, Sur les images homomorphes d'un demi-groupe ordonné, Bull. Soc. Math. France, **92**, 1964, 101–115.

48. D. G. Fitz-Gerald, On inverses of products of idempotents in regular semigroups, J. Australian Math. Soc., **13**, 1972, 335–337.

49. D. J. Foulis, Baer *-semigroups, Proc. Amer. Math. Soc., **11**, 1960, 648–654.

50. R. Freese, J. Ježek and J. B. Nation, *Free lattices*, Amer. Math. Soc. Surveys and Monographs, vol. 42, 1995.

51. L. Fuchs, *Partially ordered algebraic systems*, Pergamon Press, Oxford, 1963.

52. N. Funayama, On the completion by cuts of distributive lattices, Proc. Imp. Acad. Tokyo, **20**, 1944, 1-2.

53. N. Funayama and T. Nakayama, On the distributivity of a lattice of lattice congruences, Proc. Imp. Acad. Tokyo, **18**, 1942, 553–554.

54. B. Ganter and R. Wille, *Formal concept analysis*, Springer, Berlin, 1999.

55. A. M. W. Glass, *Ordered permutation groups*, London Math. Soc., Lecture Notes Series, **55**, Cambridge University press, 1981.

56. A. M. W. Glass and W. C. Holland, *Lattice-ordered groups – advances and techniques*, Kluwer Academic Publishers, 1989.

57. V. Glivenko, Sur quelques points de la logique de M. Brouwer, Bull. Acad. Sc. Belgique, **15**, 1929, 183–188.

58. G. Grätzer, *General lattice theory*, second edition, Birkhauser, 1998.

59. P. Grillet, *Semigroups*, Marcel Dekker, New York, 1995.

60. J. Hashimoto, Ideal theory for lattices, Math. Japon., **2**, 1952, 149–186.

61. J. Hashimoto, Congruence relations and congruence classes in lattices, Osaka Math. J., **15**, 1963, 71–86.

62. F. Hausdorff, *Grundzuge der Mengenlehre*, Leipzig, 1914.

63. H. Hermes, *Einfuhrung in die Verbandstheorie*, Springer, Berlin, 1935.

64. P. M. Higgins, *Techniques of semigroup theory*, Oxford University Press, Oxford, 1992.

65. J. V. Hion, Archimedean ordered rings, Uspechi Mat. Nauk., **9**, 1954, 237–242.

66. O. Hölder, Die Axiome der Quantität und die Lehre vom Mass, Math.-Phys. Cl. Leipzig, **53**, 1901, 1–64.

67. J. M. Howie, *Fundamentals of semigroup theory*, L.M.S. Monographs, **12**, Oxford University Press, Oxford, 1995.

68. M. F. Janowitz, Baer semigroups, Duke Math. J., **32**, 1963, 85–96.

69. M. F. Janowitz, A semigroup approach to lattices, Canad. J. Math., **18**, 1966, 1212–1223.

70. M. F. Janowitz, Regularity of residuated mappings, Semigroup Forum, **42**, 1991, 313–332.

71. C. S. Johnson Jr, Semigroups coordinatising posets and semilattices, J. London Math. Soc., **4**, 1971, 277–283.

72. J. A. Kalman, The triangle inequality in ℓ-groups, Proc. Amer. Math. Soc., **11**, 1960, 395.

73. G. Kalmbach, *Orthomodular lattices*, London Mathematical Society Monographs, **18**, Academic Press, 1983.

74. B. Knaster, Une théorème sur les fonctions d'ensembles, Annales Soc. Polonaise Math., **6**, 1927, 133–134.

75. A. G. Kurosh, Durchschnittsdarstellungen mit irreduziblen Komponenten in Ringen unt in sogenannten Dualgruppen, Mat. Sbornik, **42**, 1935, 613–616.

76. H. Lakser, The structure of pseudocomplemented distributive lattices, Trans. Amer. Math. Soc., **156**, 1971, 335–342.

77. H. Lakser, Principal congruences of pseudocomplemented distributive lattices, Proc. Amer. Math. Soc., **37**, 1973, 32–36.

78. P. Lorenzen, Abstrakte Begrundung der multiplicative Idealtheorie, Math. Z., **45**, 1939, 533–553.

79. H. M. MacNeille, Partially ordered sets, Trans. Amer. Math. Soc., **42**, 1937, 416–460.

80. D. B. McAlister and T. S. Blyth, Split orthodox semigroups, Journal of Algebra, **51**, 1978, 491–525.

81. D. B. McAlister, Regular Rees matrix semigroups and regular Dubreil-Jacotin semigroups, J. Australian Math. Soc., **31**, 1981, 325–336.

82. D. B. McAlister and R. McFadden, Regulae semigroups with inverse transversals, Quart. J. Math. Oxford, **34**, 1983, 459–474.

83. F. Maeda and S. Maeda, *Theory of symmetric lattices*, Springer, Berlin, 1970.

84. I. Molinaro, Demi-groupes résidutifs, J. Math. Pures et Appl., **39**, 1960, 319–356; ibid., **40**, 1961, 43–110.

85. E. H. Moore, *Introduction to a form of general analysis*, New Haven, 1910, 53–80.

86. K. S. S. Nambooripad, The natural partial order on a regular semigroup, Proc. Edinburgh Math. Soc., **23**, 1980, 249–260.

87. J. von Neumann, *Continuous geometry*, Princeton, 1960.

88. J. von Neumann, *Lectures on continuous geometries*, Princeton, 1936–37.

89. O. Ore, On the foundations of abstract algebra, Annals of Math., **36**, 1935, 406–437; ibid. **37**, 1936, 265–292.

90. O. Ore, Structures and group theory, Duke Math. J., **3**, 1937, 149–173; ibid. **4**, 1938, 247–269.

91. C. S. Peirce, On the algebra of logic, Amer. Jour., **3**, 1880, 15–57.

92. J. Querré, Contribution à la théorie des structures ordonnées et des systèmes d'idéaux, Ann. di Mat., **66**, 1964, 265–389.

93. P. Ribenboim, *Théorie des groupes ordonnés*, Monografías de Matemática, Universidad Nacional del Sur, Bahía Blanca, 1959.

94. F. Riesz, Atti del IV Cong. Int. dei Mat., vol II, 1909.

95. F. Riesz, Sur quelques notions fondamentales dans la théorie générale des opérateurs linéaires, Ann. Math., **41**, 1940, 174–206.

96. D. E. Rutherford, *Introduction to lattice theory*, Oliver & Boyd, Edinburgh, 1965.

97. Tatsuhiko Saito, Regular semigroups with a weakly multiplicative inverse transversal, Proc. 8th Symposium on Semigroups, Shimane University, 1985, 22–25.

98. Tatsuhiko Saito, Construction of regular semigroups with inverse transversals, Proc. Edinburgh Math. Soc., **32**, 1989, 41–51.

99. Tatsuhiko Saito, Naturally ordered regular semigroups with maximum inverses, Proc. Edinburgh Math. Soc., **32**, 1989, 33–39.

100. V. N. Saliĭ, *Lattices with unique complements*, American Mathematical Society Translations, vol. 69, 1988.

101. E. Schröder, Algebra der Logic, 3 vols, Leipzig, 1890–95.

102. R. Sikorski, *Boolean algebras*, Ergebnisse der Mathematik und ihrer Grenzgebiete, band 25, Springer, Berlin, 1964.

103. M. H. Stone, Boolean algebras and their relation to topology, Proc. Nat. Acad. Sc. U.S.A., **20**, 1934, 197–202.

104. M. H. Stone, The theory of representations for boolean algebras, Trans. Amer. Math. Soc., **40**, 1936, 37–111.

105. H. H. Stone, Topological representations of distributive lattices and brouwerian logics, Časopis Pěst. Mat., **67**, 1937, 1–25.

106. M. Suzuki, *Structure of a group and the structure of its lattice of subgroups*, Ergebnisse der Mathematik und ihrer Grenzgebiete, band 10, Springer, Berlin, 1956.

107. G. Szász, *Introduction to lattice theory*, Academic Press, New York, 1963.

108. Xi Lin Tang, Regular semigroups with inverse transversals, Semigroup Forum, **35**, 1997, 24–32.

109. A. Tarski, Zur Grundlegung der Boole'schen Algebra I, Fund. Math., **24**, 1935, 177–198.

110. A. Urquhart, Lattices with a dual homomorphic operation, Studia Logica, **38**, 1979, 201–209; ibid., **40**, 1981, 391–404.

111. M. Ward, The closure operators of a lattice, Annals Math., **43**, 1942, 191–196.

112. M. Ward and R. P. Dilworth, Residuated lattices, Trans. Amer. Math. Soc., **45**, 1939, 335–354.

113. L. R. Wilcox, Modularity in the theory of lattices, Ann. of Math., **40**, 1939, 490–505.

114. M. Yamada, Regular semigroups whose idempotents satisfy permutation identities, Pacific J. Math., **21**, 1967, 371–392.

115. M. Yamada and N. Kimura, Note on idempotent semigroups II, Proc. Japan Acad., **34**, 1958, 110–112.

Index